FLOWER PRODUCTION AND GARDENING

About the Editors

Dr P. K. Yadav, born in Bulandshahr district of Uttar Pradesh is currently Professor and Head, Department of Horticulture, S. K. Rajasthan Agricultural University, Bikaner. He received his Post Graduation and Doctoral degree from Agra University Agra. He has been actively involved in the teaching profession for more than 25 years in S.K. Rajasthan Agricultural University Bikaner. During his long illustrious carrier he has guided numerous M.Sc. and Ph.D. students. He has published around 75 Research papers in national and international journals, and has also written around 50 popular articles. He is the author of two internationally and two nationally published books and has complied three Compendiums as editor of Winter School and Training. He wrote two practical manual and five technical bulletins. He has also handled various research projects and organized one Winter School and has delivered several Radio and TV talk shows for farmers. He received Scientist of the Year Award 2012.

Prof. Raksha Pal Singh, a renowned agricultural scientist, academician and research manager, has been working as Vice Chancellor, Swami Keshwanand Rajasthan Agricultural University, Bikaner, Rajasthan, India since August, 2019. Besides Ph.D. in Agricultural Extension (1996) from B. R. Ambedakar University, Agra, UP. Prof. Singh was conferred with the Post P.G. Diploma in Mass Communication and Journalism from Kendriya Hindi Sansthan, Ministry of Human Resource Development, Govt. of India, Agra, UP. He has expertise in areas of Socio-economic-psychological issues in agriculture and rural development, information communication technology (ICT), human resource management, participatory development & leadership, farming system research etc. He has published >200 publications including research papers, books, book chapters etc. Prof. Singh has been conferred with various prestigious awards and recognitions as Best Administrator Award by Academy for Environment and Life Sciences, Agra and Life Time Achievement awards one by Indian Society of Extension Education on the occasion of ISEE National Seminar 2019 by Hon'ble Chancellor and Governor of Rajasthan Shri Kalraj Mishra and another by society for Scientific Development in Agriculture & Technology, Meerut., UP.

FLOWER PRODUCTION AND GARDENING

P. K. Yadav
Professor & Head
Department of Horticulture
S. K. Rajasthan Agricultural University
Bikaner, Rajasthan

R. P. Singh
Vice Chancellor
S. K. Rajasthan Agricultural University
Bikaner, Rajasthan

NEW INDIA PUBLISHING AGENCY
New Delhi-110 034

NEW INDIA PUBLISHING AGENCY
101, Vikas Surya Plaza, CU Block, LSC Market
Pitam Pura, New Delhi – 110 034, India
Email: info@nipabooks.com
Web: www.nipabooks.com
For customer assistance, please contact
Phone: + 91-11-27 34 17 17 Fax: + 91-11- 27 34 16 16
E-Mail: feedbacks@nipabooks.com

ISBN: 978-93-90591-21-3

Composed and Designed by NIPA

Preface

The history of systematic gardening in India is as old as it's civilization, which existed between 2500 B.C and 1750 B.C. Trees related with religious civilization were *Ficus religiosa* and *Ficus bengalensis*. Aryans came to India in 1600 B.C. and appreciated the beauty of plants, lakes, mountains, forests etc. and named their children after flowers like Kamal, Rukmani, Kailash etc. A very beautiful description of Ayodhya city with wide streets, large houses, richly decorated temples and gardens was mentioned in the famous book Ramayana written by Valmiki. In Mahabharata, it is described that the pleasure gardens were maintained with flowering plants. The famous tree of this era was Kadamba (*Anthocephalus cadamba*), associated with the Lord Krishna. Lord Buddha was also associated with Asoka, Pipal, Banyan, Mango and Sal trees. (*Source*: ecoursesonline.iasri.res.in)

Presently, floriculture is a fast emerging and rapidly expanding industry through strong research and development. It covers all aspects related to commercial growing, marketing, arrangement of seeds & bulbs, plants, flowers etc. On the other hand gardening is all about the planning and planting of an area to secure a relationship between the landscape and plants to meet the human needs for beauty and function in the best way. Indian floriculture, occupies 1044 (000 ha) area with 3651 (000 MT) production and 3.49 (MT/ha) productivity (NHB, 2018). Apart from the utilities of ornamental plants in outdoor gardening and interior decoration as in-house plants, cut-flowers, potted plants and dried products also constitute as an important segment of the international trade. The scenario of floriculture is fast changing due to steadily widening export opportunities. The total export from the total floriculture products is around 20703.97 (MT) with a value Rs. 50731.22 lacs (NHB, 2018). The demand for flowers and ornamental plants in the domestic market is also growing at a fast pace and a large number of people earn their livelihood by production and marketing of flowers and ornamental plants.

The present title has been planned and designed to meet the long - felt need for a book covering major aspects of important floriculture crops. A series of courses are operational in the curriculum of UG and PG programmes. The courses in the subject are highly ramified and need specific literature for a particular matter. The students have to turn - out the pages of numerous books to cover the matter on all diverse aspects of floriculture. It is in this context, that the work of compiling various facets of floriculture has been attempted and we feel pleased to present it in the form to meet out the needs of the readers. This book would serve as a reliable source of information about all the important and relevant aspects of floriculture and ornamental horticulture for various uses including production technology of export quality flowers to the persons who are associated with or working in floriculture with the latest information for further advancement. In this book important aspects of floriculture are explained in a concise and easy to understand format using simple and lucid language. Encompassing 32 chapters, the book is a detailed summary of facts figures yet in a comprehensive way.

It is a pleasure to acknowledge with thanks the help and cooperation of the contributors. Heartfelt thanks to all our colleagues and friends who helped us in this endeavour.

P. K. Yadav
R. P. Singh

Contents

List of Contributors

Chapter 1: Importance and Scope of Ornamental Horticulture

S. S. Sindhu, *Head, Division of Floriculture & Landscaping, ICAR-Indian Agricultural Research Institute, New Delhi*

M. R. Dhiman, *Pr. Scientist (ARS), ICAR-Indian Agriculture Research Institute Regional Station, Katrain, Kullu, Himachal Pradesh*

Chapter 2: Principles and Elements of Landscape Design

Ashutosh Mishra, *Professor & Head, Department of Floriculture & Landscaping College of Horticulture and Forestry, Jhalawar, (Agriculture University, Kota)*

Chapter 3: Styles of Gardening and Different Types of Garden

S. K. Khandelwal. *Professor and Zonal Director Research, S. K. N. Agriculture University, Jobner, Rajasthan*

Chapter 4: Turfgrass Management: Princeples & Practices

S. L. Chawla, *Associate Professor & Ex. Head, Department of Floriculture & Landscape Architecture, Navsari Agricultural University, Navsari, Gujarat*

Roshni Agnihotri, *Assistant Professor, RPCAU, Pusa, Bihar*

R. D. Pawar, *Assistant Professor & Ph.D. Scholar, (NAU, Navsari), College of Agriculture, Kolhapur, Maharashtra*

Chapter 5: Cultivation and Management of Annual Flowers

Ajai Kumar *Tiwari*, *Principal Scientist, Division of Floriculture and Landscaping ICAR-Indian Agricultural Research Institute, New Delhi*

Babita Singh, *Scientist, Division of Floriculture and Landscaping ICAR-Indian Agricultural Research Institute, New Delhi*

Ritu Jain, *Senior Scientist, Division of Floriculture and Landscaping ICAR-Indian Agricultural Research Institute, New Delhi*

Gunjeet Kumar, *Principal Scientist, Division of Floriculture and Landscaping ICAR-Indian Agricultural Research Institute, New Delhi*

Chapter 6: Shrubs and Their Management

L. N. Bairwa, *Professor and Head, Department of Horticulture, SKN College of Agriculture Jobner, Jaipur, Rajasthan*

A. K. Mahawar, *Research Scholar, Department of Horticulture, SKN College of Agriculture Jobner, Jaipur, Rajasthan*

Chapter 7: Climbers

Madhu Bala, *Scientist (Floriculture and Landscaping), Department of Floriculture and Landscaping, Punjab Agricultural University, Ludhiana, Punjab*

Chapter 8: Trees

Simrat Singh, *Scientist (Floriculture and Landscaping), Punjab Agricultural University Ludhiana*

Chapter 9: Indoor Plants

R. K. Dubey, *Department of Floriculture and Landscaping, Punjab Agricultural University, Ludhiana, Punjab*

Simrat Singh, *Scientist (Floriculture and Landscaping), Punjab Agricultural University, Ludhiana, Punjab*

Chapter 10: Cacti and Succulents

Sapna Panwar, *Scientist, Division of Floriculture and Landscaping, ICAR-Indian Agricultural Research Institute, New Delhi*

Namita, *Senior Scientist, Division of Floriculture and Landscaping, ICAR-Indian Agricultural Research Institute, New Delhi*

Neelam Thakur, *Assistant Professor, Department of Horticulture, School of Agriculture, Lovely Professional University, Jalandhar, Punjab*

Chapter 11: Bonsai

Reeta Bhatia, *Senior Scientist, Division of Floriculture and Landscaping, ICAR-Indian Agricultural Research Institute, New Delhi*

Shyam Sundar Dey, *Senior Scientist, Division of Vegetable Science, ICAR-Indian Agricultural Research Institute, New Delhi-12*

Saurabh Singh, *Research Scholar, Division of Floriculture and Landscaping ICAR-Indian Agricultural Research Institute, New Delhi*

Chapter 12: Rose

S. S. Sindhu, *Head, Division of Floriculture & Landscaping, ICAR-Indian Agricultural Research Institute,New Delhi*

Namita, *Senior Scientist, Division of Floriculture & Landscaping, ICAR-Indian Agricultural Research Institute, New Delhi*

Chapter 13: Carnation

S. R. Dhiman, *Professor & Head, Department of Floriculture and Landscape Architecture, Dr. Y S Parmar University of Horticulture and Forestry Nauni-Solan, Himachal Pradesh*

Puja Sharma, *Associate Professor, Department of Floriculture and Landscape Architecture, Dr. Y S Parmar University of Horticulture and Forestry Nauni, Solan, Himachal Pradesh*

Bharati Kashyap, *Associate Professor, Department of Floriculture and Landscape Architecture, Dr. Y S Parmar University of Horticulture and Forestry, Nauni-Solan, Himachal Pradesh*

Arshi Sultanpuri, *Department of Floriculture and Landscape Architecture, Dr. Y S Parmar University of Horticulture and Forestry, Nauni-Solan, Himachal Pradesh*

Kalkame Ch. Momin, *Scientist, Department of Floriculture and Landscape Architecture, Dr. Y S Parmar University of Horticulture and Forestry Nauni-Solan, Himachal Pradesh*

Pratibha Chauhan, *Scientist, Department of Floriculture and Landscape Architecture, Dr. Y S Parmar University of Horticulture and Forestry Nauni-Solan, Himachal Pradesh*

Chapter 14: Chrysanthemum

L. N. Mahawer, *Professor, Rajasthan College of Agriculture, MPUAT, Udaipur Rajasthan*

Chapter 15: Cultivation of Gerbera

Tanya Thakur, *Assistant Floriculturist, Department of Floriculture and Landscaping, Punjab Agricultural University, Ludhiana , Punjab*

K. K. Dhatt, *Senior Floriculturist and Head, Department of Floriculture and Landscaping, Punjab Agricultural University, Ludhiana , Punjab*

Chapter 16: Orchids

N. K. Meena, *Senior Scientist, ICAR-National Research Centre on Seed Spices Tabiji, Ajmer, Rajasthan*

Ram Pal, *Senior Scientist, ICAR-National Research Centre for Orchids Pakyong, East Sikkim, Sikkim*

Chapter 17: Anthurium

Sanchita Ghosh, *Subject Matter Specialist, Dr. Rajendra Prasad Central Agricultural University, Pusa, Distt. Samastipur, Bihar*

Ajit Kumar Kapoor, *Professor Floriculture, Department of Horticulture, G.B. Pant University of Agriculture & Technology, Pantnagar, Uttarakhand*

Tripti Tewari, G.B. *Pant University of Agriculture & Technology Pantnagar, Uttarakhand*

Syed Khudus, *Research Scholar, G.B. Pant University of Agriculture & Technology, Pantnagar, Uttarakhand*

Chapter 18: Gladiolus

L. N. Mahawer, *Professor & Principal Investigator, Department of Horticulture, Maharana Pratap University of Agriculture and Technology, Udaipur Rajasthan*

Saddam Hussain, *Research Scholar, Department of Horticulture, Maharana Pratap University of Agriculture and Technology, Udaipur, Rajasthan*

A. L. Regar, *Research Scholar, Department of Horticulture, Maharana Pratap University of Agriculture and Technology Udaipur, Rajasthan*

Chapter 19: Aster

B. D. Bhuj, *Professor (Floriculture & Landscaping) Department of Horticulture, G.B. Pant University of Agriculture and Technology, Pantnagar, U.S. Nagar Uttarakhand*

Yograj Kushwaha, *Research Scholar (Floriculture & Landscaping), Department of Horticulture, G.B. Pant University of Agriculture and Technology Pantnagar, Uttarakhand*

Chapter 20: Tuberose

P.K. Yadav, *Professor & Head, S.K. Rajasthan Agricultural University, Bikaner Rajasthan*

Priyanka Kumawat, *Research Scholar, S.K. Rajasthan Agricultural University Bikaner, Rajasthan*

Chapter 21: Bird of Paradise and Heliconia

K. Nihad, *Senior Scientist (Horticulture), ICAR-CPCRI, Kasaragod, Kerala*

Chapter 22: Dahlia

Simrat Singh, *Scientist (Floriculture and Landscaping), Department of Floriculture and Landscaping, Punjab Agricultural University, Ludhiana Punjab*

R. K. Dubey, *Department of Floriculture and Landscaping, Punjab Agricultural University, Ludhiana, Punjab*

K. K. Dhatt, *Department of Floriculture and Landscaping, Punjab Agricultural University, Ludhiana, Punjab*

Chapter 23: Lilium

M. R. Dhiman, *Principal Scientist (ARS), ICAR-Indian Agriculture Research Institute Regional Station, Katrain, Kullu, Himachal Pradesh*

Chapter 24: Statice

Ranjit Singh, *Assistant Professor (Floriculture), Department of Floriculture and Landscaping, Punjab Agricultural University, Ludhiana, Punjab*

Parminder Singh, *Professor Floriculture & Landscaping, Department of Floriculture and Landscaping, Punjab Agricultural University, Ludhiana Punjab*

Chapter 25: Jasmines

Safeena S. A., *Senior Scientist (Floriculture and Landscaping), ICAR – Directorate of Floricultural Research, (Indian Council of Agricultural Reseach), College of Agriculture Campus, Shivaji Nagar, Pune, Maharashtra*

Chapter 26: Marigold

B. S. Beniwal, *Professor, Department of Horticulture, CCS Haryana Agricultural University, Hisar, Haryana*

Sonu Kumari Sheoran, *Assistant Scientist, Department of Horticulture, CCS Haryana Agricultural University, Hisar, Haryana*

Chapter 27: Lotus

P. K. Yadav, *Professor & Head, S.K. Rajasthan Agricultural University, Bikaner Rajasthan*

Kumari Lata, *Research Scholar, S.K. Rajasthan Agricultural University, Bikaner Rajasthan*

Chapter 28: Post-Harvest Handling of Flowers

Puja Sharma, *Associate Professor (Floriculture and Landscaping), Department of Floriculture and Landscape Architecture, Dr Y S Parmar University of Horticulture and Forestry, Solan, Himachal Pradesh*

S. R. Dhiman, *Professor and Head (Floriculture and Landscaping), Department of Floriculture and Landscape Architecture, Dr Y S Parmar University of Horticulture and Forestry, Solan, Himachal Pradesh*

Priyanka Sharma, *Assistant Professor, R.L.B. Central Agriculture University Jhansi, Uttar Pradesh*

Panchal Sangmesh, *Research Scholar, Department of Floriculture and Landscape Architecture, Dr Y S Parmar University of Horticulture and Forestry, Solan Himachal Pradesh*

Chapter 29: Dehydration of Flowers for Value-Addition

Bharati Kashyap, *Associate Professor (Floriculture and Landscaping) Department of Floriculture and Landscape Architecture, Dr Y S Parmar University of Horticulture and Forestry, Solan, Himachal Pradesh*

Rajesh Bhalla, *Professor & Head, Department of Floriculture and Landscape Architecture, Dr Y S Parmar University of Horticulture and Forestry, Solan Himachal Pradesh*

S. R. Dhiman, *Professor and Head (Floriculture and Landscaping), Department of Floriculture and Landscape Architecture, Dr Y S Parmar University of Horticulture and Forestry, Solan, Himachal Pradesh*

Puja Sharma, *Associate Professor (Floriculture and Landscaping), Department of Floriculture and Landscape Architecture, Dr Y S Parmar University of Horticulture and Forestry, Solan, Himachal Pradesh*

Sangeeta Kumari, *Horticulture Development Officer (H.P.), Department of Floriculture and Landscape Architecture, Dr Y S Parmar University of Horticulture and Forestry, Solan, Himachal Pradesh*

Chapter 30: Importance of Cut Foliage and Fillers in Floriculture Industry

Vanlalruati, *Scientist, Division of Floriculture and Landscaping, ICAR-Indian Agricultural Research Institute, New Delhi*

Ritu Jain, *Senior Scientist, Division of Floriculture and Landscaping ICAR-Indian Agricultural Research Institute, New Delhi*

Prativa Anand, *Scientist, Division of Floriculture and Landscaping, ICAR-Indian Agricultural Research Institute, New Delhi*

Babita Singh, *Scientist, Division of Floriculture and Landscaping, ICAR-Indian Agricultural Research Institute, New Delhi*

Gunjeet Kumar, *Principal Scientist-Floriculture, PME Cell, Directorate, ICAR-Indian Agricultural Research Institute, New Delhi*

Chapter 31: Flower Arrangements: Principles & Methods

Yesh Pal Singh, *Chief Horticulturist, Maurya Sheraton, ITC Hotels, New Delhi*

P. K. Yadav, *Professor & Head, S.K. Rajasthan Agricultural University, Bikaner Rajasthan*

Kumari Lata, *Research Scholar, S.K. Rajasthan Agricultural University, Bikaner Rajasthan*

Chapter 32: Essential Oils in Ornamental Plants

Ashish S. Ghormade, *Research Scholar, Dr. Balasaheb Sawant Konkan Krishi Vidyapeeth, Dapoli, Maharashtra*

Kumari Lata, *Research Scholar, S.K. Rajasthan Agricultural University, Bikaner*

Parag Sable, *Dr. Balasaheb Sawant Konkan Krishi Vidyapeeth, Dapoli Maharahstra*

Sumed Thorat, *Assistant Professor, Dr. Balasaheb Sawant Konkan Krishi Vidyapeeth, Dapoli, Maharashtra*

1

Importance and Scope of Ornamental Horticulture

S.S. Sindhu and M.R. Dhiman

Ornamental horticulture consists of floriculture and landscape horticulture. Each is concerned with growing and marketing of plants used to beautify the outdoor environment and with the associated activities of flower arrangements and landscape designs. Ornamental plants provide comfort and peace in our everyday lives. Floriculture is a dynamic, global, fast growing industry comprises of cultivation and trade of cut flowers, loose flowers, potted plants, foliage plants, dried flowers and value added products etc and is becoming popular in both domestic and export markets (Rikken, 2010). In the recent years, ornamental horticulture has emerged as a profitable agri-business in India due to improved living standards and growing consciousness among the citizens to live in environment friendly ambiance. Now-a- days this activity has assumed the commercial status with great potential for generating gainful self-employment among small and marginal farmers. Globalization of the Indian economy and subsequent liberalization of the Seed Act paved the way for the advent of protected cultivation of flowers in the country resulting in increase the production of quality floricultural products witnessing their presence in the international flower markets. The production and trade of floriculture has exhibited considerable increase consistently over the last decade in the country.

Flowers and ornamental plants have been associated with mankind since time immemorial, as they have been used for religious offerings and other social ceremonies. Loose flowers are mostly grown for worship, garland making and decoration. Cut flowers like rose, chrysanthemum, carnation, gerbera, anthurium, orchids and lily are mostly grown for export purposes. Some of the flowers are also edible with their nutritional value e.g. Dahlia tubers, lily, tapioca etc., mainly their roots are rich in starch and even dahlia roots are considered good to cure the cancer.

Status

Flowers have always remained an integral part of social fabric of human life. Man has traditionally used flowers for expressing or exhibiting his innermost feelings to God and deities or presenting to the beloved ones or complimenting any one or versifying any conceivable emotion. Due to continued growth in demand in the recent years, flowers cultivation has gained a status of commercial importance and is viewed as a high growth industry. Export of floricultural products has been on the rise and therefore

Indian floriculture industry has diversified from traditional flowers to cut flowers for export purposes. The Indian floriculture market was worth INR 130 billion in 2017. The market is further projected to reach INR 394 billion by 2023 at a CAGR of 20% during 2018-2023. India is one of the leading country in floriculture with an area of 3.12 lac ha producing 2058 thousand million tonnes cut flowers and 806, 000 MT of loose flowers during 2016-17 (NHB database, 2017). On the exports point, it has steadily growing over 20% annually. The country has exported 22,000 metric tonnes of floricultural products worth of Rs. 547 crore ($82 m) in 2016-17. Major export destinations were the United States, Germany, United Kingdom, The Netherlands and the United Arab Emirates. Major export destinations for Indian Floricultural products during 2019-20 are as under:

Top ten export destinations of floricultural products from India (2019-20)

S.No.	Country of destination	Quantity (MT)*	Value (Rs. Lacs)
1.	U.S.A.	3,276.08	13,902.70
2.	The Netherlands	1,377.08	7,852.26
3.	Germany	1,112.52	4,093.47
4.	United Kingdom	1,236.74	4,091.67
5.	United Arab Emirates	1,499.07	3,311.24
6.	Singapore	2,037.67	1,279.02
7.	Malaysia	843.81	1,760.07
8.	Canada	393.05	2,012.07
9.	Italy	358.80	1,613.63
10.	Japan	205.46	1,311.67

MT=Metric Tonnes (*Source*: APEDA)

The export basket of the Indian floricultural products comprises of dry flowers (77.1%), cut flowers (6.1%), bulbs and rhizomes (0.8%), cut foliage (0.02%) and others (15.9%).

State wise area and production of flower crops in India (2018-19)

State	Area ('000' ha)	Production of flowers	
		Loose (Metric Tonnes)	Cut (Lacs, Nos.)
Andhra Pradesh	28.04	302.53	0.00
Arunchal Pradesh	0.00	0.00	0.00
Assam	5.199	34.892	58.341
Bihar	1.41	10.13	0.39
Chhattisgarh	13.06	54.94	214.81
Gujarat	20.50	195.86	0.00
Haryana	5.96	72.85	2.64
Himachal Pradesh	0.64	12.35	10.56
Jammu & Kashmir	0.28	27.00	4.45
Jharkhand	0.81	4.45	4.17
Karnataka	24.756	178.033	75.211
Kerala	53.26	0.08	44.84

State	Area ('000' ha)	Production of flowers	
		Loose (Metric Tonnes)	Cut (Lacs, Nos.)
Madhya Pradesh	31.42	375.62	0.00
Maharashtra	11.36	57.61	0.11
Manipur	0.05	0.02	0.27
Meghalaya	0.01	0.00	0.02
Mizoram	0.17	0.00	2.36
Nagaland	0.04	0.00	24.40
Odisha	6.61	24.91	48.79
Punjab	2.09	13.07	0.00
Rajasthan	4.10	7.60	0.00
Sikkim	0.24	16.50	0.09
Tamil Nadu	39.80	521.78	17.15
Telangana	4.60	20.71	10.96
Tripura	0.00	0.00	0.00
Uttar Pradesh	21.33	46.70	65.58
Uttarakhand	1.55	2.55	12.02
West Bengal	29.10	77.42	208.82
Others	6.54	1.12	0.59
All India total	312.93	2058.72	806.55

(*Source*: National Horticulture Board)

Importance

India is a home of vast wealth of genetic resources of wild species of ornamental plants spread over various habitats like forests, gardens, hilly tracts and wild habitats, which are of great floricultural importance and significance. Large collections of beautiful plant species like orchids, rhododendrons, primula, camellia, *etc.,* are found wild in the Himalayan and other mountainous regions. The Kashmir valley and the valley of flowers are also rich sources of indigenous flowers. A large number of these native Indian species have been extensively used in the improvement of various flowers in other countries. Some of these Indian floras have been compiled and described. These basic genetic resources need to be conserved and effectively utilized in the development of floriculture in India. Economic aspects of ornamental horticulture are as important as the aesthetic ones. The floricultural product of commercial importance mainly consists of cut flowers followed by flowering pot plants, tree and nursery crops, flower bulbs, dried flower products, propagation material and other value added products etc. Developed countries account for more than 90 percent of the total world trade in floricultural products. Netherlands continues to be the world leader in the export of floricultural products. The liberalized policies of the Government both at centre and state level have resulted in mushrooming of floricultural units in the country over the years.

1. Major components of floriculture industry

1.1 Traditional flowers

The domestic consumption of loose flowers especially marigold, china aster, desi gulab, jasmine, crossandra, grallardia, barleria, *etc.* has been increased tremendously. It is clear that area under traditional flowers has increased significantly. This sector, inspite of its potential, is still an unorganized and often does not get proper importance.

Research is required on developing high-yielding varieties, year round production of chrysanthemum, china aster, marigold etc. and in promotion of crops like annual chrysanthemum, desi rose, *etc.*

Among the traditional crops grown for loose flowers, the largest area is under marigold, grown all over the country. In most parts of the country only local varieties are grown for generations. African marigolds occupy more area as compared to the small flowered French types. Jasmine flowers in view of its scent are also very popular as loose flowers and for use in garlands and *Veni* (ornament for decoration of hair by women). The major areas under this crop are in Tamil Nadu, Karnataka in South and West Bengal in East. The varieties are mainly improved clones of *Jasminum grandiflorum, J. auriculatum* and *J. sambac*. The chrysanthemum, particularly the white varieties are much in demand as loose flowers during the autumn period of October-December when other flowers like jasmine, tuberose are not available for use in garlands etc. Among other traditional flowers grown in large areas are crossandra in southern states of Tamil Nadu, Karnataka and Andhra Pradesh and aster in Maharashtra.

1.2 Cut flowers

The cut flowers constitute more than 45 percent share of the total world trade in floricultural products. Rose is the principal cut flower grown all over the country, even though in terms of total area, it may not be so. The larger percentage of the area in many states is used for growing scented rose, usually local varieties akin to the Gruss en Tepelitz, the old favourite to be sold as loose flowers. These are used for offerings at places of worship, for the extraction of essential oils and also used in garlands. For cut flower use, the old rose varieties like Queen Elizabeth, Super Star, Montezuma, Papa Meilland, Christian Dior, Eiffel Tower, Kiss of Fire, Golden Giant, Garde Henkel, First Prize etc. are still popular. In recent times, with production for export gaining ground in the country, the latest varieties like First Red, Grand Gala, Konfitti, Ravel, Tineke, Sacha, Prophyta, Pareo, Noblesse. Virsilia, Vivaldi etc. are also being grown commercially.

Chrysanthemum and Lilium is another most important cut flower crop in the country. Earlier it was considered a crop for temperate regions and its growing was restricted to the hilly areas, particularly in the north eastern region, which still continues to supply the planting material to most parts of the country. However, with improved agronomic techniques and better management, the northern plains of Delhi, Haryana, Punjab, Uttar Pradesh, as well as Maharashtra and Karnataka have emerged as the major areas for production of these cut flowers in India. Both single and double flower varieties of tuberose are equally popular and also sold loose in some areas for preparing garlands and wreaths. The other main cut flower item is orchid. Its production is restricted mainly in the north-eastern hill regions, besides parts of the southern states of Kerala and Karnataka. The main species grown are Dendrobiums, Vanda, Paphiopedilums Oncidiums, Phalaenopsis and Cymbidiums.

1.3 Foliage plants

Another group of ornamental plants of considerable commercial importance are foliage plants. Its production has gained much importance in the recent years in places like Bangalore, Kolkata, Delhi, Rajamundry, Trivandrum, Agra, Pune and in many other cities and towns of the country. Several nurserymen have taken up large –scale multiplication of the foliage plants to cater to the growing needs of indoor decoration at homes, offices and hotels. In the metropolitan cities particularly, there has been a great awareness in recent years of the usefulness of house plants for home decoration. With the growing population, lack of open space and development of multi-storeyed building systems, people have to depend largely on indoor plants for decorating their surroundings. Some exports of the foliage and other ornamental plants have started in the last few years to the Gulf and Middle-East countries.

1.4 Flower seeds

The demands of flower seeds are increasing worldwide because of rise in socio-economic status of urban people and awareness towards the healthy environment. Quality seed production is labour intensive. In floriculture, demand of quality seed of annual flowers is increasing. Maintenance of genetic purity of seeds is most essential. Four types of flower seeds can be produced–Mixture or open pollinated seeds, Inbred or pure-lines, F_1 hybrids and F_2 strains. Flower seed industry is another field, with great potential for development in India with the ban on import of flower seeds, during the World War II, when the indigenous seed industry developed. However, its growth was slow and on a small-scale only. In recent years, though a few seeds men have started producing high quality F_1 hybrid seeds of certain seasonal flowers for export, there is still much scope for development of this industry in India. An average home gardener is often not able to find reliable quality seeds at reasonable price. This has also affected the development of our gardens as well as the flower industry, to some extent. In India, with a great variability in soil and climatic conditions, it is possible to produce seeds of all types of flowers. There have also been instances of large-scale production of corms/bulbs of different bulbous crops like gladiolus, lilium, dahlia, crinum, crocus, calla lily, amaryllis to meet the growing demand both in the domestic as well in the international market.

1.5 High value flowers

High value flowers like Lilium, Asiatic ginger lily, protea, heliconia, orchids, bird of paradise, etc. are being cultivated in comparatively smaller area but these fetch higher returns per unit area owing to high flower value. Such high value flowers are being cultivated in southern states and hilly region of our country where climate is favourable and remains constant as per crop requirements.

1.6 Essential oils and perfumes

In India, flower crops grown for essential oil production are limited and include mainly rose, jasmine, tuberose, etc. (Singh, 1990; Irulappan, 1994).The important aspect of commercial floriculture is the production of flowers for extraction of essential oils and perfumes. There exists a great demand of these both in domestic and international markets. Their need in the perfumery trade, in the cosmetic market as well as in the food industry is continuously increasing. The two most important flowers which find their use in the perfumes are rose and jasmine. In India, extraction of essential oils from rose and jasmine has been practised for a very long time. Large areas in Uttar Pradesh around Hathras and Aligarh and Haldighati in Rajasthan are put under production of *Rosa damascena*, the Damask rose, for the extraction of rose oil. There are plantations at Kannauj (UP) and Amritsar (Punjab) and Himachal Pradesh, where rose attar, rose water and gulkand are manufactured from Damask rose and the Edouard rose (*Rosa borboniana*). Similarly, growers around Coimbatore produce large quantities of jasmine, particularly *Jasminum grandiflorum* and *J. auriculatum* for extraction of jasmine concrete. Processes for the extraction of these oils have been indigenously developed. A good R&D work has been carried by Tamilnadu Agricultural University, Coimbatore.

2. Ornamental gardening

This includes, landscape planning and designing, establishment and maintenance contracts, etc. This is highly money spinner business owing to the growing industries, population along with general eco-friendly awareness and need for green view. There is regular demand for systematic avenue plantation, traffic island landscaping, landscaping of public places and societies.

a. Aesthetical Importance: Establishment of gardens helps in maintaining the ecological balance by producing large quantities of bio mass. Ornamental trees help in fighting environmental pollutions and also add beauty to the environment known as avenue plantation and thus, they are considered as the best component of social forestry. Waste land, salt affected areas, arid semi-arid regions can be utilized for cultivation of suitable trees for improvement of surroundings as well as for high returns. The arboriculture and plantation of seasonal flowers improve the ambience of the locality. Now a day's, many corporate houses have set up their offices in India and they are giving due emphasis to the landscape to improve their ambience and public health. Even on the road side and as an avenue trees, many fruits plants like lemon, aonla, jamun, tamarind and mango are planted as avenue on road side. They provide shade, aesthetic value and serve a purpose of avenue plantation for healthy environment. Apart from trees, many topiary works, hedges, shrubs and seasonal flowers add value of colour and design in landscape.

b. Environment

Ornamental horticulture has also played a significant role in improving the environment. Besides the important parks and gardens described earlier in the post-independence era, much emphasis has been laid on development of gardens and public parks in many cities, towns, residential and industrial areas, river bank sides and alongside the highways. The greater use of ornamental plants certainly contributes to the improvement of the environment. For instance, some trees have been found to be useful in preventing air and noise pollutions in urban areas. The trees provide shade and create better microclimate. There is enough scientific evidence to indicate that the environment does affect the behaviour of man. These gardens have also served as good recreation spots.

c. Economy Benefit

Economy of Himachal Pradesh largely depends upon apple and flowers cultivation. Grapes a have flourished greatly in Punjab, Delhi, Maharashtra, Andhra Pradesh, Karnataka and Tamil Nadu. In Uttar Pradesh, production of mango in the plains and temperate fruits in the hills has increased substantially in the last few decades. Production of coconut and cashew nut is showing sharp upward trend due to new plantation in Orissa, north-eastern states and rejuvenation of old ones in Kerala. Fruits cultivation is highly profitable as the net return per unit area of land is maximum compared to many other crops. They are one of the potential foreign exchange earners. Cultivation of fruit crops generates employment for rural masses. The horticultural crops besides providing livelihood to a large number of small farmers contribute significantly to economic security. Covering 6.1 % of the gross cropped area, they contribute 18.8 % gross value of the agricultural output and 14.5 % of foreign exchange earned through agriculture.

Ornamental plant is today's lucrative business where we have a wide range of such plants based on whole sale or retail, indoor/shade loving plants, tissue culture plants, commercial flowering plants, shrubs, climbers and tree seedlings for landscaping, annual plants, bulbous flowers, etc. are flourishing well throughout the country. Besides, there is high demand of lawn species for landscaping as well as plant rental services in the nurseries.

A nursery is a managed site, designed to produce seedlings grown under favorable conditions until they are ready for planting. All nurseries primarily aim to produce sufficient quantities of high quality seedlings to satisfy the needs of users and earn a livelihood.

Globalization has improved the chances of export of quality planting material to other countries. Special techniques and care is required for exporting the nursery material. Similarly, great care is necessary while importing nursery material from outside. Ornamental and floricultural crops are numerous and are propagated vegetatively, like gladiolus, carnation, roses, lilies etc. There is a large group of ornamental plants, which is propagated by seeds and seedling such as Asters, Marigolds, Salvias, etc. There is an increasing demand for potted plants for gifts such as bougainvillea, roses, dahlia, cactus and succulents, etc. Specialized plants like ficus, cycas, roses, and bougainvillea etc can be multiplied in nursery site. It provides employment opportunities for technical, skilled, semi-skilled, and unskilled labour as there is a huge demand of skilled professionals for grafting, budding, potting, re-potting and other nursery operations. Therefore, nursery can itself be a very remunerative enterprise in the changing national scenario.

d. Phyto-remediation

Large areas of soils have been contaminated by heavy metals, which are deleterious to the existence, reproduction, and development of living organisms including plants, animals and microorganisms. This phenomena has been threatened the health of ecosystems and human being themselves. Because soils contaminated by toxic heavy metals have important characteristics such as concealment, delay, accumulation, regionalism and irreversibility. Soil remediation has not only received more attention in environmental science and engineering, but also becomes global problems to be solved urgently. Phytoextraction which makes use of harvestable parts of plants to remove pollutants represents green and environmental-friendly tools for cleaning metal polluted soils and waters compared with conventional chemical and physical remediation technologies which are generally costly and often harmful to soil ecosystems.

Ornamental plants are an important type of higher plants apart from those in the food chain, and are quite crucial if they have hyper-accumulation properties and can be applied to remediation of contaminated soils. Up to now, there was no document on ornamental plants that had been applied to phytoremediation, which can remedy contaminated environment and beatify it at the same time. Liu *et al.*, (2008) reported that three ornamental annuals e.g. *Calendula officinalis, Althaea rosea* and *Impatiens balsamina* had higher tolerance and accumulation ability to cadmium (Cd) and Lead (Pb).

3. Nursery

Nursery is place where plants are grown for transplanting, for use as stock for budding and grafting, or for further sale. Commercial nurseries produce and distribute woody and herbaceous plants, including ornamental trees, shrubs, bulb crops, seasonal plants and all other allied activities of ornamental horticulture. India has a rich diversity of ornamental wealth and many native plants can be grown as pot plants and can be traded in domestic and export markets. Genetically pure planting material is essential for healthy and vigorous plant growth. The planting material should be satisfactory in quantity and quality and easily available for further multiplication.

Nursery is consequently the basic need of horticulture. Plant propagation techniques and practices is the core of horticulture nurseries. The planting materials for horticultural plantations are raised from seeds and vegetative parts. Role of Mother Plants is very primary and important. The fate of nursery depends on quality and truthfulness of mother plants. A good nursery entrepreneur does not depend on others for procurement of mother plants. Mother plants are required for both stock and scion. Mother plants should be selected on the basis of its genetic traits and other factors like availability and adaptation in the growing environment. The various types of nurseries can be classified based on various criteria. They include, Sale: Retail nurseries which sell to the general public. Wholesale nurseries; offer opportunities to sell plants on larger quantity to the other nurserymen or buyers. Some retail and wholesale nurseries sell their product online like sale seeds, pots and garden tools.

There is sudden increase in the demand for certain commercial plants. It is not possible to fulfil this requirement by ordinary or common nursery practices. There is necessity to have special techniques and methods to meet the demand and only Hi-tech nurseries can satisfy this type of demand. These nurseries grow plants in greenhouse, building of glass or a plastic tunnel, designed to protect young plants from harsh weather, while allowing access to light and ventilation. Modern greenhouses allow automated control of temperature, ventilation, light, watering and feeding. Some also have fold-back roofs to allow "hardening-off" of plants without the need for manual transfer of plants to the outdoor beds.

Among the ornamental shrubs, a large number of cultivars have been developed in bougainvillea, croton, hibiscus and jasmine, which are being used to beautify the gardens in many parts of the country. The development of trade in decorative and house plants is of comparatively recent origin in our country. The plants produced at some nurseries are, however, of very good quality and are being exported. In India, there is a lack of scientific information on the production of flower crops and other plans under covers, glasshouses or similar suitable environment controlled conditions. This needs more attention, particularly in view of strict quality requirements of cut flowers for export purposes, which is difficult to obtain in cultivation under open field conditions.

4. Services

Various services are being provided in ornamental horticulture sector include maintenance of landscape site; providing seasonal requirements of seeds, pots, garden tools, machinery; shifting of old trees impart technical knowhow to various stakeholders; to let nursery plants (rental services); shifting of trees; protected cultivation and accessories. Import of bonsai and various specialized kind of ficus species are being important from China and Thailand in large quantity to use in ready landscape project.

5. Tissue Culture

Demand for flowers is increasing globally. India is expected to emerge as a strong player in the consumer market of biotechnology products in future. There is ample opportunity in ornamental horticulture especially for micro-propagation of flower crops such as gerbera, lilium, spathophyllum, carnation, *etc.* This is very big business and provides an ample opportunity especially for qualified young youth and girl to create the job in their locality and to earn foreign exchange through export of tissue culture plants abroad.

Tissue culture is the *in vitro* aseptic culture of cells, tissues, organs or whole plant under controlled nutritional and environmental conditions often to produce the clones of plants. The controlled conditions provide the culture an environment conductive for their growth and multiplication. These conditions include proper supply of nutrients, pH medium, adequate temperature and proper gaseous and liquid.

Plant tissue culture technology is being widely used for large scale plant multiplication. Apart from their use as a tool of research, plant tissue culture techniques have in recent years, become of major industrial importance in the area of plant propagation to produce millions of tree to type plants. A single explants can be multiplied into several thousand plants in relatively short time period and space under controlled conditions, irrespective of the season and whether on a year round basis.

Commercial production of plants through micro propagation techniques has several advantages over the traditional methods of propagation through seed, cutting, grafting and air-layering etc. It is rapid propagation processes that can lead to the production of plants virus free.

It has an ample opportunities to set up a small to big unit and finance can be arranged from the banking sector including subsidy available under MIDH - centre - sector scheme and NHB also provide such financial support.

References

Altaf Hussain et al., 2012. Current Status and opportunities. Plant Tissue Culture: an open access and peer reviewed chapter. DOI: 10.5772/50568.

IFTS (International Floriculture Trade Statistics) (2004) Pathfast Publishing. http://www.pathfastpublishing.com/docs/ITS98101.html.

Irulappan I, 1994. Breeding jasmines of high yield of flowers and floral oil. In: Prakash J and Bhandary KR, editors. Floriculture – technology trades and trends. Oxford and IBH Pub. Co. Pvt. Ltd., New Delhi, pp. 258–268.

Khan H. 2018. Top 10 flower producing countries in the world. www.worldblaze.in.

Liu, Jia-nv., Zhou, Qi-xing., Sun, T., Ma, L.Q. and Wang, S. 2008. Growth responses of three ornamental plants to Cd and Cd-Pb stress and their metal accumulation characteristics. J. Hazardous materials, 15: 261-267.

NHB database published National Horticulture Board, Govt of India, Gurgaon.

Sindhu, S.S. (2016): Ornamental Horticulture, Published by New India publishing House (NIPA), Pitampura, New Delhi.

Xia Y, Deng X, Zhou P, Shima K and da Silva JAT. 2006. The world floriculture industry: dynamics of production and markets. Ornamental and Plant Biotechnology: Global Science books, UK, p336-347.

2

Principles and Elements of Landscape Design

Ashutosh Mishra

A garden is a place where ornamental plants are grown to put on display various forms of plant life which are purposely intended for ornamental or practical use or both. Now days we often use the term landscape gardening, Landscape architecture and Landscape design. Landscape gardening can be defined as arrangements of trees, shrubs, climbers and variety of other plants in concert with the building, walks, drives, artificial and natural features for the use of humanity while Landscape architecture is the art of arranging land and landscape for human use convenience and enjoyment.A good landscape design determines the characters of objectsto serve a certain purpose known in advance.

The subject garden planning is diverse and complex and very difficult to symbolize all ideas of wholeness. In modern living consciousness of interior decoration including furniture,woodwork,wall furnishing, carpets painting also needs careful setting and the services of specialist are considered necessary for this purpose. In order to achieve a rewarding result a garden should be designed carefully as it is semi-permanent feature and frequent change in design is not possible.

Principles of design steer designers to organize elements for a visually pleasant landscape. A melodious composition can be achieved through the principles of proportion, repetition, order and unity which are alllinked and application of one principle assists to achieve the others. By using these principles, two important concepts in garden design i.e. physical and psychological comfort can be achieved. Landscape that has order and repetition creates more psychologically comfortable feeling. Psychological comfort is also affected by the sense of satisfaction that a spectator perceives from a harmonious landscape. Users feel more physically comfortable and secure in a landscape with proportions compatible to human scale. Garden design combines the aesthetic beauty, artistic knowledge and at least the principles of gardening. In order to layout good garden some fundamental principles to be followed

Principles of landscape design

The principles of landscape design serve as guidelines that govern the association of the design elements and materials in accordance with the laws of nature. Landscape designers employ following principles of design to create landscape design that are functional and aesthetically pleasing.

1. Simplicity

Simplicity is the essence of garden design. It is the concept of reducing or eliminating unnecessary to avoid a messy look and brings clarity and purpose to the design. This can be achieved by considerately removing features from a design while still preserving its integrity.Simplicity indesign is also achieved through either repetition of same species for similar form (silhouette), habit or colour and construction material (hard cape) or by avoiding too many species in a small place.

2. Proportion and Scale

Proportion is the definite relationship in masses or association that exists among the components of landscape. Proportion, in a sense is similar to ratio whileabsolute proportion is the scale or size of an object. A chief absolute scale in design is the human scale (size of the human body) because the magnitude of other objects is measured relative to humans therefore; plant material, garden structures, and ornaments should be measured relative to human scale (Hansen, 2019) Other important relative proportions include the dimension of the house, lawn and the area to be planted.

Proportion can be established in plant material relative to people, the surrounding plants, and the house and when all three are in proportion, the composition feels balanced and harmonious. A feeling of equilibrium can also be achieved by dividing the whole design in equal proportions of open space and planted space. Planting of markedly different size of plants help to achieve dominance (emphasis) through contrast with a large plant. Likewise, planting of similar size of plants in equidistance also help to achieve rhythm through repetition of size. Tall trees are planted near a tall, narrow house or buildingto maintain proportions. A small house surrounded by large trees appears smaller than it actually is while a large house surrounded by small trees appears larger than it actually is.

Garden features are most functional for people when they are approachable. Patios, chairs, tables, pathways, arbors, and gazebos work best when people can use them easily and feel comfortable using them. The hardscape should also be proportional to the house—a deck or patio should be large enough for entertaining but not so large that it doesn't fit the scale of the house.

Human scale is also important for psychological comfort in open spaces. People feel more secure in smaller open areas, such as patios and terraces. An important concept of spatial comfort is enclosure as most people feel at ease with some sort of overhead condition that implies a ceiling. The enclosure does not have to be solid; in fact, an implied enclosure, such as tree branches, serves as a good psychological enclosure that still allows light and views of the sky.(Hansen,2019)A design is in proper proportion and scale when a pleasing relationship exists among and between each component and the design as a whole.

3. Balance

Balance is a design principle defined in terms of weight. It is the concept of equalization of visual weight and attraction, usually around a real or imaginary central axis. Balanceis affected by form, color, size, and texture of the plant. It may be symmetrical, asymmetrical, or perspective.

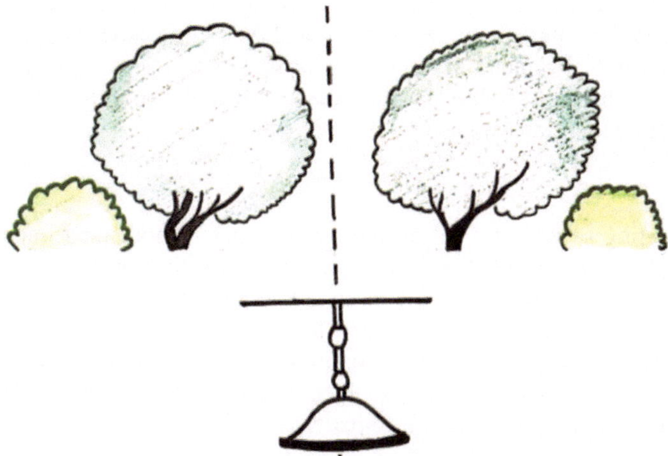

Symmetrical balance is a formal balance and also referred to as bilateral symmetry. Symmetrical balance is achieved when the same objects (mirror images) are placed on either side of an axis e.g. the same trees, plants, and structures on both sides of the axis. This type of balance is used in formal designs and is one of the oldest concepts and many historic gardens are organized using this concept. In this the mind naturally divides space by assuming a central axis and then seeks an even distribution of objects or mass (visual weight).

Asymmetrical balance is an informal balance and implies equal weights on either side of the central axis. This type of balance is usually achieved by masses of plants that appear to be the same in visual weight rather than total mass. This can be achieved by groupings of trees and structures that are approximately equal in visual weight on either side of the axis. The mass can be achieved by combinations of plants, structures, and garden ornaments. To create balance, features with large sizes, dense forms, bright colors, and coarse textures appear heavier and should be used sparingly, while small sizes, sparse forms, gray or subdued colors, and fine texture appear lighter and should be used in greater amounts.

Perspective balance is concerned with the balance of the foreground, midground, and background. The objects in front usually have greater visual weight because they are closer to the viewer. This can be balanced, if desired, by using larger objects, brighter colors, or coarse texture in the background. In most cases, either the foreground or background should be dominant.(Hansen,2019)

4. Repetition

Repetition involves repeating of an identical or similar component more than once through a design which helps to establish and add order & unity to a design. It is created by the repeated use of elements or features to create patterns or a sequence in the landscape. Repetition of line, form, color, and texture creates rhythm in the landscape. However,excessive repetition creates monotony and too little can create confusion. Simple repetition is the use of the same object in a line or the grouping of a geometric form, such as a square, in an organized pattern.

Repetition can also be made more interesting by using alternation (minor change in the sequence on a regular basis), inversion (selected elements are changed so the characteristics are opposite the original elements), gradation (gradual change in certain characteristics of a feature like height,texture and size). Repetition does not always create a pattern; sometimes it is simply the repeated use of the same color, texture, or form throughout the landscape. Repetition and pattern can be made most obvious in the hardscape because duplication is easiest with built materials that are manufactured to exact dimensions.

5. Order and Unity

Order is the overall organization and arrangement of a design whileunity is the harmonious relationship among all elements and characteristics of design. Order and unity areemotional and visual reactions to the overall formation and organization of the design elements.To achieve unity, consistency is used by fitting different elements of a landscape together to create a common unit or theme. This can be achieved by using dominance, interconnection and simplicity to assemble colors, textures, and form. Although hardscapes and plants can be amalgamated by the combination of analogous characteristics but some diversity is also important to create curiosity. The simplest way to create unity is through the use of a design theme or a design style as design themes and styles have a well-defined set of features that have maintained their fame over time because they are visually pleasing to many.

6. Focal point or Emphasis

Focal point is the point or area of landscape that attracts the viewer's eyes and is created to break the sequence and flow of the landscape. Emphasis is the property of a plant or object that attracts and grasps concentration, making the object an important feature. Focal points are used to draw attention to a particular location, move the eye around the space, or guide circulation. Through the use of emphasis,eye movement is directed towards a centre of interest that takes a position of prominence in the landscape. Focal point should be physically powerful and effective which creates as visual break in the sequence and flow of landscape but do not incorporate too many focal point otherwise, their effect will be lost. If there is more than one focal point and colour is the elements to be used as focal point in a site, than blues or whites are used for minor focal point while warm, dominant or bright colours are used for main focal point. Focal point could be a single specimen tree, a beautifully designed musical fountain, a piece of sculpture, or may be similar plants forms of varying size. Most landscape usually contains visual focal points such as plants and structures. However, running water and rustling leaves are effective as focal points, capturing the sense of sound.

The focal point in a landscape may also be focal area, which is more popular in present time, which also attracts the viewer's eye in the same manners as focal point as it consists of more than single elements or components. Focal area consists of a group of parts or components.

7. Rhythm and Sequence

Rhythm is are petitious use of sculpture in landscape to reduce monotony. This can be expressed through the placement of plants parks furniture etc. either individually or as group. Rhythm and sequence describe the dynamic unity or the related, orderly movement that implies continuity. Order and repetition help to establish rhythm and sequence in design. Rhythm and sequence characterize continuity and connection from one part of the design to another part and group the components together. Effective use of sequence is often times employed to create visual movement in the landscape. E.g. sequence could be an orderly natural combination of plant material in which low objects would appear in the foreground, intermediate objects in the middle ground and tall objects in the background.

8. Interconnection

Interconnection is a design principle for producing unity in design. Various components in the design are physically linked together. Although all features are linked to other features, the key is to make the linkage seamless so that the features blend or fit together. Hardscape is important to interconnection because it typically serves to organize and link spaces in the garden. Continuation of a line, such as a path, the edge of a built object, or a defined edge of a plant bed, can create unity through interconnection. A designer may incorporate interconnection into the entire design or into only a small space within the design.

Elements of garden design

The elements of design refers to the basic visual and physical features of the plant and hardscape components used to establish a design. These elements include line, form, texture and colour. A landscape designer uses these elements to enhance the design

1. Line

Line is one of the most important fundamental elements of landscaping. Line work as bones or the framework of the landscape. It creates a sense of direction and physical or visual movement and is able to direct the attention of the viewers to a focal point. Line is incorporated in landscape by using contrasting plant material and by forming pattern with similar plant materials in a repetitive sequence. In landscaping, lines exist in soft cape (garden plants) as well as hardscape (structures). There are four types of line in landscaping.

a. **Axis line:** This is an imaginary straight line in any garden around which garden is created striking balance. Straight lines are used to represent formality or a contemporary concept.

b. **Curved lines:** The curved line create an informal, natural, relaxed and flowing effect. Curved lines are the line of beauty and are smooth, give the sense of progress, moving and natural feeling. Good examples of curved lines in landscape are internal pathway, pathway inside big lawn, and curved edges.

c. **Meandering lines:** These are extended curved lines made in garden to mimic a flow of river or a stream. Small garden pathways are given naturalistic form by developing them with meandering lines

d. **Vertical line:** Vertical lines are purposely used to make a large space feel compact. These lines in landscape move the eye up which creates more excitement in the garden. Hardscapes such as Poles, pillars, statutes and other tall ornamental structures creates vertical line. Soft capes such as pendulous tall trees, bottle palm, Phoenix and coconut palm creates vertical lines in landscape.

2. Form

Form relates to the natural shape of the plant or is the two or three dimensional shape of Form relates to the natural shape of the plant or is the two or three dimensional shape of an object or space. Form is line surrounding mass; either it is two or three dimensional. The air space created by two plant materials set side by side is also an expression of form. The shape of trees and areas of lawn surrounded by edge are examples of form expressed in a landscape. All the components have a peculiar and natural form in a landscape. Mostly form of the plants contributes to the total design composition. The basic form of each plant depends on the natural growth habit of plant. Some of the common forms of landscape plants include conical, oval, weeping, and horizontal and fastigiated. Plants come in many forms including round, upright, spreading, etc. A round plant following a curved bed line gives a landscape a flowing and natural feel. In contrast, an upright or columnar plant placed along a linear walk takes on a more rigid and formal feel. Landscape plants may also change form when they are grouped together in a mass.When a design composition is viewed as a whole, the individual forms become subordinate to the overall mass of the composition. In nature most deciduous trees and shrubs have a rounded form. A conical form is characteristics of many evergreen trees while evergreen shrubs have more of a horizontal form. Incorporation of form into a landscape is done to manipulate a person's emotions.

Horizontal or spreading forms are used to add visual width to all structure while vertical forms for strong accents and for adding height. In design, addition of weeping or drooping or pendulous form is done to create soft lines and to provide a transition to the ground plane or as interesting accents in the garden. A weeping tree is a great choice for a focal point in the garden because the unusual drooping growth habit makes it a real eye catcher. Round or Oval plant forms create large masses and are effective as border and enclosures. The same kinds of forms, when used repeatedly, create a "visual theme" that provides consistency, harmony, and order in the landscape.

Forms (left to right) columnar, oval, vase, weeping, pyramidal, rounded

3. Texture

Texture is defined as the representation of the structure of the surface of leaves and bark, and is determined visually. It is a combined look of all the surface elements of a garden working together or it is the surface quality of any plant material or structure in the garden.Texture is relative and it relates to the coarseness or fineness of leaf, smoothness or roughness of bark and trunk, heaviness or lightness of leaves, usually defined by fine, medium and coarse or bold textured. Texture is also a feature of the aggregation of minor units that makeup the plant or structure. It is also linked with the sense of touch and sight e.g. physical surface of plants (smooth, rough, dull or shiny) which creates visual texture in landscape. Landscape plants and flowers possess a wide range of textural attributes, including leaf, flower shape and surface, bark and stems. All of these can greatly enrich the visual quality of a landscape composition. Actually texture in garden depends upon the distance from which the plant is viewed by the observer. Fine textured plants (Casurina, Thuja, Junipers, Gulmohar, Russelia etc.) and elements are usually whisky and delicate, look better in broad groups and swaths than alone. It enables certain emotions by the viewer e.g. soft textures create a sense of comfort, while rough textures (Kadamb, Teak, Kanakchampa, Palash, Spathodea etc.) add excitement and heighten attention. It creates a sense of space and depth when rough textures are placed in front and soft textures are found in the distance, the landscape appears larger. It provides contrast to emphasize or highlight parts of the landscape e.g. a soft lawn leading up to the course texture of a pillared-building of a university. It can create lines e.g. if a soft lawn is bordered by a course-textured hedge, it creates a distinct line for the eye to follow. A nice equilibrium can usually be achieved by blending about ⅓ fine textured with ⅔course and bold texture.

Fine Medium Coarse

4. Colour

Color is one of the most prominent elements in landscape design. Colour theory is very complex and very personal matter that expresses individual taste and feelings. Light is the source of colour which consists of several visible wavelengths. Colour and light are inseparable as each wavelength produces characteristic colour. White is the accumulation of all light while black is the absence of light. Colour wheel is useful to study colours and their relationships in the arrangement of annuals in beds. Red,Blue and Yellow are referred as the primary colours which cannot be made from any colour. Combination of these two primary colours produces a secondary colour and combining equal parts of primary colour and an adjacent secondary colour produces tertiary color. There are six tertiary colours are in colour wheel. Generally red, orange and yellow are warm / hard or advancing colour, as associated with colours of sun or fire. Blue, violet and green are soft or cool colour which express restfulness and have a receding visual effect. The black, grey and white are called as neutral colour. Color helps to emphasize, complement, heighten the effect of, or disguise other aspects of

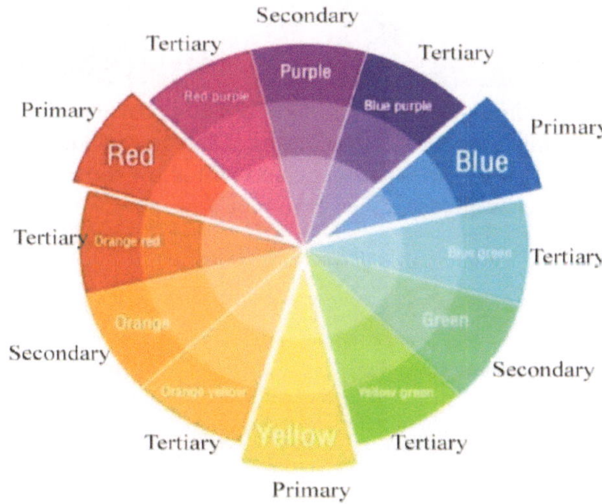

a landscape's design. Color also unifies the landscape and creates consistency (which our brains like) throughout the entire design. This colour wheel is useful for the arrangements of annuals in annual bed/herbaceous border and shrubbery border. The basic arrangement of colours in annual bed may be classified as follows.

a. **Monochromatic:** It is arranging of different tones of same colour either of the same species or in combination with different species e.g. group of blue petunias of different tones. In a monochromatic arrangement of yellow colour flowering annuals of different species such as marigold with yellow calendula or nasturtiumcan be grown in different height.

b. **Analogous:** This can achieved by using closely related or harmonious colour such as violet with blue and light blue with green, yellow with green or yellow with yellow- orange.

c. **Complementary or Contrasting:** In this two opposite or contrasting colour producing flowering annuals are arranged such as blue and orange, red and green and violet and yellow. In this triads arrangements is also possible i.e. one dominant with two complementary colours (violet pink and yellow)

The use of color and texture will simultaneously add variety and consistency to the landscape, preventing it from seeming too busy or too boring. The color will highlight the focal points, lines, and forms found in the landscape and will allow the viewer to feel the intended mood, the designer wanted the viewer to feel.

References

Hansen, G. 2019. Basic Principles of Landscape Design. https://www.edis.ifas.ufl.edu.

Misra, R.L. and Misra, S. 2012. Landscape Gardening. Westville Publishing House, New Delhi

Nambison, K.M.P. 1992. Design Elements of Landscape Gardening. Oxford & IBH Publishing Co. Pvt. Ltd., New Delhi, India

Tiwari, A.K. 2012. Fundamentals of Ornamental Horticulture and Landscape Gardening, New India Publishing Agency, New Delhi.

Whiting, D. 2014. Principles of Landscape Design, CMG Garden Notes, Colorado State University.

3

Styles of Gardening and Different Types of Garden

S.K. Khandelwal

Garden lover has to invent their own styles of gardening commensurate with their climate of site, budget and taste. Gardens fall under three categories when it comes to style – formal, informal and wild.

1. Formal style gardens

Formal style gardens are beautiful in the view and have been popular for thousands of years. Granted, it is not to everyone's taste but it creates a certain magnificence and persistence that will probably have meaning and importance in garden design for many more years in the future. A formal garden is laid out in a symmetrical or a geometrical pattern. Symmetry is the main characteristic of this style of gardening. Geometrically designed landscapes with trimmed hedges, paired flower beds, straight line planting and equal length of the sides of the area are the main features of formal or symmetrical gardening. The formal design is stiff as all garden features in a straight and narrow way. In the formal design, the land is forced to fit in the plan. In this type, an axis is drawn in the centre of the garden which divide whole of the garden in two equal halves, one half being the exact replica of the other, for every feature, element or adornment. The outline of the whole garden as well as the outline of different parts like flower beds, paths, hedges and lawns are of geometrical shape. These shapes are normally square, rectangular,oval or circular.

The Persian gardens, Mughal gardens and Italian gardens were formal in design. Geometrical formalism also influenced the French and British gardens of the pre-Industrial Revolution.

Key features

- A formal garden is laid out in a symmetrical or geometrical pattern. The formal garden design is stiff as all garden elements in a straight and narrow way. In such gardens, all garden features are laid out in straight lines.
- Roads and paths are straight and cross to each other at right angles.
- Bilateral symmetry; one side of the garden is the mirror image of the other. Hardscape, plant material, and shape of the garden beds are similar on either side of an axis so that the appearance of perfect balance is created. If there is tall

tree on one side of the garden, there will be a similar tree in the same position on the other side of the garden.

- Trimmed formal hedges, edges, cypress, Ashoka trees and topiary are typical features of a formal garden.
- Flower beds, borders and shrubbery are harmoniously arranged in geometric designs.
- Clipped shrubs, hedges, and lawn with everything appearing to be in its placc and under complete control.
- The arrangement of trees and shrubs is necessarily geometrical and kept in shape by trimming and training.
- Surrounded by a visual barrier such as a wall, clipped hedge or dramatic elevation change of the ground to give it a clear separation from the surrounding natural landscape.
- Other features like fountains, water pools, cascades etc. are used for further attraction
- Usually, gardens of public parks, university/ library gardens, or government buildings are designed in formal style garden.
- Formal gardens have no 'secrets' and the element of surprise is lost. However, attractive focal points at terminal and intersecting points of paths and roads are provided to make the formal garden effective.

2. Informal Style Gardens

This style of garden is sometimes referred to as naturalistic or asymmetrical design. It may draw inspiration from natural habitats. On the whole an informal garden design is extrovert type. It has asymmetrical balance even though the objects are not arranged around a central axis unlike the formal designs. However, on either side of the vertical axis, the dissimilar objects are placed without any similarity but equated to give the occult appearance of balance.

Informal garden is an exotic attempt to mimic the nature. It is a landscape casually designed with few straight lines, often depending on curving, organic lines and having a nice mixture of foliage, colors, textures, heights and varieties. Informal gardens act as a temporary sanctuary that offers a relaxed ambience from day-to-day stressful life. An informal garden is naturalistic and usually includes asymmetrically placed design elements, naturally shaped plants and beds, curved spaces, rustic-looking furnishings, and more casual paving, like decking and even wood chips. It can be flowery and colorful or low key to suit the personality of the gardener. This type of gardening is aimed to imitate nature and strives to produce a natural effect in a closed area. A completely informal type of gardens provide a natural appearance but are not usually suited to very small sites. In an informal garden, the whole design looks informal, as the plants and the features are arranged in a natural way without following any hard and fast rules. However, the work has to proceed according to a set and well-through-out plan; otherwise the creation will not be artistic and attractive. Planting is often with a mixed nature and there is a complete absence of set lines. Smooth, curvaceous outlines are more appropriate in this design rather than rigid lines. Informal gardens are laid out with open large lawns, bordered by clumps of shady trees or shrubs. The shape of the garden may not be square, rectangular or circular. Straight paths are avoided and irregular beds are introduced. Shrubbery and herbaceous borders are so designed as to fit the periphery of the lawn. In this gardening design, nature is preserved in an artificial way. In informal gardens plants are often positioned in clumps (a circular group of three or more plants) or drifts (an elongated grouping of plants). Clumps and drifts are most often planted with an odd number of the same kind of plants to give the appearance of a more natural grouping. As noted, informal gardens are often naturalistic, echoing the surrounding landscape. The Informal gardens tend to be large to medium sized gardens. Usually implemented on uneven landscape and even sloped to accommodate the uneven landscape.

Ex: English and Japanese gardens

Key features

- Informal gardens are characterized by gentle curving, asymmetrically arrangements of garden features and spaces. Plants are allowed to grow in their natural shapes.
- In the informal garden, the underlying framework is almost entirely disguised by planting, and the garden should look as grown up naturally.
- Informal gardens are harder to design than formal ones, as they are ruled by irregularity and natural look of plant growth.
- A site that is not geometric shape can be a sound basis for an informal garden.
- Plan is asymmetrical according to the land available for making the garden. The garden plan is made to fit the land.
- Some hardscaping structure needs to exist, or everything will relapse into an untrammelled wilderness.
- Diagonal or curved path work well, and boundaries between the various areas of your garden are often formed using native mixed hedging.
- In general, planting in informal gardens includes tall shrubs and trees to add the vertical dimension, and these will often hide the edges of the plot to create a feeling.
- The garden should revel in colour and should look as natural as possible. If you would like to introduce water, natural-looking ponds, and pools are ideal features.
- This style reflects naturalistic effect of total view and represents natural beauty.
- It is contrast to formal style.
- Water bodies are more irregular in shape.
- Features such as hillocks, waterfalls, lakes, islands, cascades, rocks, *shola* and rustic hutments are provided to create rural effect.

- Plants are appropriately grouped and they are not trimmed, so as to avoid geometrical arrangements.
- The most typical features are more flowering shapes, less regulated and controlled planting and soft curved patterns.
- Asymmetrical design is adopted with non-geometrical beds and borders.
- The plant material includes mostly the local or native plant species predominant in the region and landscaped in such a way as to merge with the surrounding nature.
- In an Informal garden, the whole design looks informal, as the plans and the features are arranged in a natural way without following any hard and fast rules. But here also the work has to proceed according to a set and well-through-out plan; otherwise the creation will not be artistic and attractive.

3. Wild style gardens

A comparatively recent style of gardening, namely, "Wild Garden" was expounded by William Robinson in the last decade of the nineteenth century. His idea was revolutionary and found many admirers to follow this. The concept of wild garden is not only against all formalism but it also breaks the rule of landscape styles. His main idea was to naturalize plants in shrubberies. He also preached that grass should remain unmowed, as in nature, and few bulbous plants should be grown scattered in the grass to imitate wild scenery. He also suggested that passages should be opened in the woodland, and trees, shrubs, and bulbous plants should be planted among the forest flora to fulfill his idea of a wild garden. His other idea was to allow the creepers to grow over the trees naturally imitating those of the forests. Before someone starts to venture into designing a garden it will be wise to get an idea about the major gardening styles of the world.

Mughal Garden

The gardens that were laid out in the mughal period are known as mughal garden and are famous till today for their own architectures and beauty. Significant uses of rectilinear layouts are made within the walled enclosures. Mughal gardens design derives primarily from the Islamic garden, although there are influences that come from the Persian and Timurid gardens. The most important feature of Mughal garden is *Charbagh* system. The Mughals were obsessed with symbol and incorporated it into their gardens in many ways. Quranic references to paradise were in the architecture, layout, and in the choice of plant life. Numerological and zodiacal significances connected to family history or cultural significance, were often used. The numbers eight and nine were considered auspicious by the Mughals and were found in the terraces or in garden architecture e.g. octagonal pools. The design symbolizes 4 flowing rivers of Paradise and reflects the gardens of Paradise derived from the Persian *paridaeza,* meaning **walled garden** abundance with 4 rivers flowing from a central spring or mountain, separating the garden into north, south, east and west.

From 1494-1707 A.D. quite a few garden were created by Mughal Emperors in different parts of the country. Babar (1494-1531), the mughal emperor, introduced gardening techniques of central Asia & Iran with slight modifications in the architect and plant material that later on was called Mughal style of gardening. Baber laid the foundation of the mughal empire in India in the year 1526 A.D. He was a great lover of flowers and gardens. Immediately after his victory over Ibrahim Lodi at Panipat (1526) he laid out a garden there. There after he laid a garden at Agra, now known as Ram Bagh. From the Baber Nama, memories of Babar, we find the names of following flowers, oleander, hibiscus, white jasmine, and keora (screwpine). Babaris credited with the introduction of Persian rose. In India Akber (1556-1605) built a new capital at Fatehpur Sikri, completed with gardens, trees and flowers. He was first Mughal to enter Kashmir and establish garden, Nasim bagh, close to Dal Lake. The tomb garden of Akber at Sikendra is an example of gardening art during his period.

Jahangir (1605-1627AD) was also a great admire of garden and flowers, and so was his famous wife Nur Jahan. The garden of Shalimar, Achhabak and Varinag in Kashmir were created by him. The garden at Itimad –ud –daulah in Agra, in memories of his father-in-law, was also his creation. Shah Jahan(1627-1658AD) also establish several beautiful gardens. He built Red fort in New Delhi, which has beautiful garden. The garden around Taj Mahal in Agra in the memory of his beloved queen Mumtaj Mahal was also his creation. The small but pretty garden at Chashma Shahi in Kashmir was laid by one of his Governor. The best garden created by Shah Jahan was the Shalimar Garden in Lahore (Pakistan). Sher Shah Suri (1540-1544 A.D.) who ruled for a short period after defeating Babur, constructed the famous Grand Trunk Road and planted avenue trees on both sides of the road. The Nawabs of Avdh also established several gardens in their empire especially in Lucknow. One of these was Sikandar Bagh which was established by the last Nawab Wajid Ali Shah (1847-1856AD), now known as National Botanical Research Institute.

The Rajput Rajas in India also established several gardens. The garden at Amber fort near Jaipur was started by Man Singh (1590-1615AD) and completed by Jai Singh II (1699-1743AD). This beautiful garden is laid in three terraces and has lake. The Mandore garden near Jodhpur is really beautiful garden still today and was built by Raja Abhai Singh (1724-1749). The city of Jaipur was founded by Jai Singh II (1727 AD) who built a palace in the heart of the city with beautiful gardens. The garden palace at Deeg in the erstwhile Bharatpur state was founded by Suraj Mal with fountains water sources and other features. It is one of the most beautiful gardens ever created by any Rajput King. The gardens and the places in Chittor, the capital of Mewar rulers was also famous. The rulers of Bundi in Rajasthan also were founder of gardening.

Mughal garden are strictly formal and has straight flowing steam with fountains as the attraction of garden. Stone edges and path, terraced lawn and trees of formal shape (*Juniprus, Thuja* etc.) on either sides of the steam after stone path are planted. Few flowering, ornamental and sweet scented shrubs like roses, jasmine and ocimum are also planted for fragrance. Whole garden area are enclosed by high protected walls and a huge gate and building at the entrance. Fruit trees like mangoes, citrus pomegranates etc. grown after lawns of either side to act as wind break and provide cooling effect. Sweet smelling fragrance in fresh air is produced when they are in bloom and these plants also provide fruits.

Feature of Mughal Garden

Site and design

Mughals gave a prime importance to the site for the construction of their gardens. Mughals were very choosy about the selection of site and always preferred a site on a hill slope with a perennial rivulet or along the bank of a river. Mughal gardens are generally rectangular or square in shape and different architectural features are the main stay of the design. Its architecture is influenced by the climate of the place, its political, social and economic conditions.

High wall

The Mughals created the gardens not only for pleasure and recreation but also as forts and residences surrounded by 15-20 feet high stone walls. The purpose of the high walls was security from the enemies and shelter against hot winds, rodents and wild animals. In the days of insecurity the emperor use these gardens as a short-term safe residence. The gardens were a place of peace for the Emperors to enjoy with their wives and concubines.

Huge gate

A large main gate was made of stone to match the high enclosing wall. It was made with beautifully carved stones to produce the greatness to the architect of Emperer's garden. Doors are huge and strong wooden structure studded with bold iron nails and pointed iron spikes. It was also to let the Emperor reach to palatial building while sitting on the elephant.

Palatial building

Mostly just after the gate a palatial building is made with large decorative rooms on I[st] or second floor for the Emperor. The whole building is made of beautifully earned costly stones. It is generally open and the rooms have decorated glass windows which open towards the garden to have a full view of the garden.

Open platform

In front of the main building towards the garden a large platform of costly stone pieces of geometrical shapes is made for strolling, open darbar and dances etc. in full moon night for Emperor and his friends. It has beautifully carved stone fencing of low high about 2 feet all around it.

Swings

In some gardens high gate like stone structure is made at suitable place especially close to palatial building. It is very beautifully carved which is used to hold the swing for entertainments.

Running water

It is said that water is present in the garden symbolically and physically as a source of life. The style for having running water by constructing canals and tanks borrowed from the Persians. The fascination for water came from the Muslim faith which says that the promised paradise is the place "where cooling waters flow". It is always in the center of garden to divide it in two equal halves. It has uniform width and flows straight from the base of the palatial building upto the end of the garden. Broad water canals were made in overall proportion to the area. The central water course was strong and dominant. In the centre of these canals, there were fountains which make for the central axis also. The cross formed in the centre by the intersection of water channels has been the symbol of the meeting of humanity with god. A fall at each terrace is made to maintain the uniform flow of water. It may have one or two small stone platforms under which the stream flows as in Taj garden. In some other gardens mainstream have straight side streams also which may open in geometrical shaped ponds as in Deeg garden. The base and sides of the streams are pitched and made with large plain square or rectangular stone pieces. These are firmly joined together to make the stream waterproof and to give in uniform geometrically shape the stream. The water canals were paved with tiles (or marble) of blue colour to create the illusion of depth. The course of water used to be manoeuvred in various ways taking advantage of each slope, however small it may be, to break up the flow into artificial falls and ripples. The water canals used to have fountains to throw up the water high in the air. Running water in the form of cascades and 'chadars' made the garden splendid. There were different types of cascades and 'chadars'. These white waters shawls (Chadars) are characteristic features of the Mughal garden. To create a pleasant site at dusk, tiny lamps placed in the niches behind the artificial water falls were lighted, and the diffusion of light through the water cast shimmering reflections.

Grills

Narrow horizontal wavy building lines are made on the base lining of the fall through out the vertical length of it so that the water when moves over these projected lines may produce tickling sound to provide lively musical effect in peaceful nights.

Baradari

Baradari, a canopied building with twelve open doors on all sides, is another feature of some of the Mughal gardens. The baradari was generally used by royalty to observe the play of the water or 'chardar' in the twilight. Sometimes one or two small open building are also on the stream at the center of the garden or at the end of the stream to rest for a while when roaming in the large garden. Baradari is evolved by Hindus to sit and enjoy fresh breeze and to watch the dark, clouds during the rainy season. But Mughals used these baradaries to sit with their slaves and listen to the songs of the dancing girls and watch pillars of baradari with their favourite design of bouquets of flowers in the vases and furnished them with thick carpets and cushions.

Stone path

The either sides or stream and firmly and beautifully paved width rectangular and squire stone pieces which are used as a garden walk. Such stone paths are also made all around the terraced lawn in the garden for convenient movements all around the green lawns. Generally, paths ran parallel to the canals so that a person walking on them could see and admire the beauty of the colourful paved marble reflected in the crystal-clear water and enjoy the beauty of the spray from the sparkling fountains by moonlight.

Stone cut work

At a little distance to the stone paths in the lawns a series of stone cut work is layed all along on one side of each path. These are generally star shaped having 6-8 angular points. The inner cutwork space of such stones is full of green grass. This stone cutwork gives a beautiful site as border to the path and stream and also adds architectural beauty in the center of the garden. Now a day some flowering annuals are grain on with in the space that hampers the ture mughal style however addition of flower improves the beauty in traditional English style. The permanently built flower beds were geometrical patterns, with one or two exceptions, where the natural shapes of flowers of various types were adopted for the flowerbed designs.

Terrace

The Mughals came from the hilly terrains and so they were fond of terraces in the gardens. For this reason they used to select the location of gardens near hill slopes. To maintain the proportion of land in view of the garden, the basic plan extended into terraces, irrespective of the topography of the land. The presence of seven, eight or twelve terraces in the garden symbolized plants, paradise and the zodiac, respectively. On either sides of the water fall on each terrace stone stairs of the width of the path are made. Such stairs are also made at the sides of platform from where the stream flows under neath. The gardens were divided into two or more terraces with a small brick-paved canal of running water flowing down from the tanks provided with fountains. In many gardens, the entrance is at the last terrace, with the exception of the Pinjore Garden, where the entrance is on higher ground and the seventh terrace is the lowest. Terraces were also made on flat ground by dividing land into terraces. Their fascination for terrace was so intense that even in the plains of Punjab they created artificial terraces.

Terraced lawns

Square and rectangular lawns are made in terraces in the garden. These are very much symmetrically planned on either sides of the stream. Lawn are regularly mowed and neatly cut to maintain velvety effect.

Water Ponds

These are made of Hexagonal or Octagonal shapes and are connected with flowing streams coming from main stream. The fountains are so arranged that they produce a dome shaped water structure while in flow. The Mughals generally preferred a garden site on a hill slope with perennial rivulets, an unbroken site of water or along the banks of a river.

Fountains

Small fountains are arranged in straight line at the center of the stream or at inside edges of the water pond. These are made of stone and beautifully carved in pointed conical shape. The fountain body is hardly visible above the water surface except the water jet coming out of it. Such jet fountains are also fixed at the either edge of main path as Sahelion kibadi. Generally such fountains are made have in nozzles and not at all visual except water jet coming out from the surface of the path. Use of such jet water fountains give additional beauty and also make the garden atmosphere humid and cool. Height of water of such fountain was low in old days as the gravitational system was used for fountains and not the power pumps.

Lake

At the end side of the garden mostly a huge water pond or natural lake or river was utilized for receiving water from the flowing stream. It also act as a source of water supply to the flowing stream by lifting water for it through adjusted channels and lifters.

Trees and shrubs

The trees were selected with careful planning and thought, as each tree symbolized something, like life, youth, death, etc. Fruit trees were considered symbols of life and youth, Cypress represented death and eternity. White kachnar (Bauhinia alba) represented the youth of life. The traditional planting of alternate cypress and flowering trees along the waterways represented immortality and the renewal of life, respectively. The Mughals had bias for spring flowering trees and flowers. Plant materials were selected carefully for formal effect in the garden. Selection of plants was dependent upon the local climatic conditions. Fruit trees, flowering plants, shrubs and colourful annulas were planted in the garden to give a new life to it. The seasonal flower beds were of geometrical pattern and constructed along the water canals or near the main buildings. The favourite flowers were rose, jasmine, carnation, hollyhock, delphinium etc. Fragrant flowers were the main attraction of the garden. Fragarant flowering shrubs like harsinghar, queen of the night, jasmine, rose and screwpine were commonly planted. Herbaceous plants like narcissus, daffodils, iris, carnation, hollyhock, larkspur, stock, sunflower, amaranthus and coxcomb were dominant. In Kashmir, the Mughals planted chinar, white poplar, cypress, apple, peach, plum, sour or sweet cherries, quince, pomegranate and fig. in the plains. The gardens were planted with fruit trees like mango, garpe, fig, orange, citron, peach and plum.

Trees by the side of stream: Symmetrically shaped trees like Juniprus, Thujas etc. were grown in the lawn at equidistance on either side of streams path. These being evergreen do not shed leaves and do not make water of the stream and the garden dirty. Being cylindrical in shape and less in width do not interfere with the over all view of the garden and very much match with the formal designs of the garden.

Shrubs and bushes: Some small flowering shrubs like jasmines Ocimum, Marigold, Harsingar, Champa, roses and other bushes were used at the edges of terraced lawn for flowers etc. In those days most of the present time beautiful shrubs were not available.

Trees on edges of the garden: Mostly trees like Mangoes, Citrus, Pomegranate etc. are planted in between the terrace and the boundary walls to fill the gap. These trees hides the unrightfully view of the walls and produce green background all either sides of the garden. Such trees produce sweet fragrance when they are in full bloom and fruits for consumption. Besides this the trees made the garden cool and change the microclimate of the garden, which helped in growing other flowering and tender plants in good condition. Garden in temperate and subtropical regions were planted with Apple, Pear, Peches, Litchi and other suitable plants of the region.

Mughal garden now a days are more beautifully decorated because of the introduction of more and more number of spps, of flowering trees, shrubs, hedges, annuals, climbers etc. which are planted in beds and suitable places in the garden in English time. Some garden adornments are also included like Arches, Standard, Pergola and conservatory etc. in mugul garden.

Japanese Garden

The art of gardening is believed to be an important part of Japanese culture for many centuries. The Japanese are the great lover of plant and flower and infact no nation has such deep passion for flower as the Japanese. The Japanese garden designs were based on ideas of heaven. Shinto, Buddhism and Taoism were used in the creation of different garden styles in order to bring a spiritual sense to the gardens and make them places where people could spent their time in a peaceful way and mediate. Core values of Japanese gardening have been largely shaped by Chinese culture and tradition. Japan is known for its gardens, tea ceremony, ikebana and bonsai. Japanese gardens are famous in the world for their unique style, natural and spiritual beauty, quietness and calmness. Japanese gardens are also known as 'nature in miniature'. Japanese gardens were influenced by the Chinese philosophy of Daoism and Amida Buddhism. These gardens have been made by Zen Buddhists.

Another strong basis of the Japanese garden is its immutability, i.e., except some seasonal changes in the deciduous trees the Japanese garden hardly goes through any other strong visible changes during different seasons. The immutability is achieved also because rather than a grand mixture of flowers, shrubs and trees more emphasis is placed on natural elements such as a simple path, a group of rocks, stepping stones, streams, waterfalls, bridges, stone lanterns, and so on. Water, stone, and plants are three most important elements to achieve these objectives.

These gardens have passed through different development phases under the influence of religion, culture and likings of rulers of different era. Main historical periods can be summarized as follows:

(i) **Asuka period (538-749 A.D.):** During this period Nara was the capital of Japan. The influence of Chinese culture was prominent. Gardens were designed to capture the landscape in its natural form. Artificial hills or rockeries were constructed which were 100 feet high with running water. The stones were selected for interesting shapes and kept near running water. eg., Gigantic Buddhist, Todaijii, kofukiji and Biodo inn at uji temples are best for this style.

(ii) **Heian period (794-1185 A.D.):** The capital of Japan was shifted from Nara to kyota. The gardens made during this period were highly decorative. Aristocratic style of gardens where created in front of the mansion with artificial ponds and islands. Big shrines were made having large lakes. Bridges, stepping stones. eg., Heianji temple is famous for its style.

(iii) **Kamakura and Muromachi period (1186-1399 A.D.):** During this period stone, hills and lakes were essential feature. A lot of importance was attached to the selection of stones with religions and moral attributes. Various names were given to different islands like 'Hill island', 'Wood island', 'cloud island, mistry island, Dry beach island eg., Ryoanji temple of Kyoto.

(iv) **Ashikaga period (1400-1600 A.D.):** Yoshimitsu Ashikaga, the 3rd Ashikaga shogun got the construction done of very important garden along with three storied golden pavilion called kinkakuji in the year 1397. Another important garden of this era is silver pavilion (Ginkakuji).

(v) **Edo period (1615-1867 A.D.):** During the Edo period, the making of small gardens dominated and no major change or feature was brought to the garden. The garden could be enjoyed from inside of a house. These gardens were similar to tea garden. eg., Katsura imperial Villa (1642) lay kobori Enshu (a master of tea ceremony).

(vi) **Meiji period (1876-1912 A.D.):** The capital of Japan was shifted from Kyoto to Tokyo in 1867 by emperor Meiji. The Meiji period saw the modernization of Japan and the re-opening to the Japan to the west. Many of the old private gardens had been abandoned and left to ruin in 1871, a new law transformed many gardens from the Momoyama and Edo periods into public parks.

Types of Japanese gardens: The Japanese gardens are further classified based on positions, shape, and purpose. The important types are:

1. **Hill garden:** This Japanese style garden represents a miniature of natural scenery that includes ponds, hills, stones, trees, fishes, bridges, moss, paths, flowers, small plants and streams. This style is known in Japanese as Tsukiyama niwa or Tsukiyama-sansui meaning hills and water because main features of such gardens are hills, streams and ponds along with other features. The hill garden is made up of one or more hills designed with earth mounds and exposed weathered stones. The other features of this garden area include water in the form of a stream or a pond or waterfalls or all the three with or without islands

and also bridges, lantern, stones, and trees. The important points in the garden are decorated with stones and selected trees. But pine trees may be planted to give the effect of being swept by wind. Untrimmed stepping stones are placed over the walks. An island is generally a usual feature in a hill garden.

2. **Flat garden:** As the name implies, *Hira-niwa* or flat gardens are laid out in flat ground without hills or ponds. Flat gardens are supposed to represent a mountain valley or a meadowland. Such gardens lacks ups and down and are devoid of hills, streams or ponds. These gardens were popular during the era of Muromachi (1392-1573). A flat, garden is not necessarily as flat as a pan -cake. Since it stimulates a mountain velly, low rounded hill designed with the help of stones or earth mounds or both will look quite appropriate in a flat garden eg., Ryoanji temple. Mostly '*Moore*' type gardens are developed to create scenic beauty, other adornments like stones, wells, water basins, trees, *etc.,* are used. Water current of an ocean effect is produced by covering the land surface with pure sand. The rocks or pebbles are so arranged that they give an effect of diversion or rush or water.

3. **Tea garden:** In Japanese culture, the tea ceremony has gained the status of national ceremony and has been intricately woven with life style. Tea garden is a type of beautiful Japanese garden that has a tea ceremony house along with the garden. The tea garden is laid out based on certain principles and customs of the Japanese tea ceremony and hence needs a considerable space of at least about 200 square meters, for its designing. Since the performance of the tea ceremony needs an atmosphere of intimacy, it is essential that the garden be enclosed by a fence. But the fence should be rustic in mature, with a gate made of very light material such as bamboo.

The present day Japanese tea garden is sectioned into three areas *viz., sotoroji* (outer section) *machi* (middle section) and *uchiroji* (inner section)

Outer section: The outer tea garden is comparatively a narrow area, with a waiting place where the guests are supposed to wait until the master of the house appears to welcome them. The guests are supposed to wait after removing shoes. Paths will be provided with stepping stones to lead to middle section. Stone benches of irregular size are provided and the area is not planted with many kinds of plants except grasses.

Middle section: Stone troughs with water are kept for the guests to clean themselves before making entry into the inner section. Here also the stepping stones and naturally looking objects are located in a improper manner.

Inner garden: The inner garden contains the tea house. The tea house of the classical time was nothing but a small straw hut with an outside waiting place, a small side room for washing the utensils, and the main ceremonial tea house itself having a capacity to accommodate only five persons.

The most important feature at the entrance of the tea house is a water-basin or a well or both for the visitors to rinse their face before entering for the tea ceremony. To illuminate the water basin and resting place stone lanterns are set in appropriate places. The selection of water-basin and stone lanterns is done with scrupulous care so that both the purposes of utility and beauty and elegance are fulfilled.

As far as tree planting is concerned, a contrast is maintained between the outer garden and the inner garden. The outer garden will have simple plantings and stone groupings. The trees, shrubs, annual and grasses are put in an informal manner. Hedge walls are provided to look natural. The selection of trees is such that when the outer garden is exposed to the light the inner must be darkened by shady evergreen trees.

4. **Passage garden:** The passage gardens, the Roji niva, are those which are laid in narrow passage, as for example a narrow space between two houses or approaches to buildings. As such areas are generally narrow, the garden layout should be simple and not over crowded. In such gardens there should be hardly any ornaments such as lanterns, basins or other man-made features. The common feature of a passage garden are a few key rocks, slabs of stones, and only a sample of types of plant. Bushy shrubs and trees are unsuitable in a passage garden, plants with open form and slender shapes are selected.
5. **Sand garden:** It is the simplest style of gardening, though not liked by many as it is totally devoid of plants. Instead of using water in this type of garden, sand or gravel is used to represent river or sea. The main feature of this style of gardening is to arrange a few vertical and prostrate stones in groups of 2 or 3 and to fill in the gap between the stones with fine white gravel. This style of gardens looks pleasant and effective only when confined to a limited area. The gravel is raked in most simple patterns simulating the ripples of flowing water. The raking has to be repeated often to keep the garden in its best shape. This style of garden looks pleasant and effective only when confined to a limited area.

English Garden

English garden is a style of landscape garden which emerged in England in the early 18th century and spread across Europe, replacing the more formal, symmetrical 17th century principal gardening style of Europe. The English garden represent an idealized view of nature. It drew inspiration from paintings of landscapes by Claude Lorrain and Nicolas Poussin and in the Anglo-Chinese garden from the classic Chinese of the East, which had recently been described by European travelers. English garden is the most beautiful garden amongst all European gardens. Due to high rainfall, the natural groundcover in the English countryside is grass. British garden architects Repton and Capability brown advocated the concept that British garden should look like countryside.

The reasons are

i. The UK enjoys the typical grassland climate with well distributed rainfall. This is favourable for the growth of herbaceous perennials.
ii. Since they were ruling many countries, valuable plant collection was possible from diverse areas.
iii. Gardening is a high cost technology and British were rich enough to afford luxury.
iv. English people are very fond of flowers.

The history of gardening in England dates back of fourteenth century. When monks and priests conceived the idea of gardening and they started with kitchen gardens with the objective to supply the fruits and vegetables with a sole purpose of getting food for monasteries and grew herb gardens for medicinal purpose. Gradually, the spirit of gardening was imbibed among the people who realized the value of residing in pleasant surroundings. By the middle of the sixteenth century the English gardens saw flower beds, topiary, and terrace gardens. Gardens consisted of symmetrical walks of clipped evergreens, statutory and masonry metal stages, statues of Greek gods were scaled on architectural fantasies. Introduction of various exotic plants though new trading companies. Flower garden got a special stimulus. It is formal in design with topiary and carpet garden.

The predecessors of the landscape garden in England were the great parks created by Sir John Vanbrugh and Nicholas Hawksmoor at Castle Howard, Blenheim palace and the Claremont landscape garden at Claremont house. These parks featured vast lawns, woods and pieces of architecture, such as the classical mausoleum designed by Hawksmoor at Castle Howard. At the centre of the composition was the house, behind which were formal and symmetrical gardens in the style of the garden with ornate carpets of floral designs and wall of hedges, decorated with statues and fountains.

The new style that became known as the English garden was invented by landscape designers William kent and Charles Bridgeman.

- William Kent (1685-1748) was an architect, painter and furniture designer who introduced Palladian style architecture to England and compose gardens look like landscape paintings. Kent is recognized as the Father of the English landscape garden and a pioneer of picturesque. Kent's inspiration came from Palladio's architecture of the houses he built.
- Charles Bridgeman (1690-1738) was the son of a gardener and an experienced horticulturist. He collaborated with Kent on several major garden, providing the botanical expertise which allowed Kent to realize his architecture visions.

In the middle of the 18th century, the gardens were laid out with more emphasis on architectural features. The main features of the gardens of this period were; curved paths, informal groups of trees, rivulets or streams, waterfalls and clipped hedges. The flowering annuals, the main stay of an English garden, came into prominence during the nineteenth century which subdued the architectural features. The English climate suit admirably well for the growth of herbaceous annuals. This prompted them to evolve a large number of hybrids of annuals as well as to collect the best flowering plants from all over the world. Most of the flowering annuals that we see today in the Indian gardens, with the few exceptions of amaranthus, balsam, gomphrena, marigold, etc., were brought here by the British.

The English garden usually included a lake, sweeps of gently rolling lawns set against groves of trees and recreations of classical temples, gothic ruins, bridges and other picturesque architecture, designed to recreate an idyllic pastoral landscape.

1. **Tree groves:** were spread throughout the landscape with paths that allowed the gardens users to wander in and out of the groves and provide a view of rolling against mass tree plantings.

2. **Ha ha wall:** The purpose of ha ha wall was to separate garden from the grazing land but was invisible from the distance. Cattle could be kept securely in their fields without the use of hedge or fence to interrupt the view.
3. **Grottos:** Grottos were used as romantic hide outs. They were manmade but build to resemble a dark natural forming cave.

The Key features of English gardens in India are:

1. **Lawn:** Lawn is a principal feature of English garden. A lawn can be of any shape and it may be plain or undulating. It can be laid by seed, turfing or plastering.
2. **Rockery:** It is the second important feature of an English garden. There are various types of rockeries.
   - Cold weather rockery of annuals like Alyssum, Calendula, Candytuft, Dianthus, Ageratum, Gaillardia, Verbena, Zinnia.
   - Rockery of succulents for dry areas.
   - Rockery of ferns in humid areas.
3. **Herbaceous border:** Flower beds or borders are characteristic feature of a British garden. The height of the plant, time, duration and overall colour scheme are taken into consideration.

References

Arora, J.S. 1999. Introduction to Ornamental Horticulture. Kalyani Publishers, Ludhiana, India.

Bose, T.K, Maiti, R.G., Dhua, R.S. and Das, P. 1999. Floriculture and Landscaping. Naya Prakash, Calcutta, India.

Randhawa, G.S. and Mukhopadhyay, A. 1986. Floriculture in India. Allied Publishers Pvt. Ltd., New Delhi.

www.HortiAgri.com

www.uahs.in

4

Turfgrass Management: Principles and Practices

S.L. Chawla, Roshni Agnihotri and R.D. Pawar

Turf is an English word, which traces its origin to India from the *Sanskrit* word *"Darbha"* means "tuft of grass" or "*Kusha*" in hindi. *Kusha* has been mentioned numerous ties even in Hind mythologies (Turgeon, 1985). Turf, is a grass which is smooth, has quality texture and a soothing colour surface that is used in the golf courses, lawns, fields, nurseries, gardens, *etc.* Turf is a piece of earth of about 5 cm thickness with grass thickly grown over it. Turfgrass has become increasingly important in today's society as people's needs increase for recreation, comfort and aesthetics.

Turfgrasses are considered as major component of landscape industry which provides aesthetic value. They provide arena for various functional, recreational and social activities.

Turfgrasses enhance the appearance of land and increase the utility of sports fields, lawns, parks, golf courses and other greenbelt areas in and around urban as well as sub urban centres. They improve the environment of urban centres by reducing erosion, temperatures, noise, dust and glare. And, turfgrasses improve the safety of sports fields and playgrounds by adding resiliency to the playing surface. Turfgrasses are increasingly being used by National Highway Authority of India for reducing dust pollution and decreasing vehicle glare by planting on central verge and sides. Only turfgrasses are used by airport authorities for planting besides runway. The grasses being of perennial nature also help in carbon sequestration thus helping combat climate change. The grasses which grow well on sandy soil will help to reduce the dust and control erosion and dust pollution on river banks.

The turfgrass industry is growing rapidly due to strong demand for residential and commercial property development, rising affluence, and the environmental and aesthetic benefits of turfgrass in the urban landscape. Increased area under aesthetic and utility turfs is mainly because of increasing awareness about the sports, degrading environment and the increase in real estate value. The economic value of the turfgrass industry in India is yet to be determined as there is no single government agency handling or collecting the authentic data on turfgrass cover, use or import. All the agencies dealing with turfgrasses are either private companies, nurseries or corporates. But, according to the scarce data gathered by golf associations, sports experts and Association of Indian Nurserymen, turfgrass industry covers an area of more than

30000 acres, in which around 6000 acre is under turfgrass cultivation (sod production). States leading in turfgrass cultivation and marketing are Karnataka, Andhra Pradesh, Telangana, Uttar Pradesh, Haryana, Maharashtra, *etc.* The total number of international cricket grounds in India are 50 and 8 are under construction, more than 520 domestic cricket grounds and athletic fields are also in operation. Indian turfgrass industry is growing around 10 *per cent* annually. Landscape business contributes 2-3 *per cent* of total construction and turf has 50 *per cent* of landscape business.

Management practices often have an overriding influence on turfgrass selection and quality. If a site cannot be watered, a drought tolerant grass is required. If a turf is not mowed for longer time, only the low maintenance grasses are suitable, a high level of maintenance implies frequent mowing, regular watering, several applications of fertilizer each year and a pest control program. In contrast, a low level of maintenance suggests no supplemental watering, monthly mowing, perhaps annual fertilization and no chemical pest control. Selecting the best turfgrass for a site does not guarantee success. But, planting a grass not adapted to a site insures failure.

Characteristics of Ideal Turf

Selection of Turfgrass

Selecting the best adapted turfgrass for a site improves the quality of a landscape and reduces the cost of turf maintenance. Knowledge of the adaptation, use and maintenance requirements of turfgrasses is essential for selecting the best grass for a site thus there is an urgent need to do an evaluation trial of different grass species which does well in different agro-climatic zones of India and also takes into consideration the prevailing edaphic factors.

Characteristics of ideal turfgrasses are as follows:

1. It should remain fresh and green throughout the year.
2. It should be cold or drought resistant.
3. It should not look patchy.
4. Quick growing and should persist regular mowing and traffic.
5. Soft to touch.
6. Not giving fowl or bad odour.
7. Tolerance to diseases and pests.
8. Tolerance to smog, salinity, compaction, cold traffic (wear) and other environmental adversities.
9. Adaptability to wide range of climatic conditions.

Classification of Turfgrasses

Turfgrasses can be classified on the basis of type of shoot growth, function, maintenance requirements, and quality and temperature requirement

Based on Temperature Requirement: Selection of turfgrass species for a site is dependent upon many factors but priority should be given on the basis of climatic condition.

1. **Cool Season Grass:** Cool season species grow well in cool temperature range between 15° to 24°C. A typical growing season starts with a flush of growth in the spring, then slowed (sometimes to dormancy) in the summer, followed by another flush of growth in the fall. Cool season areas have cool winters with temperatures that fall below freezing and having warm/hot summers. These grasses turn brown during the hot seasons. They maintain their green color during colder times. Their leaf types and textures are compatible with each other. Typical cool season grass types include bent grasses (creeping bent grass, colonial bent grass), blue grass, fescue, rye grass and many different genera. Such grasses are not utilized in India because majority of part comes in tropical and subtropical climate. But high altitude areas *viz.,* hilly parts of J&K, Himachal Pradesh, Uttarakhand and eastern states having temperate climate are suitable for cool season grasses.
2. **Warm Season Grass:** Warm season grasses are of tropical origin and can thrive during the scorching summer heat. Warm season turfgrass species are believed to be more diverse in geographical origin and ecological habitat than cool season species. The optimum temperature range for better growth and development of the grasses is 27° to 35°C. They are tough, resistant to drought, disease and insect attacks but susceptible to the cold spells of winter. The grass thins out over time and has a low tolerance to many weed control herbicides. These grasses cover maximum part of the country. Bermuda, Bahia, Buffalo, Carpet, Centipede, Saint Augustine and Zoysia grasses are some of the warm season grasses.

Based on Type of Shoot Growth

1. **Bunch type:** It spreads by forming tillers. At sufficient seed rate, it forms a uniform turf while in case of lower seed rate it forms small clumps.
2. **Rhizomatous type:** It spreads by below ground shoots called rhizomes. It forms uniform turf with upright aerial shoots.
3. **Stoloniferous type:** It spreads by below ground lateral shoots called stolons. The examples may include St. Augustine and creeping bents.

Based on Function, Maintenance and Quality

Frequently, turf grasses are divided into three groups based on function, maintenance requirements and quality.

1. **Utility Turfgrasses:** Turfgrasses that serve mainly as erosion control, receive minimal maintenance and have maximum quality are termed as utility turfgrasses. Pasture-type tall fescues, fine fescues, common Kentucky blue grasses and buffalo grasses in cool season group while St. Augustine, centipede and shade loving grasses under warm season group fall into this category.

Grasses from this group are commonly used on roadsides, industrial areas, low-use park areas, and other low-maintenance settings. The prime functions of this type of turf is soil stabilization means it has a binding effect of an interconnecting system of fibrous roots and aerial shoots to prevent soil erosion from air and water. The aerial shoots also provide a cooling effect in warm weather. On roadside, it absorbs toxic emissions from vehicles. Along airport runways, it reduces dust and prolongs the engine life of aeroplane. It also reduces the formation of water puddle.

2. **Lawn Turfgrasses:** Turfgrasses grown for their aesthetic appeal, moderate maintenance and moderate to high quality are placed in the lawn turfgrass category. Lawns serves decorative function also as it is called as heart of the Garden. It enhance the beauty of a landscape. It can also act as mulch. Mowing leftover can be fed to cattle. Common and improved types of Kentucky blue grasses, perennial rye grasses, turf-type tall fescues, fine fescues, Bermuda grasses, Zoysia grasses and buffalo grasses are included in this group. These turf grasses are commonly planted on home lawns, commercial areas, park settings, and other landscape areas requiring turf having a good appearance.
3. **Sports Turfgrasses:** Sports turfgrasses function as athletic or recreational play surfaces, generally receive high maintenance, and are of high quality. Improved Kentucky blue grasses, perennial rye grasses, creeping bent grasses, turf-type tall fescues, Bermuda grasses, and zoysia grasses are used as sports turf grasses in various parts of the mid west. These grasses are often grown on golf course greens, tees, fairways, and roughs, baseball, football and other sports sites.

Turfgrasses Grown in India

Different grasses are chosen based on the utility, climate and traffic expected. Generally, varieties of *Cynodon dactylon* like 'Selection-1' (strain of Calcutta doob), 'TifDwarf', 'Calcutta doob', *Zoysia japonica* 'Nilgiri', *etc.* are employed in landscaping of lawns due to their wear tolerance and minimum maintenance. Huge area under recreational turf is in golf course which requires immaculate selection, establishment and maintenance under constant supervision. On greens of golf courses, varieties of Cynodon like 'TifDwarf', 'TifGreen', 'Mini Verde', 'TifEagle', *etc.* are preferred which has been released from University of Georgia (USA); a world leader in turfgrass research. For other parts of golf course like tees, fairways and roughs, 'Selection-1' of *Cynodon dactylon* is preferred. *Cynodon dactylon* is widely grown in the country (Chawla *et al.*, 2016 and Chawla *et al.*, 2018).

A wide range of environmental conditions occur all over India. Mostly extremes of temperature limit the adaptation of several species in many areas when combined with other climatic and edaphic factors. There are many abiotic factors involved. In the temperate climates of Himalayan states, cool season grasses such as bluegrass, bent grass, rye grass and fescue are grown whereas in coastal areas, *Paspalum vaginatum* (salt tolerant) is preferred.

Table 1: Commonly used perennial turfgrasses for turf, sports and amenity purposes

S.N.	Common name	Botanical name	Climate	Special traits	Use and utility
1.	Kentucky Blue grass	Poa *pratensis* (L.)	Temperate	Broad adaptation,apomixis	Lawns, sports, golf (fair ways, rough and tees)
2.	Tall Fescue	*Festuca arundinacea* Schreb.	Temperate	Broad adaptation, drought tolerance	Lawns, golf (far rough)
3.	Perennial Rye grass	*Lolium perenne* (L.)	Temperate	Rapid establishment, broad adaptation	Lawns, sports, golf
4.	Creeping Bent grass	*Agrostis stolonifera* (L.)	Temperate	Low growth habit, aggressive spreading	
5.	Bermuda grass	*Cynodon dactylon* (L.)	Tropical/ sub-tropical	Aggressive spreading, drought tolerance, narrow leaves	Lawns, sports, golf, low maintenance areas
6.	Zoysia grass	*Zoysia japonica,* *Z. tenuifolia* *Z. matrella*	Tropical/ sub-tropical	Dense sod, slow growth rate, narrow leaves	Lawns
7.	Centipede grass	*Eremochloa ophiuroides* (Munro) Hack.	Tropical/ sub-tropical	Low-fertility, acid-soil, and drought tolerance, medium leaves	Lawns, low maintenance areas
8.	Seashore paspalum	*Paspalum vaginatum* Swartz	Tropical/ sub-tropical	Salt and drought tolerance, medium leaves	Lawns, sports, golf, low maintenance areas
9.	St. Augustine grass	*Stenotaphrum secundatum* (Walt.) Kuntze	Tropical/ sub-tropical	Aggressive spreading, shade tolerance, broad leaves	Lawns, low maintenance areas
10.	Bahiagrass	*Paspalum notatum* Flugge	Tropical/sub-tropical	Drought and low-fertility tolerance	Low maintenance areas

Planting

Land Preparation

Establishing a durable turfgrass/lawn requires soil testing and proper site preparation. The initial investment in site preparation should be considered over the many years that a field provides recreational use. For example, improper root zone construction and grading inhibits the movement of water away from the surface and consequently increases the duration of wetness of the lawn particularly during wet weather. Lawns can be grown in wide varieties of soils but for a better establishment, the soil should be fertile and well drained with good water holding capacity. Ideal pH range is 5.5-7.0 but it is grown even at high pH of 8.2-8.4. Proper liming is essential for vigorous and healthy root and shoot growth of turf. Applied fertilizers and other soil nutrients are used more efficiently by the germinating turfgrass seedlings or newly transplanted sod when the soil having proper pH (6.5 to 6.7). Soil should first be solarised in summer for killing microbes, pests and weeds. Before planting the lawn, ground should be levelled to give a gentle slope for drainage of extra water. Generally, 2-3 cm slope is necessary for every 50 meters towards the natural source of drainage. Removing debris such as rocks, lumber, concrete, *etc.* should be followed by tilling and addition of organic matter such as well decomposed manure, well decomposed sawdust to improve soil structure. Soil should have adequate organic matter. Rotary tilling should be done to mix 1 to 2 inches of organic matter into 3 to 4 inches of top soil. The tilling is followed by grading the seedbed to provide drainage away from the house, walks and driveway. Application of water is done as often as needed to allow the soil to settle and form a firm seedbed. It is advised not to overwater the point of runoff and soil erosion. Thereafter, the soil is raked to remove any large clods and fill depressions. Glyphosate herbicide eliminates most of the undesirable weedy grasses and broad-leaved weeds. It moves or translocates down in plants to kill underground portions.

Different Turf Establishment Methods

After land preparation, it is mandatory that turfgrass should be established or planted as quickly as possible. Lawn grasses can be propagated by both sexual and asexual means but asexual methods are widely adopted as it takes less time to establish properly. Different methods of turfgrass establishment are as under (Chawla *et al.*, 2013 and Chawla *et al.*, 2019)

1. Seeding

Many turfgrasses that can be established from seed include common Bermuda grass, centipede grass, carpet grass, bahia grass, buffalo grass and all the cool season grasses. When establishing grass from seed, seed quality, seeding rate, seeding date and seeding method must be considered.

Seeding rates depend on grass species, the purpose for the turfgrass and the time available to develop a cover. Grasses with a creeping type growth habit are generally seeded at lower rates than bunch-type grasses. Seedling survival, a function of environmental conditions and management practices during the germination period, also influences seeding rates. Generally, 25-30 kg seeds required for one hectare area. Any factor that increases the time required for germination or the time the grass is in the seedling stage, also increases seedling losses. Extremely high or low soil temperatures, drought or excessively wet soils, low soil fertility, improperly sown seeds (covered too deep, left uncovered on the surface or non-uniform distribution), excessive seedling densities and the presence of some pre-emergence herbicides all increase seedling losses. Thorough seedbed preparation, proper planting techniques, correct planting date and proper post planting care all can reduce seedling losses.

Date of planting is critical when establishing turfgrass from seed. Soil temperature and the expected duration of favourable soil temperatures are the primary considerations for selecting the planting date. Cool season grasses should be planted in the fall when soil temperature is low and months of favourable growing conditions remain. If cool season grasses are planted too early, high soil temperatures create stressful conditions during the seedling stage and result in high seedling losses.

Late spring and early summer are ideal seeding time for warm season grasses. Soil temperatures are favourable by late spring and a long growing season is ahead for the warm season grasses. Late summer is another suitable time for seeding warm grasses since soil temperatures are favourable.

Seeding methods have a tremendous influence on the successful establishment of grasses. Ideally, grass seed should be uniformly distributed at recommended seeding rates, covered with an inch of topsoil and rolled to firm the field. Grass drills, or planters, are made to perform all these tasks in a single pass over the site. Other options include broadcasting the seed uniformly over the surface and covering seed with straw or mulch.

Seeding Methods

i. **Broadcasting:** It is the easiest method but is not advisable owing to non uniformity and higher seed cost.

ii. **Seed spreader:** To obtain an uniform turf, seed should be evenly distributed over the prepared soil. A drop-type spreader is ideal for distributing seed on small areas. Seeding in two directions resulted in fewer skips and a more uniform application. Thus, it is required to calibrate the spreader to deliver half of the desired amount of seed and cover the entire lawn area twice in opposite directions. Starter fertilizer should be applied just before or just after seeding. Starter fertilizer will provide sufficient nutrients to ensure rapid establishment of the new seedlings. Using a light weight roller to roll the entire area is advisable as rolling presses the seed into the soil.

    On small sites, seeds can be broadcasted with a hand seeder or fertilizer distributor over the surface of a lightly raked seedbed. After seeding, firm the soil with a roller and lightly water the site. If watering is limited to light, frequent applications, this method works well. But, excessive watering will wash the seeds and result in a non-uniform stand of grass field.

iii. **Hydroseeding:** Hydroseeding is a method of lawn seeding that has become popular in recent years. It involves preparing a mixture of seed, water, fertilizers, and mulch in a large tank and pumping the mixture through a hose or gun onto a prepared ground. Although lawn/turf of any size can be established through hydroseeding, this method is most efficient for large sites or steep slopes.

Most hydroseeding equipments consist of a large tank equipped with an agitation system to keep the mix of seed, water, fertilizer and mulch in suspension. The hydroseeder operator moves the hose or gun back and forth until the entire area is covered with the prepared mix.

Advantages of hydroseeding over other methods of lawn establishment from seed are that large areas can be seeded quickly and sloped areas are easier to seed. Disadvantages include the high cost of hydroseeding equipment, the fact that seed is placed on the soil surface and may not be in close contact with the soil (sometimes resulting in seedlings drying out), and a hydroseeder takes some experience to operate. Hydroseeding was developed for planting sites that were difficult to reach with conventional equipment. Steep slopes, rocky slopes, reclamation sites and other inaccessible areas were well suited for hydroseeding. Since its development, the hydroseeder has found much broader use in grass establishment and is used today for seeding lawns, golf courses, sports fields and other sites.

2. Sprigging

All for the Bermuda grasses, the zoysia grasses and seashore paspalum are readily established by sprigging. St. Augustine grass and centipede grass can also be established from sprigs, but the risks are greater. For large plantings such as golf courses and sports fields, sprigging provides the most practical method of establishment. A grass sprig consists of a stem or rhizome segment with at least one node or crown(multiple nodes) with some leaves and roots that might be attached to the node. Usually, 3 or more nodes are found on a single sprig.

Sprigs are harvested by shredding sod, by rototilling sod and raking, by verticutting or by a sprig harvester. Sprigs consisting of rhizomes, crowns and only a few green leaves are most desirable. Such sprigs will transport and store much better than green, leafy sprigs. Rhizomes are also more drought tolerant and survive several days without water; whereas, a green leafy sprig without a crown or rhizome segment may die within hours if not kept moist after planting.

Bermuda grass, zoysia grass and seashore paspalum are the only vegetatively propagated warm season grasses with rhizomes. Thus, those grasses are best suited to sprigging. St.Augustine grass, centipede grass and buffalo grass sprigs that consist only of stolons are much more susceptible to drought stress after planting. Such sprigs are also more likely to overheat and deteriorate during shipment. Sprigs that appear mouldy when they arrive at the planting site should not be planted. Such sprigs desiccate rapidly after planting and replanting is often required.

The quality of sprigs, like that of seeds, is critical for successful establishment. Just as weed seeds are undesirable in a lot of grass seed, off-type grasses are undesirable in a bushel of sprigs. Common Bermuda grass sprigs can cause serious problems in a bushel of hybrid Bermuda grass sprigs. Likewise, Bermuda grass sprigs in zoysia grass or other turfgrasses lead to serious weed problems for the turf manager. In addition to purity, sprig vigor is another aspect of sprig quality. Sprigs harvested by digging are usually the most vigorous since they include rhizomes, crowns and the large stolons found near the soil surface. Such sprigs are more vigorous than the leafy sprigs harvested above the soil level. The more vigorous sprigs tolerate stress and develop a root system rapidly than the less vigorous sprigs. Consequently, sprig survival is much greater for the vigorous sprigs.

Sprigging rates depend on grass varieties, planting method, sprig quality and the time available for a complete cover. Bermuda grass sprigs spread faster than all other grasses and can be planted at lower rates. Planting method also influences sprigging rates. Most sprigs are broadcasted over the site with a distributor or hydroseeder at rather high rates of sprigs. However, row planters are available for sprigging that use much less planting material. If the row planter functions properly, it places the sprig in a narrow furrow, covering 50 to 80% of the sprig with soil. If water is applied immediately (within 30 minutes), a very high percentage of sprigs survive. In contrast, broadcasting sprigs over the surface and depending on an irrigation system to keep them moist results in high sprig losses. Covering broadcast sprigs with mulch helps to reduce their losses.

Sprigging rates can also be reduced by planting high quality sprigs. Sprigs harvested by digging are more vigorous and have a higher survival rate; thus, lower rates of planting can be used. Where only the "tops" of Bermuda grass are planted, twice the quantity of planting stock may be needed when dug sprigs are used.

Finally, the time available to develop a complete cover will significantly influence planting rates. The higher the rate of sprigging, the faster a cover will develop. However, the cost of sprigging also increases with increase in planting material.

3. Stolonizing

Stolonizing is a form of sprigging that involves broadcasting sprigs or stolons over moist soil. The sprigs are then pressed into the soil, again leaving about 25% of the sprig's length exposed. This process can also be used with a hydromulcher, where sprigs are broadcasted in a spray with mulch onto the soil surface, often referred to hydrosprigging.

4. Plugging

Plugging refers to the setting out of small sod blocks or plugs (2 to 4 inches in diameter) to establish a turfgrass. The plugs need to be rolled so that the soil on the sod block, or plug is levelled with the soil surface. Rooted plugs are available which have been grown in about 2 inches of soil or organic matter. When rooted plugs are planted, a hole about 2-inches deep should be dug. Do not simply place the plug on the soil surface.

St. Augustine, zoysia and centipede grasses are frequently established from plugs. St. Augustine grass plugs will cover the site in about 2 months while Centipede and zoysia grasses need one growing season for same spacing. Although plugging requires lesser planting material than sodding, the labor requirement for plugging is greater than that for sodding. Rooted plugs, however, require less attention after planting than sod since their root system is developed prior to planting.

5. Sodding/Turfing

Sodding is usually the most expensive method of establishing turfgrasses. However, it is the quickest way to achieve a perfect lawn. Sodding is the laying of blocks or rolls of sod directly over a prepared ground to provide an instant lawn. Solid sodding obviously requires more planting material than other methods, but it eliminates many of the problems associated with seeding, sprigging or plugging. When sodding a site, lay sod blocks or sod rolls like bricks on a smooth surface that has been firmed by rolling or settled by rainfall or irrigation. The ground needs to be moist, but not wet, at the time sod is laid. After the sod is laid, roll it lightly to eliminate air pockets and to develop firm contact with the soil. In addition to avoid many problems with other methods of establishment, sodding can be done any time of the year if irrigation is available at the site. Frequent mistake on commercial sites is poor seedbed preparation prior to sodding. Although, sod will cover up a lot of problems, the turf manager is left to deal with them. All of the steps of good seed bed preparation should be followed prior to sodding a site.

Advantages of sodding over seeding

- The quickest way to establish turfgrass
- Gives immediate soil erosion control
- Eliminates problems with dust and mud
- Minimizes the need for weed control after planting
- Can be planted nearly year-round
- Can be used sooner than other methods
- Available for some of the best turfgrass varieties
- Can be used for a total turf installation or the repair of smaller areas

Synthetic Turf (Astro turfing)

The development of synthetic fibers and their integration into a carpet to try to simulate the conditions of natural grass was first successfully achieved by the Monsanto Company in the United States of America in 1964. It is a synthetic lawn popularly used in developed countries in roof gardens as well as in play grounds. It dispenses the normal maintenance usually required for normal lawns.

High-quality and consistent surface (even ball roll/bounce), no requirement of water, relatively low maintenance requirements, tolerance to adverse weather conditions, potential to generate significant rental income, intensive programming, potential of multi-use, recyclable content and reusability, reduction in pesticide and fertilizer use, increased predictability of sports fixtures and events and ability to be programmed intensively both in terms of time bandwidth (day and night), as well as subdividing field space and increasing participation are its merits over the natural grass. However,high initial capital cost, requirement of high fencing to protect the facility, costly to repair if damaged, heat retention and reflection in certain circumstances, need of upgradation or replacement every eight-fifteen years, variability of performance across the range of high value to low value surfaces and the difficulty for new adopters to differentiate are major drawbacks of Astroturf along with the perception that it is more expensive to establish, maintain and upgrade, increased injury potential in sports fields, negative environmental impact and negative health impacts.

Mangement of Turfgrasses

Care After Turf Planting

Irrigate newly planted turf areas regularly. Water lightly and frequently to prevent the surface from drying. As the young seedlings develop or as the sprigs or sod begin to take root and spread, reduce the frequency of watering and increase the amount of water applied per watering. This permits the development of a deep root system and ultimately reduces the amount of water needed.

The time to mow will depend on the species planted. The area is mowed as soon as the grass is 1 to 2 inches high. It is done frequently enough to prevent removing more than half of the growth.

1. Nutrient Management for Turf

The term 'nutrient management' refers to additions of fertilizer in efforts to positively influence plant growth. Chemical fertilization is the most common means of supplying turfgrasses with their nutritional requirements as inorganic fertilizers are the sources of nutrients available instantly to the plants. Inorganic fertilizers have been used in various forms to improve the productivity of the growing soil.

Careful consideration of the nutritional requirements of the turf, based on soil fertility, expected quality of the turf, use of the turf, suitability of the growing environment, grass species and varieties present, and available management resources is must for proper nutrient management. The necessary nutrients needs to be added at proper timing (late summer, late spring for cool season, late spring to midsummer for warm season) proper application rate, proper material selection and proper placement. In today's world, need of the hour is reduction of fertilizer application to the lowest possible level to maximize efficient use of nutrients by the plants in the turf system while eliminating waste and minimizing nutrient losses. Awareness of the potential for adverse impact from nutrient contamination on precious natural resources, particularly water, from off-site movement of nutrients due to factors such as improper application, runoff, erosion and leaching. In general, top-dressing refers to application of thin layer of manure or fertilizers, but in case of turf management, it refers to the distribution of a thin layer of soil over turfgrass area. After spiking, raking or scrapping, the turf should be top-dressed with the mixture of well decomposed weed freee FYM or neem cake powder and sand in a proportion of 1:1 of 1 cm thickness. After top-dressing, sprinkling with a hose is advisable. Top dressings are often applied on newly established turfgrass.

Nitrogen application for turf grass is divided into two categories of fertilizers viz. quick release and slow release. Quick release fertilizers are those that are soluble in water, relatively inexpensive, can be applied in spray and granular form in any season; most important is that they provide quick response to turfgrass. Such fertilizers are ammonium nitrate, ammonium sulphate, potassium nitrate, *etc.* Slow release N fertilizers are those that are not readily soluble, from moderately to highly, with one or two exceptions can only be applied in granular form and provide a slow and constant N release over a long period. Examples of slow release fertilizers are methylene urea, sulfur coated urea, isobutylidenediurea (IBDU), neem coated urea (NCU), *etc.*

Fertilizing the turf thrice a year is adequate to maintain turf in low maintenance condition. Application of urea or ammonium sulphate @ 1 kg / 50 sq. m. during February - March, June - July and October - November is quite beneficial. At the same time, well decomposed compost @ 1 kg/sq.m. area will be sufficient as top dressing.

After the emergence of seedlings, surface applications of N are applied when decline in shoot color, growth and vigor is evident.

Mowing practices related to nutrient management are leaving grass clippings and other organic materials in place when feasible for *in-situ* utilization of nutrients. Recycling removed turf clippings as a nutrient resource (compost, mulching, organic matter supplement, *etc.*). It is not required to mow the turf during periods of severe plant stress or intensity of disease. Mowing frequency should be increased during rapid, vigorous plant growth and decreased during dry stressful periods.

Methods of fertilizer application:

There are two methods for application of fertilizers

a. By sprayers: for liquid application of soluble and suspended fertilizers
b. By spreaders: for dry application of granular fertilizers

2. Irrigation Management

Irrigation is the artificial application of water for the growth of plant. On the basis of location, species, weather and maintenance, tropical turfgrasses use from 2-3 mm water per day and that is why proper irrigation is necessary to provide an adequate supply of moisture for turfgrass growth. Turf is also irrigated to wash out fertilizers and some pesticides following application, maintain sufficient surface moisture to promote germination of inter-seeded turfgrasses along with modification of soil temperatures. Lesser frequencies of irrigation required for deeply rooted turf while irrigation frequency is higher in shallow rooted turfs. Moreover, the amount of water required per irrigation depends on the volume of the turfgrass root zone, method of irrigation and canopy resistance (shoot density, leaf orientation, leaf area growth rate). Shallow-rooted turfs require more frequent but less intensive irrigations than deep-rooted turfs. Various environmental factors and cultural practices influence rooting depth. Poor soil aeration due to compaction or waterlogged conditions, close mowing, excessive fertilization, and frequent irrigation are sometimes associated with shallow turfgrass rooting. Irrigation requirement of turfgrass changes with the undergoing changes in the growing season.

Irrigation Frequency

Frequency of irrigation depends on climatic conditions, growth stage of turf and soil type. High disease incidence, reduced wear tolerance, low vigor, and high susceptibility to injury from climatic stresses are identified as problems associated with frequent irrigation of turf. As irrigation frequency is reduced, turfgrasses typically improve until inadequate moisture limits growth. First principle of irrigation management is deep (field capacity) and infrequent. The standard rate is 10-15 mm/3-4 days.

Irrigation Timing

Irrigation can be performed at any time of the day or night as long as the rate of application does not exceed the infiltration capacity of the soil.

The best time to irrigate is early morning hours which avoids interference to play or use in moist situations. In morning hours, water loss through evaporation is minimum and wind disturbance is least. Always maintain appropriate soil moisture for better growth and development as per turf species, soil type, weather, *etc.*

At night, the absence of solar radiation results in much less evaporative losses of irrigation water thus more efficient use of the water for satisfying turfgrass moisture requirements but there are more chances of attack of fungal diseases.

Irrigation Systems

The function of an irrigation system is to supply water in sufficient quantity. The system should supply water in the entire field uniformly and appropriate soil depth. An irrigation system's capacity to perform its function is limited by its design, construction and operation. Turf is irrigated by four main systems:

***Portable irrigation systems*:** It can be easily moved and shifted from one place to another place in turf field but it is not much efficient.

***Travelling irrigation system*:** These systems consist of rotating sprinkler attached to hose. The sprinkler continue propelling and winding itself.

***Quick coupler system*:** In this system, a series of underground pipes with quick couplers permanently installed in the ground and valves are constructed. Hose pipes and sprinklers can also be attached directly to the system.

***Installed irrigation system*:** This is most efficient system of turf irrigation. It is automatic and reliable system. To irrigate large turf area, pop up sprinklers are installed. Soil moisture sensors may be installed.

Irrigation management to prevent nutrient loss as runoff is also a sustainablepractice. It is required to irrigate lightly after application to wash fertilizer from surface of turf. Irrigation intensity should be low enough to allow infiltration. Irrigate turf based on weather and soil water availability (as needed) and not on a calendar schedule.

Irrigation controller technology should be optimized to ensure efficient irrigation patterns for efficient use of fertilizers.

3. Weeding

Weeds are the biggest menace to a beautiful lawn. Weeds deter the quality due to competition for nutrients, light and water. Weeds also detract uniformity and beauty of lawns due to the distinct contrast in colour and texture between the desired grass plants and weeds. Weed proliferation should be avoided by purchasing pure planting material free of weed seed, by pre-treating the lawn area with herbicides (Glyphosate) before planting but after planting, regular hand weeding is better way. Although, some herbicide like Metsulfuron-methyl (8 g/100 liter water) is used to control broad leaved weeds in standing turf.

4. Mowing & Rolling

These are most important practices for a visually pleasing and well trimmed lawn. Mowing improves turfgrass appearance and uniforms its surface by periodic removal of a portion of leaves or tillers. The objective of light mowing is to help the grass anchor itself firmly and keep the surface levelled. Accordingly, proper cutting height and mowing frequency must be defined in order to maintain an attractive and vigorous turf. Never remove more than 1/3 of the grass blade length at any time. A healthy lawn can survive an occasional close cut. Repeated close mowing produces a brown lawn and has several harmful side effects. Cutting height may extremely vary (from $\approx$ 0.3 up to 10 cm) on the basis of the intended use of turfgrass, species/cultivar and the period of the year. In addition, mowing frequency should be tailored to the shoot growth rate of the turf and environmental conditions by avoiding single removal of more than 30–40% leaf tissue. Mowing interval is expressed as number of days between successive mowings and is dependent on the increase in culm length of the turfgrass. Very frequent mowing can result into less rooting, reduced rhizome and shoot growth, depleted carbohydrate reserve and increased shoot density and succulence. There are many patterns by which a lawn can be mowed.

Rolling is process in which cylindrical roller is used to make soil compact and remove air gaps in newly established turf and to correct minor variations in turf surface caused by traffic. It is important for rhizomatous roots to get proper anchorage on the ground and also in levelling.

5. Aeration

Aeration or aerification refers to the process of mechanically removing small plugs of thatch and soil from a turf area to improve soil aeration (coring, spiking and slicing). The aeration process is also commonly called core aeration or simply aeration. It helps in improved air exchange between the soil and atmosphere, enhanced soil-water uptake, improved fertilizer uptake and use, reduced water runoff and puddling, improved turfgrass rooting, reduced soil compaction, enhanced heat- and drought-stress tolerance, improved resiliency and cushioning and enhanced thatch breakdown. Equipment with hollow tines removes soil cores. Equipment having solid

tines penetrates the soil surface. The turf should be spiked, raked or scrapped during February or September. Annual aeration is beneficial for most of the grasses. Lawns growing on heavy clay or sub-soils, and lawns exposed to intense use benefit from more than one aerations each year. After proper watering to turf, spiking can be done by inserting fork about 10 cm deep at 10-15 cm interval. Tines at least 1/2 inch in diameter or more are suggested for severe compaction or thatch problems. Using a garden rake, raking of turf should be done after mowing and removing the grass clippings. Applying fertilizers after aeration helps the lawn compete against weeds. Water the lawn carefully after aeration, particularly in areas where drought and high temperatures are common.

6. Thatch Control

Thatch control is most important practice for smooth and even looking turf. Thatch is a layer of living and dead grass stems, roots and other organic matter developed due to imbalance of organic matter produced by the turfgrass and the rate of its decomposition. It depends on growth rate and growth habit of the turf along with its maintenance. Any species of grass that spreads by above ground stolons or below ground rhizomes develops thatch. e.g. *Cynodon* and Zoysia are prolific thatch producers. Some amount of thatch is necessary for some places like sport field but excessive thatch is always undesirable. Development of thatch is influenced by grass species, fertilization, soil pH, mowing height, pesticide application, frequency of raking and soil type. High fertilization and intensive maintenance develope moderate to heavy thatch.

Efficient use of fertilizers, pesticides and irrigation water helps in reducing production of more roots and leaves. When thickness layer of thatch exceeds 1.5 cm, it can become problem and should be reduced by dethatching. One of the most common methods of thatch control is vertical mowing, sometimes called 'verticutting'. The blades of vertical mower cuts vertically into the turf canopy, severing lateral stems and removing thatch that develops on the surface of the soil, then it is raked and removed. Dethatching should be performed at times of vigorous growth and should be avoided before anticipated periods of environmental stress such as drought. After dethatching, light fertilizer and irrigation should be applied for speedy recovery of turf.

Power rake – It is used for thatch removal. This is a lawn- mower type machine with tines instead of blades that rip the thatch out of the ground. Hand raking is then needed to remove the debris.

Vertical mower – Also called a verticutter which is similar to a power rake but cuts down through the thatch into the soil. A verticutter is better than a power rake if the plan is to overseed after thatch removal. It cuts deeper into the soil and therefore provides better seed-to- soil contact. Raking up the debris is necessary before sowing the seed.

7. Turf/Lawn Renovation and Total Re-establishment

Renovation refers to the repair and improvement of an existing lawn without complete re-establishment of the turf cover. Renovation practices reduce the labour and cost of completely re-establishing the lawn. In most cases lawn renovation is preferable over total lawn re-establishment because it is less labour intensive, leads to minimum erosion, some foot traffic can be tolerated during the process and generally results can be observed more readily. Where lawn quality is unacceptable, complete renovation might be necessary. Renovation involves planting grass into a pre-existing lawn area, often adding new grass varieties to repair damage or increase tolerance to drought, shade or wear. Renovation usually isn't necessary until approximately 20-25% of the lawn is bare or covered with weeds. Knowing the cause for lawn failure can make awareness of changes in cultural practices such as mowing, watering and fertilizing that may be necessary.

Diseases and Pests

The major diseases of turfgrasses and their management practices are as below (Janakiram *et al.*, 2015):

Dollar Spot: It is a fungal disease caused by a *Sclerotinia homoeocarpa.* The grasses such as annual blue grass, bahia grass, Bermuda grass, fine-leaf fescue, colonial bent grass, creeping bent grass, perennial rye grass, Kentucky blue grass, creeping and spreading fescue are its important host plants. The symptoms of this disease include bleached or blighted areas of turf with silver colour and size of dollar appears initially which further may merge to form large areas. Tan lesions appear with a dark border girdle blades. It is prevalent in nitrogen deficit turfgrasses. The management of this disease includes maintaining adequate nitrogen level, reducing thatch accumulation; avoid over or late irrigation and application of systemic or contact fungicides such as COC, propiconazole, triadimefon, *etc.* It can also be controlled with a bacterium called *Enterobacter cloacae.*

Fairy Rings: Many fungi such as *Marasmus oreades, Lepiota* spp.*, Agaricus* spp.*, Lycoperdon* spp.*, Agrocybe pediates, Bovistra plumbea, etc.* are the cause of this disease which hosts all the turfgrass species. The initial symptoms of this disease are the appearance of circular patterns of dark green grasses or growth of mushrooms around the perimeters that tend to grow in a circle through the organic matter in the soil, mat and thatch. It is due to the fungi that break down organic matter. A ring of dead grass occurs due to the interference of water permeability by the dense fungal mycellial growth present in the soil. It often appears after rains or heavy irrigation and is a more serious problem on golf greens. For the management of this disease apply sufficient nitrogenous fertilizers, apply sufficient irrigation to sufficiently wet the root zone, frequent mowing, application of soil fumigants to eradicate the fungus and spray application of triademefon, azoxystrobin, flutolanil.

***Pythium* Blight:** It is a fungal disease caused by *Pythium aphanidermatum* that commonly hosts annual bluegrass, Kentucky bluegrass, rough bluegrass, velvet bent grass, tall fescue, fine-leaf fescue. It is very common in cool and wet conditions. Symptoms of this disease include round to irregular water-soaked "greasy" sunken patches which turn yellow to brown before collapsing. Damage may appear in streaks following drainage or mowing patterns. In pre-emergence attack, the seed rotting takes place. While in post-emergence infestation we can see the thin stands of grasses with irregular dying grass patches. To control this disease improve the drainage, avoid excess irrigation and high seeding rates, avoid creating excessively lush turfgrass and apply strobilurin fungicides such as pyraclostrobin, azoxystrobin.

Major insect-pests of turfgrasses are as under:

Ants: It is a pest of turfgrasses which burrow into the soil and form the colonies or nests in the soil thus destroying the turf. To control the ants, trace the line of march of ants to the nest location and apply chlorpyriphos @ 2.5 ml/l or methyl parathion dust @ 20 kg/ha for their control.

Termites: Termites like ants burrow into the soil and form the colonies underground causing massive damage to the turf grounds thus destroying the turf. To control the termites, destroying the colonies along with queen is the quickest method. It can also be controlled by drenching with chloropyriphos @ 0.05%.

White Grubs: The C-shaped larvae feed on the roots of turfgrasses while the adult stage of this pest feeds on the grasses thus destroying the turf above and below ground. The root feeding damage can be identified by the brown areas development or rolling up like a carpet. The attack is high during the dry summers. The damage can be identified by lifting the grasses that breaks away easily at root level. To control the white grubs, spray chloropyriphos @ 25 ml/l or malathion 50EC @1 ml/l or apply chlorpyriphos @ 50 ml/100 m^2 area by mixing with soil and sprinkle over the turf followed by a heavy irrigation.

References

Chawla S.L., Sudha Patil, Patel M.A., Patel R.B. and Patel R.M. (2013). Turfgrass Management. Manual published by ASPEE College of Horticulture and Forestry, Navsari Agricultural University, Navsari, Gujarat.

Chawla, S.L., Roshni Agnihotri; Patel, M.A.; Sudha Patil and Shah, H.P. (2018). Turfgrass: A Billion Dollar Industry. Paper presented in National Conference on Floriculture for Rural and Urban Prosperity in the Scenario of Climate Change during February 16-18, 2018 at CAE and PHT, CAU, Ranipool, Silkkim.

Chawla, S.L., Roshni Agnihotri; Sudha Patil, Patel, M.A.and Shah, H.P. (2019). Turfgrass Establishment Techniques. Paper presented in National Conference on Ornamental Horticulture to Uplift Rural Economy during January 11-13, 2019 at MPUAT, Udaipur, Rajasthan.

Chawla S.L., Roshni Agnihotri, Sudha Patil and M. Sree Devi (2016).Prospects of Turfgrass Industry in India. Paper presented in National Conference on Floriculture and Landscaping for Urban and Rural Prosperity February 28-29, 2016 at ICAR-IARI, New Delhi during.

T. Janakiram, Namita and Ritu Jain (2015). Introduction to Turfgrasses. Westville Publishing House, New Delhi

Turgeon, A.J., 1985. Turfgrass Management. Reston Publishing Co., Inc., Virginia

5

Cultivation and Management of Annual Flowers

Ajai Kumar Tiwari, Babita Singh, Ritu Jain Jain and Gunjeet Kumar

Annual flowers are a cluster of herbaceous plants which grow from seeds, produce flowers, set seeds and complete their life-cycle within one year or one season. They offer a gorgeous display of colours within the garden. Whether it is a small homemade garden or a giant public garden, it is incomplete without beds of annual flowers. They enhance the ornamental value of a garden within a small span of time. At their blooming time, one feels elated when rainbows of colours appearin the garden.

Classification of Annuals

Flowering annuals are grown in beds and pots. They're used for several purposes in the garden. In beds, they are grown individually with or without perennial plants in borders. They are developed in hanging baskets, window boxes or rock gardens, for training on walls and trellises and for planting in the form of edges, borders or ground covers. Dried flowers of some annuals are similarly used for interior decoration. Some annuals provide sweet fragrance.

1. Based on season: the annual flowers are divided into 3 groups—winter, summer and rainy season.

Winter season annuals

These annuals are grown in Sept. -Oct. and transplanted in Oct. -Nov. and they yield flowers in winter season. Majority of annuals grown in this season. This group includes a large number of seasonal flowers. The important ones are acroclinium, agrostemma anchusa, annual carnation, annual chrysanthemum, antirrhinum, arctotis, bells-of-Ireland, brachycome, calceolaria, calendula, campanula, candytuft, China aster cineraria, clarkia, cornflower, daisy, delphinium, dianthus, dimorphotheca, echium eschscholzia, gazania, godetia, helichrysum, helipterum, hollyhock, gypsophila limonium, linaria, linum, lupin, marigold, matricaria, mesembryanthemum mignonette mimulus, myosotis, nasturtium, nemesia, nicotiana, nemophila, nigella pansy, petunia phlox, pimpinella, poppy, rudbeckia, salvia, saponaria, scabiosa, schizanthus, senecio stock, sweet alyssum, sweet pea, sweet sultan, sweet william, venidium, verbena viola, wall flower etc.

Summer season annuals

These flower can grow on higher temperature during summer. The seed are sown in Jan. -Feb. and transplanted in the month of Feb. -March. The common warm season annuals are coreopsis, cosmos, gaillardia, kochia, marigold, portulaca, sunflower tithonia, zinnia etc.

Rainy season annuals

The rainy season annual are tolerating high heat and humidity and can stand in rains. This group includes all summer season annuals and others like amaranthus, balsam celosia, gomphrena, coleus and torenia etc.

2. Based on Utility: These can be categorized for different purposes for which it has been used.

- **For cut Flowers:** Antirrhinum, annual chrysanthemum, annual carnation, bells-of-Ireland, cornflower, China aster, Coreopsis, gypsophila, gaillardia, larkspur piminella, statice, stock, sweet sultan and sweet pea are grown commercially for cut flowers for interior decoration.
- **For Loose flowers:**Marigold, Annual Chrysanthemum, Marigold, Gomphrena Zinnia, Gaillardia, Straw flower, Paper flower,
- **For hanging baskets**: Nasturium, Partulaca, Daisy, Verbena, Phlox, Sweet Alyssum Petunia *etc.*
- **For pots**: Annual Carnation, Antirrhinum, Pansy, Ice Plant, Candytuft, Balsam Marigold, Stock, Salvia, Calendula etc
- **For Dry flowers**: Limonium, Helichrysum, Acroclinum, Nigella etc
- **For Shade**: Alyssum , Begonia, Cineraria, Impatiens, Nicotiana , Phlox, Salvia Torenia, Verbena, Viola (Violet),
- **For Fragrance**: Alyssum, Lemon Bergamot, Tobacco, Nasturtium, Moonflower Vine(*Ipomoea Alba),* Phlox, Stocks, Sweet Williams, Sweet Sultan, Sweet Peas etc.
- **For Rock gardens**: Ice Plant, Verbena, Phlox etc
- **For Screening**: Kochia, Hollyhock, Sweet pea, Ipomoea
- **For Border plants:** Almost every annual can be used for this purpose
- **For edging:** Ageratum, Alyssum, Dianthus, Candytuft, Brachycome, Marigold Portulaca, Pansy, Phlox, Petunia, *Zinnia* etc.
- **For Cut greens:** Kochia, Amaranthus, Coleus, Cineraria and Celosia

3. Based on height: Annuals can be categorized into 3 groups viz. , dwarf, medium and tall.

- **Dwarf Annuals:** Acroclinium, Ageratum, Sweet Alyssum, Arctotis, Daisy Brachycome, Cacalia, Calceolaria, Campanula, Wallflower, Cuphea Cynoglossum, Dianthus, Echium, Eschscholzia, Felicia, Gaillardia Gazania, Godetia, Candytuft, Linaria, Linum, Lobelia, Matricaria, Stock,

Mesembryanthemum, Mimulus, Nemesia, Nemophila, Nierembergia, Pansy Petunia, Phacelia, Phlox, Portulaca, Mignonette, Salpiglossis, Scabiosa, French Marigold, Torenia, Nasturtium, Venidium, Verbena and Viola.

- **Medium-Tall Annuals:** Agrostemma, amaranthus, anchusa, antirrhinum calendula, celosia, sweet sultan, cornflower, cineraria, clarkia, coreopsis cosmos, larkspur, carnation, digitalis, dimorphotheca, gomphrena, gypsophila balsam, kochia, limonium, lupin, bells-of-ireland, nicotiana, nigella, garden poppy, rudbeckia, salvia, schizanthus and african marigold.
- **Tall Annuals:** Hollyhock, Chrysanthemum, Sunflower, Helichrysum Heliotrope, Pimpinella and Zinnia, Lupin, Malvika, Salvia.

4. Based on Colour Scheme: These can be classified into monochromatic, analogus and complementary colour scheme.

- **Monochromatic Colour scheme:** In this colour scheme, only one colour or its different shades and tinges are used. For instance, if blue colour is to be used annuals like blue corn flower, blue larkspur, ageratum, anchusa, etc. can be used. For yellow colour annuals like yellow dahlia, yellow antirrhinum, yellow calendula, yellow nasturtium, yellow marigold, yellow annual chrysanthemum yellow coreopsis etc. can be used.
- **Analogous colour scheme:** In this colour scheme, the annuals of closely related colours are used as per colour wheel. Since, the arrangement harmonizes with the surroundings beds hence called a harmonious colour scheme. The arrangement of colours will follow the order as white, creamish yellow, light yellow and finally deep yellow. In case of red colour the order of colours used will be light pink, deep pink, light red, deep red and crimson, or vice a versa.
- **Contrast colour scheme:** In contrast colour scheme, the opposite colour of colour wheel are used and plants of contrast colours are planted accordingly e. g. red salvia against the background of green or next to blue corn flowers creates a good contrast. For yellow colour, violet, blue and red are divergence. Yellow marigold, red phlox or petunia and white brachycome can be planted for a creation of contrast. The width of the border is roughly divided into three parts in proportion of 5:3: 2 or 6:3: 1 for tall, medium and dwarf plants.

5. Based on Colour: These can be classified into array of colours viz. , white, red, pink, yellow, blue, purple, mauve, violet etc.

- **Annuals with White Flowers:** *Ageratum houstonianurn* (Mexican Ageratum), *Alyssum maritimum* (Sweet Alyssum), *Antirrhinum majus* (Snapdragon), *Arctotisgrandis* (Arctotis), *Argemone grandiflora* (Pricklepoppy), *Brachycome iberidifolia* (Swan River Daisy), *Browallia demissa*, *Campanula* (Annual Canterbury Bells), *Centaurea* (Royal Sweet-sultan, Cornflower, etc.), *Centranthus macrosiphon* (Spur-valerian) *Chrysanthemum* , *Clarkia* , *Cleome spinosa* (Spiderflower), *Clintoniapulchella Collinsiabicolor*, *Crepis barbata*rubra (Hawkweed), *Datura fastuosa Dimorphotheca aurantiaca* (Cape-marigold), *Echium Plantagineum Hybrids Gilia tricolor*, *Godetia amoena* and *G. grandiflora*, *Gypsophila elegans*

Helichrysum bracteatem (Strawflower), *Iberis amara* (Candytuft), *Impatiens* (Balsam), *Lathyrus odoratus* (Sweet Pea), *Lavatera trimestris* (Tree mallow), *Limonium sinuaturn* (Statice or Sea-lavender), *Lobelia erinus* and *L. tenuior Lupinus mutabilis* (Lupine), *Malcomia maritima* (Virginian-stock), *Matthiola incana* (Stock), *Nemesia strumosa, Nemophila maculata, Nicotiana* (Flowering Tobacco), *Oenothera americana* (Evening primrose), *Papaver somniferurn* (Opium Poppy), *Petunia hybrida, Phlox drummondii, Portulaca grandiflora, Saponariavaccaria* (Soapwort), *Scabiosa atropurpurea, Schizanthus pinnatus* (Butterfly Flower), *Senecio elegans* (Purple Groundsel), *Verbena erinoides* (Moss Vervain), *Vinca rosea* (Periwinkle), *Viola tricolor* (Pansy), *Xeranthemurn annuum* (Immortelle), *Zinnia elegans.*

- **Annuals with Red, rose, and pink Flowers:** *Abronia umbellata* (Sand-Verbena) *Alonsoa acutifolia* (Maskflower), *A. warscewiczii Amaranthus caudatus* (Love-lies bleeding), *Antirrhinum majus* (Snapdragon), *Calandrinia grandiflora, C. speciosa, Callistephus chinensis* (China Aster), *Campanula* (Annual Canterbury Bells), *Celosia* (Cockscomb), *Centaurea eyanus, C. imperialis C. moschata Centranthus macrosiphon, Clarkia, Collinsia bicolor* (Chinese-houses), *Collomia coccinea, Cosmos bipinnatus, Crepis barbatarubra Cupheaignea* (Fiery Cuphea) *Diascia barberm* (Twinspur), *Echium creticurn E. plantagineurn* hybrids, *Emilia flammea* and *E. sagittata, Eschscholtzia californica* (California Poppy), *Gilia, Godetia Gypsophila muralis, Helichrysum bracteaturn Helipterum manglesii* and *H. roseurn, lberis umbellata* (Candytuft), *Impatiens* (Balsam), *Lathyrus Odoratus* (Sweet Pea). *Lavatera trimestris* and *L. rosea* (Treemallow), *Leptosiphon hybrida, Limonium Suworoni* (Statice or Sea-lavender), *Linaria bipartita* (Toadflax), *Linum grandiflorurm* (Flowering Flax), *Lupinus hartwegii* (Hartweg Lupine), *Lychnis* (Rose-o f- heaven) *Malcolmia maritima* (Virginian stock), *Myosotis dissitiflora grandiflora* (Mallow wort) *Mathiola* (Stock), *Mimulus* (Monkeyflower), *Myosotis dissitiflora* (Forget-me-not) *Nemesia strumosa, Nicotiana sanderc* and *N.* sylvestris, *Papaver rhocas* (Shirley or Field Poppy), *P. somniferum* (Opium Poppy), *Petunia hybrida, Phlox drummondii, Portulaca grandiflora* (Portulaca) *Rehmannia angulata, Salvia splendens, Silene armeria* (Sweet-William) *Saponaria calabrica* (Calabrian Campion), Tropeolum (Nasturtium), *Scabiosa atropurpurea, Schizanthus pinnatus* (Butterfly-Periwinkle), *Senecio elegans* (Purple Groundsel).
- **Annuals for yellow flowers:** Marigold, Dimorphotheca, Zinnia, Coreopsis
- **Annuals for blue/purple/mauve coloured flowers:** Ageratum, Anchusa Browalia, clitoria, Corn Flower, helipterum, ipomoea, Larkspur, Linaria lobelia, petunia, salvia, verbena.
- **Annuals Orange colour flowers:** calendula, dimorphotheca, California poppy Helichrysum, marigold, nasturtium, zinnia

5 Based on Hardiness: These can be categorized as hardy, half hardy and tender annuals

- **Hardy annuals:** Hardy annuals are the most cold-tolerant of all of the types of annuals. The term hardy, as it applies to plants, refers strictly to the ability to withstand cold. Hardy annuals may be planted in early spring or in the fall allowing them to bloom again the following spring. Examples: Pansies and Johnny jump-ups (Viola tricolor).
- **Half-hardy annuals:** Half-hardy annuals will tolerate periods of wet or cold weather but may be damaged or killed by frost. Along with the hardy seasonals these are the plants that we typically consider of as spring annuals. Examples: Calendula and Nasturtiums
- **Tender annuals**: Tender annuals need warm soil temperatures and should only be planted after all danger of frost has passed. Examples: Celosia, Zinnias and Petunias

6. Based on Day Length: The annuals can be classified based on the requirement of sunlight for the flowering as short day, long day, day neutral and intermediate annuals.

- **Short day annuals:** Cosmos, Salvia, Amaranthus
- **Long day annuals:** Antirrhinum, Petunia, Carnation
- **Day Neutral annuals:** Balsam, Gomphrena
- **Intermediate annuals:** Ornamental Coleus

7. Annuals for problematic soils: Based on this annuals can be classified as

- **Annuals for very poor soil:** *Mentzelia aurea* (Bartonia), *Eschscholtzia californica* (California poppy), *Coreopsis tinctoria, Papaver rhoeas, Celosia plumose* (Cockscomb), *Mirabilis jalapa* (Four-o'clock), *Gaillardia lorenziana* (Gaillardia), *Impatiens balsamina* (Garden Balsam), *Godetia grandiflora* (Godetia), *Amaranthus caudatus* (Love-lies-bleeding), *Tropeolum majus* (Nasturtium), *Petunia hybrida* (Petunia), *Portulaca grandiflora* (Rose Moss) *Cleome spinosa* (Spiderflower), *Alyssum maritimum* (Sweet Alyssum) *Amberboa moschata* (Sweet-sultan).
- **Annuals for a neutral or alkaline soil:** *Impatiens, lberis, Papaver, Phlox Reseda odorata* (Mignonette), *Tropeolum majus, Zinnia elegans.*

8. Based on blooming time

- **Early blooming:** Ageratum, Alyssum, Amaranthus, balsam, Celosia, *Coreopsis drummondii*, cosmos, gomphrena, marigold, salvia, Stock and zinnia
- **Late blooming:** Anchusa, Hollyhock, Salvia, Snapdragon, Sweet Sultan, Sweet William, Wall flower
- **Annuals with a short season of flower:** Several Sowings should be made for Succession of Bloom. *Coreopsis tinctoria* (Coreopsis), *Iberis umbellate* (Candytuft), *Dimorphotheca aurantiaca* (Cape-marigold), *Centaurea cyanus* (Cornflower), *Myosotis* spp. (Forget-me- not), *Gypsophila elegans* (Gypsophila) *Nigella damascena* (Love-in-a-mist), *Reseda odorata* (Mignonette), *Phlox drummondii* (Phlox), *Papaver* spp. (Poppy), *Alyssum maritimum* (Sweet Alyssum).

Climate

The annual flowers are grown successfully in all parts of the country. In north Indian plains, winter season annuals are grown during winter, while summer and rainy ones during summer and rainy seasons respectively. In northern hills, all annual flowers are grown during summer season. Under warmer temperature of south India, where there is no distinct wintertime, common winter annuals, i. e. carnation, sweet pea antirrhinum, pansy, viola etc. do not grow well. In the milder climate, particularly in Bangalore and Punc, almost all types of annuals can be grown.

Soil

Annuals thrive best in well-drained, sandy loam soil rich in organic matter. The soil should be neither too acidic nor too alkaline. The soil pH should be between 6. 0 and 7. 5. In acidic top soil, lime should be added for making it neutral. In clay soil, organic matter is added to make it porous. In sandy soils, add sufficient organic matter for improving its texture.

Propagation

Almost all annuals can be propagated by seeds. Marigolds can be propagated by cuttings also. Their sowing time depends upon climate. It varies in different parts of India. In the northern plains, winter season annuals are sown during September–October. However, late-flowering types—China aster, carnation and cineraria are sown during August–September. Summer season annuals are sown during January–February, while rainy season annuals during May–June. In south India, September–October is ideal time for sowing annuals. In northern hills, March–April is most appropriate time for sowing annual flower seeds, whereas March–May in southern hills.

The seeds of annual flowers are sown in nursery beds, earthen pots, seed pans or wooden seed trays. The seeds of a few annuals like sweet pea, morning glory, lupin nasturtium and hollyhock which have bold seeds can be sown directly at permanent places. Mixing of fine sand in very small seeds is advisable for even sowing. The seeds of echium, lobelia and flowering tobacco do not germinate unless first exposed to sunlight, while seeds of nigella and cineraria germinate only in dark. The seeds recently harvested from the plant, although given required conditions, fail to germinate. This may be due to physical condition or chemical reaction of seed coat of seeds. Seeds of clianthus need stratification or scarification for germination. Some seeds require after ripening period for germination.

Location and Layout of seed bed

For selecting an area for establishing seed bed it is worthwhile to consider the following points:

- Seed bed should be prepared in such place where no water stagnation is experienced, and have good drainage system.

- Land for seed bed should be well drained and located at on a high level.
- The soil for seed bed should be sandy loam and normal in pH (around 6. 5-7. 0).
- The plot for seed bed should be selected near to a water source.
- Seed bed plots should be chosen near the farm building so that frequent supervision can be made easily.
- Seed bed should be away from the shady places.
- Seed bed plots should be selected at one side of the field to isolate the other fields for doing cultural practices easily.
- Site should be safe from stray animals and excessive diseases and pest attacks.

Methods

Soil preparation for seed bed

Seed bed preparation is an important step in crop management because it largely affects crop stand and its performance at field level. Therefore, soil should be worked to a fine tilth by repeated ploughing and spading. Dead plant parts which are seem to be occupiers of pathogens and pests should be collected, removed and burnt. Well decayed organic manure @ 40-50 kg/10sqm, should be mixed thoroughly in the soil. The seed bed compost should consist of one part each of garden soil, coarse sand, farmyard manure and leaf-mould. For making the nursery beds, the topsoil should be turned up thoroughly and sufficient farmyard manure should be mixed in soil. Raised nursery beds of convenient size (normally 60cm wide and 15cm high) should be prepared. If soil is heavy, some amount of sand may be added. It is better if the soil of nursery bed or earthen pots is sterilized with 2% formalin. For this, soil is drenched with formalin solution and is protected with polythene sheet for 45hr. afterwards, the polythene is removed and soil is dried before sowing the seeds. Before sowing, the seeds should be treated with Cerason (0. 2%) and Captan or Bavistin (0. 2%) to avert the seedlings from damping off disease. The seeds should be sown delicately and evenly as thick sowing causes damping off of seedlings.

Soil treatment

Soil treatment is an essential step in a successful Seed bed management because it the base for seedlings stand, source of nutrition and pathogens. In humid region damping-off caused by soil borne fungi like *Pythium, Rhizoctonia, Phytophthora* etc, is a common disease in the nursery beds. In addition, pests like snails, cutworms and termite and mites are also damages young seedlings. There are numerousmethods for soil treatment like soil solarization, chemical treatment, biocontrol treatment etc. Soil solarization can be completed with transparent polythene of 25-100 mm thickness during the hot and dry periods. For this soil should be moist before mulching because it increases latent heat and thermal sensitivity for resting of soil borne pathogens, harmful pests and weeds which can be reduced to a sustainable level. Chemical treatment of nursery beds can be completed by Formalin solution (1:100: Farmaline: water) @ 5 lt/sqm. The treated area should be covered with polythene sheets for 7-8 days and after that it should be remained open for next 7-10 days for facilitation of formalin emission. Besides, the seed beds and seeds can be treated with some fungicides like Ceresan or Bavistin @ 2g/litre before sowing. The insect-pests can be controlled by treating soil with Chlorpyriphos or Sewin dust powder (20-25g/sqm) before seeding and or at the time of nursery preparation. Presently a large number of bioagents like *Trichoderma, Pseudomonas* and *Bacillus formulations* are available for soil as well as seed treatment which performs as antagonist to harmful microorganisms. Selected botanicals like NSK, Neem cake, Karanj cake etc. also can be used for nursery bed treatment.

Seed bed preparation

In nursery beds, the seeds are sown in rows spaced 6 cm apart. Then, they are covered with finely sieved leaf-mould. Watering is to be done with a watering can having a fine rose both in beds and pots. In beds, when sprouting is over, water is given for proper moistness. Thereafter, the beds should be kept weed-free. Aoccupied area of 60 cm between the beds is necessary. This facilitates ease in sowing of seeds, weeding, watering, spraying and lifting of seedlings. Irrigation channels are to be laid out conveniently.

Once sowing seeds the beds should be levelled and pressed gently to make it firm. Nearly 15-20 cm raised beds width are always preferred for raising nursery. Beds of 1-meter width of any convenient length are to be made. However, its length should be made according to the requirements or size of plots but should not exceed 5-6 m. In between beds, drains of about 30-45 cm width are prepared and connected to the main drain for removal of excess water during heavy pour. This space facilitates easy movement during intercultural operations and acts as physical barrier for inoculums spread. The drains are flooded during dry period to modify microclimate of nursery beds in favour of seedlings. In recent years various advancements have been made in Seed bed management for bed preparation to avoid possibilities of pathogen spread like use of soil less media, plug tray technique, perforated poly trays etc.

Cultivation

Planting

Their seedlings are transplanted 25 days after sowing at 4-leaf stage. Before transplanting, seedlings are hardened off by withholding water for 1 or 2 days or by exposing them gradually to sunlight. Transplanting is, usually, done either on a cloudy day or in the evening. Transplanting in sunset is good as the night cool temperature is beneficial for the establishment. Light irrigating every day in early morning or late in the afternoon is required for about a week for proper establishment of the seedlings.

Dwarf annuals should be planted at a distance of 30cm × 30cm, whereas medium-tall annuals should be planted at a spacing of 45cm × 45cm. The plants of tall annuals should be planted 60cm apart.

Manuring and fertilization

The farmyard manure or compost @ 3kg/m^2 is mixed in the soil. Chemical fertilizers @ 20g urea, 60–120g superphosphate and 30–60g muriate of potash/m^2 should also be added. Half quantity of urea and full of superphosphate and muriate of potash should be applied at the time of bed preparation. The left over quantity of urea must be applied one month after transplantation. Spraying plants with 2% urea twice or thrice is beneficial for good growth and flowering. Fertilizers should on no occasion come in the direct contact of the foliage since they cause scorching. Fertilizers should never be applied in the pot-grown annual flowers. However, some readymade pot-mixtures can be used. The pot-mixture should consist of 2 parts of garden soil and one part each of coarse sand and farmyard manure. Instead of fertilizers, it is better if pot-grown plants are given liquid feeding. The liquid manure is prepared by fermenting 1–2kg each of fresh cowdung and oil cake in 10 litres of water in a drum for one week. It is diluted to tea shade and sieved with the help of a muslin cloth. It is applied @ 500–1000 ml/pot at 7–10 days intervals.

Growth and flowering

Environmental causes and various cultural conditions affect progression and flowering of many annuals. The *Cosmos bipinnatus, Nicotiana* sp. and *Callistephus chinensis* are classic short-day plants for vegetative growth and flowering, while *Dianthus barbatus* and *Nigella damascene* are classic long-day plants for vegetative growth as well as flowering. On the opposite hand, some seasonal need long days for flower initiation followed by short days for flower development. An imbalance of N either hastens or delays flowering. Clarkia, candytuft and salvia flower too early if the soil is deficient in N. If N level is high in soil, sunflower, lupin, flowering tobacco and African marigold flower hastily.

Growth delaying or stimulating substances play a foremost role in getting dwarf plants or higher flower yield. Growth retarding chemicals—CCC (1, 000–2, 000ppm), B-Nine (2, 000–5, 000ppm) and SADH (1, 000–3, 000ppm)—are used to control growth and flowering in hollyhock, arctotis, sweet sultan, coreopsis, cosmos, phlox and pansy. These chemicals retard plant tallness, produce more number of leaves and branches and improve flowering, Similarly, GA_3 (100–400ppm) gives beneficial results in respect of growth and flowering in African marigold, China aster and antirrhinum. Use of these growth substances is more effective at vegetative stage.

Aftercare

After transplanting, beds are cleared, hoed and sprayed regularly. As soon as plantlets are established in beds, pinching is done for creating the plants bushy. Pinching is not accomplished in antirrhinum, lupin, larkspur, stock and hollyhock. Sweet pea, carnation, morning glory and nasturtium, have weak, slender or straggling stems. They want support when they are 15–20cm tall. The stakes prepared from split bamboos are painted green so that they can be accorded with the foliage colour of the plants. Annuals which benefit from pinching are *Ageratum houstonianum, Calendula officinalis, Chrysanthemum, Petunia, Phlox, Dianthus chinensis, Salpiglossis sinuate Schizanthus, Piñatas, Antirrhinum,* Verbena and Zinnia.

Occasionally, seedlings of carnation, marigold, China aster, cosmos and zinnia produce flower buds at an initial stage. These buds should be removed as soon as they seem. The number of buds/ stem is reduced by disbudding the axillary buds, if large blooms are wanted.

Irrigation

Slight water is desirable everyday up to 7–10 days after transplantation. When the plantlets start new growth, plentiful watering once or twice a week is required in beds. Later, regularity and quantity of irrigating depend upon soil and season. In lighter soils, added frequent irrigation is needed than that in heavy soils. The season of planting also regulates the frequency of irrigation. During hot season, irrigation should be done at weekly breaks in beds, while at 10–12 days intervals in winters. Irrigation throughout rainy season depends upon prevailing weather conditions. Potted plants need daily watering during summer, whereas on alternate days in winter.

Harvesting and postharvest management

Most of the annual flowers are grown for garden demonstration purpose in various ways. However, marigold, China aster, antirrhinum, gypsophila, statice, gaillardia, annual carnation, annual chrysanthemum, cornflower and bells-of-Ireland are grown commercially for cut flower or loose flower purpose. Their flowers are reaped when they are fully open and are sold in the local shops. Antirrhinums are cut when one-third of the florets are unfurled. China aster flowers are cut along with their stems when they develop their unique colour. Marigold flowers are gathered when they are fully open. Gypsophila flowers are cut when these are open, but not over mature. The flowers, in general, are cut either late in the evening or very early in the pre-lunch. After harvesting, cut flowers should be put in a bucket of liquid filled up to one-fourth of the volume as it helps in their recovery from the shock of being cut away from the plant. As far as possible, the recently opened flowers should be cut as freshness improves their shelf-life. In African marigolds, produce of 20 to 22 tonnes of fresh blossoms is obtained from one hectare crop, whereas in French marigolds 10–12 tones/ha of fresh flowers is attained. China aster gives the produce of 10–12 tonnes/ha of fresh flowers.

Some annuals are used as cut flowers. Therefore, suitable postharvest managing is necessary for lengthening their vase-life. The flowers are sorted according to stem length, flower size, flower shape, flower colour and freshness. The cut flowers/loose flowers of most of the annuals are sold in local markets. However, cut flowers of gypsophila, bells-of-Ireland and Limonium are traded in the international marketplace.

Woody-stemmed flowers should have their stems split or crushed to expand the surface in contact with water. Eliminate all foliage from stems which are underneath water otherwise they deteriorate and clog up the xylem vassels of stem. It is better to cut stems obliquely because it exposes more surfaces to water.

If flowers are not sold the day they are harvested, to stock them in a cold storage is imperative. Antirrhinum flowers can be kept for short duration at a temperature –0. 6°C to 1. 7°C. For keeping stock cut flowers, 4°C temperature is optimum. Gypsophila flowers can be stored for 1–2 days at 4°C. Statice cut flowers can be stored for 2–3 weeks at 2°C. A temperature of 0°–2°C is suitable for the storage of cut carnations.

Herbaceous border or bed

The establishing of annuals in the border of a plot is called as herbaceous border. It is developed in flower beds to provide bulk effect of different colors. Borders are uninterrupted beds of more length than width containing plants of one type only.

Due to varied range of flowers obtainable in winter it becomes a site of attraction in the garden. The concept of herbaceous border has been familiarized in Indian gardens by Britishers and now it has become a common feature.

Type of Herbaceous Border

The herbaceous edge can be of two types i. e. Single and Double face.

Single face Herbaceous Border: Single face herbaceous border is prepared when border is positioned on one side of plot and having certain background like wall or shrubbery border. Annuals are organized in the border according to their height i. e. big plants in back; mediocre in center and dwarf in front line.

Double face Herbaceous Border: Double face border is more absolute when border is to be prepared in between a big plot. In such border, there is no background and tall annuals are planted in the middle; medium and dwarf on both sides in descendent direction so that attractiveness can be watched from both sides.

Site Selection

The selection of spot for herbaceous border is very vital and desired picturesque effect may not be obtained if place is not selected appropriately. The site should be original sunny and fronting towards south side so that plants get adequate sunlight for maximum hours of the day. If this way is not available, it should face east and last choice is for the west.

Back and Fore Grounds

To make an actual herbaceous border, there should be appropriate back ground as well as fore ground. Back ground may be of dwarf flowering trees, shrubbery border or back ground prepared by training of sweet peas, brick wall in arrangement of both.

Fore ground is usually of turf grass which may or may not have other garden decorations like lily pool, sun dial, statue etc. Green turf in opposite of herbaceous border is a living green carpet which serves a place for enjoying beauty of flowers while resting.

Size of the Border

The size of border depends upon the obtainability of space. The border can be of every length but width should be about 1-1. 5 m or more subjects upon the area of fore ground vacant. Therefore, the width can be attuned accordingly.

Colour Scheme of the Border

The preparation of these annuals according to colour scheme expresses the mood and spirits of the gardener. Dominance of a specific colour in border is suggestive worth of that colour. Hence, the prearrangement of these annuals should be done very carefully. In environment green colour being the colour of foliage dominates in garden all over the year. Warm colour alike red dominants in winter whereas, yellow colour in spring.

There are three main colour schemes according to which these annuals can be arranged

1. Monochromatic scheme
2. Analogous or harmonious colour scheme, and
3. Contrast colour scheme.

Monochromatic colour scheme hampers the use of single colour or its shadows which may be obtainable in the same or altered annuals. For instance, if blue colour is to be used, annuals like blue corn flower, blue larkspur, ageratum, anchusa, etc. can be used. For yellow colour annuals alike yellow dahlia, antirrhinum, calendula, nasturtium marigold, annual chrysanthemum, coreopsis etc. can be used.

The trouble may be in receiving the seeds of pure colours. According to analogous colour scheme, annuals are organized according to the adjacent wave length of colour. This preparation harmonizes with the neighbouring beds and, hence, it is called a harmonious colour scheme. The arrangement of colours will follow the order as white, creamish yellow, light yellow and finally deep yellow which will follow the similar but in descending order.

If red colour is to be used the order will be light pink, deep pink, light red, deep red and crimson and colours will incline in the same order. In contrast colour scheme, the conflicting colour of colour wheel are used and plants of contrast colours are planted consequently e. g. red dahlia in contradiction of the background of green or next to blue corn flowers makes a good contrast. For yellow colour, violet, blue and red are divergence.

Yellow gamolepis, red phlox or petunia and white alyssum can be planted for aintense contrast. The width of the border is unevenly divided into three parts in proportion of 5:3:2 or 6:3:1 for tall, medium and dwarf plants. Similarly, the plots are made sideways.

The front line of border is not a conventional line, but it is made curvaceous and thus plots become irregular in form. In these plots planting is also done irregularly at an estimated distance so that the border looks natural at the time of blossom.

6

Ornamental Shrubs and Their Management

L.N. Bairwa and A.K. Mahawar

Shrubs are defined as woody, semi woody or herbaceous perennial plants, branches arises from the base of the plants and grow up to height of about 0.5 to 4.0 meters. These plants are commonly erect and bushy. Most of the shrubs are very hardy, require little attention and grow in all types of soils. Although some plants flower almost throughout the year, in most cases it is seasonal and selection can be made in such a way that some plants in the shrubbery will be in bloom in different seasion of the year. Shrub flowers regularly every year and produce flowers of almost all the important colors white, light yellow, golden pink, scarlet, crimson, violet, blue, and some are sweet scented.

In gardens, shrubs are usually planted in three ways:

(1) In the form of shrubbery, on the long strip of land where taller shrubs are grown at the back with medium height in middle and dwarf once at the fronts. Colors of flowers and time of flowering are also considered so that shrubbery remains colorful throughout year with bloom and foliage.

(2) In small garden or to break monotony of open space shrubs are often grown in groups and in this case few plants of the same variety or of different species are selected.

(3) Shrubs are also grown as a specimen plant or in only one row along the border of the garden or a wall.

Shrubs are used in gardens for decorative purposes as these shrubs are found more in ornamental and provide a variety of interests. They are evergreen and best against noise and dust. Any perennial woody, perpetual and thick plant of relatively low height that branches into a few stems or trunks arising from the ground base is known as shrub. They are littler than a tree but taller than herbaceous plants. They are also called as bushes when they are much-branched and dense. They grow upwards up to 0.5-4.0meter and may be flowering and ornamental. Shrubs play important role in permanent landscaping *i.e.* formal decoration which contributes pattern, colour, fragrance and utility. They may be either evergreen or deciduous in nature. Some shrubs have attractive foliage, flowers and attractive berries. Shrubs are planted as borders on the sides of paths and at the corners of lawn in a curving line. Shrubs are used for back ground, boundary wall, screen, shade, wind breaks and foundation planting. Hence, knowledge on the form, growth habit and habitat is essential to select the shrubs for planting. Shrubs are important part of landscaping or landscape designing is an art of modifying the overall appearance of a given piece of land using different ornamental plants. It is an artificial process. A garden or national and international playgrounds are the example of artificially landscaped area. Ornamental plants are a group of different flowering, foliage and stylish plants. These plants are used for the beautification of selected part of a site or other places.

A shrubbery is an area planted with various types of shrubs and a shrub border is one where only one kind of shrub is used. The shrubs which are used for topiary work require frequent pruning and trimming. Tall growing shrubs can be used to screen the displeasing object and backyard. Handsome shrubs can also play an important role to make attractive pot plants for indoor and outdoor decoration.

Shrubs have different types of shape, size, colour of flowers and fragrance according to their growth habit. They form the chief features of landscape gardening placed in front of tall trees and joining the spacious lawn etc. Shrubs which are amenable for frequent training are chosen for topiary work. Tall growing shrubs generally serve as screen purposes. Hardy dwarf Shrubs like *Gardenia Jasminoides*, *Lantana Camara Depressa*, *Malpighia Coccigera*, *Plumbago capensis*, *Russelia juncea* are also grown in rockery to add colour. They are usable as a single specimen in the lawn. They can be trained to make standards *viz.,* trained to single stem and assent to branch out and form a handsome head only above a particular height *e.g. Bougainvillea*, *Ixora*, *Murrya exotica.*

Characteristics of a shrub

- Lack of a trunk is the main feature of shrubs. They branch near the ground hence do not have a well-defined trunk and rather have several narrow stems.
- Shrubs are smaller than trees usually less than 1-3 m tall as they are less rigid than a single thick trunk.
- Shrubs are usually broad leaved plants.
- Shrubs are hardy plants that can be grown in all kind of soils.
- Shrubs are seasonal, since they generally thrive in a sunny environment conditions well.

Functions of shrubs

- Shrubs play an significant role in preventing soil and water erosion.
- They are a source of fuel wood.
- They are good source of foodstuffs such as fruits, leaves, tuber shoots and roots.
- Shrubs are also used in the landscapes.
- They can be grown as backdrops, garden beds and in public gardens.

Purpose of planting of shrubs

- To glorify overall appearance of surrounding.
- To provide wind break.
- To provide Shelter for birds in summer.
- Used as live boundary.
- For avenue planting.
- Rookeries.
- Cut flower and foliage.
- Terrace garden.
- Hedge and edge.

Ideal site for planting shrubs

Avoid shade of tree for planting of shrubs. Shrubs should planted in sunny situation and they are not produce attractive flower in shade. Planting of shrub should be done in South-East direction. They require availability of ample space.

Classification of Ornamental Shrubs: The Ornamental shrubs can be divided into two groups:

i) Hedge

ii)Edge.

Hedges: When the tall hedges are planted at the border of nursery, crop field or around a building or garden, it is called hedge.

A garden should be appended by a good hedge or fence. It protected from cattles and provides shelter from wind and privacy. The best plant material for forming a hedge would be quick growing and hardy shrub with attractive foliage, handsome flowers and drought resistant. It should also stand trimming to different shape and capable of being easily and quickly raised from seed or from cuttings to fill up the gaps promptly.

The decorative internal hedges are formed of small growing or under shrubs which have handsome foliage and in some cases bear handsome flowers as well. The usual height for an ornamental internal hedge varies from 30 to 65 cm. The main objective is to that justly divide the garden into a number of parts and each part containing the distinctive feature of it sown as a rosary or flower bed or collections of bulbs etc. Boundary hedges or screen hedges or the ornamental hedges are meant to be permanent features of a garden and proper care should also be exercised in planting them. Trenches are made in the size of 30-40 cm wide and 40-45 cm deep and refilled with the layer of top soil, farmyard manure and red soil. Rainy season is the best time for planting the hedges either with the seed or cuttings. Shoots of hedge should be tipped as they grow and induce to branch out and side shoots should be cut back to the desired dimensions as possible. If, possible, a compact or thick hedge is trimmed as often as necessary, cutting back the overgrowing shoots strictly to the desired dimensions (Arora, 2007, Randhawa and Mukhopadhyay 1986, Pal and Krishnamurthi 1967) .

According to height hedge can be classified into two groups.

1. Tall ornamental hedge
2. Dwarf ornamental hedge
3. Tall protective hedge
4. Dwarf protective hedge

Tall ornamental hedge: These type of shrubs growth to a height of 1-3 metres and are characterized by having attractive foliage or flowers. Following are the examples of tall ornamental hedge:

S. N.	Scientific Name	S.N.	Scientific Name
1.	*Polyalthia longifolia*	6	*Casuarina equisetifolia*
2.	*Murraya peniculata*	7	*Tecoma stans*
3.	*Thuja orientalis*	8	*Poinsettia pulcherima*
4.	*Bougainvillea* spp.	9	*Calliandra hematocephala*
5.	*Narium indicum*	10	*Ficus retusa*

Dwarf ornamental hedge: These type of hedges growth to a height of 1 metres and they possesses attractive foliage or flowers. Following are the examples of tall ornamental hedge:

S. N.	Scientific Name
1.	*Russelia juncea*
2.	*Russelia floribunda*
3.	*Acalypha spp.*
4.	*Pedilenthus*
5.	*Lantana camera*
6.	*Agave spp.*
7.	*Opuntia* spp

Tall Protective hedge: The height of such edge is about 1-3 meters. Such hedge is characterized by having thick growth that possesses thorns. examples

S. N.	Scientific Name	Remarks
1	*Bougainvillea spp.*	Native to S. America
2	*Inga dulcis*	Native to laconic coast
3	*Carissa carandus*	Fruits are used to make pickles
4	*Acacia farnensiana*	Vachellia farnensia var. farnensia is wild species
5	*Lawsonia alba*	Source of dye hina

Dwarf protective hedge: Dwarf shrubs grow about 1 metre and posses thorns.

S. N.	C. N.
1.	*Opuntia*
2.	*Agave*

Plants suitable for ornamental internal hedges

S.N.	Scientific Name	S.N.	Scientific Name
1	*Acalypha* sp.	7	*Cupressus macrocarpa*
2	*Barleria* sp	8	*Duranta plumieri*
3	*Bougainvillea* sp.	9	*Eranthemum* sp
4	*Hamelia patens*	10	*Hibiscus rosa-sinensis*
5	*Lantana camera*	11	*Malpighia coccigera*
6	*Meyenia erecta*	12	*Pedilanthus tithymaloides*

Plants suitable for boundary or tall hedges

S.N.	Scientific Name	S.N.	Scientific Name
1	*Acacia farnesiana*	6	*Agave Americana*
2	*Acalypha*	7	*Aralia*
3	*Bougainvillea*	8	*Caesalpinia pulcherrima*
4	*Casuarina equisetifolia*	9	*Carissa carandas*
5	*Pithecelobium dulce*	10	*Punica granatum*
6	*Tecoma stans*	11	*Thevetia nereifolia*

Edge: When low growing dwarf perennial plants are employed at the border of plant nursery, path, roads, flower beds, lawns or other place of demarcating spaces and beautification, it is called edge.

Edging are two types

Formal edging: It is also called as mechanical edging. Formal edging is done by using following materials *viz*. bricks, tiles, rough or irregular stones, broken bottles and bamboo pieces.

Informal Edging: It is also known as live edging. It is created by using grass verge and other low growing plants *viz*. Eupatorium, Coleus, Echeveria, Iresine, Pilea, miniature rose etc. The edging plants should be dwarf growing in habit with handsome foliage and responsible for regular trimming. They are perennial, hardy, easily propagated and should have lasting foliage or flower or both. Frequent trimming is necessary operation to maintain proper shape and height. The height of an edge should be about 15 to 20 cm. A strip of grass at the side of road, track or path, originally in a garden is known as grass verge. The usual size of grass verge for edging is about 60 cm.

Examples of dwarf ornamental edge: These have dense foliage *i.e. Lantana sellowiana, Linustrum ovalifolium*

Examples of dwarf protective edge: These edges are dwarf in nature and are protective by thorn. *Berberis* spp. , *Opuntia* spp.

Foliage plants suitable for edging: These plants are low growing (less then 50cm) example *Alternanthera, Coleus* sp., *Eupatorium cannabinus, Iresine* sp., *Santolina chamaecyparissus*, Echeveria, *Cineraria maritime*

Flowering plants suitable for edging: Low growing shrubs and some small annuals are suitable for this prpose are: Alyssum, Amaryllis, Gerbera, Lobelia, *Gazania splendens*, *Plumbago capensis*, Zephyranthes.

Shrubs can be classified easily according to the beauty of plant parts

1. **Flowering shrubs:-**These shrubs produce very attractive and spectacular flowers which are grown for mass effect and carpetting purchases. Suitable shrubs are *Hibiscus rosa sinensis, Hibiscus mutabilis, Bougainvillea* sp., *Ervatamia divericata, Nerium indicum, Caesalpinia pulcherrima, Ixora parviflora, Callindra* sp., etc.
2. **For foliage:-**These shrubs are grow in the garden for handsome and richly variegated foliage e.g. *Acalypha tricolour* – Medium shrub and colour full foliage, *Manihot variegata* – Pale yellow and green variegated foliage, *Codium variegata* (Croton) – Variegated leaves, *Hamelia patens* – Variegated.
3. **For flowered and foliage:-**Some shrubs have foliage and produce good flower also. They add beauty in the garden tremendously. *Bougainvillea* sp. (Thimma and Dr. Rao Varieties) Tall shrub and all type colour flower, *Hamelia patens* – Tall shrub and red type flower, *Hibiscus variegated* – Tall shrub and red colour flower, *Nerium odoratum* – Tall shrub and white, red & pink colour flower.
4. **For fragrant flowers:-** These shrubs produce fragrant flowers eg. *Cestrum diurnum* (Din Ka Raja) – Medium shrub, white colour flower, *C. nocturnum* (Raat Ki Rani) – Medium shrub, white colour flower, *Murraya panniculata* (Kamini) – Tall shrub, white colour flower, *Jasminum sambac* (Mogra) – Dwarf shrub, white colour flower, *J. grandiflorum* (Chameli) – Dwarf shrub, white colour flower and *Buddelia asiatica* – Tall shrub, white colour flower.
5. **Specimen Shrubs:-** These shrubs are planted singly or in group in the beds on the lawn to create a picturesque effect. These shrubs have elegant foliage and produce beautiful flowers. e.g. *Bougainvillea sp.* Tall shrub all type colour of flower, *Hibiscus rosa sinensis* – Tall shrub, red colour flower, *Hamelia patens* – Tall shrub and red colour flower, *Tecoma stans*- yellow colour flower and *Thevetia peruviana* – Tall shrub and yellow colour flower (Pili Kaner).
6. **Scented flowers:** *Cestrum nocturnum, Gardenia jasminoides* syn. *G. florida, Hamiltonia suaveolens, Jasminum sambac, Nyctanthes arbortristis, Tabernaemontana divaricata* syn. *Ervatamia coronaria.*

Some popular shrubs produce attractive different colour flower

White	Yellow	Red
Abelia grandiflora	*Allamanda cathartica*	*Caesalpinia pulcherrima*
Citrus aurantium	*Acasia farnesiana*	*Cordyline terminali*
Cestrum nocturnum	*Berberis vulgaris*	*Hibiscus rosa sinensis*
Erenthemum nervosum	*Cassia alata*	*Hamalia patans*
Gardenia resinifera	*Tecoma stans*	*Lagerstroemia indica*
Feijoa sellowiana	*Thevetia peruviana*	*Malvaviscus arboreus*
Hiptage benghalensis (syn.H. madablota)		*Punica granatum*
Ixora arborea(syn.i.parviflora)		
Jasminum sambac		
Ligstrum indicum		
Murraya paniculata		
Roupellia grata		

Various Colour	Pink	Purple	Crimson
Bougainvillea spectabilis	*Buddleia lindleyana*	*Beloperone chrysophloe*	*Acalypha sanderiana*
Camellia japonica	*Callistemon citrinus (syn.C.lanceolatus)*	*Malpighia glabra*	*Ixora coccinea*
Rhododendron indium (syn.azalea indica)	*Tamarix troupii (syn.t.galllica)*		*Russelia juncea*
Hibiscus rosa- sinensis			
Lantana camara			
Salvia spledense			

Shrubbery Border

The area of the garden devoted exclusively to shrubs is called as shrubbery border. In the garden to secure privacy and provide partition, formal hedges are used. The labour and cost of maintaining hedges is much more than keeping to well planned shrubbery. Therefore, the shrubbery is preferred which also adds beauty to the garden. In planting shrubbery, consideration of foliage and colour effect should be kept in mind before planting. Prior to laying out shrubbery border, the following points should be kept in mind.

- By keeping shrubs in front of the trees, a very pleasing effect will be obtained.
- Best results achieved by facing east or south.
- Allow greater width where greater height is desired.
- The shade of tree should be avoided as per as possible.
- Shrubbery is also used to limit the garden especially in case of large gardens

Generally border are planted by two methods:

Single face: In this type of border shrubs are planted in one side of a lawn. First tall shrubs are planted then medium and dwarf shrubs are planted in descending order.

Double face border: Double face shrubbery border is preferred to get pleasing or natural effect of the shrubbery border from both the sides. In this type, tall shrubs are planted in the centre followed by medium and dwarf shrubs both the ways.

Planning of Shrubbery Border

Site selection is very important step to establish shrubbery. After deciding the site, first of all a paper plan should be prepared which will help in final planting.

Colour Scheme for Shrubbery Border

The best colours to pacify are blue, purple and to a lesser degree, green and white, can also be helpful. So for example, if you have a yellow shrub, in a border that is predominantly pink if you add blue or purple foliage or flowering plants between the yellow and the pink, it will help avoid the clash. Pacifying colours work because they help blend strong colours. They are also colours that go with everything, so they are the bridge between other strong colours.

Creating really good colour schemes is definitely an art. Choosing plants that have great flower and foliage combinations can be tricky to achieve. Borrowing colour schemes that talented colour experts like fashion designers and artists have already come up with, is a great way to get colour combinations that work.

Growing of shrubs

Soil- Any garden soil is suitable for growing shrubs, if soil is bad, top layer up to 45 cm should be removed and refilled with good fertile and well drain soil. The shrubs can be grown in varied pH of soil. If soil is highly alkaline or acidic in nature, it should be reclaimed by using soil amendments. Salt tolerant shrubs are *Thevetia peruviana, Bougainvillea* sp.*, Caesalpinia pulcherrima, Nerium indicum* etc. Salt sensitive shrubs are *Buddleia asiatica, Lagerstroemia indica, Barleria cristata* and *Hamalia patens.*

Propagation of Shrubs- The shrubs are propagated by seeds, cutting (Hardwood and semi hardwood cutting) and air layering.

By Seeds

Seeds are collected when fruits are fully ripe. They are dried in the shade and store in airtight bottles in dry places. In rainy season, nursery is prepared and seeds are sown. Seedlings are transplanted in prepared ground. Following shrubs are mainly planted by seed are S*tenolobium stans*, *Thevetia peruviana*, *Caesalpinia pulcherrima*, *calliandra* sp. etc.

By Cutting

Most of shrubs are propagated by cuttings. February- March is the best season of propagation by cutting. The size of cutting should be 15-20 cm long, which is planted in pot or beds containing ample of sand. The plants raised from cutting should be planted in next planting season. Shrubs are mainly planted by cuttings are *Hibiscus rosa-sinensis*, *Achania malvaviscus*, *Jasminum sambac*, *Hamelia patens*, *Bougainvillia*, *Cestrum diurnum*, *Castrum nocturnum* , *Lawsonia alba* etc.

By Layering

Some shrubs are propagated by air layering which are not easily propagated by seeds or cutting. Layering is done in rainy season by removing the bark of 2.5 to 3.0 cm long and wrapping by garden soil and sphagnum moss. The shrubs which are propagated by layering are Bougainvillia, *Ixora* sp., *Jasminum sambac*, *Jasminum multiflorum* (ground layering) ete.

Planting of Shrubs

The best season of planting of shrubs are rainy season. Planting can also be done in the month of February-March depending upon irrigation facilities available during summer. The shrubs should be planted according to height and spread. The distance of planting of tall shrubs may be 150-180 cm. The medium shrubs should be planted in one or more lines between tall and dwarf shrubs at the distance of 90-120 cm. The dwarf shrubs in front line should be planted at least 30 cm away from the inner edge at the distance of 45-90 cm in a group of 4-6 plants in the length of 3.0 to 4.5 meter.

Planting and Maintenance of Shrubs

The best time of panting of shrubs is during rainy season. When irrigation facility is available. Planting can also be done during February-March. Digging a pit of 75 x 75 x 75 cm size for the larger shrubs and 45 x 45 x 45 cm size for dwarf shrubs is sufficient. The digging of pit is done during summer and left exposed to sun for at least 15 days for sterilizes the soil and destroying of harmful micro-organisms of the soil. The pits are filled by using 10 kg FYM, 10 kg soil and 10 g Endosulfan dust. The pits are filled 5 cm above the ground so that it may settle down properly after application of irrigation water. The shrubs are planted in the centre of the pit and it is watered. The shrubs are planted at a distance of 1-3 metres depending upon nature of growth of plants.

Regular watering is done to shrub plants during summer at five days interval and during winter at 10-15 days intervals. Deficiency of irrigation water leads to wilting, defoliation and ultimately dyeing of the plants. Time to time pruning is necessary to maintain the shape of shrubs. Pruning is done after February – March when danger of frost is over. Pruning of dormant shrubs is done during winter season. Best time of pruning in tropical regions is before the rain and just after flowering is over. Training of shrubs is done to removing of dead, damaged, disease infected, twiggy and overgrown branches. Spraying of 0.02 per cent Malathion and 0.2 per cent Dithane M-45 is used to control insect-pest and diseases of shrub plants. To control the termite, imidachlorepid should applied with irrigation water.

Pruning of Shrubs

Pruning of shrubs is the necessary practice which creates a better look to the garden. Winter flowering shrubs are pruned at the end of summer season whereas others are pruned at the close of winter season. All the superfluous growth, old or week branches are best cut at ground level or as low as possible without leaving the stub.

Important shrubs

S. No.	Common name	Botanical name	Family	Flower colour	Remarks
(A)	**Tall shrubs (above 1.8 meter)**				
1.	Malti (chandni)	*Ervatamia divericata*	Apocynaceae	White	Shrub only to look on a moon-light.
2.	Kamini	*Murraya peniculata*	Rutaceae	White	Scented flower
3.	Harshringar (Night jasmine)	*Nyctanthus arbotristis*	Oleaceae	White, Orange	Night flowering plants.
4.	Shravani	*Lagestromia indica*	Lythraceae	White, Red, Purple	Medium flower and small leaves
5.	Hamelia	*Hamelia patens*	Rubiaceae	Red	
6.	Cup and saucer	*Holmskioldia sanguinea*	Verbanaceae	Red	
7.	Abutilon	*Abutilon indicum*	Malvaceae	Yellow	
8.	Pili kaner	*Thevetia peruviana*	Apocynaceae	Yellow, white, orange	Fruit is an odd-shaped berry. "Thevetin," a strong poison obtained from bark.
9.	Doembeya	*Doembeya spectabilis*	Sterculiaceae	Pink	
10.	Lady of the night	*Brunsfelsia Americana*	Solanaceae	Yellow	
11.	Gudhal (China rose)	*Hibiscus rosa sinensis*	Malvaceae	Scarlet, pink, white and blue	Heart-shaped leaves. Large ever-green shrub. Single bell-shaped flowers.
12.	Dancing lady	*Fuchshia hyrida*	Onagraceae	Red, purple	
13.	Gardenia (Cape jasmine, Gandharaj, Gardenia)	*Gardenia jasminoides*	Rubiaceae	White	Jasminoides means "resembling Jasmine" and Radicans means "resting"
14.	Brilliant gardenia	*Gardenia lucida Syn. C. resinifera*	Rubiaceae	White and fragrant	Used in ulcers and sores,
15.	Tithonia	*Tithonia tagetiflora*	Astraceae	Red-orange	

S. No.	Common name	Botanical name	Family	Flower colour	Remarks
16.	Genista	*Spartium junceum*	Leguminoceae	Yellow	
17.	Basant	*Rainwardtia indica*	Linaceae	Yellow	
18.	Dalina	*Osyris arboria*	Santalaceae		
19.	Jacobinia	*Jacobinia carnia*	Acanthaceae	Red, pink	
20.	Exocaria	*Exocaria bicolor*	Euphorbiacea	Yellow, red	Leaves are variegated
21.	Buddlia	*Buddlia asiatica*	Loganiaceae	White	Large flowers
22.	Ixora (Kangan)	*Ixora perviflora* *I. rosea, I. chinensis, I. fulgens, I. lutea, I. coccinea*	Rubiaceae	White, pink orange	Also known as Torch Tree.
23.	Poinsettia (Poinsettia, Flame Leaf)	*Euphorbia pulcherrima* *Syn. Poinsettia pulcher-ima*	Euphorbiaceae	White, red	
24.	Lal kaner (Kaner, Oleander)	*Nerium indicum*	Apocynaceae	White, red, pink	Flower throughout the year, but best during the rains.
25.	Peacock flower (Dwarf Poinciania)	*Caesalpinia pulcherima*	Leguminoceae	Red, yellow	Ink is made from the charred wood. Pulcherrima means "most beautiful".
26.	Azalea	*Azalea* spp.	Ericaceae	Yellow, pink, orange, red and white	
27.	Jatropha	*Jatropha multifida*	Euphorbiaceae	Red	
28.	Anar (ornamental flower)	*Punica granatum*	Lythraceae	Red	Fruits are very nutricious
29.	Tecoma	*Tecoma stans*	Bignoniaceae	Yellow	
30.	Hemiltonia	*Hemiltonia suaveolens*	Rubiaceae	Mauve	
31.	Ficus	*Ficus pandanus*	Moraceae	Foliage	
32.	Snow bush	*Phyllanthus nervosus*	Euphorbiaceae	Foliage	

S. No.	Common name	Botanical name	Family	Flower colour	Remarks
33	Duranta (Sky Flower, Pigeon Berry)	*Duranta plumeri*	Verbanaceae	Blue	Flowering period-January-February, Fruits appeared soon after flowering
34.	Lolypop plant	*Pachystachys sp.*	Acanthaceae	Golden yellow	
35.	Weigela	*Weigela florida*	Caprifoliaceae	Pink-red	
36.	Aralia	*Schefflera spp.*	Araliaceae		
37.	Moutan peony	*Paeonia suffruiticosa*	Rananculaceae		
38.	Privet	*Ligustrum ovalifolium*	Oleaceae		
39.	Gamelina	*Gamelina asiatica*	Verbanaceae	Yellow	
40.	Cassia	*Cassia glauca*	Leguminoceae	Yellow	
41.	Bougainvillea (Paper Flower)	*Bougainvillea sp.*	Nactaginaceae	White, pink and all colours	
42.	Candle bush	*Cassia alata*	Leguminosae	Deep yellow, cone shaped	Craw-craw plant, acapulo, ringworm bush, or ringworm plant.
43.	Cassia	*Cassia auriculata*	Leguminosae	Deep yellow	
44.	Raat ki rani/ night jasmine	*Cestrum nocturnum*	Solanaceae	Small white and scented	
45.		*Euphorbia pulcherrima* syn. *Poinsettia pulcherrima*	Euphorbiaceae	Yellowish flower	Use as a foliage flower

S. No.	Common name	Botanical name	Family	Flower colour	Remarks
(B)	**Medium shrubs (1.2 to 1.8meter)**				
1.	Night queen (Rat ki Rani)	*Cestrum nocturnum*	Solanaceae	Cremish white	Fragrance during night
2.	Day king (Din ka Raja)	*Cestrum diurnum*	Solanaceae	White	Fragrance during day time
3.	Kund (bela)	*Jasminum multiflorum*	Oleaceae	White	
4.	Tecoma	*Tecome capensis*	Bignoniaceae	Red, yellow	Tecoma is Mexican name and stans means "erect"
5.	Thunbergia	*Thunbergia eracta*	Acanthaceae	Violet	
6.	Acalypha	*Acalypha spp.*	Euphorbiaceae	Colorful foliage	
7.	Achania	*Achania malvaviscus*	Malvaceae	Red	
8.	Ghanari	*Lantana camera*	Verbanaceae	White, red, yellow	Most troublesome weeds in India
9.	Mussaenda	*Mussaenda luteola*	Rubiaceae	Yellow , pink	
10.	Stachytarpetha	*Stachytarpetha indica*	Verbanaceae	Violet	
11.	Croton	*Croton tiglium*	Euphorbiaceae	Colourful foliage	Oil is extracted from seeds.
12.	Camellia	*Camellia japonica*	Ternstroemiaceae	White, shade pink, red or varigated	
13.	Blue Sage	*Eranthemum nervosum*	Acanthaceae	Blue	
14.	Honeysuckele	*Lonicera japonica*	Caprifoliaceae	White changing to yellow	Rambling shrub
15.	Escobillo	*Malpighia glabra*	Malpighiaceae	Pale purple	
16.	Chandani/ crape jasmine	*Tabernaemontana di-varicata* syn. *Ervatamia coronaria*	Apocynaceae	Daxxling white	

S. No.	Common name	Botanical name	Family	Flower colour	Remarks
(C)	**Dwarf shrubs (0.6 to 1.2meter)**				
1.	Mogra	*Jasminum sambac*	Oleaceae	White	Scented
2.	Chitra	*Plumbago auriculata*	Plumbaginaceae	Blue	
3.	Weeping merry	*Russelia juncea*	Scrophulariaceae	Red	
4.	Barleria	*Barleria cristata*	Acanthaceae	Violet-blue	Funnel shaped flower, Flowering season -November to January
5.	Galphimia	*Galphimia gracillis*	Malphigiaceae	Yellow	
6.		*Brunsfelsia pauciflora*	Solanaceae	Blue	
7.	Red bell bush (Dawa, Santha)	*Woodfordia fruticosa*	Lythraceae		Spreading, leafy and deciduous shrub. A red dye obtained from flower (contain much tannin) which is used to dye silks.
8.	Allamanda (Golden Trumpet)	*Allamanda cathartica*	Apocynaceae	Yellow	It is a fairly large climber

References

Arora, J.S. 2007. Introductory Omamental Horticulure (Fifth Edition) Kalyani Publishers, Ludhiana.

Chadha, K.L. and Choudhary, B. 1986.Omamental Horticulture in India. ICAR, New Delhi.

Desh, Raj, 2004. Floriculture at Glance, Kalyani Publishers, Ludhiana.

Pal, B. P. and Krishnamurthi, S. 1967. Flowering Shrubs. ICAR ,New Delhi.

Randhawa,G.S. and Mukhopadhyay, A. 1986.Floriculture In India ,Allied Publisher. New Delhi.

Swarup,V.1997. Garden Flowers. National Book Trust of India, New Delhi.

7

Climbers

Madhu Bala

Climbers are the group of ornamental plants which possess weak stem and climb up the support with the help of certain modified organs to reach the top and receive sunlight and air. Climbers enhance the beauty of any garden provided they are selected carefully depending upon the availability of space and the orientation of the building. The growth habit, mass effect of foliage and diverse flower colour give charming appearance to the climbers for their characteristics.

Ornamental climbers are commonly used to train on suitable supports such as arches, pergola, trellis, along the walls and trees. These are most suited in congested regions prone to scarcity of space especially in urban environment. In addition to adding beauty to the surroundings, the climbers also provide or fulfill other need-based requirements in landscaping such as:

- Beautify the indoors and outdoors
- Create mass effect and colour contrast in a garden
- Add greenery and mimic nature in miniature
- Screening of the undesirable premises or from the adjacent houses
- Maintaining privacy
- Break monotonous view in gardens
- Camouflaging the bare walls that otherwise look dull.
- Providing habitat for terrestrial creatures such as squirrels, ants and birds etc.

Nature has equipped these climbers with different modified organs like:

(i) **Tendrils:** Twisted thread like segments developed by the plant from leaf axils or stems. These tendrils coil around the support, hold it tightly and the climbers continuous to grow indeterminately.

E.g. *Antigonon leptopus, Adenocalymma alliaceum, Bignonia gracilis, Clemetis panniculata, Pyrostegia venusta etc.*

(ii) **Thorns**: Certain climbers possess spine like structures called thorns that help them against foraging by wild animals and lend support against walls, wire-mesh or barbed wires. Such thorny climbers are best to use on exposed sites to avoid tresspassing and invasion of wild animals.

E.g. *Bougainvillea species, Climbing roses, Euphorbia milii, Persea equileata etc.*

(iii) **Roots and rootlets**: Some climbers possess root and rootlets that arise from their nodes. These climbers release gum like exudates or sticky substances from growing points that help them to anchor themselves especially on vertical bare walls. Such climbers (or creepers) are most suitable for covering un-plastered bare walls.

E.g. *Campsis grandiflora, Ficus repens, Ficus pumila,Hedera helix, Hedera nepalensis, Hoya carnosa etc.*

(iv) **Hooks:** These are also modified organs resembling claw of a cat. These hooks are present along the succulent stem at specific nodes. These hooks possess sharp claws that anchor themselves in minute pores on the wall. eg. *Bignonia gracilis etc.*

(v) **Clingers:** *Passiflora caerulea, Passiflora quadrangularis, Hydrangea petiolaris etc.*

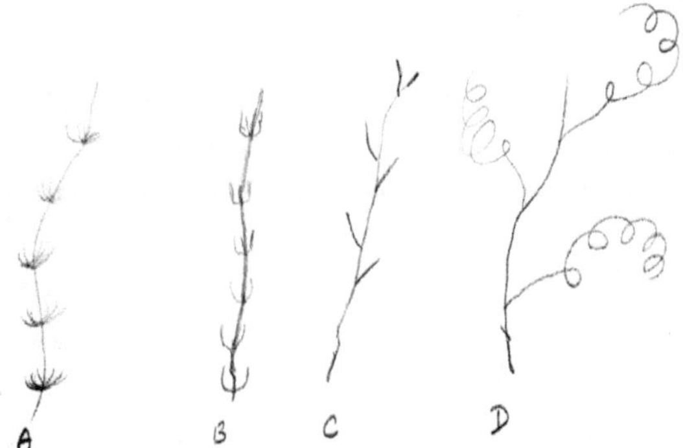

Fig. 1: Modified organs for climber to climb the support.

A. Root and rootlets; B. Hooks; C. Thorns; D. Tendrils

Twiners: These are also the group of climbing plants which do not have any modified organs but climb over a support or a plant by twining themselves spirally around such supports. After sufficient growth, these twiners look beautiful producing mass effect of their delicate and colourful foliage. Depending upon the growth habit and kind of support, these twiners also cover the ground with their partially hanging foliage. Often a times, such plants require periodic pruning to check their growth and also for rejuvenation.

E.g. *Ipomoea purpurea , Lonicera japonica, Hiptage benghalensis, Clerodendrum splendens etc.*

Ramblers and stragglers: Plants which climb up a support by using modified organs and produce large clusters of small flowers several time a year. These plants fail to climb but manage to support themselves over the trunks, stems, or branches of other plants by outgrowing and adding their additional weight over the support that come along their way. e.g. *Quisqualis indica*

Supporting Structures supports for Climbers

Climbers require a suitable support to climb over the top to gather sunlight and space. These supports could be in the form of natural objects such as tree stumps, wooden logs etc. often climbers in their natural habitat are seen climbing on walls, boulders with irregular rough surfaces lending them a soft touch to otherwise harsh and hard landscapes. Such scenic views are generally observed on hill stations where native climbers grow indefinitely refreshing the surroundings with their elegant foliage and cooling the ambience by regulating the micro-climate near their vicinity.

However, due to rapid urbanization and scarcity of space in urban environment, there is need to lend a support to the climbers which are otherwise domesticated outside their native range. Therefore, artificial supporting structures are provided varying in design, size for the climbers to climb. In addition to lend a support, these supporting structures also help in providing utility services such as shading a walkways, providing natural enclosures, shading for vehicles etc. Depending upon the growth habit of climbers, the supporting structures are classified as follows:

1. **Arches:** These are rectangular or semi- circular structure erected over gate at the entrance or made as a highlighted feature in a garden. The climbers are planted alongside the foundation of the arches and are tied with wires to help them climb. These arches vary in sizes and design as shown in Fig 2.

Fig. 2: Diagrammatic sketch of arches for the climbers

2. **Pergola (syn Arbour):** Extension of an arch or a series of arches that are joined together with a framework of wires over the top. The resultant structure is strong and durable enough to bear the weight of the climbers. Such type of structures add utility services such as shading walkways that provide a respite from heat especially during the summers for the pedestrians. Pergolas also form a sort of enclosures in a private home garden or may also be useful to park the vehicles for providing shade and enclosures (Fig 3).

Fig. 3: Diagrammatic sketch of Pergola supported with pillars

3. **Trellis:** A framework of light wooden or metal bars usually made from an open framework or lattice of interwoven or intersecting pieces of wood, bamboo or metal that is normally made to support and display climbing plants. Trellises are typically fastened against a wall or now a days custom made trellis are available that can be anchored at a suitable place in a garden for dividing the garden into two parts or may be use as an enclosure to screen out the other undesirable sites.

Selection of Climbers

Climbers can be selected and planted according to the orientation of building that provide different exposures (like sunny, partial shade and shady areas) at a particular place , nature of growth, leaf retention characteristics, colour of foliage and flowers, individual's choice and availability. The climbers suitable for different situations and purposes are as follows:

I. Based on Exposure of Site to Sunlight

1. **Sunny situation:** Sites oriented towards South or West facing receive full sun. Such sites may be provided shade by placing a suitable support depending upon the growth habit a climber. Climbers such as *Clerodendrum thomsoniae, Bougainvillea* spp, *Campsis grandiflora, Quisqualis indica, Pyrostegia venusta, Antigonon leptopus, Ipomoea indica, Adenocalymma alliaceum* are considered hardy and are most suitable for planting for the sites exposed to full sunshine. etc.
2. **Partial shade**: Certain sites within the house premises may receive full sun light for 3-6 hours daily and for the other period they may remain under shade. For such conditions climbers such as, *Trachelospermum jasminoides, Petrea volubilis, Lonicera japonica, Clerodendrum splendens, Thunbergia grandiflora etc.*are suitable that may appear with more brighten foliage and flowers under partial shady conditions.

3. **Shady areas:** Sites facing North remain comparatively cooler and soils hold moisture for longer duration. However, the intensity of sunlight received may not be adequate for all climbers to sustain their growth. Therefore, under such conditions, climbers such as *Ipomeas* , *Trachelospermum jasminoides, Lonicera japonica* can be planted.

II. Based on Nature of Growth

Climbers are appreciated for their mass effect of foliage and beautiful flowers. These form a dense network of intermingled branches that continue to grow inderminatly adding more foliage. The selection of climbers based on their nature of growth should be done according to available space, area of site to cover, whether the site is exposed or under management etc. All these factors determine the selection of climbers that may be classified as heavy or light climbers.

1. **Heavy climbers:** These climbers grow very vigorously and produce luxuriant growth displaying mass effect of their profuse foliage and flowers. These climbers are suitable to cover bigger areas and training along the fences to screen the exposed sites or to avoid trespassing especially into the private property. Once established, such climbers continue to grow and cover the available space in a shorter period of time.

   E.g. *Pyrostegia venusta, Adenocallyma alliaceum, Clerodendrum splendens, Antigonon leptopus, Bignonia magnifica, Bignonia gracilis, Wisteria sinensis, Clerodendrum splendens, Campsis grandiflora, Beaumontia grandiflora, Bougainvillea* spp. *Hiptage benghalensis, Quisqualis indica, Wisteria sinensis, Petrea volubilis etc.*

2. **Light climbers:** These climbers exhibit sparse foliage and are relatively less dense than there lighter counterparts. These are most suitable for small areas or those that have a limited volume of soil, such as pots. These climbers remain light in growth and spread.

   E.g. *Allamanda cathartica, Lonicera japonica, Nastutium, Gloriosa superba, Trachelospermum jasminoides, Clitoria ternatea etc.*

III. **Fragrant flowers:** The flowers of certain climbers have very pleasant odour that fill the ambience with their fragrant volatile compounds diffusing into surrounding air. Such climbers are best suitable for panting in public parks giving a mass effect. The fragrance of these flowers calms the mind and also possesses potential therapeutic properties that may relieve stress and strain.

E.g. *Jasminum grandiflorum, Jasminum humile, Lonicera japonica, Trchelospermum jasminoides, Hiptage benghalensis, Clematis panniculata etc.*

IV. **Climbers for pots:** These are best suited for small areas where there is no provision to plant climbers on the ground. Under such conditions, earthen or plastic pots of preferably 14-18 inch size are best for growing climbers. These pots may be placed at a suitable location and once climbers began to grow, these must be supported with wire and left undisturbed to let them create mass effect. A climber such as *Epipremnum aureum Syn. Scindapsus aureus* is most suitable for indoor rooms under bright indirect light and well ventilated conditions.

E.g. *Bougainvillea* spp. *Philodendron* spp. *Tristellatia australis, Adenocalymma allicea, Clitoria ternata, Bignonia purpurea, Clerodendrum thomsoniae etc.*

V. **Making hedge:** Like shrubs, certain climbers also can be maintained as dense hedge, once planted, they continue to grow indefinitely and may or may not be pruned depending upon the site characteristics. In sites such as public park or private garden, the climbers can be pruned as desired depending upon the available space. In other exposed sites, these may be left un-pruned and allowed to grow them profusely to cover maximum area to avoid trespassing or invasion of wild animals.

Eg. *Clerodendrum inermi and Bougainvillea* spp. *etc.*

VI. **Indoor decoration**: Certain shade loving climbers can be planted in beautiful planters and placed indoors in the rooms, corridors and office at suitable locations. Care should be taken to ensure availability of bright indirect light and area should be well ventilated to allow adequate gaseous exchange essential for foliage quality and general vigour of climber.

Eg. *Epipremnum aureum Syn. Scindapsus aureus, Monstera deliciosa, Philodendron* spp. *Asparagus* spp.

VII. **Screening purpose:** Climbers when planted at suitable locations and trained as per the situation may present a picturesque effect. The passage, balcony or multistorey building may be naturally beautified by planting climbers and train them so as to create a screening or camoflauging effect.

Eg. *Vernonia elegnaefolia, Pyrostegia venusta, Jacquemontia violaceae, Passiflora edulis, Antigonon leptopus, , Clerodendrun splendens, Thunbergia etc.*

VIII. **Annual climbers:** There are climbers, which are seasonal in nature and can be grown annually from the seeds in pots, in hanging baskets or in low trellises. For climbers to raise in pots, the pots may be custom-designed with the provision of in-built trellis attached at the pot periphery. Care should be taken that such pots should be durable and be able to bear the support of the trellis as well as the climber. These pots must not be placed on windy sites that may cause toppling resulting in breakage of pot and harming the aesthetics of climber.

E.g. Sweet pea, trailing nasturtium. *Ipomea purpurea, Clitoria ternatea* .

IX. **Climbers for arches, bowers, trellis and pergolas:** Supporting structures such as arches, bowers, trellis and pergolas present a durable and robust platform for the climbers to climb. These structures can be custom made with durable metal iron rods. In addition to lending a support to the climbers, these also provide utility services such as shady walkways or site enclosures in private garden. In a typical rural setting, a well designed pergola may be a feasting room providing shade and coolness for the cattle. The evaporation from the exposed leaf surface creates a micro-climate cooling the surroundings.

Eg. *Passiflora caerulea, Lonicera sempervirens, Allamanda cathartica, Antigonon leptopus, Aristolochia elegans, Solanum seaforthianum* etc.

X. **Climbers with attractive foliage:** Certain climbers become a centre of attraction in a garden landscape for their peculiar characteristics of leaves – for their shape, colour, and arrangement. These also possess beautiful flowers but may remain attractive even when not in flowering due to their peculiar foliage features.

   Eg. *Asparagus spregeri, Ficus repens, Hedera helix, Epipremnum aureum Syn. Scindapsus aureus, Passiflora racemosa Monstera deliciosa* etc.

XI. **Drought and salt tolerant climbers:** Such climber are most suitable for panting in drought prone and neglected sites. Certain hardy climbers are resistant to varied edaphic factors and harsh conditions lacking management. The purpose of planting may be to enclose a boundary or to avoid trespassing into private property.

   Eg. *Bougainvillea* spp.

Selection of Soil

The climbers can grow well in any soil provided it is rich in organic matter, fertile, having good water holding capacity and proper water drainage. The soil should be devoid of any hard pans and free from stones and pebbles at subsoil depth of about 3-4 ft. The soil preferably be under cultivable range with optimum pH (6.5-7.5) and EC ($<1 dsm^{-1}$).

Planting of Climbers

The area should be prepared well ploughing 3-4 times and leave the soil exposed for sterilization under hot sun rays. The pits of 30 cm x 30 cm x 30 cm should be prepared well in advance (at least one month) prior to planting of climbers during the onset of monsoons. The dug pits must be refilled with soil mixture comprising of 10-15 kg of well rotten FYM. The pits should also be drenched with chloropyrephos @ 4-5 ml/l of water to guard against termite attack. The pits should be watered and soil be allowed to settle. The depression caused due to settling of soil must be filled with mixture of Soil + FYM till ground level. This operation is necessary to avoid ponding of water near the crown portion of the climber to avoid rotting.

Planting Time

The best planting time for evergreen climbers is July-August and February-March (provided availability of water is there). The winter deciduous climbers are planted during their dormant stage during Jan-Feb before spring flush. The climber should be watered as per need and prevailing climatic conditions. Proper support should be provided at the time of planting. Training is necessary in first year to make the climber climb over the support and to have the desired effect and shape depending upon the nature of support and growth characteristics of climbers.

Intercultural Operations

Although, the beauty of climbers is in their mass effect of foliage and flowers. The more they grow profusely, the more they are of worth appreciation. However, like other indoor ornamental plants, they also need adequate care and maintenance for maintaining their vigour, flower quality and elegant display. The cultural operations such as training and pruning must be done taking into consideration the nature of growth and leaf retention characteristics of the climber.

1. **Training:** At the time of planting, the main branches of climbers must be trained over the supporting structures. It is not advisable to use wire to train them that may cause girdling over the branches that may interfere with the uptake of water and nutrients. Therefore, always use thick plastic or nylon thread and gently tie at strongest node. Care should be taken not to cause bruising to stem or leaves of climbers while training.
2. **Pruning:** Climbers require corrective pruning to contain their growth within the availability of space and nature of support provided. Light trimming of climbers is done in the initial 1-3 years, however, the method of pruning depends upon their form and age and growth habit. Climbers need pruning after they are well established. Winter flowering climbers e.g. *Pyrostegia venusta* and *Clerodendrun splendens* are pruned at the end of their flowering in summer season, whereas, others are pruned in the end of winter season but before emergence of spring flush.

   The winter deciduous climbers are pruned in the second fortnight of January before new sprouting of shoots in the spring season. The heading back of old climbers helps in rejuvenation and improving the vigour and flowering. All the superfluous growth, old or weak branches are best cut at ground level or as low as possible without leaving the stub. Evergreen climbers are pruned only after their flowering is over or prior to start of their active phase of growth.
3. **General maintenance:** After planting the climbers, regular watering for proper establishment. Cultural operations such as weeding and hoeing should be done to avoid competition from weeds for space and sunlight. Drastic pruning of climbers is sometimes essential depending upon their growth to keep them in a desired shape. General pruning consists of removal of dead wood, diseased branches and excessive growth. Heavy climbers need drastic pruning, whereas, light climbers require light pruning. Care should be taken to keep the climbers healthy and free from insects, pests or diseases by periodically spraying the recommended pesticides or fungicides as per the severity of pest or disease infestation.

Table 1: Description of growth characteristics and landscape use of climbers

Sr. No.	Botanical Name	Common name	Family	Origin	Flowering Time And Colour	Method of Propagation	Growth Characters and Its Landscape Use
1.	*Adenocalymma alliaceum*	Garlic climber	Bignoniaceae	Brazil	Oct-Nov. Purple, bell shaped	Cuttings	Hardy climber with shinning leaves produce garlic like smell on rubbing. Suitable for pergolas, arches and planting against walls.
2.	*Allamanda cathartica*	Golden trumpet	Apocynaceae	Tropical America	May-Sept. Yellow, purple, peach	cuttings and layering	Leaves are light green, narrow and pointed arranged in whorl of four. It can be planted as shrub or climber.
3.	*Antigonon leptopus*	Coral vine or Loves chain	Polygonaceae	South America	August - Nov. Light pink	Seeds and cuttings	Fast growing and vigorous creeper with weak branches. Used to train along the fences, arches and pergolas.
4.	*Aristolochia elegans*	Duck climber	Aristolochiaceae	South America	Rainy season White with purple brown blotch	Seeds	Evergreen twiner with heart shaped pale green leaves giving poor smell, can be trained on trellis, peculiar duck shaped flowers.
5.	*Banisteria laevifolia Syn. Banisteriopsis laevifolia*	--	Malpighiaceae	Tropical America	Winter months Yellow	Cuttings and layering	Generally uses to train on pergolas and trellis.
6.	*Beaumontia grandiflora*	Nepal Trumpet creeper	Apocynaceae	Eastern India	Feb-March white	Seed, cutting and layering.	Suitable for climbing on tall trees and very heavy and vigorous growing creeper.

Sr. No.	Botanical Name	Common name	Family	Origin	Flowering Time And Colour	Method of Propagation	Growth Characters and Its Landscape Use
7.	*Bignonia magnifica*	Glow vine	Bignoniaceae	Clumbia and South America	October- November, Funnel shape, large purplish violet in colour	Cuttings and layering	This climber is suitable for planting against walls and arches.
8.	*Bougainvillea* spp.		Nyctaginaceae	Brazil, introducing in 20th century in India	Almost throughout the year, various colours	Hard wood cuttings	Hardy creeper with multiple uses. Can be trained like standard shrub, bush, climbers, hedges, pot plant, bonsai and ground cover on sloppy lands.
9.	*Buhinia vahlii*	Foot creeper	Leguminosae	Indian , common in Western Ghats	April-May, Creamish white	Seeds	It is fast growing climber with luxuriant growth. Suitable for bigger areas, parks and can be trained on trees.
10.	*Campsis grandiflora*	Trumpet climber	Bignoniaceae	China and Japan	May-October , Orange	Cuttings	Very hardy creeper, can be trained on Arches and pergola as it climb up with the help of roots and rootlets.
11.	*Clerodendrum splendens*	Flaming glory bower	Verbenaceae	Tropical Africa	December- March, Scarlet deep red flowers	Root cuttings and suckers	Suitable for shady and semi-shady locations. Leaves are beautiful, round and dark green in colour.
12.	*Clematis panniculata*	-	Renunculaceae	Himalayan region	July- August, Fragrant white flowers	Cuttings	It is a medium to heavy climber, can be grown on trellis and as pot climber.

Sr. No.	Botanical Name	Common name	Family	Origin	Flowering Time And Colour	Method of Propagation	Growth Characters and Its Landscape Use
13	*Clitoria tarnatea*	Butter fly pea creeper	Leguminosae	India	May-Sep Bluish purple	Seeds	It is one of the annual climbers, Slow growing, can be grown as a pot climber.
14	*Dolichandra unguis-cati Syn. Bignonia gracilis*	Cat's claw trumpet	Bignoniaceae	South America	March-April and July-August, trumpet shape yellow flowers	Seeds and layering	Evergreen, vigorous climber. Seed, germination rate is very high. Suitable to train against high walls to create impressive effect.
15	*Ficus repens*	Indian Ivy, Chipkni bel	Moraceae	India	-	Stem tip cuttings and root suckers	It climbs up with help of rootlets and producing sticky substances. It fixes its roots in the cracks and crevices of the walls. Best for growing in shady and semi-shady situations.
16	*Gmelina hystrix*	Ghanti vel/ Parrot's beak	Verbenaceae	-	Year round Yellow	Seeds and stem cuttings	Hardy, evergreen rambler, can be grown near the walls as a protective climber and as thorny shrub.
17	*Hiptage Benghalensis*	Madhvilata	Malpighiaceae	India	February- March, Creamish white scented	Seeds and layering	Evergreen, heavy climber with shining leaves with sweetly scented flowers.
18	*Ipomea cairica Syn. Ipomea palmata*	Railway creeper	Convolvulaceae	Tropics of hemispheres	Year round, purple flowers	Seeds	Quick growing, evergreen vigorous creeper, can be used as a ground cover to cover large area and on boundaries.

Sr. No.	Botanical Name	Common name	Family	Origin	Flowering Time And Colour	Method of Propagation	Growth Characters and Its Landscape Use
19.	*Jacquemontia violacea*	Jacquemontia	Acanthaceae	Tropical America	Aug-Sept Bluish purple	Terminal cuttings.	Light climber, can be used for screening in sunny and shady areas and as pot plant also.
20.	*Jasminum humile*	Pili Chameli/ Yellow Jasmine/ Italian jsmine	Oleaceae	Afganistan, Nepal, Himalayas, Burma, South west china	March Yellow	Cutting/ layering	Medium sized, bushy evergreen climber, can be grown on fences.
21.	Jasminum grandiflorum	Spanish Jasmine	Oleaceae	South Asia, Arabian, Africa	Nov - May White	Cutting/ layering	Light Climber, flowers are used for oil extraction, making veni, garland, bracelet etc.
22.	*Lonicera japonica*	Japanese Honey suckle	Caprifoliaceae	Japan and China	Winter Creamish white	Layering and cuttings	Evergreen creeper, best for growing on trellis or on small trees in semi sunny situation.
23.	*Passiflora laurifolia*	Wrist watch climber or Phoolghari	Passifloraceae	South America	September Purple	Seeds and layering	The flowers are very interesting and look like Rakhi, should be planted near the entrance or from where you can see the flowers closely.

Sr. No.	Botanical Name	Common name	Family	Origin	Flowering Time And Colour	Method of Propagation	Growth Characters and Its Landscape Use
24.	*Petrea volubilis*	Purple Wreath	Verbenaceae	Tropical America	April-May, Purplish-blue flower	Layering, cuttings and suckers	Creeper with woody stem and leaves with very rough surface. Suitable for training against walls or for shrubbery border.
25.	*Pyrostegia venusta*	Golden shower	Bignoniaceae	Brazil	February- March, Bright orange	Ground layering	Most beautiful climber, woody climber. Used to adorn buildings, houses and to train along the walls, pergola, trellis, boundary wall and screening to cover large areas.
26.	*Quisqualis indica* Syn. *Combretum indicum*	Jhoomeka bel or Rangoon creeper	Combretaceae	Malaya	Year round pink and white	Cuttings and root suckers	Moderately scented flowers, planted on long pillars, walls, pergola, trees and trellis etc.
27.	*Rosa* spp.	-	Rosaceae	Northern Hemisphere	Several flushes in a year White, pink, red etc.	Budding, grafting, cuttings	Rose creepers are mostly grown along the pillars, arches and trellis. Because of their thorns these should not be grown near the path.
28.	*Tarlmounia elliptica* Syn. *Vernonia elaegnifolia*	Cutain creeper, Parda vel	Compositae	South America	--	Cuttings	It is most suitable for creating weeping effect on porch, balcony, boundary and against pillars, walls and for scrrening purpose.

Sr. No.	Botanical Name	Common name	Family	Origin	Flowering Time And Colour	Method of Propagation	Growth Characters and Its Landscape Use
29.	*Thunbergia grandiflora*	Sky flower	Acanthaceae	Eastern India	Feb-March, July-September, Blue	Seed, cutting and layering	Leaves are pointed at tip and broader at base, and easily recognized as heart shaped. It can cover walls, trellis, pergola and trees etc.
30.	*Trachelospermum jasminoides*	Star jasmine	Apocynaceae	China and Japan	May-June, White	Semi hardwood cuttings	Evergreen climber. It is best for growing in shady situation. Bears moderately fragrant flowers.
31.	*Wisteria sinensis*	Grape flower vine	Leguminoseae	China and Japan	March, White and Purplish blue	Cuttings	It is a winter deciduous creeper. Can be planted along the walls or on pergolas, fences and trellis.

References

Arora, J S. 2017 Introductory Ornamental Horticulture. Kalyani Publishers, New Delhi

Bhattacharjee, S K. 2006 Landscape gardening. In Advances in Ornamental Horticulture. Vol 5 pp: 5-24.

Pal, B.P. 1972 Beautiful climbers of India. Indian Council of Agricultural research (ICAR), New Delhi. PP: 1-105.

Randhawa, G.S. and Mukhopadhyay, A. 2004. Floriculture in India. Allied publishers Pvt Ltd.

Trivedi, P.P. 1996. Home gardening. Indian Council of Agricultural research (ICAR), New Delhi. P: 129-146.

8

Trees

Simrat Singh

Trees are amazing nature's creations that are not only important for mankind but also sustain ecological balance in nature. They are the source of livelihood and provide valuable raw material for manufacture of household utility products. They are important from spiritual aspect and have been exhaustively described in ancient Indian texts with their significant Ayurvedic and other potential health benefits. Botanically a "tree." is a large, long-lived (perennial) woody plant that attains a height of at least 6 m (20 ft) at maturity. A tree possess certain characteristic features such as a trunk (single main self-supporting stem) which gives off spreading branches, twigs and foliage to make a crown. Since the diameter at breast height (dbh) of trees is determined internationally at 1.35 m (4.5 ft) above the ground, a tree must be un-branched at least up to 1.5 m (5 ft) from the ground.

Importance of Trees

Trees are the important components of terrestrial ecosystem and are sacred precious gifts of nature. The welfare of all living beings on earth is affected not only by density and diversity of trees, but also by their direct and indirect ecological services that help in maintaining the sate of equilibrium. The plural word "TREES" has been described (Seth, 2002) with a logical meaning owing to their value and eco-system services they provide as:

T Timber, the first and the foremost use of trees

R Restoration, reclamation and rejuvenation of denuded and disturbed soils

E Ecological, eco-developmental and environmental services

E Educational, art, culture and bio-aesthetic planning,

S Source of sustenance; i.e., food, fuel, fodder etc.

Festival of Trees

Indian people are well known for celebration of different occasions. Trees hold a special relevance to Indian people. To bring awareness among the masses for planting and caring trees, a week long festival called "Vanamahotsava", literally translated as great festival of forests is celebrated annually. It is celebrated during first week of July annually with the onset of monsoons. The concept of Vanamahotsava was initiated by (Late) Kanaiyalal Maneklal Munshi (then Union Minister for Food & Agriculture) during 1950's. He was a great politician and environmentalist and always encouraged the people to plant and care trees to improve the depleting quality of environment and make it congenial for the mankind and other living creatures.

Trees Indeginious to Indian Subcontinent

Indian subcontinent is one among the 17 mega –biodiverse countries in the world. According to International Union for conservation of Nature (IUCN), India accommodates 2.4% of earth's surface area and accounts for 7-8% of the world's bio-diversity. There are estimated to be over 45,000 species of plants in India, and nearly one-third are endemic. A wide variety of trees, plants and wildlife can thrive in India due to the distinct climatic zones that exist in the country. Trees have always been associated with wisdom and immorality in India. Hindu literature describes a celestial tree as having its roots in the heaven and its branches in the underworld that unites and connects beings of every kind. The majestic Banyan tree (*Ficus bengalensis*) is regarded as the National Tree of India. The trees that are native to the Indian sub-continent and are most commonly planted as an avenue or for their aesthetic and functional utility in a landscape have been described in tabular form in Table 1.

Selection of Trees

To observe a tree requires a great skill especially when these are observed for their beauty and other qualitative characteristics. Selection of trees for beautification of a landscape is based on several factors including morphological growth features of trees, site characteristics and climate of a region. A careful selection of tree species suiting to edaphic and prevailing environmental conditions is of utmost importance. Therefore, for beautification of a landscape after undergoing necessary site analysis (beyond the scope of this chapter), a selection of trees is made depending upon their classification discussed as follows

i. Based on Height

Knowing the average height of tree at maturity help us in selection of tree species for planting at a particular site. The selected tree species should be ensured adequate availability of above and below ground space so as to help them proliferate and flourish for next 15-20 years. Presence of above ground electricity wires, below ground sever pipes, adequate space for expansion should be determined that is also a part of site analysis. However, the trees based on height can be classified into three major groups.

a **Large trees:** The trees that attain height more than 10 m (35 ft) at maturity are classified as large trees. Such trees are most suitable for planting in larger compounds or playgrounds where there is ample space for them to grow. Such trees provide shade and shelter to birds, insects and other terrestrial vertebrates and sometime also a source of food for them.

Eg. *Ficus infectoria, F. retusa, Bombax malabaricum, Azadirachta indica, Millingtonia hortensis, Peltophorum roxburghii, Cassia siamea, Cassia nodosa.*

b. **Medium trees:** The trees with an average height ranging from 6-10 m (20-35 ft) at maturity are classified as medium trees. Such trees can be planted as avenue (discussed later in this chapter) or in public parks in groups for their mass effect.

   Eg. *Cassia fistula, Lagerstromia thorelli, Melia azedarach, Plumeria* spp., *Saraca indica*

c. **Dwarf trees:** The trees with an average height ranging from 3-5m (10-20 ft) at maturity are classified as dwarf trees. Such trees are most suited for comparatively smaller areas such as private home gardens, commercial establishments etc. The advantage of their dwarf architecture help in accommodating these where there is a scarcity of space.

   Eg. *Parkinsonia aculeate, Plumeria rubra, Bixa orellana, Brownea grandiceps, Cordia sebestena, Erythrina blackei, Wrightia tomentosa.*

ii. Based on Form of Canopy (Growth Habit)

The shape or the pattern of growth formed by a network of tree branches and leaves is termed as canopy. Depending upon the growth habit, the tree can exhibit canopy of varying shape and size that tend to increase with the age of the tree. The shape and spread of tree canopy help us in determining the requirement (number) of trees to be planted at a known dimensions of a site. Ideally, the distance between the same tree species will be approximately twice the length of one-half uniform spread of canopy on either side. This will ensure only required overlapping of canopies of two trees planted adjacent to each other. Relatively closer planting may be done to form a dense screen or as per the requirement depending upon the landscape. Based on form of tree canopy, the trees can be classified as follows:

a) **Weeping or drooping shape**: The branches of such trees droop naturally and appear to convey a gesture of welcome or regards. Such trees are planted at the entrance of a house to greet the visitors that also get attracted provided the selected tree species are in flowering.

   Eg. *Callistemon lanceolatus, Sallx babylonica, Salix alba, Putranjiva roxburghii, Tecomella undulata*

b) **Round or umbrella shape**: Such trees are planted where there is availability of ample space without obstruction of overhead wires. Certain trees with well defined round canopy (*Mimusops elengi*) are most suitable for planting at parking sites. Trees with spreading umbrella shaped canopies such as *Delonix regia, , Chakrassia tabularis* and *Chorisia speciosa* accommodate larger space and the requirement of such trees is relatively less compared to trees that have conical or pyramidal canopies.

c) **Conical shape/ Pyramidal**: Such trees usually have an upright growth habit with less spread of primary and secondary branches. These trees form a dense enclosure especially if planted at a closer distance at the outer boundary of a garden or farmhouse.

   Eg. *Grevillea robusta, Araucaria cookie, Polyalthia longifolia, Sterculia foetida, Pinus roxburghii.*

d) **Columnar shape**: Such trees are valued for their peculiar columnar shape and can be planted as a single specimen in a garden.

Eg. Polyalthia longifolia var pendula, Juniperus chinensis, Euclayptus robusta

e) **Spreading**: Trees with spreading canopies accommodate larger space for providing shade.

Eg. *Cassia siamea,* Fic*us lacor, Delonix regia, Thespesia populnea,* F*icus glomerata*

iii. Based on Leaf Retention Characteristics

The classification of ornamental trees based on leaf retention period has been generally limited to evergreen and deciduous species. A more comprehensive way of expressing the leaf retention characteristics of trees has been proposed (Singh and Kushwaha, 2005) based on the intensity of leaf retention depending upon the changing seasons. The functional traits (such as deciduousness/leaflessness and evergreen) of trees observed during phenological studies have been viewed from the broader perspective to precisely depict the extent and intensity of leafless and leaf-flush duration. Various functional types have been described in the following table that will help in categorizing ornamental trees observing their phenological stages with respect to changing climate.

Functional type	Description
Evergreen	Species which maintain high level of canopy (> 75%) throughout the year
Semi-Evergreen	Species with a short deciduous stage
Leaf-exchanging	Young leaves emerge immediately after leaf shedding
Brevideciduous	Species in which synchronous bud break of vegetative buds generally occurs after mid-March and is preceded by a period of deciduousness which varies from several weeks to a few days.
Deciduous	Species in which leaves are shed early in the dry season followed by a strong dehydration of stems
Fully Deciduous	Species in which each individual loses its leaves for a period of at least one month during dry season
Semideciduous	Species maintain relatively low canopy (15–50%) throughout the year
Spring flushing	Species showing synchronous bud break induced by increasing photoperiod.

(Singh and Kushwaha, 2005)

iv. Based on Adaptability

Only few of the ornamental trees have been found indigenous to the Indian sub-continent. Majority of the trees have been introduced from other countries (such as Madagascar, South America, South Africa and East Asian countries) with varying climatic conditions. These introductions are termed a exotics in India. Due to differences in intensity, duration and seasonality of climate, these trees also exhibit varying response outside their native habitat. These trees have undergone certain structural modifications to sustain under hostile conditions comprising drought, salinity and heat. These trees have adapted themselves with modifications such as deep root system, leaf shedding in summers to conserve moisture and other xerophytic characteristics to help them sustain, flower, develop fruits and seeds to perpetuate for generations. The trees that have been native or introduced from other geographical areas and are survive under harsh climate and varying soil conditions have been mentioned as follows:

1. **Sandy soils (Light textured):** *Acacia tortilis, Prosopis cineraria, Albizia lebbek, Azadirachta indica, Leucaena leucocephala, Grass species: cenchrus ciliaris, Cenchrus setigerus, Acacia nilotica, Thespesia populnea*
2. **Clayey soils (Heavy textured):** *Lagerstromea flossreginae, Jacranda mimosaefolia*
3. **Shallow soils:** *Acacia tortilis, Prosopis juliflora, Dichrostachys glomerata, Ailathus excelsa, Zizyphus mauritiana, grass species: Cenchrus ciliaris, Dichanthium annulatum, Panicum antidotale etc.*
4. **Drought prone sandstone rocky sites:** *Albizia lebbek, Boswellia serrata, Cenchrus ciliaris and Cenchrus setigerus, Acacia senegal, Cassia auriculata, Butea monosperma, Butea frondosa, Tecomella undulate, Casurina equisetifolia, Cratavea religiosa, Morus indica, Melia adedarach*
5. **Saline areas:** *Acacia spp., Azadirachta indica, Tamarix auriculata, Prosopis juliflora, Prosopis tamarugo, Salvadora oleoides, Chenopodium, Chorisia species, Phoneix dactylifera, Butea frondosa, Phyllanthus emblica, Bassia latifolia*
6. **Shifting sand-dunes:** *Prosopis juliflora, Prosopis cineraria, Acacia senegal, Albizia lebbek, Tamarix articulata, Cenchrus ciliaris, Sacharum munja.*
7. **Edible trees for arid areas:** *Zizphus mauritiana, Psidium guajava, Phoenix dactylifera, Aegle marmelos, Annona squamosa.*
8. **Trees for wetlands Swamps and marshes:** *Salix babylonica* (water willow), *Alnus cremastogyne* (long peduncled alder), *Alnus trabeculosa* (trobeculate alder), *Morus alba* (mulberry), *Taxpdoium distichum* (swamp cypress), *Taxodium scandens* (pond cypress), *Musa paradisica* (Plantain).
9. **Trees for noise reduction**: *Juniperus chinensis, Madhuca indica, Alstonia scholaris, Pterospermum acerifolium, Puttranjiva roxburghii.*
10. **Dusty and smoky sites:** *Alstonia scholaris, Ficus benjamina, Terminalia arjuna, F. bengalensis, F. religiosa, Bombax ceiba, Pongamia pinnata, Madhuca indica*

v. Air Pollution & Dust Tolerance

Trees also play a vital role in sequestering or adsorbing gaseous pollutants and VOC's from the air. Trees are often projected as "lungs" of an ecosystem because they absorb carbon dioxide and give oxygen. Trees also act as an ecosystems "liver" too, filtering atmospheric pollutants like SO_2 and N_2O through their leaves. Trees are particularly effective in removing particulate matter (PM). Industries are the major source of pollution and in such areas deciduous trees or trees having thick shining leaves will be more successful e.g. *Morus sp., Populus hybrida, Plumeria acutifolia, Alstonia scholaris, Terminalia arjuna, Albizia lebbek, F. bengalensis, Butea monosperma, F. benjamina, Acacia auriculiformis, Pongamia glabra,* etc.

Characteristics for Air Pollution Tolerances

- Deciduous trees with compound leaves
- Bigger canopies can trap more particles than smaller ones
- larger leaves can trap more pollutants than small ones
- Trees with rough, rugged and hairy surfaces that act as the "best filters"
- Tree leaves with epicuticular waxes and leaf hairs (trichomes).

Classification Based on Landscape Use

1. **Based on Beauty (Specimen):** Such trees are planted singly for their attractive shape, beautiful foliage or flowers or certain peculiar characters such as drooping branches (indicating a gesture of humbleness). eg *Araucaria cookii Salix babylonica, Plumeria alba, Magnolia grandiflora* etc.
2. **Fragrant Flowers:** These trees produce colorful flowers and are planted for their beautiful fragrant flowers. eg *Pterospermum acerifolium, Plumeria alba,Citrus sp, Magnolia grandiflora, Citrus etc.*
3. **Foliage Beauty:** These trees are valued for their peculiar leaf characteristics such a leaf shape, leaf texture, leaf variegation. Such trees can be planted singly as a specimen. Eg: *Erythrina indica, Acacia auriculiformis, Thespesia populnea, Ficus lyrata, Grevillea robusta.*
4. **Fruit & Berry Bearing**: Some of the ornamental trees are values for bright colour of their fruits and can also be planted as a specimen in garden. Eg: *Citrus spp., Courouptia guanensis, Kigellia pinnata, Dillenia indica.*

Based on Purpose of Planting

I. Avenue Plantation

Selection of trees for roadside depends upon width of the road, vehicular traffic and the available space alongside the roads for plantation. The trees alongside the road can be planted in a single row or in 2-3 rows accommodating dwarf and medium sized trees on the foreground of tall trees that act as a supporting background. There are four planting schemes followed in avenue plantation depending upon the selection of trees based on flowering colour, flowering time, canopy form and leaf retention habit. These planting have been described as follows:

1. **Shady tree spp. planting on both sides of road**: Trees along side avenues or roads are generally planted for shade or flowers. Commonly planted trees are *Cassia fistula, Grevillea robusta, Jacaranda acutifolia, Ficus infectoria* etc.
2. **Single tree spp. flowering on both sides of road:** When an avenue of single flowering tree spp. is planted on both the sides of the road, the road looks very beautiful and elegant posing a welcoming gesture to the commuters. However, rest of the period, the roadside view looks monotonous as the trees remain devoid of flowers and may or may not remain in vegetative phase depending upon their nature of growth.

a. *Eg. Cassis fistula, Delonix regia, Lagerstromia speciosa, Cassia nodosa, Cassia javanica etc.*

3. **Two tree species flowering same season both sides of road:** In this scheme, the trees are planted alternately on both the sides of the road. The overall effect appears pleasing as there is variation oin the flower colour, shape and size of the trees whch avoids monotonous effect.

   a. *Eg. Grevellia robusta* [coppery yellow flowers in April-May]- *Jacranda acutifolia* [blue flowers in April-May]- *Grevellia robusta*

4. **Two tree species flowering in different season both sides of road:** Following this planting scheme, the roadside will look colourful for extended season. The commuters can enjoy flowering and foliage beauty of either of the tree species during different seasons of the year.

   a. **Scheme I.** *Bauhinia blakeana* [rosy purple flowers in Jan-Feb] - *Spathodia companulata* [red flowers in April-May]-*Bahunnia blakeana*

   b. **Scheme II.** *Delonix regia* [Scarlet flowers in April-June]- *Chorisia speciosa* [bright pink flowers in Oct-Nov]

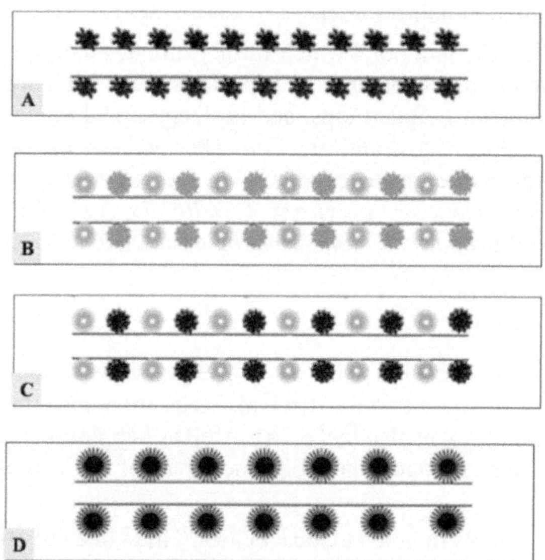

Source: Ramesh Kumar, JS Arora and BM Bharadwaj

Fig. 1: Pictorial representation of different avenue planting schemes

A. One kind of flowering tree on both sides; B. Two kinds of flowering trees blooming at one time on both sides; C. Two kinds of flowering trees blooming at different time on both sides; D. Shady trees only on both sides

i. **Specimen:** Such trees are planted singly for their attractive shape, beautiful foliage or flowers or drooping branches (sign of humbleness). eg *Araucaria cookii, Salix babylonica, Plumeria alba, Magnolia grandiflora* etc.

ii. **Shade:** Shady trees are planted in chosen spots of large public garden which provides place for picnic and relaxation. Such trees are also planted along the borders of roads as avenue for giving shade. Such trees have round canopy or umbrella like crown. Leaves are large and dense so that no or very little sun is allowed underneath them. eg. *Azadirachta indica, Alstonia scholaris, Ficus religiosa F. intectoria, Mimusops elengi, Morus* spp., *Eugenia jambolina, Pongamia pinnata, Chakrassia tabularis* etc.

iii. **Screening & Wind Break:** Tall upright trees are planted very close to give a look of a curtain or screen. Such trees are planted to hide ugly views or to crate partition. eg. *Grevillea robusta, Eucalyptus* spp, *Populus spp, Polyalthia longifolia*

iv. **Home Gardens:** Certain trees with confined canopy and dwarf stature can be planted in limited spaces such as residential home garden. The selection of such trees should be carefully done taking into consideration the available space, allergic problem caused by pollen shedding, fragrant flowers and leaf retention characteristics (evergreen or deciduous). Eg: *Plumeria spp.* (*P. rubra* with red flowers and winter deciduous; *P. alba* with white flowers with evergreen growth habit) can be planted as specimen in home gardens.

v. **Bonsai:** A bonsai is literally a "tree in a pot," which further imitates nature in miniature (the projection of an old tree in nature). Old specimen trees in nature, unlike juvenile trees, have compact rounded tops, and horizontal or drooping branches, which make them, appear aged and graceful. The methodology of preparing Bonsai is beyond the scope of this chapter. However, trees such as *F. benjamina, F. religiosa, Cupressess semipervirens, F. deltoidia, Adansonia digitata, Chorisia speciosa, M. azedarach, Juniperus chinensietc.*are suitable for making bonsai.

Scared Grove Plantation (Religious)

Indians have been worshipping trees since time immemorial to offer gratitude to God's creations as lot of tribals in India depend upon tree for survival. Trees are believed to posses consciousness like humans and they can also feel pain as well as happiness like us (Wohlleben, 2016). So trees and their products have become a part of our rituals and ceremonies.

Sacred Groves are patches of forests usually devoted to local deities. These tree groves are protected by tribal communities over generations because of their religious beliefs and traditional rituals. Sacred Groves are (small or large) patches of vegetation of various sizes and may consist of a multiple tree species depending on the history of the vegetation and associated beliefs. With passage of time different trees like Neem (*Azadirachta indica*), Banyan (*Ficus bengalensis*), Bel (*Aegle marmelos*) and many more have been added to the religious sanctity. Even various Gods and Goddess have been associated with different trees like Bel, Rudraksa seeds (*Elaecarpus ganitrus*) are close to Lord Shiva, peepal (*F. religiosa*) to Lord Visnu, Mango *(Mangifera indica)* to Lord Hanuman, Asoka (*Saraca indica*) to Kamadeva.

Topiary / Sculpturing

Topiary is art of training perennial plants by clipping the foliage and twigs into characteristic geometric shapes. The plants used for making topiaries must be evergreen, woody and have small leaves for producing dense foliage. The trees for making topiary should respond to frequent clipping and must not shed leaves or susceptible to foliage disease. Certain coniferous trees such as *Cupressus sempervirens, Juniperus chinensis, Puttranjiva roxburgii, Buxus sempervirens, Taxas baccata* are widely known for their beautiful shapes maintaine as topiaries. The hanging gardens of Bombay (Kamla Nehru park) is famous for topiary work.

Source of Sustenance

Trees in addition to providing shade and other utility services also are excellent source of food. Certain trees have edible leaves, edible flowers, and edible pods with nutritive value. Eg. Asian Persimmon (*Diospyros kaki*), Jujube (*Ziziphus jujube*), *Moringa oleifera, Bahuinia variegata, Phoenix dactylifera* etc. can be grown at home garden for their aesthetic and nutritional value.

Importance of Trees in Bioaesthestic Planning

Professor Lancelot Hogben coined the term 'Bio-aesthetic planning' defined as planning of flora and fauna with the object of beautifying the country. Bio-aesthetic planning was thus recognised as the aesthetic branch of horticulture, which deals with planting of ornamental plants to create a picturesque effect. Bio-aesthetic planning can play important role in environmental amelioration of wastelands, urban and industrial areas along with their beautification.

Dr. M. S. Randhawa was the flag bearer of Bio-aesthetic planning and strived hard for the cause of environmental amelioration by planting of flowering trees. He promoted the idea that rich colourfulness of nature in the form of ornamental trees should be conspicuously noticeable in the form of avenues along streets and roads in towns and even in home gardens. He advocated garden suburbs around commercial city centres and created garden city of Chandigarh in collaboration with Le Corbusier of France, which has greatly inspired future generations of town planners and architects.

Importance of trees in bio-aesthetic planning

- Reduction in the heat island effect
- Improvement in Air Quality Index (AQI)
- Noise abatement
- Carbon sequestration
- Reduce glare and reflection
- Deflection of strong wind currents
- Maintain evolutionary processes
- Support urban wild-life
- Provide congenial social environment for people

- Trees enhance life of public amenities
- Enhance the biodiversity and beautify the road landscape

Propagation of Trees

Sexual propagation (Seeds)

Trees saplings can either be bought from a reputed ornamental nursery or tree seeds may be collected directly from ripen fruit pods. It is essential to remove that any fleshy mesocarp / endocarp surrounding the seed before planting because this often slows or prevents germination. It is essential to check the viability of the tree seeds prior to sowing, especially if these are stored for longer periods.

The best time of sowing seeds is during the onset of monsoons (Mid July – August). Under North Indian plains, the seeds can also be sown in Feb-March under assured irrigation. Places with milder climate, the nursery sowing of seeds can be done round the year except during hot and dry months of the year. Sometimes, the tree seeds may need scarification *(Cassia fistula, Delonix regia)* or stratification (temperate stone fruit trees) to break the dormancy. Pre-treatment of seeds with hard impermeable seed coat should be done by soaking the seeds in conc. H_2SO_4 (30 min), hot water or cold tap water to soften the hard seed coat and promote imbibition of water. Seeds of trees such as *Acacia auriculiformis, cassia nodosa, Peltophorom pterocarpum, Schleichera oleosa, cassia nodosa* etc are subjected to scarification prior to sowing.

The main factors controlling germination are temperature and moisture. For most of the tree seeds to germinate, the acceptable temperature range is wide and only becomes important if it is exceptionally cold or hot. There needs to be enough moisture for the seed to absorb (imbibe) to hydrate the contents and burst the seed coat for emerging out roots and shoots. Trees usually raised from seeds are *Mimusops elengii, Jacaranda acutifolia, Milletia ovalifolia, Pongamia pinnata, Polyalthia longifolia, Cassia fistula, Delonix regia* etc.

Asexual propagation (Vegetatively)

There are several ways of propagating the ornamental trees vegetatively. Trees can be propagated through grafting, air-layering, hardwood cuttings or through tissue culture for multiplication of rare and elite germplasm.

Cuttings: Propagation by cuttings is the most convenient and cheapest method and usually preferred. For example: *Erythrina indica, F. benjamina* (var. nuda, starlight and curly top), *Bauhinia blakeana, Ficus bengalensis, Ficus retusa, Ficus elastica* are propagated through cuttings. Both terminal and stem cuttings can be utilized for propagation. Cuttings of desirable size (4-6") and thickness are taken from healthy shoots during onset of monsoons, especially if the tree are evergreen. For deciduous trees, the cuttings are made during dormant phase during January. The cuttings are treated with IBA / seradex powder available for soft, semi-hard and hard to root cuttings. The basal portion is dipped in rooting hormone and cutting are inserted into sand beds taking care of their polarity. Sand beds are kept uniformly moist but not wet.

The cuttings are then transplanted in polybags of desirable sizes filled with garden soil and FYM (2:1). The saplings should be monitored regularly for any insect or disease infestation and suitable measures must be taken. The tree saplings must be regularly inspected for their roots that may coil or penetrate deep into soil surface. Periodic shifting of old stock plants must be done to avoid development of taproot and encourage fibrous root system in perforated biodegradable fiber-based tree grow bags.

Air layering: Air-layering (marcottage) is a type of vegetative propagation whereby the new saplings are made *in-situ* over the stock plants (or trees) by gently removing a piece of bark (4-5 cm) from the pencil thickness healthy branch during active phase of growth (July–August). The excised bark portion is covered with moist sphagnum moss holding in a cellophane paper. The moss grass is tied securely with durable thread from both the sides and is left undisturbed for formation of roots. The new baby saplings are ready for transplanting in polybags within 45-60 days depending upon the nutritional status and prevailing climatic conditions. Trees such as *Ficus elastic, Bauhinia blakeana, Citrus chinensis, Ficus krishnae, Magnolia grandiflora and Terminalia metallis* are propagated through air layering.

Table 1: Arboriculture practices for care & management of newly planted tree saplings.

Arboriculture practices	Specification
Pit excavation	Pit measuring 3x3x3 ft dia be excavated and subsoil be mixed with well rotten FYM (2:1).
Soil treatment	Drench chloropyriphos@4ml/litre water against termite infestation.
Planting & Staking	Insert and tie stakes 6-12 inches apart from tree trunk, taking care not to bruise or disturb the root system.
Fencing	To protect young tree saplings from grazing by animals. To protect from strong winds. Iron tree guards or thorny bushes/ bamboo enclosures may be placed alongside to avoid physical injury to newly planted saplings.
Irrigation	Need-based watering is done for the first 3 years fro proper establishment of tree saplings. Water should not touch the base of the trunk that may cause collar rot or fungal infection. The water should contain within the tree berm made alongside 3-4 ft periphery alongside the tree trunk.
Gap filling	Important in mass tree plantation campaigns. The dead, dried and diseased trees should be removed and gap-filled with saplings preferably of same age and size within 3-4 years of planting.
Training and pruning	Done during initial stage of tree growth for proper framework of scaffold branches. Undesirable branches should be removed making a sharp cut and Blitox paste (mixed in linseed oil) must be applied at cut ends to avoid fungal infection. Corrective pruning may be done as required.
Weeding	Periodic removal of weeds alongside the berms helps tree gather required sunlight, space and soil nutrients for its proper growth & development. Mulching the exposed surface alongside berms with paddy straw helps regulate soil temperature, check weed growth and reduce evaporation from the ground surface.
Protection from hot and cold climate	Newly planted saplings can't withstand hot winds during summers and severe frost during winters. Adequate precautionary measures may be taken erecting thatches over young saplings keeping SW side open during winters and shading the saplings by sowing green manure (Sesbania asculata) and tying few plants over top of the saplings to protect from hot winds.

Table 2: Description of ornamental trees and their landscape use

S.No.	Botanical Name/ synonymous	Common or English Name	Family	Origin	Description / Landscape use
1.	*Acacia auriculiformis*	Australian wattle	Fabaceae (Mimosaceae)	Arid region of Australia	An evergreen tree with pendulous branches and phyllodes leaves. The tree is medium sized and bears yellow flowers during Aug-Sept. A. cyanophylla (Golden acacia) is a dwarf species having brittle branches.
2.	*Alstonia scholaris*	Devil's tree	Apocynaceae	Indo-Malaya	It is tall, spreading and shady tree. Leaves are shiny and dark green. The flowers are greenish yellow, small in size appear in Oct-Dec.
4.	*Agathis robusta*	Queensland kauri pine	Araucariaceae	Australia.	Tall upright coniferous tree, leaves with parallel venation, planted as a specimen tree, can grow under semi shade and prefer moist soils.
5.	*Araucaria cocki*	Monkey puzzle (Krishmish Tree)	Coniferae	South America	Horizontal spreading branches with needle like leaves, mostly dioecious, planted as specimen and as pot plant. Tree began to decline later on.
6.	*Azadirachta indica*	Neem	Meliaceae	Indo-burma	Tall tree with pinnate and serrated leaves. Flowers are small white sweat scented, appear end of April to Mid May. Planted as an avenue and for shade.
7.	*Bauhinia blakeana*	Kachnar	Leguminosae	India	A small sized tree, bears rosy purple flowers during Jan – Feb. propagated through air layering. Planted as specimen and group planting.
8.	*Bauhinia variegate*	Kachnar	Leguminosae	India	A medium to tall tree. Flowers are pink or rosy emerge during mid March, when the tree is leafless. The flowers are used for making curry. Planted as specimen or in groups or as avenue tree alongside roads.

Contd.

S.No.	Botanical Name/ synonymous	Common or English Name	Family	Origin	Description / Landscape use
9.	*Bombax ceiba*	Simbal, semal, red cotton silk	Bombaceae	Temperate and tropical Asia	A large deciduous tree, stem is straight and undivided, branches appear in whorls. The young stem is covered with sharp prickles, bears large red/ orange flowers during Jan –March when tree is leafless. Fallen flowers cause nuisance on roads.
10.	*Butea monosperma*	Dhak or flame of forest	Leguminosae	India	A small deciduous tree having irregular branches. It bears bright orange red flowers from Feb - April when the tree is leafless. It is drought resistant tree and can be grown in alkaline soils
11.	*Cassia fistula*	Amaltas, golden laburnum	Leguminosae	India	A medium sized tree bears pendulous racemes of large bright yellow flowers in May-June, when tree is leafless. It is drought-tolerant and can tolerate quite a high salt concentration.
12.	*Cassia javanica*	Java ki rani, Java cassia	Leguminosae	Java, Malasyia	A medium sized tree with spreading crown. Bears clusters of pink flowers intermingled with the foliage in May-June. Leaf tip is blunt unlike its pointed in Cassia nodosa. Branches are brittle.
13.	*Cassia nodosa*	Pink mohar	Leguminosae	Indo-Malaya	Tree larger than Cassia javanica. It bears flowers in pink clusters giving appearance of bunches of roses on long branches during May-June. Planted in parks and as avenue.
14.	*Cassia siamea*	Kassod tree	Leguminosae	Indo-Malaya	A medium sized shady tree. It is drought resistant and can tolerate lot of dust. It is suitable for planting along roadside and bears yellow flowers from Nov. to Feb.
15.	*Callistemon lanceolatus*	Bottle brush	Myrtaceae	Australia	Tree with pendulous branches but leaves are golden-greenish in colour. Scarlet flowers appear borne on long spikes during March-April and Aug-Sept. Best suited for planting alongside ponds or pools.

Contd.

S.No.	Botanical Name/ synonymous	Common or English Name	Family	Origin	Description / Landscape use
16.	*Callistemon salignus*	White or willow	Myrtaceae	Australia	Tree having very narrow leaves when crushed giving smell like lemon, bear white flowers in April.
17.	*Cedrella tuna*	Tun, Toona	Meliaceae	Indo-Malaya, Australia	Large and quick growing tree having shinning foliage. It is easily propagated by seed. It is planted in gardens and roadside for shade.
18.	*Chakrassia tabularis*	Bastard cedar, White cedar	Meliaceae	India and Tropical SE Asia	A tall growing, spreading tree with evergreen foliage. Reddish brown foliage appears on the top of shoots.
19.	*Chorisia speciosa*	Mexican semal	Malvaceae	Mexico and Brazil	It is beautiful tree with a bottle shaped green trunk thorny. It bears bright pink flowers in Oct – Nov. when there is no other tree in bloom.
20.	*Crataeva nurvala*	Barna, Varuna	Capparaceae	India, SE Asia	A medium sized deciduous tree. It bears white or pale-yellow flowers during April - May. It is hardy and drought resistant also.
21.	*Cupressess semipervirens*	Saru	Coniferae	West Asia , South Europe	Leaves are dull green, tree erect, columnar growth. Planted as specimen and propagated through seeds.
22.	*Dilenia indica*	Elephant apple	Dilleniaceae	China / tropical Asia	Tree is evergreen small to medium-sized, growing to 15 m tall. flowers are large, 15–20 cm diameter. fruits main source of food for elephants, monkeys and deer. Seed dispersal through herbivorous animals. Planted in groups or as specimen.
23.	*Delonix regia*	Gulmohar,	Leguminosae	Madagascar	A large tree with umbrella like crown and feathery leaves. It bears scarlet flowers during April to June.
24.	*Erythrina variegate*	Indian Coral tree	Leguminosae	India	A quick growing tree. It bears scarlet flower during March - April when the tree is leafless. The flowers are shaped like the beak of a parrot, hence commonly called tota tree.

Contd.

S.No.	Botanical Name/ synonymous	Common or English Name	Family	Origin	Description / Landscape use
25.	*Eucalyptus spp.*	Safeda, Blue gum tree	Myrtaceae	Australia	Eucalyptus is adapted to fire and tree stumps or seeds re-sprout after fire. Fast growing and have timber value. Provide nectar for honeybees, planted as avenue tree alongside canal roads.
26.	*Eugenia jambolina*	Jamun , Malabar plum, Java plum	Myrtaceae	India and SE Asia	Slow growing, dense foliage provides shade, wood is water resistant, used to make railway sleepers, start flowering from March to April. The flowers are fragrant and small, about 5 mm in diameter. Planted as a shady tree in fields.
27.	*Ficus benjamina (var. nuda, starlight, curlytop)*	Golden Fig	Moraceae	Indo-Malaya	An evergreen quick growing large shady tree. Leaves are small, glossy and dark blackish green, curly and shiny in var. curly top, or variegated (var. starlight). Elegant crown giving drooping look, bark is grey in color and smooth, suitable for planting as avenue or even as a specimen tree in big park.
28.	*Ficus benghalensis*	Bohar or Banyan tree	Moraceae	India	Tall and shady tree. It provides shelter and food for birds.
29.	*Ficus religiosa*	Pipal	Moraceae	India	Tall, evergreen, shady tree. It is considered as sacred tree also.
30.	*Ficus glomerata*	Gulhar	Moraceae	India, Australia, SE Asia, Malasyia	Tall and shady tree. It bears fruiting bodies, which attract birds. Suitable for planting in large compounds for shade.
31.	*Ficus benghalensis var. krishnae*	Makhan katora, Krishna butter cup	Moraceae	India	Leaves are folded and appear like folded cup. Associated with lord Krishna. Natural mutant of F. benghalensis. Medium evergreen tree and planted as a specimen or in groups.

Contd.

S.No.	Botanical Name/ synonymous	Common or English Name	Family	Origin	Description / Landscape use
32.	*Ficus retusa*	Chilkan	Moraceae	Indo-Malaya and China	Evergreen shady tree, medium size, leaves are shining green. Good for growing in parks and along roads.
33.	*Ficus elastic*	Rubber tree	Moraceae	Indo-Malaya	Evergreen tree with shining bright blackish green leaves. Good for shade and can also be used as a pot plant for ornamental purpose.
34.	*Ficus panda*	Ficus panda	Moraceae	Asia	Small tree with new leaves of lemon yellow color. It can be used for hedge or bonsai making.
35.	*Ficus lyrata*	Bingo / Fiddle leaf fig	Moraceae	Western and central Africa	Slow growing tree with curved large dark green leaves
36.	*Ficus lacor*	Pilkhan	Moraceae	Indo-burma	A large deciduous, low crowned thick shady tree, 35-40 ft. high with greenish-grey smooth bark, bending down aerial roots. In April, it is covered with delicately tinted copper colored foliage, which look very attractive. It is suitable where shade is required.
37.	*Grevillea robusta*	Silver oak	Proteaceae	Australia	An evergreen tree with a conical crown. The leaves are fern like and silvery from the lower side. It bears coppery yellow flowers in April.
38.	*Heterophragma Adenophyllum*	Maror phalli, Monkey tail	Bignoniaceae	Indo-Malaya	A large upright growing tree with medium spreading, large sized leaves, for landscaping.
39.	*Inga dulcis*	Jungle jalebi	Leguminosae	India	Tall spreading thorny, suitable protected area, hedge
40.	*Jacaranda acutifolia*	Neeli Gulmohar	Bignoniaceae	Brazil	Medium-sized tree, with feathery bipinnate leaves, bears mauve flowers in March-April, quick growing.
41.	*Juniperus chinensis*	Chinese juniper,	Cupressaceae	China, Taiwan, Myanmar	Leaves are green and needle like, adapted to coastal climate planted as specimen or in groups over mounds, require regular trimming to keep in shape

Contd.

S.No.	Botanical Name/ synonymous	Common or English Name	Family	Origin	Description / Landscape use
42.	*Kigelia pinnata*	Jhar fanoos/ Sausage tree, Balam Khira	Bignoniaceae	Africa	A large tree with good spreading crown. Flowers are cup shaped deep maroon and are borne at the end of long pendulous stalks in April-May, which develop in to sausage like fruits.
43.	*Lagerstroemia spp.*	Jarul, Pride of Inda	Lythraceae	India and SE Asia	Medium size tree, winter deciduous, bears white (spp. parviflora), pink (spp. floribunda) and purple (Spp. speciosa) flowers. Flowers in mid April till August. Planted as a specimen or in groups. Lagerstroemia thorelii is a dwarf tree with pink or mauve flowers from July to September.
44.	*Madhuca indica*	Mahua	Sapotaceae	India	A large deciduous tree. It bears cream colored flowers in Feb to April when the tree is leafless. The leaves are borne in clusters near the end of the branches and are coppery colored when young. The flowers are used for making vegetables and wine.
45.	*Melia azedirach*	Persian lilac, Dharek	Meliaceae	Persia	Fast growing, winter deciduous, compound leaves, flowers first forth-night of April. Planted for shade or at the periphery of the property
46.	*Mimusops elengi*	Molsari	Spotaceae	Indo-Malaya and Sri Lanka	Slow growing tree with round canopy and dark green foliage.
47.	*Michella champaca*	Son Champa or sone champa	Magnoliaceae		A medium sized evergreen free. Bears fragrant flowers of light yellow colour in mid May
48.	*Magnolia grandilora*	Bara champa	Magnoliaceae	USA	A medium sized, evergreen tree with beautiful foliage, large scented white flowers appear in April-May.
49.	*Millingtonia hortensis*	Akash neem, cork tree, tree jasmine	Bignoniaceae	Burma-Malaya	A tall evergreen tree which bears sweet scented white flowers in April-June and November-December. The tree is quick growing and the wood is very brittle.

Contd.

S.No.	Botanical Name/ synonymous	Common or English Name	Family	Origin	Description / Landscape use
50.	*Milletia ovalifolia*	Molumein rosewood	Leguminosae	Burma	A medium sized tree. Bears profuse mauve color, small flowers in March when the tree is leafless
51.	*Moringa oleifera*	Soanjana, Drumstick	Moringaceae	India	Round canopy, bright green foliage, creamish white flower, pods are used for pickle making, leaves sticks are used in curry making. Branches brittle. planted in home gardens for its pods and leaves to make nutritious vegetables.
52.	*Neonauclea purpurea*	Kadam	Rubiaceae	Indo-china and Sri Lanka	The tree is beautiful and columnar. It is growing and bears golden ball like flowers during June and August-September
53.	*Phyllanthus emblica*	Amla	Phyllanthaceae	India	Tree is small to medium in size, dried and fresh fruits of the plant are used in ayurved, tolerant to saline soils
54.	*Pinus roxburghii*	Chir/pine	Pinaceae	Across Himalayas	Evergreen tree. Tree height varies from 30-50 m. Tree grown in hills. It can be grown as specimen tree. Leaves are fine needle shaped.
55.	*Plumeria spp.*	Pegoda/ Gulachin	Apocynaceae	Mexico and Gautemala	Dwarf trees with large dark green leaves with acute apex (spp. acutifolia/rubra), or leaves with obtuse apex (spp. obtuse), bear white (spp. alba), red (spp. rubra) fragrant flowers from May –Sept. planted as a specimen or avenue for street roads.
56.	*Pterospermum acerifolium*	Kanak Champa	Sterculiaceae	Bengal, Assam and Burma	Medium sized evergreen tree with beautiful foliage and creamish yellow scented flower which appears from April to June.
57.	*Putranjiva roxburghii*	Jiva-pota	Euphorbiaceae	Indo-Burma and Sri Lanka	An evergreen, slow growing shady tree. Branches are drooping which give beautiful effect, very good for planting in gardens and along the roads, can be used as tall hedge when closely planted.

Contd.

S.No.	Botanical Name/ synonymous	Common or English Name	Family	Origin	Description / Landscape use
58.	*Polyalthia longifolia*	Ashoka	Anonaceae	Sri Lanka	Tall evergreen tree with straight stem and slender branches which are at right angles to the main stem which form a symmetrical pyramidal crown, suitable for group and avenue plantation.
59.	*Sapium sebiferum*	Makhan tree	Euphorbiaceae	Eastern China and Taiwan	Medium size, fast growing, and a good shade tree tree with round and shiny leaves. Prized for its abundant and spectacular autumn foliage. Tree grow well in urban areas.
60.	*Sapindus mukorossi*	Ritha, soapberry	Sapindaceae	Coastal India	Deciduous tree. Flowering in summer. tolerant to reasonably poor soil
61.	*Saraca indica*	Sita Ashoka	Leguminoceae	Sri Lanka	Dwarf tree with drooping branches, bear scarlet red flowers in April – May
62.	*Schleichera oleosa*	Kusum tree	Sapindaceae	Indo-Burma and Sri Lanka	Medium sized evergreen tree with beautiful foliage. It is ideally suited as specimen or shade or as avenue tree.
63.	*Spathodea Companulata*	Fountain tree	Bignoniaceae	Africa	A medium sized tree, bears red colored flower in April-May. Flower on pressing gives out water like a fountain.
64.	*Sterculia alata/ Pterygota alata*	Buddha's coconut	Malvaceae	India	Habitat mostly in Moist and deciduous forest, leaves are big and tree has upright growth and suitable for planting as avenue for narrow width roads
65.	*Sweetinia Mahagony*	Mahagony	Meliaceae	Tropical America	Tall and evergreen, spreading crown, suitable as a shady tree in park or alongside roads, shed leaves during frost
66.	*Tabebuia rosea*	Mauve Tabebuia	Bignoniaceae	Mexico	Large tree which bear pinkish flowers in March
67.	*Thespesia populnea*	Tulip tree	Malvaceae	Pahari Pipal	A medium sized evergreen tree. It bears yellow flowers which turn red.
68.	*Terminallia arjuna*	Arjun tree	Combretaceae	India and Sri Lanka	Large evergreen shady tree. Flowers are not showy. Suitable for planting along roads and in groups for shade.

Contd.

S.No.	Botanical Name/ synonymous	Common or English Name	Family	Origin	Description / Landscape use
69.	*Terminalia belerica*	Bahera	Combretaceae	India and Sri Lanka	Medium size tree with medicinal properties
70.	*Terminalia chebula*	Harar	Combretaceae	South Asia including India	Medium size spreading evergreen tree, medicinal properties
71.	*Tecoma argentea*	Golden Bell, Silver Trumpet Tree.	Bignoniaceae	South America	Yellow flowers appear in dense clusters that are funnel in shape appear in April, winter deciduous, specimen and also planted as avenue, in parks and home gardens.
72.	*Tecomella undulate*	Rugtora	Bignoniaceae	India	It is dwarf tree, yellow orange colour flowers appear in April – Aug.

References

Arora J S. 2017. Introductory Ornamental Horticulture. Kalyani Publishers, New Delhi

Bhattacharjee S K. 2006. Advances in ornamental Horticulture. Vol 5 (Amenity Horticulture, Biotechnology and postharvest technology). Chap. Landscape gardening. P: 5-24.

Harris R W., Clark JR and Matheny N P. 2003. Arboriculture: Integrated management of landscape trees, shrubs and vines. 4th Edition. Prentics Hall.

Hirons, A D., Thomas, P A. 2018. Applied tree biology. Publishers- John Wiley & Sons Ltd, UK.

Randhawa, G S and Mukhopadhyay A. 2004. Floriculture in India. Allied publishers Pvt Ltd.

Randhawa, M S. 1996. Flowering trees. National Book trust, New Delhi. Series-India: The land and the people.

Santapau, H. 1966. Common trees. National Book trust, New Delhi. Series-India: The land and the people.

Seth, M K. 2004. Trees and Their Economic Importance. The Botanical Review 69(4): 321-376.

Singh, K. P. and C. P. Kushwaha. 2005. Emerging paradigms of tree phenology in dry tropics. Current Science, 89(6): 964-975.

Wohlleben, P. 2016. The hidden life of trees. Penguin random house India Pvt. Ltd. Pp: 320.

9

Indoor Plants

R.K. Dubey and Simrat Singh

Growing foliage and flowering potted plants in the home is a great way to develop an intimacy with the nature. Caring and attending the plants gives us immense pleasure and satisfaction, especially retiring after day long working schedule. Now a days, wide variety of houseplants are available that can accommodate almost any indoor location in terms of available space, light and management factors. Research carried at National Aeronautical Space Agency (NASA) has shown that having plants in the home or office, especially in a closed air-conditioned environment, can be good for your health and well-being.

Our indoor environment

It is surprising to know that our comfort zone environment may have 2-5 times the pollution levels than outside air. Although, the magnitude may vary and could potentially reach 100 times depending upon the size of building, occupants, provision of ventilation, volatile organic compounds (VOC's) emitted from upholstery and paints. The quality of outside air also determine the indoor air quality that may get mixed-up resulting in elevating the levels of carbon-dioxide (CO_2) and other harmful gases beyond the permissible limit. It has been revealed that the indoor CO_2 levels are substantially higher (3-5 times) than those recorded outdoor. The CO_2 being a narcotic cause drowsiness and loss of concentration leading to decline in performance and productivity of employed staff. This sudden upsurge in CO_2 and other toxic gases evolved from the volatile compounds from paints likely raise the Air Quality Index (AQI) that affects the health and working efficiency of building occupants.

It has been scientifically proved that the equipments installed for ventilating closed environment of residential or commercial buildings are not efficient in removing gases and thereby, to some extent filter only the particulate matter (PM). As per the prescribed standards of World Health Organization (WHO) and American society of heating, refrigeration and air conditioning engineers, the maximum permissible limit for indoor CO_2 is 1000 ppm. The term "Sick Building Syndrome" (SBS) is a very common disorder associated now a dayswith the people working in closed environments, especially that lack ventilation.When the CO_2 levels reach 1500 ppm, symptoms such as sore throat, nasal flow and cough are quite commonly observed among the people, thus affecting their general health and well being. Poor indoor air quality has been a major factor linked to SBS, reducing productivity of employees and also cause impaired learning in children in schools.

Plants as potential carbon sinks

It has been well scientifically proved that indoor foliage plants are potential sinks of carbon and substantially reduce CO_2 from the indoor environment, besides assimilating harmful VOCs affecting the general health and well being of occupants. The regulation of gaseous exchange and increase in ambient air humidity through transpiration from leaf surface significantly affect the indoor environment and provide required comfort for an ideal living. The research undertaken by NASA claim that photosynthetic parts of plant and their associated microorganisms had potential to neutralize harmful gases, bacteria, viruses and organic chemicals that are assimilated for forming new tissues during plant growth and development. For past five decades, the selection of plants, their associated physiological processes and underlying mechanisms to remediate indoor air pollutants has been progressed from conventional approaches to state of the art analysis in testing the photosynthetic efficacy of plants inside fully environment regulated closed chambers.

Indoor plants: Origin and History

Indoor plants, suitable be called as 'foliage plants' due to significant worth of trade associated with around 500 species of foliage plants, with different colours, textures and forms available and traded worldwide. These foliage plants are grown in containers under shade, and are used primarily as living specimens for interior decoration or interior-scaping (Chen et al. 2002). Majority of the indoor plants or more precisely 'Foliage plants' have been a native of tropical and subtropical countries.

Many foliage plants are adapted to climate within these regions, mainly characterized by temperature range of 12-38°C due to variation in elevation, varying intensity of light and uniform distribution of rainfall.The foliage plants that are native to tropics have characteristic adaptations for their tolerance of low light, sensitivity to below zero temperatures, photoperiod insensitivity and lack of dormancy.

Foliage plants originating from subtropics are more adaptable to heat, drought, and chilling temperatures and may also exhibit dormancy during winters. Plants adapted to xerophytic environment, such as succulents and cacti have unique adaptive features such as thick succulent foliage capable of storing large amount of water, varied shapes to escape heat and drought stresses (Arora, 2017). Few of the foliage plants are native to the cold temperate regions such as English ivy (*Hedera helix)* is probably the most notable foliage plant of temperate region widely used as either potted or hanging-basket foliage plants.

The use of plants for keeping indoors began 3,500 years ago during rein of Sumerians and ancient Egyptians (Randhawa and Mukhopadhyay, 2004). The ancient Chinese people were also very fond of exotic plants for decorating their interiors. Europeans also imported plants from Asia and the East Indies during Crusades. The aristocratic families of France and England during sixteenth century also had developed a desire for exotic plants. It is estimated that nearly 5,000 species of exotic foliage plants had been brought into Europe. Foliage plants gained popularity in the second half of 19th century in (Europe Pal, 2006). The grand palaces and guest rooms of Victorian houses had exotic collections of foliage plants, especially palms and ferns. Apopka,a place in Florida is known as the foliage capital of the world, with the production and trading of exotic and indigenous foliage plants since 1960s.

Temperature

Temperature is important factor that can affect the growth and development of potted indoor plants, especially those that are of tropical origin that cannot tolerate freezing temperatures. The ideal temperature range for most indoor plants is a daytime temperature of 18°C to 24°C and a nighttime temperature about 10 degrees cooler than the day. Cooler nights are important because it enables plants to recover from daylong stress and build new tissues. Generally, plants are grow in very well areas having temperature 15-30 °C. Plants that are grown under consistently high temperatures suffer from leaf scorching subjected to photo-oxidation of pigments thereby fail to photosynthesize and decline. Soil temperature should 18-21°C soil.

Propagation of Indoor Plants

Propagation is the practice of reproducing or multiplying the plants either through seeds or vegetative plant parts or through a more specialized method of biotechnology involving culturing of explants through tissue culture (David *et al.*, 2012). Most of the indoor plants can be easily propagated and the methods involved are easily accessible to nurseryman or amateurs. Most of the indoor pot plants are propagated vegetatively that include all asexual means such as cuttings, layering, division, air layering, offsets etc. the methods for propagating different indoor plants have been summarized in tabulated form (Table 1).

There are different vegetative methods of propagation are broadly classified as follows:

Seeds: Propagating indoor plants through sexual means (seeds) is done to evolve variation for novel colour, either in flower or foliage. The indoor plants such as cactus and African violets can be raised from seed. However, the plants grown from seeds have long juvenility and may take longer time to reach maturity. The seeds should be collected from mature and ripened pods. The factors that determine the germination of seeds are age of seed that in turn affects its viability, seed germination media, moisture, temperature and oxygen. Seeds should be provided with the proper moisture and temperature conditions for germination. Germinating media useful for seeds include sand, sphagnum moss, peat moss, vermiculite, perlite, and a mixture of these media. The easiest method is to purchase commercial soil mixtures that are specifically for germinating seed. The foliage plants that are raised through seeds are *Araucaria, Podocarpus, palms, Brassaia* and *Coffea.*

Cuttings: Cuttings are severed parts of a mother or parent plant that produce roots and/or shoots when these are placed in a suitable rooting media to form a baby or a new and independent daughter plant. Vegetative plant parts such as stems (apical or middle portion), leaves, leaf-buds, or roots may be used. Certain consideration should be kept in mind while taking cuttings from the stock plants. The mother plants should be healthy free from disease and pests and be at their active stage of growth except when these have shed leaves and the plant is dormant. The basal cut should given round with a sharp knife. The basal portion is given a dip treatment with arooting hormone for encouraging rooting on difficult-to-root cuttings. It is advisable to keep some leaves over the cuttings for better translocation of rooting hormones, carbohydrates for better development of callus and for initiation of roots.

Care of Cuttings: As cuttings method of propagation is most commonly followed. Care should be taken to never allow the cuttings to dry during the rooting, or establishment phase. Although, majority of succulents and cacti have stored large amounts of water within their fleshy tissues, these should be kept relatively dry compared to cuttings from other foliage plant species. There should be adequate humidity in and around the container, holding the cuttings that may help reduce the risk of drying out.

Several methods of propagating foliage plants by cuttings, cuttings can be rooted in water, sand, perlite, or vermiculite. Some plants are easy to root in water (*Drascaena, Scindapsusaureus, Hedera helix, Tradescantia*), but perlite or vermiculite can also be used as an amendment mix for providing aeration to the developing cuttings. The basal portion of cuttings should be pushed about 1 inch deep in the medium that is kept moist but not soggy. The propagation though cuttings preferably be done with the onset of monsoons when the summer heat temperature drops with rise in humidity.

Types of Cuttings

1. **Tip and Stem Cuttings:** Tip cuttings (taken from the tip or actively growing regions of plants) are used to propagate common house plants such as the *Velvet plant, Aglaonema, Diffenbachia* and jade plant. Tip cuttings generally are 3 to 5 inches long and are removed from the parent plant at a point just below a leaf bud. *Swedish ivy, pilea, and fittonia* can be propagated by stem cuttings (cuttings made from sections of stems with leaves attached). The cuttings should have three or four leaves for best rooting.
2. **Cane Cuttings:** Cane cuttings are used for propagating dumbcane, Chinese evergreen, and similar plants that produce cane-like or leafless stems. The canes are generally short stems containing buds are cut into small pieces 2 to 3 inches long and placed horizontally, slightly below the surface of the rooting medium. A bud will sprout and form a new stem.
3. **Whole Leaf Cuttings:** This method is used for mass propagation of foliage plants such as *Begonia rex, Peperomia, Sansevieria, African violets* and *Bryophyllum* with limited availability of stock plants. Whole leaf cuttings are prepared from leaves with or without their petioles. Sometimes, the leaves are cut into pieces and basal excised portions are inserted in rooting media. Roots and leaves eventually will form at the crease of the leaf.
4. **Leaf Bud Cuttings:** Foliage plants with vine-like growth habit are mass propagated through this method. Leaf bud cuttings consist of removing a single leaf attached to a piece of stem (1- to 1½-inch). The shoot and branches eventually arise with the sprouting of dormant buds, located near the leaf petiole. The leaf bud cutting should be inserted in the rooting medium with the

excised portion containing bud below (about one-half inch) the surface. *English ivy, succulents* or *cactus* are generally propagated by this method. The rooting media should be allowed to dry between subsequent watering depending on species and size of leaf buds. The dryperiod will aid in formation of callus at cut edges,thatwill prevent excessive wetting of cut portion.

Division: Division of foliage plants from their crown portion containing roots,is the quickest way to multiply plants. A major advantage of division is the appreciable size of daughter plants. Plants can be divided at any time, however, it is generally advisable to divide mature stock plants during monsoons or at dormant period, before spring flush. Foliage plants such as *Sansevieria, Calathea, chlorophytum, Spathyllium* are multiplies through division

Runners: Certain foliage plants grow with prostrate creeping shoots such as found on strawberry geraniums, can be multiplied by excising the leafy clusters. The cut ends are placed in rooting media for growth of small rootlets that are already present. Foliage plants such as *Chlorophytum, Saxifraga* can be multiplied through this method.

Air Layering: Multiplying plants by air layering is useful in 'pruning' or containing the growth of certain plants which become "leggy," due to maturity or over-growth such as *Ficus elastica* (the rubber plant). Although this method is highly laborious the advantage of getting adequate plant size with developed root system makes this method a popular among nurserymen. Other foliage plants such as *Dracaena Monstera, F. benjamina, crotons* etc. can be propagated through air layering. Care should be taken to wrap up the excised portion of stem properly ensuring packing with moist sphagnum moss. This is practiced during rainy season and the roots generally appear pink taking almost 3 months to transplanting after cutting from mother plant.

Micro-propagation: This is method is increasingly becoming popular to mass multiply exotic foliage plants that otherwise fail to survive into new environment or multiply through conventional ways. Explants comprising meristems, shoot tips, leaf tips, small root segments, excised embryo are taken from healthy disease free stock plants and are cultured under aseptic conditions in petri-plates containing MS media. Foliage plants such as exotic varieties of *syngonium, philodendron, spathiphyllum fittonia, peperomia* etc have been mass multiplied through tissue culture and traded worldwide.

Acclimatization

Acclimatization is defined as a process of making the plant adaptable to its new environment. It is essential to increase the chances of survival of plants that have been introduced into the new environment. Usually, it's not uncommon to see a plant declining after it has been moved indoors from the garden center. This generally happen as a natural response of plant for adjusting in new environment. The drastic change in environment factors such as light intensity, humidity, and other factors usually result in decline of plant with typical symptoms of yellowing of foliage followed by shedding and drop from the plant. Most of the plants need time and will gradually adapt to the new environment. Therefore, to reduce the shock of entering into different environment, care should be taken to furnish the ideal growing conditions that are conducive for the plants to sustain themselves.

However, some ornamental nursery retailers and growers acclimatize their plants before they are brought to the display or sales area. It would be favourable if nurserymen had such provision for acclimatization. It is a wise step to ask for the plants that have been acclimatized to ensure their chances of survival in indoor environments. Plants that have already been acclimated to typical interior conditions usually do perform and grow better when these are placed in home or at office.

Rule of thumb is never to place the indoor plants suddenly outdoors that may cause leaf scorching due to excessive light intensities. Always ensure the gradual exposure of plants to outside environment. Indoor foliage plants must be placed outside for a few hours each day for several weeks preferably in the morning sunlight. Again the exposure of plant can vary depending upon their sensitivity.

Selection of containers

The container selection is an important decision on the part of customer taking into consideration the placement of container i.e. site (outdoor/indoors), container material, size (volume) and cost. Now a days, wide variety of containers are available in market with different colours and sizes (3" to 36"), shapes (round to squarish), material (clay, plastic, ceramic, fiber) and cost (Rs. 10-2500 per piece) depending upon the material and the size.

The container must fits for the size of the plant that should be consistent with the container circumference and height. The container should have optimum volume that can hold required quantity of potting mix to anchor the plant firmly besides providing and holding required moisture and nutrients for plant growth. A general rule for choosing a proper container is that the diameter of the container should be about one-third the height of the plant (at maturity) from the top of the foliage.

The most widely used indoor plant containers are made of clay and plastic. Clay pots are porous and allow moisture to evaporate through the sides of the container. This causes the soil to dry out faster, so more frequent watering is required. However, homeowners overwatering plants raised in clay pots will likely escape rotting due to adequate gaseous exchange through the small micro-pores and drainage hole. Whereas pots made from plastic, glass, metal, wood, or ceramic retain their moisture longer than clay pots that may likely cause rotting of plants if overwatered. Although, pots apart from clay look clean, fancy and more decorative but it is important to ensure good drainage and avoid overwatering in such pots.

Generally, the container adds to the aesthetic display of foliage plant, but the management practices, essentially watering (over or under watering) is most potential cause of decline of foliage plants within the limited pot volume of containers. The containers must possess drainage holes at the bottom for drainage of excess water, provided the potting media possess requisite pore space to facilitate downward movement of excess water. Clay pots (terracotta pots) are porous and require more water than nonporous glazed or plastic pots. The size of the container can be selected depending upon the type of plant to be grown. Plants in large containers (more volume) would generally require less frequent watering than smaller pots (less volume) made of similar material that would require more frequent watering.

Growing media: Selection of a good potting mix is a critical factor for successful growth and development of foliage plants. The foremostability of growing media to anchor a plant will determine the aesthetic display of plant as a specimen. Other factors such as providing the plants with essential nutrients, moisture, and gaseous exchange are also essential that in turn depends upon the ingredients of making a potting mix. The individual physical and chemical properties of respective growing media likely influence the resultant property of media mixture. Therefore, the selection of ingredients for formulating a potting mix is also important cultural practice to safeguard the rhizosphere environment against moisture stress besides other considerations discussed above. The media which contain 2 part garden soil: 1part sand: 2 part leafmould :1 part farmyard manure is very useful for pots. There are many different agri-wastes (composted FYM, sugarcane bagasse ash, rice husk – burnt and un-burnt, paddy straw compost, composted green waste, spent compost, biochar, biogass slurry, press mud etc.) that can potentially be utilized as a potting media after though processing of these media amendments (for uniformity and stability) before utilizing as an ingredients for formulating a potting mix (Bhattacharjee, 2006). However, an ideal potting mix should possess following characteristics:

- Locally available growing media components
- Fertile containing most of the macro and micro-nutrients
- Well drained and light in weight (Bulk density below 0.75 g cm-3)
- Adequate water-holding capacity (20-60% by volume)
- Adequate porosity (5-30% after effective drainage)
- Adequate C:N ratio (<20:1)
- Good Cation Exchange Capacity (CEC) 2-5 meq/100g
- Desirable pH (5.5-6.5)

Foliage plants and growing environment

Growing indoor plants in pots and suitable planters is not difficult if they are raised in proper well drained organic matter-based growing medium with adequate aeration and water holding capacity. Besides, other environmental factors such as light, temperature, general care and managementrespective of each plant are met. The most important factors affecting the growth and development of potted plant are light, temperature, water, humidity, soil, fertilization and other need-based cultural practices that have been discussed below:

Light: Light is regarded as the single most important factor in successful indoor gardening. Plants use light as energy to synthesize their food that is translocated to different parts. The green pigment'chlorophyll' in plant leaves utilized light to combine carbon dioxide, water, and nutrients to make food for growth through the process called photosynthesis.

Even though all indoor plants require adequate amount of light, but the intensity of light (Fig 1) varies with the plant species and its nature of origin, whether temperate, sub-tropical or tropical climatic conditions.

The amount of light a plant receives is called light intensity. The light intensities of indoor areas are generally described as being high, medium, or low and so the plants are classified (Table 1).The light intensity is measured in foot-candles. High light intensity represents those areas receiving 1,000 or more foot-candles of light; medium represents 500 to 1,000 foot-candles; and low represents 50 to 500 foot-candles. Usually, if a particular spot within a house is receiving less than 50 foot-candles, it will not sustain plant life for very long. Plants are unable to utilize low light intensities to carry out photosynthesis and manufacture their food.

The suitable locations within the house can well be identified for placing the indoor plants without measuring the intensity of light. The following guidelines can help determine the light level of various areas and provide guidance for placement of indoor plants:

- Areas nearest windows and glass doors generally offer the best light for plant growth. The size of the windows, transparency, and intensity of tinting also affect the amount of available light. Obviously, more clear and larger the glass pane, more will be available light.
- Orientation (north, south, east, or west) of a window sill or balcony also affects available light. South,east and west facing sites offer bright light and full sun for several hours each day round the year. In general, the south direction of a building offers bright light, while the north facing windows offer low light.

Supplemental lighting

Roof overhangs, porches, drapes, sheers, and blinds also drastically reduce the amount of light available to plants. Artificial light can be used to supplement or replace natural light. Cool-white fluorescent lights alone provide effective supplemental lighting. LED grow lights are color-optimized for plant growth. They have low heat output and a very long lifespan. Plants grown under supplemental lighting need about 12 to 14 hours of this light per day for good growth. A timer will ensure that the lights turn on and off at the proper time each day. It is also important that plants must be kept within 1 to 3 feet from the artificial light source to avoid leaf scorching due to heat.

Water

It has often been observed that irrigating number of different plant species at home is often found confusing. It is well supported fact that every physiological process of plant growth requires water that accounts for 80 to 90 percent of plant biomass. It is not the under-watering but over watering that is often stated to be the main cause of decline in indoor pot plants.

Factors determining water requirement

Some plants require more water than others because of their origin from different climate. Certain modifications in their tissues and growth habits also determine the water requirement. For example, cacti require much less water than ferns. Usually, the brighter the light and warmer the temperature, the more water a plant will need. A plant placed in well lighted place with adequate temperature will require more water than the same plant placed in lower light and comparatively cooler environment.

The presence of moisture in the air also affect a plant's requirement for water. The plants release water into the atmosphere through a process called transpiration through tiny openings (stomata) on the leaves. High humidityslow water loss through transpiration. A relatively low humidity cause more waterloss through transpiration. The centrally air conditioned buildings have typically 10 to 30 percent humidity that may not be sufficient for growth and development of indoor plants especially native to tropical origin.

Ways to raise the humidity

Under such conditions, the humidity can be raised by placing plants on a shallow tray filled with water and pebbles. The pots placed on pebbles avoid complete submergence and help regulate air-water balance at the base of pot. Excess water that drains out of the container and gets collected in drainage tray. This water evaporated gradually and help provide continuous humidity. Homes are often around 20% to 60% relative humidity. Such a range is acceptable, although most species thrive near 80% relative humidity.

Another very effective way to maintain a micro-climate is grouping several plants together. This will help in increasing humidity in addition will create a picturesque display through mass effect of these plants, especially relevant for single species.

Occasional misting the plant foliage with water can also be done to maintain a constant moisture over the exposed leaf surface. However, misting should be done taking into consideration the susceptibility of plants to certain foliar diseases due to presence of high moisture.

Scheduling irrigation for potted plants

Watering, especially to potted plants is an art and science. Any rigid watering schedule usually will not meet the requirement of plants that are grown in a limited pot volume and being under the influence under varied environmental factors. Following a rigid schedule will cause some plants to be overwatered while others may be left dry. It is always advisable to water plants on an individual basis depending upon their species, nature of growth and origin.

The time to irrigate a plant can be determined in several ways. Although there is no foolproof method, but certain thumb rules may give an idea to water plants as follows:

- Inspect the individual plant's foliage for signs of wilting
- Feel the soil for moisture
- Judge color: Dry soils are lighter than moist soil mixtures.
- Decide as per the prevailing climate
- Stage of growth
- Soil moisture indicator
- Drying soil will shrink away from the sides of the container

It is advisable to let the soil within a pot dry out between each successive watering. This is essential to ensure adequate air-water balance desirable for health of the root system and thus the growth of whole plant. It is also important to take care that the soil might be adequately moist near the middle or lower root zone, even if the soil near the top is dry. Certain other pants require more even presence of moisture and will not tolerate drying. Such plants should be watered frequently and lightly to maintain the moisture in the root zone.

Watering the plants – Art and Science

Watering potted plants is a skill that may take a test of your sense to determine the time, frequency, amount and method to irrigate the potted plants. The most convenient and efficient method of watering is to pour water on the soil surface and allow water to move evenly through the limited volume of pot. Usually this method is better than sub-irrigation, a process of watering plants by placing the pot in a shallow amount of water in a tray and allowing the water to rise through the rooting media by capillarity.

It is always advisable to use water for irrigating indoor plants at room temperature. Drinking tap water is usually suitable for houseplants. However, cold water should never be used as it may cause a sudden and extreme fluctuation in soil temperature, which may cause root injuries. Watering should be done gently taking care not to splash the potting soil on the leaves or out of the pot. At each watering, make sure to wet the entire depth of soil thoroughly. Excess water will drain out of the container through the drain opening, which should always be provided of reasonable size and be covered with suitable concave shaped crock to enable excess water to drain.

Sometimes, a growing media in a soil dries too much and will shrink away from the sides of the container. As a result, the water applied may quickly run through the cracks and out of the container and the volume of the pot media may never get wet. The easiest way to wet the soil under such conditions is to submerge the container in water. As per conveyance, submerging potted plant containers in the sink or bathtub may also help to gradually soak the water and wet the whole medium.

Nutrition of foliage plants

It is well documented that plants need 16 essential elements to make their food for growth. The most important elements are nitrogen (N), phosphorus (P), potassium (K), calcium (Ca), magnesium (Mg), sulfur (S), and iron (Fe). The minor elements are carbon, hydrogen, oxygen, manganese, copper, boron, zinc, molybdenum, and chlorine.Some of these elements are required in large quantities, while others are needed in small amounts. The amounts needed depend primarily on the plant's growth rate. Indoor houseplant that flower should be fertilized with fertilizers, slightly higher in amount of P (1-3-1, for example). Those plants that are valued for their foliage should be fertilized with fertilizers with slightly higher proportion of N (5-3-3). The fertilizer ratio may also be balanced in proportion of nutrients (such as 19:19:19). Plants that grow rapidly demand more nutrients than plants that grow slowly. Remember that most indoor areas do not provide the ideal growing conditions for plant growth. Therefore it is necessary to fertilize such plants with balanced dose of nutrients to ensure optimum growth at all stages.

Growing mixtures often lack major nutrients, so these indoor plants will not grow well unless you fertilize them. Keep in mind that it is not always necessary to continually fertilize indoor plants. Fertilizer encourages leaves, stems, and roots to grow if all conditions are ideal, but having a larger, lush plant is not always desirable. The fertilizer nutrition to indoor potted plants depend upon several factors including type of plant (foliage or flowering), climatic conditions, stage of growth etc. There are many forms and strengths of fertilizer materials at garden centers. Some are liquids, while others are granules or powders. All forms are suitable if used properly and need-based.

Slow-Release Fertilizers

Some manufacturers offer specially prepared fertilizer materials that are called "slow-release formulas." Most are granules that are designed to release nutrients gradually and evenly over a long time depending upon the indoor temperature, and watering practices. They offer a consistent and uniform availability of nutrients to the plant. Fewer applications are needed with slow-release fertilizers, but these may not able to deliver the nutrients when the plant needs them.

Fast-Release Fertilizers

Fast-release fertilizers are useful because their nutrients are immediately available to plants and create an immediate response. Fast-release fertilizers are typically available as liquid, soluble crystals, or granular fertilizers. These materials disperse all of their nutrients at once. Fast-release fertilizers, however, are subject to rapid depletion from the soil because of leaching. If fast-release fertilizers are applied too heavily, the plant can be damaged, usually by burning.

Applying Fertilizers

A plant that is growing rapidly needs nutrients more often than a plant that is growing slowly or resting. For plants in an active growing condition, apply a balanced dose of fertilizers once every 2 or 3 months. During the winter months or under other conditions of low light, less frequent applications are needed. If you move indoor plants to the porch, patio, or balcony during the summer, these plants will begin to grow rapidly. Under these conditions, the plants may need fertilizers once every 2 to 3 weeks. Care should be taken that before fertilizing the potted plants, the soil should be moist to avoid burning of plant root due to evolution of heat from the applied fertilizer.

Over-fertilizationis very dangerous because the excess salts can damage plant roots. The plants will suffer because a damaged root system is not healthy and efficient. The typical signs of over-fertilization and excess salts are:

- Appearance of a white or yellowish, crusty spots of salt on the surface of growing media
- Roots began to rot and give offensive smell
- Yellowing foliage that soon drops off the plant

With the appearance of above symptoms, applications of fertilizer should be reduced and excess salt build-up must be washed by thoroughly watering the plant and excess water should be drained out below the bottom of pot.

Repotting

With the passage of time during a life cycle of a pot plant, the root system will fill all the available soil and space in the container and become root- or pot-bound. When this happens, the plant's growth will be restricted unless you provide more space for root growth by repotting the plant into a larger container. A simple examination of the root system is the best way to determine whether or not your plant is root-bound and ready for repotting. When under all favourable conditions the potted plants seems show decline in vigour, irrigate the plant as usual and wait for the potting medium to reach at field capacity. Gently hold the pot upside down and remove the earth ball by tapping the rim of the inverted pot against the hard surface. A typical symptoms of root coiling around the root ball indicates that the plant has become pot bound and need re-potting.

Method of Repotting

Repotting preferable be done during periods of high humidity during monsoons to lessen the chances of transplanting shock. Select a container slightly larger than the existing container. The new container should generally be only 1 or 2 inches larger in top diameter of pervious container. Add a layer of soil mix and firm it with your fingers. This layer of soil should be enough to bring the top of the root mass to within 1 inch of the container top. By allowing this 1-inch buffer near the top, you can water the plant more easily. Gently place the plant with root-ball intact and fill around the space between root ball and pot periphery with soil mix, and firm it with your thumbs packing all loose soil devoid of any pore space. Then water thoroughly and allow the potting soil to drain well.

Place the newly potted plant in indirect light until it recovers from the shock of repotting, usually about 7 to 10 days. This shock is caused by a disrupted root system, which affects a plant's ability to absorb water. Be very careful not to overwater newly repotted plants. Also, don't fertilize the newly repotted plant until the root system has become successfully reestablished. This usually takes about 2 to 4 weeks; after this time, you can resume regular fertilizer practices.

Grooming

As we are domesticating the plants outside their natural environment that may likely invite undesirable and unusual problems that can affect the optimum growth and development of plants. The most common of the problem is dust accumulation on their leaves. A layer of dust on a plant's foliage can interfere with the plant's ability to photosynthesize and function properly; dust also makes plants unattractive.

It is best to clean plants every 2 to 3 months with a soft cloth moistened with warm water. Be sure to wipe clean both the top and bottom surfaces of the leaves. Another way to clean plants is to place them in the shower, sink, or outdoors and wash them gently with fine spay of water.

Turning plants to expose all sides to available light is also important. Turning the plant promotes the development of a uniform shape. Periodically remove the dead, diseased, or discolored foliage. Use scissors to trim dried brown leaf tips and margins. For best results, follow the shape of the leaf when trimming to maintain a natural appearance.

Diseases

Diseases of houseplants are relatively rare. Many problems attributed to diseases are actually caused by unfavorable growing conditions, mostly resulted from overwatering. Prevention is the best way to control indoor plant diseases. Inspect the plant carefully before these are brought from garden centre and also periodic inspection should be carried at home for an indication of diseases, wilting, or leaf spots or algal growth as mentioned in Table 2.

Fungi and bacteria cause most houseplant diseases. Bacteria usually produce angular lesions frequently surrounded by yellow halos or water-soaked areas. Fungi produce a wide variety of symptoms including root rots, stem lesions, leaf spots, mildews, and rusts.

Table 2: Diagnostic symptoms of common indoor plant problems

Possible causes /	Foliage							Growth			
	1	2	3	4	5	6	7	8	9	10	11
Excess light			X			x	x				
Insufficient light			X	x				x	x		
Day length									x	x	x
High temperature			X	x			x	x	x		
Low temperature		x				x				x	x
Lack of water	x			x	X	x	x		x	x	x
Poor drainage / oer watering	x		X	x	X		x	x	x	x	x
Excess fertilizer	x		X	x	X		x		x	x	x
Insufficient fertilizer	x		X	x				x	x		
Compact soil			X	x			x	x	x		
Container defect		x					x		x	x	
Low humidity	x	x	X	x	X				x	x	x

Symptoms associated with foliage
1 – Brownish tips or margins
2 – curling or drooping
3 – yellowish green
4 – old leaves drop
5 – all leaves drop
6 –spots on leaves
7 –wilting

Symptoms associated with growth

8 –Weak thin & soft
9 –new leaves small
10 –none leaf develop
11 –plant died

Low Light intensity plants

Scientific name	Related species	Common name	Family	origin	Propagation	Distinguished features
Aglaonema spp.	A. commutatum, A. costatum, A. modestum A. roebelenii	Silver Evergreen, Chinese Evergreen	Araceae	Tropical regions	Division of basal shoots, terminal stem cuttings	Leaves green to greenish grey, leathery, ovate to spear shaped
Aspidistra elatior	A. lurida, A. punctata	Cast-Iron Plant	Liliaceae	China	Division	Leaves arise from the base, oblong to lanceolate , broad and leathery
Chamaedorea elegans	Chamaedorea seifrizii, C.erumpens	Parlor Palm	Palmae	Mexico and Gautemala	seeds	Slow growing, feather shaped leaves suited for pot culture
Scindapsus aureus	S. pictus, S. aureus var. tricolor, S. aureus var. Marble queen	Marble Queen Pothos	Araceae	Solomon Islands	Stem cuttings, leaf bud cuttings	Suitable for hanging baskets, totem pole planters, grows well in water, require bright indirect light for bright foliage
Sansevieria trifasciata	Sansevieria cylindrica, S. stuckyi, S. ehrenbergii	Snake Plant	Liliaceae	South Africa	Division of suckers and leaf cuttings	Succulent, erect stiff sword like leaves, variegated, do not need frequent watering
Raphis excelsa	R. humilis, R. flabelliformis	Lady palm	Palmae	China and Japan	suckers	Susceptible to leaf scorching, good for pots and shade gardens

Medium Light (75 to 150 foot-candles)

Scientific name	Related species	Common name	Family	origin	Propagation	Distinguished features
Asparagus densiflorus 'Myers'	Asparagus setaceus	Foxtail Fern	Liliaceae	South Africa	Division of tuberous roots	Dense needle like foliage soft to touch. Used as cut greens, possess tuberous roots
Calathea makoyana	C. zebrine C. mangifica	Peacock Plant	Marantaceae	Tropical America	Division	Small plants prized for their attractive foliage
Caryota mitis	C. furfuraceae	Burmese Fishtail Palm	Palmae	Burma , Malaya	Division of suckers and seed	Produce suckers freely, suitable as pot specimen. Greenish to purple flowers develop that drop down the plant axis resulting in plant death.
Chlorophytum comosum 'Variegatum'	C. picturatum, C. comosum vittatum	Spider Plant Spider Ivy	Liliaceae	South Africa	Division of crown or plantlets borne on cascading branches	Production of flowers or plantlets stimulated by short days, tolerant to low temperature, sensitive to fluoride and low humidity resulting in browning of the leaf tip.
Dieffenbachia amoena	D. bowmannii, D. bausei, D. picta	Giant Dumbcane	Araceae	Tropical America	Stem cuttings, air layering and suckers	Suitable for warm and dry places, sensitive to cold. Water little during inactive growth. Chewing leaves may cause temporary speechless.
Dracaena deremensis 'Warneckii'	D. fragrans lindenii, D. godseffiana, D. goldieana D. snaderiana	Corn plant	Agavaceae	Guinea	Cuttings/air layering	Prefer warm and moist medium, can be grown in water. Keeping under low light will cause loss in yellowish variegation and broadness of leaf.
Ficus benjamina	Ficus benjamina var. starlight, curlytop, nuda	Weeping Fig	Moraceae	India , SE Asia	Stem cuttings and air layering	Shed leaves when placed in shade. Grow profusely in warm and bright indirect light

Contd.

Scientific name	Related species	Common name	Family	origin	Propagation	Distinguished features
Ficus elastica 'Decora'	Ficus elastic var. Doescheri, Rubra	India Rubber Plant	Moraceae	India , SE Asia	Air layering, leaf bud cuttings	Plant enjoys full sun and tolerant to low light levels as well
Gynura aurantiaca	G. sarmentosa G. bicolour	Velvet Plant / purple passion vine	compositae	Java	Stem cuttings	Flowers smell offensive and removed at bud stage. Frequent pinching is done to induce business. Foliage gets subdued under poor light
Hedera helix	Hedera helix var. Baltica, Scutifolia	English Ivy	Araliaceae	Europe, West Asis	Stem cuttings	Can be grown in plain water. Pinching is done to induce bushiness. Good for AC rooms
Howea forsteriana	H. belmoreana	Kentia Palm	Palmae	Australia	seed	Slow growing, best for cooler and darker environment, too dry or wet conditions cause leaf tip browning.
Kalanchoe blossfeldiana	K. daigremintiana, K. eriophylla	Kalanchoe	Crassulaceae	Madagascar	Leaves or stem cuttings	Fleshy leaves need plenty of bright light, porous soil and air.
Livistona chinensis	L. Australia, L. humilis, L. rotundifolia	Chinese Fan Palm	Palmae	East Asia (Taiwan, Ryukyu Islands, SE China)	Seeds, suckers	Hardy species, prefer partial shade, moderately drought tolerant, can be planted as buffer strips in parking lots

Contd.

Scientific name	Related species	Common name	Family	origin	Propagation	Distinguished features
Monstera deliciosa	M. acuminate, M. friedrichsthalli, M. leichtlinii M. oblique, M. pertusa	Cut-Leaf Philodendron	Araceae	Mexico, Central America	Leaf bud cuttings, tip cuttings, stem cuttings, air layering	Leaves develop perforated foliage with deep sinuses at maturity. Need hard pruning for rejuvenation when lower leaves drop.
Nephrolepis exatata 'Bostoniensis'	N. acuminata N. cordata N. davallioides N. tuberose	Boston Fern	polypodiaceae	Tropical regions	Division of crown portion	Prefer constant moist conditions, leaves are called fronds, pinnately compound forming an arc, green in colour, forming pendulous character with age.
Pandanus tectorius	P. baptistii P. nelsonii P. sanderi	Screw Pine	Pandanaceae	Polynesia	Offsets / seeds	Leaves are leathery and recurving with white stripes near the margins.
Peperomia caperata	P. argyreia P. brevicepes P. clusiifolia P. elongate	Baby rubber plant	piperaceae	Tropical America / Southern Florida	Stem and leaf cuttings	Have distinct leaf shape and marking. The media should be moist and not wet, leggy growth occur during poor light
Philodendron bipinnatifidum	P. andreanum P. elegans P. dubium P. gloriosum	Tree Philodendron Heart leaf	Araceae	Eastern Mexico	Stem cuttings, air layering, leaf bud cuttings	Ephytic plant, exhibit excellent display under poor light, self heading climber, pot plant
Pilea cadierei	P. microphylla P. muscusa P. macrophyll	Aluminum Plant	Utricaceae	Vietnam	Stem ot tip cuttings / division	Shrubbery and trailing growth habit, sensitive to fluctuations in temperature , over vigorous stems should be pinched off.
Spathiphyllum x clevelandii	--	Cleveland Peace Lily / white anthurium	Araceae	Tropical America	Division of clumps and Tissue culture	Prefer warm temperature, withstand air conditioning well, can also tolerate very low light levels

Scientific name	Related species	Common name	Family	origin	Propagation	Distinguished features
Syngonium podophyllum	S. wendlandii	Syngonium, Nephthytis, arrow head vine	Araceae	Mexico to Panama	Stem cuttings with roots, leaf bud cuttings	Requires periodic pinching if vine growth is not desired. Aerial roots formed at the joints.
Tradescantia fluminensis/ Zebrina spp.	T. aibiflora T. blcssfeldiana T. fuscata T. purpurea	Inch Plant	Commelinaceae	South America	Cuttings, root easily in water	Need to groom periodically as lower leaves fall on maturity causing straggly appearance, excellent hanging basket specimen

High Light (150 to 1,000 foot-candles)

Scientific name	Related species	Common name	Family	origin	Propagation	Distinguished features
Aloe barbadensis	A. aculeate A. aristata A. brevifolia A. excels	Aloe Vera	Liliaceae	Mediterranean	Offsets, seeds, suckers	Sensitive to water stagnation, bear attractive tubular flowers on long stalks
Alternanthera ficoidea	A. amoena A. bettzickiana A. amabilis	Joseph's Coat	Amranthaceae	Brazil	Cuttings	Low growing and withstand trimming very well, used as herbaceous plant for edging
Aphelandra squarrosa	A. aurantiaca A. cristata A. roezlii A. tetragona	Zebra Plant / saffron spike	Acanthaceae	Brazil	Stem cuttings / air layering	Maintain adequate humidity, keep plants pot bound to initiate flowering, showy bracts lasts about six weeks
Araucaria columnaris	A. bidwillii A. cookie A. cunninghamii	Norfolk Island Pine	Araucaraceae	Norflok island/ South Pacific	Seeds, cuttings of upright shoot tips	Require repotting every 3-5 years to maintain good shape. Plant tend to decline with time.

Contd

Scientific name	Related species	Common name	Family	origin	Propagation	Distinguished features
Beaucarnea recurvata	B. tuberculata	Ponytail Palm / elephant foot tree	Agavaceae	Mexico	seeds	Requires repotting to accommodate the size of developing ball, suitable for small planters and develop into large specimen over years
Begonia rex	B. bicolour B. glabra B. nitida B. metallica	Elephant years, Beef-steak	Bignoneaceae	Tropical and subtropical areas	Tubers and seeds	Surface rooted, flowering during Aug-Oct. tubers are lifted soon after the leaves dry.
Chrysalidocarpus lutescens	C. medagascariensis Hyophorbe indica	-	Palmae	Bourbon	Suckers or seeds	Planted as a pot specimen or shade garden
Codiaeum variegatum	Var. Pokhran, SB lal, Sonar Bangala, Mohan Kumarmangalam	Croton	Euphorbiaceae	Souh India, Ceylon, Malaya	Stem cuttings , air layering	Require bright light warmth and humidity for good colouration and bushiness, fast growing.
Cordyline fruticosa	C. australia C. banksii C. termanilis C. rubra	Hawaiian Ti Plant, good luck plant	Agavaceae	India, Malaysia	Cane and tip cuttings Air layering	Prefer high humidity and moisture during active growth, also grows in water.
Crassula ovata	C. argentea C. orbiculata C.secunda C. elegans	Jade Plant	Crassulaceae	South Africa	Stem cuttings lead bud cuttings	Being succulent, watering is critical. Pants should be allowed to dry between subsequent watering in winters

Contd.

Scientific name	Related species	Common name	Family	origin	Propagation	Distinguished features
Euphorbia tirucalli	E. abyssinica E. bojeri E. carinata E. globosa	Milk cactus	Euphorbiaceae	Tropical Africa	Cuttings (dried for couple of days for stopping milky latex)	Can be planted as protective medium sized dense hedge on the outer boundary, also as a pot specimen
Iresine lindenii	I. acuminate I. herbstii var. brilliantissima	Blood Leaf, beef leaf, chicken gizzard	Amaranthaceae	South America	Stem or tip cuttings	Foliage colour brightens during full sun. reduce watering in winters to avoid rotting. Used as borders and edging
Monstera deliciosa	M. acuminate M. friedrichsthalii M. oblique	Split-Leaf Philodendron , Swiss cheese plant	Araceae	Mexico, central America	Leaf bud cuttings, air layering, tip cuttings	Heavy rooted climber with large leathery leaves, perforated with oblong hole at margins. Very effective to cover naked stems of large trees
Phoenix robelenii	P. rupicola P. spinosa P. sylvestris	Miniature date palm	Palmae	Canary Islands to southern China and Malaysia	Seeds	Pot specimen, valued for their pinnate arching leaves.
Podocarpus macrophylla	P. longofolia P. macrophyllus P. taxifolia	Weeping Podocarpus, Japanese yew	Podocarpaceae	china	Seeds / stem cuttings	Dense coniferous tree with needle like leaves, tough plant, endures cold, can be maintained in same pot over the years
Saintpaulia ionantha	-	African Violet	Gesneriaceae	Tanzania	Leaf petiole cuttings/ division of crowns	Flowers are violet to blue, leaves radiate from central crown, cold water damage leaf surface, use bush to remove dust over fuzzy leaves.

Scientific name	Related species	Common name	Family	origin	Propagation	Distinguished features
Schlumberger truncate	-	Thanksgiving Cactus / Christmas cactus	Cactaceae	Brazil	Cuttings	Epiphytic cactus, stem branches, bear no leaves, bright flowers arise at the end of stems during Christmas. Flower development influenced by temp and day length
Sedum morganianum	S. acre S. bellum S. adolphii S. craigii S. rupestre	Burro's Tail, Donkey tai, Lamb tail	Crassulaceae	Mexico	Leaves / stem cuttings (require removal of numerous leaves)	Require good drainage, reduce watering during resting periods, place pots where there is no disturbance. Flowers rose pink on terminal clusters.
Tradescantia spathacea	T. pexata T. purpurea T. navicularis T. fuscata	Moses-in-the-Cradle, wandering jew, Inch plant	Commelinaceae	South America	Stem cuttings	Allow to dry between watering to prevent rotting, pinch frequently for dense bushier growth, suitable as pot specimen and hanging baskets

Table: Commonly used palms for Landscaping and pot culture

Scientific name	Related species	Common name	origin	Propagation	Distinguished features
Areaca Catechu		Areca palm		Seed / suckers	Very suitable plants for and hotels room
Livistona chinensis	L. Australia L. humilis L. rotundifolia	Chinese fan palm	SE Asia including Malaysia	seeds	Fan leaved palm, commonly grown in pots. Slow growing. Trunk is ringed and covered with remains of dead leaf sheaths.

Contd.

Scientific name	Related species	Common name	origin	Propagation	Distinguished features
Dypsis lutescens (Chrysalidocarpus lutescens)	D. madagascariensis	Areca palm , Golden cane palm	Madagascar		The plam has a feathered leaf, forms a clump of slender trunks at the base. The leaf tips are drooping.
Caryota urens	C. mtis	Fish tail palm	India & Malaysia	Seeds	Quick growing, feathered leaf resembling fins of fish. Yields sap on maturity for making toddy
Chamaedorea elegans	Chamaedorea seifrizii, C.erumpens	Parlor Palm	Mexico and Gautemala	Seeds	Feathered shaped leaves, slow growing, grown as a pot specimen
Rhapis excelsa	R. flabelliformis R. kwaniwortsik	Ground rttan palm	China & Japan	suckers	Stem remain covered with fibrous base of leaves, good for raising in pots under partial shade
Washingtonia filifera	W. robusta	California palm/ desert fan palm	southwestern United States	Seeds	Sturdy, columnar trunk and waxy, fan-shaped (palmate) leaves. Leaves called fronds have long, thread-like, white fibers
Phoenix roebelenii,	P. sylvistris P. rupicola P. dactylifera	Pygmy date palm	Southeastern Asia	Suckers	Small to medium-sized, slow-growing slender trunk, leaves light feathery and arching. Suitable as pot specimen in early years
Roystonea regia	R. oleracea	Cuban royal palm or the Florida royal palm,	Mexico, Central America and the Caribbean,	Seeds	Possess barrel shaped swollen stem having ring marks of old fallen leaves. Leaves are large, pinnate with long leaf sheath stalks. Suitable as a specimen in garden and alongside driveway.
Chamaerops humilis	C. excels	European fan palm	continental Europe	Seeds	very drought-tolerant, prefer hot summers. slow-growing plant
Bismarckia nobilis	-	Bismarckia palm	Madagascar	Seeds	tolerates fairly dry and hot climates
Butia capitata	-	Jelly palm	Brazil	Seeds	Cold hardy, slow growing, leaves arch downward towards stem, bst grown as a specimen. Tolerate harsh environmental conditions
Wodyetia bifurcata	-	The foxtail palm	Queensland, Australia.	Seeds	Beautiful long pinnate leaves like a tail of fox. Suitable as a specimen in pots or as group planting and patio decoration. Hardy and can grow in full sun and partial shade.

Representation of range of light intensities in an indoor environment

Schefflera arboricola

Ficus elastica Rubber plant

Monestra oblique *Codiaum variegatum* (Croton)

Areca catechu

Dieffenbachia snowtropic

Ficus benjamina

Philodendron sp

Cycas revolute

Aglaonema crispum Silvar bay

Suggested Readings

Arora, J. S. 2017. Introductory Ornamental Horticulture. Kalyani Publishers, New Delhi

Bhattacharjee SK. 2006. Advances in ornamental Horticulture.Vol 5 (Amenity Horticulture, Biotechnology and postharvest technology). Chap. Landscape gardening. P: 5-24.

Chen J, HennyR.J. and McConnellD.B. 2002. Development of New Foliage Plant Cultivars. In: Trends in new crops and new uses. J. Janick and A. Whipkey (eds.)

David E. Lott, Extension Educator; and Dale T. Lindgren 2012. Professor Emeritus Propagating House Plants. Division of the Institute of Agriculture and Natural Resources at the University of Nebraska–Lincolncooperating with the Counties and the United States Department of Agriculture.

Pal A.K.. 2006. Ornamental plants with attractive foliages. In: Advances in ornamental horticulture. Ed. Bhatiacharjee S. K. Vinayananda S and De L. C .Vol (2) p:273-343.

Randhawa GS. and Mukhopadhyay A. 2004. Floriculture in India.Allied publishers Pvt Ltd.

10

Cacti and Succulents

Sapna Panwar, Namita and Neelam Thakur

Cactaceae family comprises of perennial shrubs or trees characterized with the presence of spines which arise from specialized axillary meristems (areoles), which in some taxa may also bear fine barbed hairs called glochids. A large family consisting of about 124 genera and 1,427 species. It is native to North and South America and widely distributed from Canada to Chile. Nowadays they are found growing in many countries like Australia, South Africa, and Mediterranean countries. They are highly adaptable to drier regions because of their characteristic leaflessness nature. Since cacti are succulent in nature that's the reason that all cacti are succulents but all succulents are not cacti. Cacti conserve water by two means, first presence of cuticle which prevents water loss and second it is CAM (Crassulacean Acid Metabolism) plant where stomata opens at night and close during day time therefore prevents water loss.

Uses

Cacti found its main use for food, fodder and medicinal purposes. Nowadays, it is also being used as an ornamental plant for landscaping purposes. Moreover, it forms an important basis for landscaping of arid areas i.e. xeriscaping because of its drought tolerant properties. Some of the species are also housed as popular indoor plants. Due to spiny nature of cacti, it also serves as an excellent protective hedge plants.

Essential Features of Cacti

1. Presence of areoles i.e. specialized organs which correspond to nodes on other plants. Mostly round in shape and consist of two buds, size varies from small to large and usually provided with spines or glochids.
2. Perennial nature.
3. Dicotyledonous in nature.
4. Fruit is one celled berry.
5. Inferior ovary.

Morphology

Most of the cacti are terrestrial in nature, but some species like *Epiphyllum, Rhipsalis Schlumbergera* are epiphytic in nature. All cacti except genus *Pereskia* are almost leafless wherein the leaves are either absent or greatly reduced or modified to spines. Cacti possess thick fleshy stems which performs photosynthesis due of the presence of chlorophyll. Roots are generally thin, fibrous, and shallow in nature. Cacti varied widely in size i.e. from small size (*Lophophora, Echinocereus*) to the big size (*Carnegiea gigantea*). Cacti mostly bear large bisexual flowers. Inflorescence is axillary solitary or rarely terminal cymes. Flowering time also varies like in some genera flowers open during day time and close by evening e.g. *Echinopsis, Astrophytum, Chamaecereus Lobivia* etc. while this is just opposite in *Selenicereus* where the flowers open after evening and close by morning. There also exists wide diversity for flower colour which range from white to pink, purple, yellow, crimson, dark crimson red, violet, and combinations with spots and streaks. The fruit is a berry which comprise of numerous seeds. Pollination is done by wind, birds, insects, or bats.

Classification

Britton and Rose divided the Cactus family into three tribes.

Tribe I. Pereskieae: It consists of only one genus *Pereskia* and assumed to be the most primitive members of cacti. They give the appearance like that of any other trees, shrubs, and vines but they are true cacti since they posses areoles. It bears large leaves all the year round except is some cases the leaves drop down during resting season. Flowers are stalked and mostly grow in clusters.

Tribe II. Opuntieae: The most characteristic feature of the tribe is the presence of glochids. It bears small leaves which fall very shortly. Some genera of this tribe are: *Pereskiopsis, Pterocactus, Nopalea, Opuntia and Grusonia.*

Tribe III. Cereceae: It is the largest tribe comprised of 8 subtribes. It is most complex in nature with plant ranging from the small to tall ones. It is characterised by the features like absence of glochids, leaves (except in the cotyledonary stage) and sheathed or barbed spines. It comprises of more than 100 genera and possibly over 1000 species. It produces beautiful and attractive flowers which in some genera last for only few days, some for only one night, and a few for an hour or two. Members of the tribe can be easily distinguished by the presence of prominent ribs which is absent in other two tribes.

Subtribe I Cereanae	Subtribe II Hylocereanae	Subtribe III Echinocereanae	Subtribe IV Echinocactaneae	Subtribe V Cactanae	Subtribe VI Coryphanthanae	Subtribe VII Epiphyllanae	Subtribe VIII Rhipsalidanae
Examples: *Cereus, Cephalocereus, Cleistocactus, Espostoa, Haageocereus Trichocereus, Myrtillocactus, Oreocereus, Pachycereus, Pseudoespostoa*	Examples: *Aporocactus, Hylocereus* and *Selenicereus*	Examples: *Chamaecereus, Echinocereus, Echinopsis, Lobivia* and *Rebutia*	Examples: *Ariocarpus, Astrophytum, Echinocactus, Ferocactus, Gymnocalycium, Hamatocactus, Lophophora, Notocactus, Parodia*	Examples: *Melocactus* and *Discocactus*	Examples: *Coryphantha, Mammillaria and Thelocactus*	Examples: *Epiphyllanthus, Epiphyllum* and *Zygocactus*	Examples: *Erythrorhipsalis, Pseudorhipsalis, Rhipsalidopsis* and *Rhipsalis.*

Climate

Cacti are highly adaptable to arid regions but they can also be grown under diverse agro climatic conditions. Extremes of high and low temperature may affect its growth and development. High temperature produces yellowing and scorching symptoms while at low temperature cacti exhibit chilling injuries. Mostly cacti require ample amount of sunlight but prefers morning sun for its overall best performance. In shady conditions, cacti exhibit lanky growth and may results in sparse flowering or no flowering. Hardy species of cacti can be easily grown under open conditions but mostly cacti grows well under glass or polyhouse or indoors. Good air circulation is required for obtaining good results in terms of both vegetative and reproductive growth.

Soils

Drainage is one of the important properties of the growing medium which helps in deciding the composition of potting mixture. Cacti grow well in potting mixture consisting of 2 parts sandy loam soils: 1 part sharp sand: 1 well rotten farm yard manure: 1 part leaf mould. Addition of small quantity of bonemeal is also beneficial. Cacti also grow well in soil mix which consist of 2 parts sand and 1 part soil.

Propagation

Cacti can be propagated through sexual and asexual methods.

Seed Propagation: Cacti can be propagated from seeds but it is time consuming method which may sometimes take more time for plant establishment for some cacti due to their slow growth rates. Cacti propagated through seeds are *Mammillaria bombycina, Parodia leninhaussi , Ferocactus latispinus, Myrthilocactus geometrizans* and *Hylocereus undatus.*

Cacti seeds are very small with size ranges from 0.5 - 5 mm. Cacti exhibit enormous diversity with respect to amount of seeds produced per fruit e.g. *Ferocactus wislizenii* (1600 seeds), *Gymnocalycium mihanovichii* (2040 seeds) and *Mammillaria* spp. (300 seeds). The most important factor for seed germination is temperature and humidity. Seeds require little water and temperature of 18-30°C for germination. Seeds can be raised in seeds pans throughout year under greenhouse conditions. Germination rates also vary for e.g. some cacti seeds germinate within a week or a fortnight, whereas in some it may take several months. Some cacti seeds also exhibit dormancy which can be easily breakdown by following physical and chemical methods. Bottom watering is preferable for the young tender plants as they cannot withstand the overhead spraying. Avoid early transplanting of small seedlings and allow them to grow for several weeks in the seed pans. Care to be taken to avoid injury of roots. Watering should be done after two days of seedling transplant.

Vegetative Propagation

Offsets: The most common method for propagation of cacti is by division. Certain cacti species produce offsets like *Echinopsis, Lobivia, Mammillaria* and *Rebutia*. The rooted offsets are separated from mother plant and new planting is done.

Stem cuttings: This method is suitable for those cacti species which possesses elongated actively growing stems. Cuttings should be taken just above a node with a sharp, clean knife and allow the cut surface to be air dried for period of several days and then planted in pots with sand as rooting medium. Avoid deep planting of cuttings. some researchers also suggested mixture of 50% peatmoss and 50% sand planted at temperature approximately 22°C for induction of rooting. Rooting in cacti will normally take more than two weeks. Examples of cacti propagated through stem cuttings are *Opuntia*, *Cereus*, *Trichocereus*, *Myrtillocactus, Hylocereus, Lophocereus*, *Echinopsis*, *Mammillaria*. etc.

Grafting: It is also one of important method for propagating cacti and is done for the following cases

1. To grow those cacti which are difficult to grow well on its root system.
2. To grow those cacti which shows abnormal growth or some chlorophyll free forms which cannot grow independently.
3. To fasten the growth in slow growing species of cacti
4. To save damaged cacti suffering from disease like root rot.
5. To raise a rare or specimen cacti in very short period.

In grafting the compatibility between scion and rootstocks is important which determine the success of grafting. Most common rootstocks in cacti are *Cereus, Trichocereus, Harrisia, Opuntia, Cleitocactus*. Mostly splice and cleft grafting is done.

Rootstock	Scion
Cleistocactus baumannii	*Aporocactus flagelliformis, Chamaecereus silvestrii, Mammillaria elongata, M. gracilis*
Cephalocereus spp.	*Cleistocactus strausii, Haageocereus bicolor*
Opuntia tomentosa	*Opuntia microdasys*
Hylocereus trigonus	*Gymnocalycium mihanovichii, G. denudatum, Chamaecereus silvestrii, Cereus cristata, Notocactus scopa, Mammillaria theresae, Sulcorebutia rauschii, Eriocactus leninghausii.*
Cereus peruvianus	*Melanocactus* spp., *Gymnocalycium mihanovichii cv. 'Hibotan', Rebutia muscula*
Trichocereus spachianus	*Cereus* spp., *Lobivia* spp., *Espostoa* spp., *Melanocactus* spp. *Echinocactus* spp.

Planting

Cacti are planted directly either in the ground or pots. Earthen pots are more suitable for its growth. Planting can be done throughout year by following the necessary precautions. However, planting can be preferably done in March or April.

Watering

Avoid excess watering in cacti as they are highly sensitive to excess water. Excess watering leads to root rotting. Winter season is slow growth period for cacti so avoid frequent watering. However, during summers, watering should be done at regular intervals. There is one thumb rule of watering cacti i.e. allow the soil to go nearly dry conditions between watering.

Nutrition

Cacti are not heavy feeder of nutrients and respond well at low doses of fertilizers. They should be fertilized during the active growth phase i.e. March to October. Application of high doses of fertilizers results in lush green plants which make them more susceptible to rotting. However, no fertilization is required during winters as cacti undergo resting phase and there is occurrence of very little growth during winter season. Application of balanced liquid fertilizer with dilution at once or twice a year is highly recommended.

Description of Important Cacti

All cacti belong to family cactaceae. There are numerous cacti all over the world; however, few important cacti are described hereunder.

Beavertail Cactus (*Optunia basilaris*): This is known as prickly pear cactus having blue-green pads. It grows up to 50 cm height and 180 cm width. Flowers are dark cherry pink in colour that glows and smells like watermelon. It is suitable plant for desert areas or gardens. It blooms in late winter to early summer.

Blue Flame Cactus (*Myrtillocactus geometrizans*): Blue flame cactus is commonly known as bilberry cactus, garambullo, or whortle berry. It grows up to 13 feet height and 8 to 12 feet wide. Crested species are usually smaller. It forms thick forests and reaches up to a height of 30 feet in its natural growing conditions. It is most valued for its blue-green color, upright candelabra shape, and purple fruit.

Candelabra Cactus (*Myrtillocactus cochal*): Candelabra cacti grow about 10 feet tall and broad. Flowers are cup shaped which open in the day time and close in the evening. Fruits are acidic and edible. Planting of candelabra cacti on slopes offers a natural landscape. It is widely used in xeriscaping and rock gardens along with other succulents. It can grow under drought stress conditions.

Claret Cup Cactus (*Echinocereus triglochidiatus*): This is commonly known as hedgehog, Mojave mound cactus and kingcup cactus. It grows to 3 feet high and up to 6 feet wide. Juicy fruits taste like strawberries and upon ripening turns bright orange. Planting claret cup cactus with sage, poppies, yucca, penstemon, and native grasses in landscaping design will have a pleasing effect. Claret cup cactus prefers a soil that contains more gravel than traditional cactus/succulent mixes.

Golden Ball Cactus (*Parodia leninghausii* or *Notocactus leninghausii*): Golden ball cactus is also known as lemon ball cactus or yellow tower. It grows up to 3 feet tall. Golden ball cactus appears in bunches and is suitable for fire resistant landscaping. Specimen planting of this cactus in containers provides pleasing effect to the viewers. It is an excellent choice for beginning cactus gardens.

Golden Barrel Cactus (*Echinocactus grusonii*): It is the most popular type of cactus grown in drought prone areas. Golden barrel cactus is easily recognizable and grows up to 4 feet tall. Planting of several cacti in a grid provides striking landscape design in a front or back yard. Younger plants prefer light shade for growth but once mature, golden barrel cactus thrives well in shade areas.

Strawberry Hedgehog Cactus (*Echinocereus engelmanii*): This cactus is also recognized as strawberry cactus, purple torch cactus, and saint's cactus. It produces free branching clusters or bundles of erect stems. It grows up to 70 cm tall. The characteristic feature of *Echinocereus* is pointed ornamental spines which compactly cover the plants. It can grow in rock and drought tolerant gardens along with other succulents and wild flowers. It can tolerate light shade in outdoors, while in indoors, it finds best sunniest spot.

Mexican Fence Post Cactus (*Pachycereus marginatus*): This cactus can grow up to 16 feet high and has stems of 7 to 20 cm in diameter. It blooms during mid-to-late spring. Mexican fence post cactus reacts well to frequent watering in hot climate. During frost, the plant can be protected by keeping styro foam cups or burlap over growing tips. This cactus is having columnar shape and upright growth due to which it is used as a living fence. Planting of this cactus near a bright colour wall or in containers with native flowers provides the dramatic effect.

***Mammillaria polyedra*:** Most of the species of this cactus are native to Mexico. *Mammillaria polyedra* grows up to 30 cm tall and 12 cm wide with 2.5 cm blooms. Growth of *Mammillaria polyedra* starts as an individual plant and then ultimately forms intense clusters. Unlike most cacti, mammillaria doesn't like more than about 4 hours of direct sun each day. Bright and indirect light is best suited for its growth.

Old Man Cactus (*Cephalocereus senilis*): Old man cactus is also called the old man of Mexico. It has tall, columnar or branching growth, often covered with long woolly hairs which are the characteristic feature of old man cactus. This can be propagated easily by seed or cuttings. It produces blooms at night in mid spring after attaining a height of 9 feet. Old man cactus can reach up to 49 feet height. It is highly suitable for drought tolerant gardens.

Orchid Cactus (*Epiphyllum* hybrids or *Phyllocactus*): In their native habitats, orchid cacti grow up to the height of 225 feet. The flowers of this cactus are admired for their stunning 10 cm blooms and are primarily grown in hanging baskets.

Peanut Cactus (*Chamaecereus silvestrii*): Peanut cactus is cylindrical, low-growing cactus that grows up to 1 foot in height. It bears vase-shaped flowers and blooms in spring. Peanut cactus prefers shallow containers and grows in xeriscapes and rock gardens. Peanut cactus is propagated by using offsets.

Prickly Pear Cactus (*Optunia indica*): Prickly pear cactus is commonly known as mission cactus, barbary fig, or tuna cactus. It has 10 cm size flowers and red or yellow coloured decorative fruits commonly known as tunas. It can grow to 15 feet height and 6 feet wide. It is used in drought tolerant gardens along with turfgrasses

Rat Tail Cactus (*Aporocactus flagelliformis*): Rat tail cactus produces beautiful tubular flowers in the spring. It can grow up to 3 to 5 feet tall. It grows as a hanging plant and require afternoon shade. It is quite easy to keep rat tail cactus as a potted plant in colder climates, moving it indoors for the colder months.

Silver Torch Cactus (*Cleistocactus strausii*): Silver torch cactus grows to be slender columns up to 8 feet tall with 7.5 cm long, tubular blooms which protrude from columns horizontally. Plant likes full sun and well-draining soil. Mixing of silver torch cacti with succulents provides an attractive display.

Star Cactus (*Astrophytum ornatum*): Star cactus is also known as monk's hood. They grow up to 30-100 cm tall and 15-30 cm wide. It is the tallest member in the genus *Astrophytum.* It is identifiable by its five to eight ribs which certainly twist into spirals. Yellow brown spines and yellow flowers forms at the centre are the main features of star cactus. Its fruits form a star pattern.

Turk's Cap Cactus (*Melocactus matanzanus*): This grows up to 9 cm tall and 9 cm in diameter. This cactus requires high maintenance and should not allow drying out. It grows well in rockery and xeriscapes with native shrubs, perennials, and wild flowers.

Succulents

Succulent refers to those plants which possess very thick fleshy foliage or stem or both meant for storage of water. Succulents differ from the cacti mainly for two main features first succulents possess leaves while cacti are leafless except genus *Pereskia* and second succulents do not have areoles which are present in cacti.

Uses

They can be used as pot plants in indoors for decorating rooms, balconies and verandas. Many succulents are highly suitable for window-box gardening, bowl gardening, trough gardening, handing baskets, rockery etc. They can be also be used as specimen plants in gardens for e.g. *Agave, Furcraea* etc. Succulents like *Agave* and *Yucca* can be used for hedge purposes while *Echeveria* spp. is excellent for edging purposes.

Climate

Succulents perform well under mild climate conditions and a relatively dry atmosphere. Most succulents prefer morning sunlight. High light intensity produces scorching symptoms in succulents and there are many species that are tolerant to shady conditions. Most of the succulents need protection from extreme low temperatures, but there are many frost tolerant succulents like yuccas, sempervivums, sedums, agaves. Proper air circulation is required for its optimum growth.

Soils

Succulents require soils with good drainage. It grows very well in mixture containing garden loam, leaf mould and sand (1 part each). Other mixture which is found suitable for most of succulents comprises of 1 part leaf mould, 2 parts coarse sand, 1/3 part well rotten FYM, 1/3 part garden loam, 1/3 part crushed brick and charcoal on volume basis.

Propagation

Seeds: Seed propagation is possible for succulents but not common method of propagation. The optimum temperature for seed germination varies from 30-35°C with relative humidity of 90%. It is followed in *kaloanchoe* family. Sedums like *S. acrem, S. album, S. kamtschatticum* and *S. reflexum* can be propagated through seeds.

Leaf Cuttings: Most succulents are usually propagated by leaf cuttings for e.g. *Kalanchoe, Sedum, Echeveria, Cotyledon* etc.

Bulbils and Plantlets: Succulent propagated through bulbils and plantlets are *Agave murpheyi, Agave vilmoriniana, Agave fourcryodes, Kalanchoe* spp., *Aloe* spp.

Offsets: Succulent propagated through offsets are *Agave* spp., *Aloe* spp., *Haworthia* spp., *Kalanchoe* spp. *Sedum* spp., *Graptopetalum* spp. etc.

Watering: During summers, it should be sufficiently watered at frequent intervals but during winters the succulents need less watering and should be limited to as and when required. Similarly as for cacti same thumb rule of watering is followed i.e. allow the soil to go nearly dry conditions between waterings.

Nutrition: Like cacti, succulents also not requires high dose of fertilizers. Fertilization is done mainly in summers just like other houseplants and done once in a month with a quarter-strength of water-soluble fertilizers. There is no need to fertilize during the winter.

Description of a Few Important Succulents

***Agave*:** Agave is commonly known as 'Century Plant'and belongs to family Amaryllidaceae. Leaves are thick, fleshy, sword shaped mostly with toothed margins and appear in rosettes. It bears small flowers on a long flowering stalk which appear after a long time gap. Propagation is done through seeds, bulbils and offsets. It is used as specimen plant in gardens. It also forms an important component for rockeries and fencing.

Some important species are *A. americana., A. angustifolia* var. *woodrowii, A. filifera, A. stricta, A. parviflora, A. sisalana, A. verschaffeltii, A. victoriae-reginae.*

***Aloe*:** It belongs to Liliaceae and it is perennial succulent flowering plants mostly stemless in nature or sometimes have short stems. It produces the soft, thick and fleshy leaves with their margins either entire or toothed appear in rosettes. In some species leaves are banded or streaked with different colours. Inflorescence arises from leaf axils which produces flowers mainly of orange, red or yellow colour. It is propagated through seeds or suckers. *Aloe variegata* is widely cultivated species and are suited for pot culture.

***Bryophyllum*:** It belongs to family Crassulaceae. The genus appears as close resemblance to *Kalanchoe*. They produce thick fleshy simple leaves on upright stems and bear pendulous flowers of yellow, orange, red, pink or purple. It is easily propagated by leaves or leaf cuttings. Important species are *B. crenatum* and *B. tubiflorum.*

***Crassula*:** It comprises mostly perennial, evergreen succulent shrubs and herbs belonging to family crassulaceae. It bears very attractive, fleshy, thick leaves with wide variations in their texture, size and shape. It is highly valued for their graceful foliage. It also bears tubular, star or funnel-shaped flowers in white, red or pink colours. Propagation is done from cuttings and seeds. *C. ovata* is commonly known as jade plant is most common indoor plant. Some of important species are *C. arborescens, C. argentea* var. *variegata, C. columnaris, C. cornuta, C. falcata, C. pyramidalis* etc.

***Echevaria*:** It is a rosette succulent which produces very fleshy leaves arranged in spiral arrangement and belongs to family Crassulaceae. It offers wide diversity in shape and size of ornamental leaves. In many of the species, the rosettes of leaves are very ornamental and appear as rose bloom. Propagation is through cuttings. Echeverias are also grown under indoors. Some species are also found suitable for carpet bedding. Some of the important species are *E. agavoides*, *E. crenulata*, *E. elegans*, *E.glauca*, *E. pallida*, *E. peacockii*, *E. secunda* etc.

Euphorbia **:** It belongs to family Euphorbiaceae. Euphorbia is characterized by two features i.e. cyathia and production of milky latex exudates when injured. Most of the species are thorny in nature and produces very small leaves which soon drop off. Propagated is done by cuttings. Two species, *E. splendens* and *E. bojeri* produces attractive flowers. Some of the other species are *E. antiquorum, E. canariensis, E. tirucalli, E inermis* etc.

Furcraea **:** These are ornamental succulent plants belongs to family Amaryllidaceae and have very close resemblance with Agave. They produce bulbils which are used for propagation. The variegated form of the genus is more ornamental and is used as specimen in lawns, also in rockery. The ornamental species are *F. gigantea, F. selloa* etc.

Gasteria**:** These belong to family Lilacecae are characterized with the presence of fleshy, thick, tongue shaped leaves spotted with white or purple colour and show some resemblance to *Aloes*. It is propagated through offsets and leaf cuttings. Important species are *G. croucheri, G. trigona, G. verrucosa* etc.

Haworthia: These are very attractive succulent plants with short or no stem belonging to family lilacecae. It produces fleshy leaves in rosettes and tubercles are mostly present on the leaf surfaces. Propagation is through offshoots, and leaf cuttings. They are available in diverse forms ranging from dwarf to elongated forms which are as follows.

Dwarf rosettes - *H. arachnoidea, H. cuspidata, H. fasciata with H. retusa* , *H. turgida, H. browniana,* etc.

Columnar habits: *H. coarctata, H.glauca, H. reinwardtii,H. viscosa, etc.*

Kalanchoe**:** *Kalanchoe* belongs to family crassulaceae and resembles with genus *Bryophyllum.* They produce thick fleshy leaves which may be either smooth or felted and often spotted. They are easily propagated from cuttings and seeds. Some important species are *K. daigremontiana, K. tomentosa, K.eriophylla, K. marmorata* etc.

Lithops**:** These are small and nearly stemless succulents which resemble like that of pebbles or stones hence they are called as flowering stones this belongs to family Liliaceae. They are available in wide diversity of colours i.e. gray, brown, rust, green and pink and are marked with different patterns of dots, lines or patches which aids them to mimic to their surroundings. Some of the important species are *L. dorotheae, L. aucampiae, L. optica, L. verruculosa, L. karasmontana, L. otzeniana, L. fulviceps, etc.*

Sansevieria**:** These are hardy plants which produce thick, stiff, leaves that arise out from the base this belongs to family *Liliacecea.* Propagation is through division and leaf cuttings. Important species are *S. cylindrica, S. trifasciata, etc.*

Sedum **(family Crassulaceae):** Mostly perennial herbaceous plants with distinctive fleshy foliage. Leaves vary widely in their form and their arrangement on plants. It produces star shaped flowers in clusters or sprays of varied colours viz. white, yellow, rose. Propagation is done through cuttings and division. Some of the important species. S. *spectabile, S. hintonii, S. morganianum, S. pachyphyllum, S. stahlii* etc.

Senecio (family Compositae): It is valued for its ornamental foliage available in beautiful shapes like that of pearls (*S. rowleyanus* to even bananas (*S. radicans*). These plants belong to family Compositae. The form varies from low growing ground covers, loose rosettes to trailing cascade plants. It produces flowers with colour range from off-white, orange, red, to purple. Propagated form cutting, seeds, division. Important species are *S. crassissimus, S. haworthii, S. pendulus,* etc.

Common Insect Pests of Cacti and Succulents

These are relatively free from pests; however they may be attacked by insects like mealy bugs, scales, white flies, mites, etc. and diseases such as soft or dry rot, cactus anthracnose, charcoal rot, etc.

Insects

Mealy bugs: Mealy bugs are one of the most common insects for succulents. They have a waxy, white and cottony appearance on the body. They move slowly in clusters along leaf veins or spines, on the underside of leaves, and in hidden areas at joints. Main symptom is stickiness or black mold on or near the plant. These damage plants by sucking plant juices resulting in weak plants having wrinkled and shriveled leaves. In case of severe infestation, mealy bugs spread and affect many parts of the plant causing withering and ultimately die. These make the plant highly susceptible to rot from bacteria and fungus.

Scale: These insects resemble small cottony or dome-shaped shells which are attached to stems and leaves but can be scraped off. Like mealy bugs, scales also suck plant juices resulting in weak plants having wrinkled and shriveled leaves.

Spider mites : Mites also damage cacti and succulents by sucking the plant juices. These insects made webs and small brown dots, especially on younger growth by damaging plant's epidermis. Spider mites are very small and look like dust when the affected area is tapped over a piece of paper.

Aphids: Aphids are species of green fly and black fly that reproduce quickly and commonly seen on garden plants. They feed on the plant sap and produce sweet honeydew which encourages black mould. Aphids are often found on the flowers, flower buds, and tender new growth of succulents and cacti.

Integrated Pest Management

- Follow the good cultural practices such as proper sunlight, water and drainage for growing cacti and succulents. Keep the pots free of dead leaves and spent flowers.
- Wash mealy bugs and spider mites off the plant with a strong stream of water.
- Remove mealy bugs and scale using cotton swabs dipped in alcohol. Or spraying the plant with a dilute alcohol solution (1:3) may also help in controlling pests that are not easily visible; however, it may damage the epidermis of sensitive cacti and succulents.

- Use insecticides such as Neem or pyrethrins or imidacloprid or acephate to control pests.
- Watering and spraying succulents will reduce mite attack

Diseases

Cactus Anthracnose (Causal organism: *Colletotrichum (Gleosporium)* spp.): It affects many cacti such as *Cereus, Echinocactus, Mammillaria*, and particularly *Opuntia.* Symptoms include spots which are small at initial stages but become enlarge in later stages covered by the small spore producing light pink coloured pustules. Remove and destroy diseased cladodes as soon as noticed. Copper fungicide spray may help in controlling the disease in order to control the disease.

Charcoal Spot (Causal organism: *Stevensea (Diplotheca) wrightii*): It is a common disease of opuntia. Initial symptoms include small spots on the plants. The spots are surrounded by a circle of small raised fruiting structures. In later stages, spots become large, but remain separated. To control the disease, remove and demolish diseased plants.

Dry Rot (Causal organism: *Phyllosticta concava* and *Mycosphaerella* spp.): Firstly small black circular spots will develop and thereafter, size will reach up to a diameter of 2.5-5.0 cm. Remove and destroy diseased plants in order to control the disease.

Soft Rot (Causal organism: *Erwinia carotovora*): The bacterium enters tissue through natural gaps and injuries. These bacteria reproduce very fast, spread to other healthy parts of the plant under high humidity conditions. Diseased portion of the plant is soft, black and deteriorate fast. Avoid wounding in plants, treat broken surfaces with copper fungicide and do not keep plants in highly humid places.

Stem and Root Rot (Causal organism: *Aspergillus alliaceus*)**:** Fungal attack or cold or damp conditions due to over watering may lead to rotting of roots and stems. The rotten tissues tend to turn red, brown, or black. The parts that have gone rotten are usually soggy, slimy, and emit a bad odour. Rotting generally starts from the root towards upper portion. It can control by cut off all the rotten parts if rotting has just started in a small specific region. Cut the stem well above the rotten part and propagate it to grow into a new healthy plant. Improve ventilation, do not overwater and do not keep succulents and cacti in a cold damp condition

Physiological Disorders

Scab: It is very common on opuntia cactus. Corky areas of rust coloured appear on the stems. Overwatering and poor ventilation cause the scab in plants. Increasing light and decreasing humidity may control this disorder.

Etiolation: Its symptoms include pale, sickly, excessive growth due to insufficient light, incorrect feeding, or overcrowding. Plants can be prevented by growing them in sunlight for at least half a day.

Sunburn: It happens in cacti when plants are suddenly exposed to strong light. Generally it occurs when plants are kept outdoors in spring after being in the house in low light all the winter. Plants can be protected from sunburn by introducing them gradually to higher light levels.

Overwatering Damage: This is the main cause of death for any cacti. Irrigating cacti plants with too much water makes them weakened and susceptible to fungus attack, collapse and turn into a watery mess. Irrigating cacti with more water after a long dry period cause splitting of the skin. It can be reduced by allowing cacti to dry thoroughly between two watering intervals. After a period of rest, enhance watering amount and frequency gradually.

Bud Drop: It occurs mostly in the epiphytic cacti due to lack of water, dry air or too high feeding of nitrogen. It may be caused by spraying very cold water which causes cold shock. After moving from one place to other place, well formed buds should face the sun in same position as in their previous position. Otherwise, buds twist themselves off and try to follow the light. Buds on desert cacti such as *Echinopsis, Rebutia, Lobivia, Opuntia*, etc. drops if watered too much in early spring. To prevent bud drop, water regularly and avoid moving plants during flowering period.

Cold Damage: During frost, cacti and succulents first appear glazed or semi-translucent, then afterwards collapse into a mush. Epiphytic cacti develop orange spots. Tender succulents such as *Sansevieria* like high temperature i.e. above 7 °C. Cacti such as *Echinopsis, Rebutia, Lobivia, Trichocereus, Opuntia*, few *Echinocereus* grow well below freezing if kept dry. Cutting back the damaged tissue of cacti can prevent from this damage.

11

Bonsai

Reeta Bhatia, Saurabh Singh and Shyam Sundar Dey

Bonsai is an astonishing art-form, combining horticultural approaches and Asian aesthetics in a distinct way, which reflects the nature's beauty (Pietraszko and Sobota 2008). The art of bonsai is considered to be originated in Chinese empire (Doyle et al., 2019) and was later adapted by the Japanese, where it has become the integral part of their culture (Ching 1990, Roth 2014). The word 'Bonsai' is derived from two Japanese words 'Bon' (Shallow pan) and 'Sai'(plant) which literally means "planted in a container". The Chinese started creating miniature landscapes in the form of Bonsai during Han Dynasty over 2000 years ago. During the Kamakura period (700 years ago) the Japanese adopted the bonsai art-form and subsequently a different Japanese style form came into existence. This art of bonsai was further redeveloped under the influence of Japanese Zen Buddhism (Doyle et al. 2019).

A bonsai is a nature's replication of plant in its miniature form, without depicting the human interference too clearly. Thus, they are normal trees and shrubs which are dwarfed by a specific system of pruning-training-wiring. Techniques such as pinching buds, pruning and wiring branches, and carefully restricting but not abandoning fertilizers are used to limit and redirect healthy growth in bonsai. They are different from the pot plants which are mainly grown for either their foliage or floral beauty, but in case of bonsai the whole artistic look of the plant in its miniature form is maintained for years. Bonsai can be grown both outdoors and indoors, especially they are ideal for display in living rooms, verandas, terraces and window-sills etc. The good bonsai specimens are usually hardy species that can be kept outdoors the year round wherever winters are mild. They can be brought into the house occasionally for appreciation and enjoyment. In Japan, they are customarily displayed in a recess area or on small tables in a living room and later returned to their outdoor bonsai stands.

The bonsai has become tremendously popular among the common people and the demand for naturally-dwarfed wild trees is increasing. The modern day bonsai reflects changing tastes and the diversity of culture from different countries. While Japanese bonsai artists still prefer to use traditional native species such pines, azaleas, and maples, bonsai growers in other countries are more open to innovation and novelty. The oldest known collection of living bonsai trees can be found in Happo-en in Tokyo, Japan, where about 40-800 years old Bonsai are maintained. The important characteristics of Japanese bonsai are: miniaturization (keeping the tree small enough to be container grown), gravitas (sense of physical weight), leaf reduction (varies across the life cycle of a bonsai), lignification (Woodiness of a bonsai trunk), nebari (visible spread of roots above the growing medium), ramification (splitting of branches and twigs into smaller ones), deadwood and curvature (curvature of the trunk that occurs between the roots and the lowest branch is known as *tachiagari.)* (New World Encyclopedia).

Classification of Bonsai

The basic principle of the art of bonsai is to imitating nature by shaping the plants grown in trays to give the impression of those found in the countryside or forest. There are different approaches to classify bonsai like its size, style, displaying area and can be more. When shaping a bonsai tree, the first step is to decide which style is best suited to the tree's natural design. While there are complex arrays of shapes and styles in which bonsai can be trained (Koreshoff 1984; New World Encyclopedia; Pietraszko and Sobota 2008).

Formal Upright (*Chokkan*)

This style of bonsai has a straight and upright trunk from base to apex. The trunk is is generally thick at base and thinner at the ends. The tapering of the upright-growing trunk must be clearly visible and hence the branching is allowed to begin at about 1/4 of the total length of the trunk. The chokkan style is frequently found in nature especially when the tree is exposed to ample of light and does not face any competition from adjacent growing trees. The formal upright style is best suited to the conifers and look great in oval or rectangular containers. Fruiting or naturally informal trees are not suitable for this style. Larches, Junipers (*Juniperus chinensis)*, Pines (*P. sylvestris, P. excelsa*) and Spruces can be easily trained in this style.

Informal Upright (*Moyogi*)

The tree trained in this style has a curved trunk representing'S' shape of the English letter. The tree trunk is thicker at base and tapering at end and at every turn branching occurs. The informal upright style looks best in an oval or rectangular container. Around 70 percent of the bonsai are trained in this style. The Japanese Maple (*Acer palmatum*), Trident Maple (*Acer buergerianum*), Beech, conifers and other ornamental trees such as the Crab Apple, Cotoneaster and Pomegranate are trained in this style.

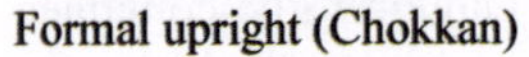

Formal upright (Chokkan) **Informal upright (Moyogi)**

Source: https://botakbonsai.com/bonsai-tree-styles/

Slanting (*Shakan*)

In this style, the plants grow upright but allowed to bend in one direction. This style signifies the nature where the tree will lean in one particular direction due to the blowing of the wind in one prominent direction or when a tree grows in the shadow. In this style, the plant is allowed to grow at an angle of about 60 - 80 degrees relative to the ground. The roots are well developed on one side to keep the tree standing. Generally, on the side toward which the tree is leaning, the roots are not well developed. The branches are uniformly arranged, and are positioned on opposite sides of the trunk. The first branch grows opposite the direction of the tree, in order to create a sense of visual balance. This style looks best when planted in the center of a round or square container. The trunk can be slightly bent or completely straight, but still be thicker at the bottom than at the top.

Cascade (*Kengai*)

This style has a trunk which bends down and tip of cascade branch usually grows below the feet of the container. The trunk grows straight for a while and then bends below till the feet of container. The crown of the tree usually grows above the rim of pot and the subsequent branches grow alternatively to maintain the balance. This style is the representation of the trees growing on a steep cliff that has bend downward as a result of snow or falling rocks. For cascade style tall and deep pots are used. Juniper and Cotoneaster because of their natural tendencies to grow close to the ground are best suited for this style. Besides, Japanese Black and White Pines, Mountain Pine, Scotch Pine and Ficus species can also be trained as cascades.

Slanting – Shakan
Source: https://botakbonsai.com/bonsai-tree-styles/

Cascade–Kengai
Source: https://bonsaiempire.tumblr.com

Semi–Cascade (Han-kengai)

The semi-cascade bonsai are easily distinguishable from cascade style as the tree does not extend below the base of the pot. Unlike cascade style, the crown of the tree is usually above the rim of pot and they never grow below the bottom of pot and. The trunk grows upright for a small distance and then bends downwards/sideward. In nature, the semi-cascade style is frequently seen on cliffs and on the banks of rivers and lakes. Most semi-cascade pots are slightly deeper than their length and width. The best shapes are round, square and hexagonal. Plants that are well adapted to the cascade and semi-cascade styles are prostrate Junipers, and few flowering plants such as Wisteria, Willows, and Star Jasmine. Japanese White Pine, Cotoneaster, Japanese Zelkova, Black Pine, Ponderosa Pine, Crab Apple, Spruce, Beech and Ficus can also be trained in semi cascade style.

Broom Style (Hokidachi)

They have straight and upright trunk which is not allowed to continue its growth till the top. It branches out at about 1/3rd of the height in all directions. In this style, the trunk is usually kept short and branches are trained in such a fashion that it will look like an upside-down broom. All the branches are trained diagonally in a fan-shape manner with no horizontal or criss crossing branches. The branches and leaves form a ball-shaped crown which gives it striking look even during winter months. This style is best suited for deciduous trees with extensive, fine branching. The Brush Cherry, Siberian Elm, Chinese Elm, Beech, Olive, Pomegranate and Crab Apple are trained in this style.

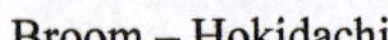

Broom – Hokidachi

Semi-Cascade – Han-Kengai

Source: https://botakbonsai.com/bonsai-tree-styles/

Windswept Style (*Fukinagashi*)

As the name suggests, this style represents the trees growing in nature that are affected by the constant blowing of wind in one particular direction. Unlike slanting style, its trunk along with the branches bent to one side. In this style, the branches grow out on all sides of the trunk but eventually bent to one side. While selecting trees for windswept style, care must be taken to select the evergreen trees. Trees like Chinese Juniper, Japanese White Pine, Mountain Pine, Scotch Pine, Red Pine and Japanese Plum are suitable for this style.

Double Trunk Style (Sokan)

This style has two trunks growing out of one root system. The trunks may split immediately above the roots, or may rise as one from the soil and then split after a few inches. In this style, one trunk is generally taller and thicker than the other, however both the trunks contribute to the artistic look of the tree. The two trunks vary in their thickness and length, the thicker and more developed trunk grows nearly upright, while the smaller trunk will grow out a bit slanted. Both trunks contribute to form a single crown of leaves/canopy. Though the double trunk style is quite common in nature, but is not actually that common in the art of bonsai. The Japanese Maples, Japanese White Pine, Pomegranate, Crab Apple, Juniper, Black Pine, Beech, Spruce and Olive are ideal for this style of bonsai.

Windswept style (Fukinagashi)

Double trunk style (Sokan)

Source : https://www.homestratosphere.com/pictures-bonsai-trees/

Multi Trunk Style (*Kabudachi*)

A multi-trunk or clump-form bonsai is same as double trunk style but has three or more odd-numbered trunks. All the trunks form one crown and are supported by a single root system. Out of all the trunks, the thickest and tallest trunk form the top, the others should be lower and thinner which give triangular or dome shape to the bonsai. Evergreen plants like *Cedrus atlantica* (Acer cedar), Beech, Japanese White Pine, Juniper, Black Pine and Ficus species are ideal for this style.

Forest Style (Yose-ue)

It is one of the most popular styles of bonsai.The forest style looks similar to multi trunk style but the only difference is that it is comprised of several trees rather than one tree with multiple trunks. While developing the forest style, the most developed trees are planted in the center and the smaller ones are planted on the sides. Large and shallow pots are used in this style. In order to provide the realistic forest appearance, planting is done in staggered pattern rather than following the straight line planting.

Most of the deciduous and temperate climate plants like *Acer buergerianum* (Trident maple), *Acer palmatum* and Japanese Beech are suitable for the forest style.

Multi-Trunk – Kabudachi

Forest Style (Yose-ue)

Source: https://www.bonsaitreegardener.net/bonsai-trees/styles

Growing on a Rock Style (*Seki-joju*)

Developing a root-over-rock bonsai is a simple but time consuming technique. This style replicates the trees growing on the rocky terrain. The roots of these trees often penetrate in to cracks and holes for search of the nutrients. Such roots remain unprotected and are exposed to sun till they reach the ground, hence a special bark grows around them. In this style the root of Bonsai grows over rock and survives with the help of soil kept below the rock. The roots of these bonsai must be grown in such a way that they grip the rock or stone securely for both the stability of the tree and the integrity of the piece. The tip of the roots penetrates in the rock and the rest part form a cage like structure around rock. Juniper, Buxifolium, Azalea and Ficus are trained in this style.

Growing in a Rock Style (*Ishisuki*)

In this style, the roots of the tree grow in the cracks and holes of the rock. Hence, there is limited space for the roots to develop and absorb nutrients. Such trees will never look healthy and appear as if they have struggled to survive. For developing this style of bonsai, the rocks are often placed in a shallow pot, which is sometimes filled with water or fine gravel. The plants trained in this style require frequent fertilization and watering.

Growing in a rock style (Ishisuki)

Growing on a Rock Style (Seki-joju)

https://www.homestratosphere.com/pictures-bonsai-trees/

Raft Style (*Ikadabuki*)

In the raft style, the trunk grows straight along the ground and represent a tree in nature that has been blown over by the wind. In this style, the main trunk falls and is survived by the branches. Sometimes when the main trunk dies, the old root system provides the branches with nutrients to survive. Eventually, new roots develop from the fallen trunk and the branches begin to grow vertically. The old branches subsequently develop as main trunks. These newly developed trunks contribute to one single canopy. The Japanese White Pine is trained in this style.

Shari Style (*Sharimiki*)

In some tree, with maturity and due to harsh weather conditions the trunk develop bald or bark less areas. These bald area usually begins at the place where the roots emerge from the ground, and grows increasingly thinner as it progress upwards on the trunk. The intense sunlight bleach these parts and imparts a characteristic feature to the tree. In bonsai, the trunks are shaved with knife and the bald area is bleached with calcium sulfate to speed up the bleaching process. The California Juniper (*Juniperus californica*) can be trained in this style.

Raft – Ikadabuki

Source: https://www.bonsaiempire.com

Mame style

Mame Style

These are very small Bonsai of less than 6 or 7 inches height but with complete character of a grown up tree. The main advantage of this style is that they are often ready for display in a shorter time than the larger forms. For developing mame bonsai, simpler styles are preferred, as due to their small size it is not possible to train them into the more complicated shapes such as Winding or Clasped to the stone. The Upright, Oblique and Cascading styles all make attractive mame bonsai, and a group planting can be most effective. Juniper, Acer, Pinus and Picea species, *Cryptomeria japonica, Chaenomeles speciosa. Pyracantha angustifolia* and *Prunus incisa* are frequently trained as mame bonsai.

Selection of Plants Species for Bonsai

Selecting a right plant species for bonsai making and giving them constant care is critical factor for any style of bonsai cultivation. A vast knowledge and experience is must for choosing right species for right style. Commonly the plants having small size leaves, flowers and fruits in which the foliage is reduced to 1/4th automatically are preferred. Plant species having vigorous growth habit and bearing flowers on leafless branches are suitable for bonsai making. The hardy plant species capable to rigorous growth even in a shallow container are ideal. The species which are able to withstand the training, pruning, wiring operations and reflect live replica of normal plants are good for bonsai art. All the target species should have a prominent natural trunk and exhibiting seasonal variations in terms of growth pattern and flowering habit are preferred. Plants must be able to grow under stress conditions and must have the ability to flower and fruit in small size containers.

Some of the examples of trees or shrubs ideal for bonsai art are as Junipers (*Juniperus chinensis*, *J. prostrate*), pines *(P. sylvestris*, *Pinus excelsa.*), Japanese maple (*Acer palmate*), *Ficus religiosa* (Peepal Tree), *Ficus retusa, Ficus benjamina, Ficus benghalensis,* Boxwood (*Buxus sempervirens*), *Ulmus parvifolia,* (The Chinese elm), Common beech (*Fagus sylvatica*), *Casurina equisetifolia,* Jade plant (*Crassula*), Bougainvillea, *Adenium obesum,* Azaleas, *Cotoneaster horizantalis, Schefflera arboricola* (Dwarf schefflera), *Fortunellla japonica, Jatropha podarica, Murraya exotica.*

Propagation

Different methods are followed for propagation of bonsai like through seeds, cuttings, layering, grafting, and buying from nursery and collection from forest, wild or fields. The natural collection of a bonsai in desirable shape is difficult to get as compared to establishing a bonsai from seeds, cuttings, layering. For collecting the plants from wild, high altitude areas are preferred where poor light or poor soil renders stunted plant growth. While, lifting from wild the care should be given not to damage the roots or rootlets. For the seed propagation high quality viable seeds are desirable for growing a good bonsai. The seeds can be collected from wild, but they are having high risk of infection. Hence it is desirable to buy the seed material from a commercial store. The hard shelled seeds need scarification and stratification for softening the seeds to enhance their germination. Since the plants raised from seeds will take more time to be trained as bonsai. Hence, vegetative propagation through cutting, layering and grafting are preferred for starting a new Bonsai.

Selection of Containers for Bonsai

The selection of the right pot for bonsai is very crucial. Generally shallow containers made up of ceramic or clay are preferred for planting of bonsai. Now a day's cement, plastic and wood containers are also gaining popularity. The bonsai pot must be large enough to stretch the roots of bonsai a little. The glazed pot of soft shades like light blue, cream, or green are preferred for the deciduous or flowering bonsai. Whereas, for conifer and evergreen bonsai, unglazed pots of an austere color, such as brown, gray or reddish clay, are chosen as they reflects the environment of their natural habitat. Selections of pots depend upon the style of the bonsai, sizes of plants, colour of fruits, colour of leaves etc. Usually the length of pot is chosen based on the overall height of the bonsai. A tall bonsai, in general, requires a long pot. The depth of the pot depends upon the thickness of trunk of the bonsai. Deeper pots are preferred for thick trunk bonsai. Ideally, two thirds of the plant height is considered as suitable length of the bonsai pots. For example if the plant height is 18" then one should select 12" length of the pot. The depth of the pot can be as per the thickness of trunk of a plant. Hence, deeper pots are preferred for the thick trunk Bonsai. For 1" thick plant trunk, 2" height/depth of pot is considered ideal. The bonsai pot should have at least 3/4 hole in its bottom portion. So that excess water can drain out. The style of the bonsai is also considered as a crucial parameter in determining the size and shape of the containers. The shape of the containers chosen for bonsai can be round, oval or rectangle. Generally, oval or round containers are preferred for single upright stem. Pots with lip are chosen for wide spreading trees. The tall slender trees look best in a simple round pot. For cascade style tall and deep pots are used. The pot should be slightly narrower than the tree spread. The colour of the container should be as natural as possible.

Growing Media

For the successful cultivation of any plants, the key component is the growing media. The well drained, coarse medium meeting basic needs of water, nutrition and oxygen is ideal for bonsai cultivation. The soil should be porous enough so as to hold the air for roots. It should also have good nutrient holding capacity. The growing media consisting of well-rottened humus/leaf mold, sand and soil in equal proportion (1:1:1) is ideal for bonsai growing. In the growing media, the sufficient quantity of humus in the top most layers is desirable. The conifer plants mostly prefer dry soil while, for the soil enriched with humus is desirable for fruit trees.

Potting

The ideal time to plant Bonsai is February-March and July-August. To prevent soil from draining out of the pot, plastic or steel wire mesh are laid over the drainage hole of the pot. Place a layer of brick pieces/pebbles at the bottom of the pot which will serve as drainage layer. Then layer of sand or small brick piecesis spread over it to cover the gaps. Finally, the well prepared soil mixtureis filled in to the pots. For planting, carefully uproot the plant from nursery container and remove 2/3rd portion of soil from the plant. Choose front position of the plant i.e. the side of the plant which looks most beautiful and people will see the plant from that side. Cut any tap thick or encircling roots before planting the plant into the pot. The position of the plant in a pot is determined by the shape of the pots and the style of bonsai. In a round or square shaped pots, planting is done in center. However, in rectangular or oval shaped pots, place the plant off the center towards the rear on any side, left or right. A straight bonsai is planted in center of the pot, whereas a cascade type should be planted on one side –the side over which it cascades/hangs. After planting, fill the container by the prepared soil. Keep the plant under shade for few days. If the plant requires some support, tie a wire from outside the pot anchoring the trunk of the plant.

Fertilization

Bonsai grows in small and shallow container where only small amount of nutrients are available for its growth. Hence, regular application of fertilizers is necessary for growth and survival of the bonsai. Organic fertilizers are considered good for bonsai as they slowly release the nutrients to the soil and support the growth of the plant. Sludge or well rotten cow dung slurry is ideal for basic nutrition. To improve the growth of leaves and stems Nitrogenous fertilizers are required. Phosphorus encourages healthy root growth and helps in the growth of fruits and flowers, while it is also protecting against disease and unfavourable winter conditions. Phosphorus encourage the growth of and roots.While, Potassium promotes the overall health of the plant. Different ratios of NPK can be used for different trees at different times of the year. Application of some of the micronutrients like Iron, Manganese, Boron, Molybdenum, Zinc and Copper also have beneficial effect on health of the bonsai. Excessive application of fertilizers should be avoided as they can prove detrimental to bonsai.

Most commercial fertilizers contain all three main nutrients plus some trace elements. By regular application of fertilizers plants get good quantity of nitrogen, phosphorous, potassium and other micro nutrients helping their growth. Flower and fruit bearing Bonsai requires more phosphorous fertilizers than others. Inorganic fertilizer-mixture can also be prepared at home by mixing Ammonium Sulphate, Potash and Super Phosphate in equal proportion by weight. This should be used after diluting in water. In 5 liter water two tea-spoonful of the above mixture is sufficient to give good result. It should be used once in every week period. Any newly potted plant should not be fertilized for at least one month.

Watering

Appropriate quantity of water and nutrients is essential for bonsai to exhibit their maximum beauty of flowers and foliage. As they are grown in small containers, these miniatures are totally dependent upon their attendant for meeting their requirement of water and nutrients. The most appropriate time for watering is morning or evening. Avoid the excessive watering to prevent the waterlogging.

Training of Bonsai

The training of bonsai is very crucial as it determines the overall look or shape of the miniature tree. The bonsai is generally trained based on its style. Most of the bonsai are trained in such a way that they have a single apex. In order to maintain the visual balance, most bonsai are trained to a triangular shape. While training bonsai, no branching should be allowed at the bottom 1/3rd part of plant. The middle 1/3rd part should bear bigger branches and top 1/3rd part of plant should have smaller branches. This will give a triangular shape to the plant and allows uniform distribution of sunlight to every part of the bonsai. The lowest branches of a bonsai must be the heaviest. While training, care must be taken to allow the branches to grow only in horizontal direction and branches crossing each other should be trimmed off. No branching should be allowed to grow directly towards the front. Bonsai are trained in such a manner that each one of them bonsai must have a front and a back. Front of Bonsai is the side from which the plant looks most beautiful. Back branches are required for depth of a bonsai. For the compact plant growth always cut growing branches leaving two sets of leaves and plant roots should be exposed on the soil surface in radial direction.

Wiring

Wiring is a relatively modern approach to train and style a bonsai. By wrapping wire around the branches one can reshape, curve and bend them. A wired branch or trunk will take the desired shape just by bending in any direction. While wiring branches, the thumb rule is to keep the wire at 45° angle to branch. It is very important to select right kind of material for wiring. Two types of wires i.e. anodized aluminum and annealed copper are commonly used for the training of bonsai. The aluminum wires are used for deciduous species, while the harder copper wire is used for training the conifers and pines. While wiring a bonsai, the utmost care should be taken to not to hurt the plant. After few months, when the wires starts penetrating into the bark, they must be removed.

Timing of Wiring

Bonsai wiring is temporal and can be done round the year for most of the species. The deciduous plants are wired in early summer after their leaves have matured and the wires are removed in autumn to avoid any damage to bark. The coniferous plants are wired during winter months because they need longer time period to become fixed in position.

How to Wire

The wires are generally wrapped around the branches in the direction in which the branch is desired to be bent. While wiring, initially wrap the wire around the trunk and then proceed with the first branch. Start wiring from the base of the branch to the tip before and then wire the other branch. The wire should be wrapped around the branches at an angle of 45 degrees. It will enable the tree to grow thicker while remaining its new shape. When wiring an entire tree, first wire the trunk and then start wiring the primary branches followed by the wiring of the secondary branches. Two branches of similar thickness located near each other can be wired using a single wire (double-wiring) and the remaining branches can be wired separately (single-wiring). The thickness of the wire depends on the thickness of branch or trunk desired to be shaped. Generally, the thickness of the wire should be about 1/3 of the thickness of the branch. The wire should be thick enough to hold the branch in its new shape. The branch should be then bent once into its final position without causing any harm to the cambium layer under the bark.

The following points should beconsidered during the wiring process:

- To give a specific shape to the bonsai, the wire has to be kept in place for several years. Hence, the rewiring has to be done at every 6-12 months interval to avoid any damage to the bark. A gap of few months should be allowed between two wirings.
- Copper wires are preferred over the aluminum wires for wiring as they remain soft even after a long use. The copper wire that has first been annealed at a low-temperature must be heated to make them pliable.
- While wiring the branches with tender bark, before wiring a cotton tape is wrapped around the branch to avoid any damage to bark.
- The wire should be wrapped rigid, but not too tight. Due care must be given to maintain the wiring tension as high tension leads to break the branch and low tension causes slip of wire.
- In case of partial damage garden tape can be used to seal the branch to avoid it from dying.
- To separate the two parallel branches or trunk a small triangular wooden piece is used as wedge and is removed after 4 months. While to bring close the two parallel branches snugging is practiced with the use of a belt or soft chord.
- The conifer species take long time of up to 12-18 months for final shaping. Wiring should not be left for very long period as they can leave scars on the bark of the branches. The wires should be removed after 6-12 months before they start penetrating into the bark .
- While removing the wires from the bonsai, always use a wire cutter to snip the wire at each turn. Never unwind the wire or use pliers to cut the wire, since this will damage the branches
- After wiring, bending and repositioning of the branches is done. Proper precautions must be taken to avoid any damage to the branches during the process of bending. The straight sections of the branches should be bent slightly so that they look more natural.

Repotting

The shifting of plants to new pots with trimming off of the excess of roots (not > 20%) to promote new root growth is known as repotting. Repotting is done to prevent the tree from becoming pot-bound and ultimately starving to death. Repotting will also supply the tree with new nutrients that it needs to grow and flourish. For an old Bonsai is should be done in every 2/3 year and in every year of new Bonsai. While repotting, only bonsai soil should be used. Bonsai should not be kept in direct sunlight after repotting and proper watering should be done from time to time. It is preferably done during rainy season. It should be done when Bonsai leaves start turning yellow in spite of regular care and watering.

Care and maintenance of a Bonsai

- Proper soil disinfection is must to avoid any microbial activity which is harmful for the plants.
- Plants should be protected from strong sunlight during summers and heavy winds.
- Keep the bonsai in neat and clean place and any leaf debris should be removed to avoid the growth of any fungal or microbial load.
- Repotting should be done in every year for new bonsai and in 2-3 year for old bonsai.
- Proper sunlight should be there for a Bonsai plant and it should not be kept inside for long time.
- During winters watering is done only when needed while in summer it should be done frequently.
- Regularly pinch or prune the new leaf buds of Bonsai to maintain overall shape and maintain the vigour or the plan.

Insect-pest and Diseases

Insects

Ants: A number of insects pests like ants, aphids, spider mites, vine weevil, caterpillars and scales are often found to attack the bonsai. Ants themselves are not harmful but they are the vector of aphid eggs or fungus like sooty mold. These insects can be easily controlled by the spray of any commercially available insecticides. The insecticides such as Thimet and 10G Furadon can be used to control the insects present in the soil.

Diseases

Like other plants, the bonsai are also infected by various types of fungal, bacterial or viral diseases infect the Bonsai as described in table 1.

Table 1: The list of the important diseases of bonsai

S.No.	Name of the disease	Causal organism	Symptoms
1.	Powdery mildew	Fungal	White floury layer appears
2.	Black spot	Fungal	Leaves are covered with black patches
3.	Mildew	Fungal	Greyish mould on underside of leaves.
4.	Rust	Fungal	Orange- brown patches are formed. Blisters spread underside of leaves.
5.	Root rot	Fungal	Caused due to contamination from poorly decomposed organic fertilizers used straight on roots.
6.	Canker	Bacterial	Swelling of the bark and callosities.
7.	Mosaic	Viral	Mottling-alternate green and yellow stripes or other shaped areas- on the leaves

Spraying of Bavistin or Blitox at monthly interval is recommended for the control of various fungal diseases of bonsai.

References

Ching YC. 1990. Japanese spatial conception. The Japan Architect. 'The general Notion of Beauty' 8404: 66-68

Doyle JF, Tisdell EJ, Palmer C. 2019. Bonsai as mediation for adult learning. In : New Directions For Adult And Continuing Education, no. 161, Spring 2019 © 2019 Wiley Periodicals, Inc. DOI: 10.1002/ace.20315

Koreshoff, Deborah R. 1984. Bonsai: Its Art, Science, History and Philosophy. Timber Press, Inc.

Pietraszko K, Sobota J. 2008. Bonsai as reflection of nature's beauty: styles and aesthetic value. WSEAS Trans Environ Dev 4: 588-604.

Roth K. 2014. Bonsai: The haiku of the tree world. Peninsula Clarion, Retrieved from http://peninsulaclarion.com/arts-entertainment/2014-04-02/bonsai-the-haiku-of-thetree-world

The Bonsai Site, A Detailed History of Bonsai. Retrieved June 12, 2008.

The New World Encyclopedia of Bonsai, https://www.newworldencyclopedia.org/entry/Bonsai

Thomas Pakenham, Remarkable Trees of the World (New York: Norton, ISBN 9780393049114).

Yuji Yoshimura and Barbara M. Halford. 1957. The Art of Bonsai: Creation, Care and Enjoyment. Tuttle Publishing, North Clarendon VT USA. ISBN 0-8048-2091-0.

12

Rose

S.S. Sindhu and Namita

Rose (*Rosa* × *hybrida* L.) is the most beautiful creations of nature and is commonly known as queen of flowers. It occupies first position in the international trade. Rose is known for its beauty and is used in worship, bouquets, cut and loose flowers, preserves, garlands, decorations, landscaping, etc. It has gained wide acceptability due to varied growth habits, beautiful shape, dissimilarity in size and form, attractive colour, pleasant fragrance and abundant varieties. These can fulfill the requirement of climbers, trees, hedges shrubs, and edges in the landscape gardening. Commercial products of rose such as rose water, rose attar, gulroghan, gulkand, rose dry petals, rose oil, etc play a vital role in international and national market. Rose hips are one of the imperative sources of Vitamin C and are used for preparation of rose tea. With the ever-increasing demand of novel varieties in the world trade, a large number of varieties are being bred every year with many novel traits. Its share is around 51 per cent of the whole of European markets. In India, 29.41 thousand ha area, producing 301.95 thousand MT is under rose cultivation in the states of West Bengal, Karnataka, Gujarat, Chhattisgarh, Maharashtra (NHB database, 2017). Cultivation of rose such as *Rosa damascena, R. bourboniana, R. centifolia* is also done for essential oil extraction from rose petals.

Origin and Taxonomy

Rose belongs to the family Rosaceae. The genus is divided into four sub genera of which sub genus *Rosa* includes nearly all the species (Gudin, 2000). The sub genus *Rosa* is further divided into different sections of which the actual number is arguable. There are about 100 to 300 species in rose based on various reports. The rose genome is complex consisting of five haploid chromosome sets each made of seven chromosomes. (Nybom *et al.,* 2005). The chromosome number of wild species varies from $2n = 2x=14$ to $2n= 8x=56$, with most of the species being diploid or tetraploid (Nybom *et al*., 2005). *Rosa* × *hybrida* L. is not a species in botanical terms, however, is a highly complex artificial species (Gudin, 2003) that may be used to describe the modern rose cultivars.

Classification, Species and Varieties of Roses

The genus *Rosa* is endemic to temperate areas of the Northern hemisphere including North America, Europe, Asia, the Middle East, etc. with the large diversity of species in Western China. Various species of rose are also distributed in the warmer areas such as Bengal, Iraq, Ethiopia, New Mexico, and Southern China (Nybom *et al.*, 2005). No endemic species are found in Southern hemisphere.

The genus *Rosa* contains about 150 species (Quest Ritson, 2003), although the number varies according to authority. However, various studies on taxonomy and cytology of roses have concluded that only eleven species namely *Rosa canina* L., *R. rugosa* Thunb., *R. chinensis*, *R. wichwaiana* Crép., *R. foetida* Herm., *R. gigantea* Colett ex Crép., *R. moschata* Herm., *R. multiflora* Thunb. Ex Murr., *R. Phoenicea* Boiss., *R. gallica* L., and *R. rubra* Blackw were involved in the development of modern roses (Hurst, 1941 ; Wylie, 1954 ; Maia and Vénard, 1976). About sixteen species and four hybrid species of rose were found growing wild in various phyto-geographical zones of India (Rathore and Srivastava, 1992). The species endemic to Himalayas are *R. brunonii, R. gigantean, R. involucrata, R. longicuspis, R. leschenaultiana, R. macrocarpa, R. macrophylla, R. moschata, R. sericea,* and R. *webbiana* whereas the species growing wild are *Rosa ecae* Aitch, *R. banksiae, R. beggeriana, R. foetida, R.* × *bourboniana, R.* × *centifolia, R.* × *alba, R.* × *damascena, R.* × *bracteata, R. laevigata, R. multiflora, R. indica, R. chinensis and R. gallica. R. bracteata* and *R. laevigata* which perform well under sub temperate climate and possess excellent vegetative and floral characters justifying their use as donor parents (Viraraghavan, 1991). Wild species of rose such as *R.* × *bourboniana* (Gulkand preparation) and Kakinada rose (loose flowers) are probable species for Geographical Indicators. Development of ecotypes *viz.*, Desi, Cheenia, Kaithali and Meeruti are the result of *R.×bourboniana* cultivation in India (Shrivastava and Chandra, 1985).

ARS Classification Scheme 2000 is broadly used to classify roses (Cairns, 2003). The ARS has defined old garden roses as those rose types cultivated prior to 1867 and are further divided into 21 types. The most popular classes described by American rose society are: (www.rose.org)

Alba: These are known as white roses. Alba roses have upright growth, climbing habit, dense and shinny foliage. They are resistant to diseases.

Ayrshire: Plants of this group have climbing and sprawling habit and originated in Scotland. They bloom once in a season.

Bourbon: These roses were developed from the Hybrid China roses. They have repeat-flowering habits. Their name was derived from the location of the first members of the class, the Ile de Bourbon in the Indian Ocean.

Boursault: These are thornless, rambling and once blooming type roses.

Centifolia: Since flowers mostly contain more than 100 petals, hence these were named as Centifolias. These roses are once blooming, winter hardy and 4 to 8 feet plant height. These are also known as cabbage roses and are marked in Dutch flower paintings during 17th century.

China roses: Plants of this group are repeat blooming, winter tender and have few thorns. Flowers borne in small clusters. China roses are the historical group of rose and originated in Southeast Asia.

Damask: These roses are famous for their strong fragrance. Some varieties belonging to this group are repeat blooming.

Hybrid China roses: Plants of this group are generally small and have repeated blooming habit. Stems are very weak and can't able to support the clusters of blooms. Flowers have spicy fragrance. Plants are sensitive to winter and need protection in low temperature.

Hybrid Gallica: These plants are small and winter hardy. Flowers have fragrance and are in brilliant colours. Flowers of this group have once blooming habit.

Hybrid Perpetual: Roses are described by repeat blooming habit, six feet plant height, fragrance and colour range mostly red and pink.

Moss: These plants are named after the mossy thorn growth on the peduncle just below the flower bud and sepals. These roses release a pine scented oleoresin when moss is rubbed between the fingers. Plants are winter hardy and few varieties are repeat blooming.

Noisette: This group was developed in the United States by Philippe Noisette of Charleston. Thereafter, he introduced these roses in France during 1817. Plants are large and rambling type. Fragrant flowers are produced in clusters.

Portland: These roses were derived from crosses involving damask, hybrid gallica, hybrid china and centifolia. They are of 4 feet plant height, have repeat blooming and short peduncles. Name of this group was given in the honour of Duchess of Portland.

Tea roses: These plants have variable height, with some of the best cultivars being Climbing Teas. The stems of these roses are weak, hence, large flowers cause drooping. They are one of the ancestors of the modern Hybrid Tea. Tea roses are winter-tender and required only light pruning for best growth.

Five broad classes of European roses such as Gallica (*R. gallica*), Alba (*R. alba*), Damask (*R. damascena*),Centifolia (*R. centifolia*) and Moss rose (*R. centifolia moscosa*) were emerged during 18th Century (Marriott, 2003). European roses were crossed with China roses during later part of the 18th century because china roses have low branching habit, recurrent flowering, crimson colour, new fragrance, distinct flowers with high centre and slender flower buds that unfurled on opening (Higson, 2007). The crossing of the China roses with the old European roses resulted in new classes such as Portland, Bourbon (Autumn Damask × Parson's Pink China), Noisette (Parson's Pink China × *R. moschata*), Hybrid Perpetual (Portland × Bourbon × Hybrid China) and Tea (China × Bourbon or Noisette) (Marriott, 2003 and Higson, 2007).

Roses bred after 1867, were considered as modern roses of the world. Modern roses were originated from different 10 species namely *R. canina, R. gallica, R. gigantea, R. multiflora, R. moschata, R. Phoenicia, R. rugosa, R. chinensis, R. foetida, and R. wichurana* (Gudin, 2000). The Hybrid Tea roses derived from crosses between Hybrid Perpetual and Tea roses were the first modern roses (Marriott, 2003). 'La France' was first recognized Hybrid Tea which was developed in 1867.

A French breeder Joseph Pernet-Ducher made crosses between Hybrid Perpetuals and *R. foetida* to introduce yellow colour during early 1900. Hybrid Tea roses are upright with single, well shaped flowers having high spiraling centres at the end of long stems. They have sturdy and shiny petals, pointed buds, large glossy leaves and strong stems. Hybrid Tea roses were used to develop many modern rose classes such as Polyantha (*R. multiflora* x Hybrid Tea), Floribunda (Polyantha x Hybrid Tea), Grandiflora (Floribunda x Hybrid Tea) and Miniature (Polyantha x *R. chinensis minima*) (Marriott, 2003).

Varieties

Some of the important varieties are listed below under different classes

Hybrid Teas

Aalsmeer Gold, American Heritage, Anvil Sparks, Bajazzo, Bhim, Blue Moon Bonne Nuit, Brandy, Careless Love, Century Two, Charles Mallerin, Christian Dior Command Performance, Crimson Glory, Doris Tysteman, Eiffel Tower, Ena Harkness First Prize, Frau Karl Druschki, Friendship, Garden Party, Ganga, Gladiator, Golden Giant, Granada, Happiness, Inge Horstmann, Jadis, June Bride, Kiss of Fire, Lady X, Louisiana, McGredy's Sunset, Mister Lincoln, Montezuma, Mridula, Mrinalini Nurjehan Oklahoma, Papa Mielland, Paradise, Pasadena, Peace, Peter Frankenfeld Sonia Meilland, Sugandha, Super Star, Virgo, Abhisarika, Anurag, Arjun, Arka Parimala, Aruna, Ashwini, Bhim, Century Two seedling, Chambe di Kali, Chitra Dil-Ki-Rani, Dr. B.P. Pal, Dr. Benjamin Pal, Dr. Bharat Ram, Dr. M.S. Randhawa Dr. R.R. Pal, Dulhan, Eiffel tower X Queen Elizabeth, Ganga, Golden Afternoon Haseena Homage, Jawani, Lal Makhmal, Madhosh, Lalima, Mother Teresa, Mridula Maharani Mrinalini, Mrs. K. B. Sharma, Nayika, Nurjahan, Nehru centenary Priyadharshini Pink Montezuma, Pusa Ajay, Pusa Arun, Pusa Bahadur, Preyasi, Pusa Garima, Pusa Mansij, Pusa Mohit, Pusa Priya, Pusa Sonara, Raja Ram Mohan Roy, Raja S.S. Nalagarh, Rajkumari, Raktagandha, Raktima, Ranjana, Ratnaar, Sahasradhara Shanti Pal, Shreyasi, Sir C. V. Raman, Soma, Sugandha, Surabhi, Surekha, Surkhab Jawahar, Shiloz Mukherjee, Indian Princess, Pusa Gaurav , Pusa Mahak, etc.

Floribundas

Africa Star, Charisma, Charleston, Else Poulsen, Flamenco, Gene Boerner, Iceberg Mercedes, Sea Pearl, Shocking Blue, Summer Snow Pusa Shatabdi, Akash Sundari Delhi white Powder Puff, Delhi Pink Powder Puff, Anitha, Arunima, Banjaran Chingari, Deepak, Delhi Brightness, Delhi Princess, Dr. S. S. Bhatnagar, Himangini Jantar Mantar, Krishna, Lahar, Loree, Madhura, Manmatha, Manasi, Mohini, Navneet Neelambari, Prema, Punchu, Pusa Abhishek, Pusa Barahmasi, Pusa Komal, Pusa Pitambar, Rupali, Sabnam, Sadabahar, Shola, Sindhoor, Suchitra, Surdas, Suryakiran Suryodaya, Tarang, Pusa Virangana, Pusa Manhar, Pusa Urmil, Pusa Muskan, Rose Sherbet, etc.

Polyanthas

Baby Faurax, Cameo, Chattilon Rose, Echo, Nartaki, Paul Crampel, Swati, Vatertag

Miniatures

Blue Mist, Chandrika, Cricri, Dazzler, Deep Velvet, Dwarf Queen, Honey Comb, Little Buckaroo, Little Flirt, Pushkala, Starina, White Madonna

Climbers

Casino, Cocktail, Delhi Pink Pearl, Delhi White Pearl, Golden Showers, Marechal Niel, Pinata, Prosperity, Royal Gold

Poly house varieties

Long stemmed varieties : Aalsmeer Gold, Bianca, Black Magic, Dallas, First Red, Grand Gala, Grand Prix, Jacaranda, Konfetti, Noblesse, Ravel, Red Velvet, Skyline, Tineke, Virginia, Vivaldi. **Medium stemmed varieties:** Europa, Flirt, Frisco, Golden Times, Kardinal, Kiss, Lambada, Ohio, Prophyta, Souvenir, Tina, Vanilla.

Fig. Recently Released Rose Varieties from ICAR-IARI

Propagation

Rose can be propagated by seed and vegetative methods like cutting, layering grafting and budding.

Seed: Seed propagation is generally followed by breeders to produce new varieties. Seed propagation is also used to multiply rootstocks like *R. canina.* Seed is harvested from mature and ripe hips. The seeds are botanically achenes with a very hard seed coat. The seed coat can be softened by following acid scarification. The seeds of rose contain an immature embryo resulting in dormancy. Stratification is an ideal and widely followed method for breaking dormancy. The seed is placed in media like sphagnum moss and wet stratified at 1.6-4.4 °C for 2-6 months depending on genotype. The treated seeds are sown in raised beds or plugs in the month of October- November.

Cuttings: Stem cuttings are used for clonal propagation of root stock. They are also used for propagation of vigorous cultivars. Roses belonging to classes of climbers, miniatures and polyanthas are propagated by cuttings. Propagation of roses by cuttings is slowly gaining prominence in many countries. IBA 1000- 2000 ppm is used for rooting of cuttings.

Layering: This method is rarely followed and limited to climbers and ramblers. Ground layering and air layering are generally followed.

Rootstocks required for budding and grafting

The rootstocks play an important role in budding and grafting of the roses. Climate, compatibility, soil, pests and diseases and utility play a vital role in the selection of root stocks. Some of the root stocks widely used are listed below

Rosa indica var. ororata

This is recommended for Northern plains. This can with stand dry, wet conditions and salinity. This rootstock is also tolerant to powdery mildew and insect pests.

Rosa bourboniana

This is widely used in North India for producing standard roses as it throws vigorous and strong shoots. It has not gained prominence because of susceptibility to powdery mildew.

Rosa multiflora

This is widely used as root stock in South and Eastern India and hills. It is resistant to nematodes.

Rosa canina

This is commonly used as root stock in Europe. This is propagated by seed and is resistant to drought and alkaline conditions.

R. manetti, Dr. Huey, *R. laxa* and Natal briar are also used as roots tocks in other countries. Natal Briar is gaining prominence as rootstock for green house roses.

Fig. Preparation of Rootstock *Rosa indica* var. *odorata*

Budding: Roses are commercially propagated by T budding or shield budding. The ideal time for budding in North India is from January to March. In case of areas with mild climate , it can be done throughout the year.

Budding in rose

Stenting: It is a quick method of propagation of roses where grafting of a selected cultivar is done on an unrooted cutting of a root stock resulting in a complete plant in 3-4 weeks. The stented plant is called a stentling. Whip grafting is used for stenting. A stem having five leaflet leaf and a single dominant axillary bud is used as a scion. A stem consisting of a single internode without buds or leaves is used as root stock. Scion and stock are held together with the help of a peg. The basal end of a rootstock is dipped in a powder with IBA at 0.4%. The rooting medium is a mixture of peat and perlite in 1:1 ratio. The relative humidity is maintained at 100% with intermittent misting. The ideal temperature is 25°C.

Climate

Production of roses is favoured by an extended growing season with a majority of sunny days. Roses in gardens should have sunshine for a minimum of six hours and prefer morning sunlight. Temperature has a significant influence on quality aspects such as stem length, bud and leaf size and stem diameter. The ideal temperature for rose cultivation is 15-27°C. Best commercial quality is obtained when the night temperature is round 16°C. Optimal rose production occurs at 12-15 mol/m^2/day. The idcal humidity is 60-65%.

Soil

Rose plants prefer well drained soil having a pH near to seven. Roses are also tolerant to pH below 7 as long as it is not below 5.5. Roses are not salt tolerant, so EC values should be less than 2.0dsm^{-1}. Sandy loam soils are preferred due to their rapid infiltration rates. Roses are susceptible to water logging and cause leaf fall under such conditions.

Bed Preparation and Planting

Planting of rose in gardens is done in beds of different shapes. The beds are prepared by incorporation of organic manure and basal dose of phosphorus in the soil. Planting should be done in well prepared beds. Planting should be done in such a way that the bud union should be 2.5-5.0 cm above the soil level. After planting, the plants are to be watered copiously. In case of bare root roses, the roots along with the stem up to the bud union should be kept in water for six hours before planting. In north Indian plains and Eastern India, planting can be done from October to February. In Southern India , planting is preferred during October- December. In temperate areas , planting is done during spring and autumn. In areas with mild climate, planting can be done throughout the year. Planting distance for gardens varies from 30-75 cm between plants and rows depending on the class, variety, soil and climate. In polyhouses, rose is planted on raised beds having one metre width, 30 cm height and length as per the structure. Roses are planted in double lines at 30 x 15-20 cm. The soil may be sterilized with soil sterilizing chemicals and thoroughly mixed with FYM , phosphatic and potassic fertilizers. Farm yard manure at the rate of 100 t/ha should be thoroughly incorporated in the soil.

Lighting: Four year old plants of rose cv. Dallas were grown at a minimum temperature of 12^{0}C with 2000 lux supplementary lighting (Total light period of 16hrs/day) from HPS lamps during the period of Nov-Feb. Flower yield increased by 18% over control with increase in quality. No further increase in yield was observed when raising the irradiance to 3000 lux.

Irrigation

Drought stress in rose field cause defoliation, sun burn of canes and may contribute to spider mite. However, over watering or poorly drained soils may cause root disease or nutritional harms. Rose production cycle may be delayed due to water stress applied at the stage of leaf primordia formation but do not have any negative effects on flower shoots or the flower buds quality. Water stress at the stage prior to petal initiation affected the quality of flower buds i.e. reduction in the number of well formed petals and height of the flower buds. Water stress applied prior to stamen initiation affects the quantity and the quality of flowering shoots and buds , whereas, water stress applied after stamen and carpel formation do not affect the number of flowering shoots or quality of flowers produced. Generally, a rose plant requires water at 8-10 ml/m^2 a day.

Carbon Dioxide Enrichment in Greenhouses: Rose plants respond to CO_2 enrichment of the greenhouse atmosphere with a decrease in the rate of lower bud abortion, an increase in lateral sprouting (Hand and Cockshull, 1975) and an increase in the fresh weight of flower stems (Mortensen & Moe, 1983). Prevention of flower bud abortion can be a result of alteration in partitioning of assimilates between upper buds. An additional possibility is that the effect of the CO_2 on the flower bud retention is due to anti ethylene action of CO_2. The increase in number of flowers and quality may be the result of CO_2 on photosynthetic process, partitioning of assimilates and the effect of CO_2 on ethylene. Increasing CO_2 levels during day time to 800-1000ppm are beneficial for rose production due to increased dry matter production.

Pruning

Rose requires annual pruning exercise to revitalize the plant before it produces a new flush of blooms in the season. Pruning encourages new growth from bud union. This removal of old growth sometimes acts as a form of dormancy in warmer climates which is otherwise lacking. Pruning slows down the plant processes for a short period of time. Pruning in rose is done for the following reasons

1. To enhance the growth of an old and unproductive rose.
2. To make the plant to produce fewer, larger blooms or many, smaller blooms.
3. To remove old and diseased shoots.
4. To shape the plant.
5. To control the blooming time in the season, so that the blooming coincides with most favourable time to get good quality blooms.
6. To encourage repeat blossoms.

Pruning Types in Rose

Hard Pruning or Low Pruning

The shoots are cut back to three or four buds from the base or bud union. This leaves short sturdy shoots. Hard pruning is suggested for newly planted bush roses of the hybrid tea, floribunda and grandiflora tribes. Hard pruning is frequently used by growers to produce show blooms for exhibition. This method is not suitable for established garden roses and should not be practiced. It may be used to rejuvenate sick and neglected plants.

Moderate or Medium Pruning

The shoots are cut back to about half of their length. Weaker stems are cut back more depending on their position on the bush. Moderate pruning is suitable for established garden roses. Floribundas, hybrid teas and grandifloras respond well to this type of pruning .

Light or High Pruning

The shoots are cut back to about two thirds of their length. In this method, after removal of unwanted wood, the remaining stems are simply tipped. Light pruning is not generally suggested as it produces spindly bushes because it results in an early blooming bush with poor quality flowers. In case of vigorous hybrid teas, climbers and shrub roses, light pruning is the only the way to cut these plants.

Hybrid Tea Roses: Hard Pruning is mandatory to develop a strong root system and to encourage the growth of sturdy, fresh shoots near to the base of the bush in newly planted roses. Moderate pruning is the best method for general garden display of roses. Hard pruning is sometimes used for show blooms. In case of hybrid teas, only 4-5 shoots are left out after pruning.

Floribunda Roses: Moderate pruning is the excellent way to prune established floribundas. But hard pruning is done in old stems within a few inches of the ground, while light pruning is carried out in new shoots arise from the base. This method ensures a long period of continuous bloom. In floribundas, many older shoots are allowed after pruning due to the habit to bear flowers on short stems in clusters.

Standard Roses: Hard pruning is recommended for standard roses. Stem length should be about 20 cm long. Moderate pruning is suggested to form an appropriately balanced head which will produce abundance flowers. Hard pruning should be avoided on established tree roses due to effect on the shape of the plant which make it further less attractive.

Miniature Roses: Pruning is not needed for miniature roses except the removal of any dead or broken shoots.

Climbing Roses: Pruning is not required for climbing roses except removal of dead canes and tips.

After Pruning

After the pruning operation is over the pruning cuts may be sealed with a proprietary fungicidal paste like Bourdeaux mixture or Blitox Paste. A dormant spray of pesticide may be given before the sensitive buds sprout. Fertlizers can be applied after 2-3 weeks.

Bending

Bending was originated in Japan. All weak shoots are bent down to fill any area full of foliage to attain a desirable leaf area index by optimizing photosynthetic potential and facilitating the transport of sugar to the developing shoots. The shoots which arise on dormant buds are usually more vigorous and produce high quality flowers. Bending is required for keeping enough leaves on the plant for production of carbohydrates.

Nutrition

A fertilizer dose of 520 kg N, 868 kg P_2O_5 and 694 kg K_2O/ha/year was recommended for high density planting (30x30 cm) of Super Star under open field conditions (Bhattacharjee and Damke, 1994). Potassium application rates affected the number and quality of flowering stems and reduced diseases. Potassium rates closer to 50 g/m^2/year are recommended to obtain maximum quantity of commercial stems regardless of the cultivar (Barbosa *et al.*, 2007). Heavy dose of organic manure like FYM at the rate of 50-100 t/ha is recommended in rose. Bhattacharjee (1994) observed that the soil application of sulphur at 10 kg/ha, magnesium sulphate at 50 kg/ha and calcium sulphate at 50kg/ha significantly improved flower yield and quality in cv. Raktagandha

Weeding

Generally, weeding in roses is done manually. Rajamani *et al.* (1992) recorded better control of monocot weeds with glyphosate 1.0 kg a.i/ha and dicot weeds with oxyfluorfen 0.5 kg a.i/ha.

Harvesting, Packing and Transport

Roses should be harvested at the tight bud stage when one or two petals begin to unfold. The stage of harvest depends on the variety, distance to market place, climate and consumer preference. Roses cut too early may develop bent neck. Flowers should be cut in the morning or evening. They should be cut leaving two 5 leaflet leaves on the stem.

After cutting, they are immediately placed in a hydrating solution to maintain turgidity. After they are cut, they are cooled or graded. Roses are stored at 2-3°C with a relative humidity of 90-95%. The flowers should be transported to cool rooms. Precooling removes the field heat and improves the post harvest life. The hydrating solution may be acidified with citric acid 300 ppm to improve the uptake of solution. The vase life can be improved by using floral preservatives in vase solution. Aluminum sulphate and citric acid @ 300 ppm improve the vase life. The flowers can be wet stored at 2-3°C. After cooling the flowers are shifted to grading room. All the inferior stems and those infested with pests and diseases are removed. The flowers are sorted to different grades manually or automatic graders. Long stemmed varieties are graded from 40 cm onwards with a difference of 10 cm whereas; short stemmed are graded from 40-65 cm with a difference of 5 cm. Then graded stems are made into bundles of 20 each. The buds are wrapped with corrugated paper and leaves are removed from lower 5 cm portion of the stems. The bunches are packed in fiber or card board boxes of 100 × 32.5 × 20 cm which accommodates approximately 80-100 cut roses of 60-65 cm length . The stems should be tightly packed to avoid movement during transport.

The optimum yield of cut roses in open field is 25-50 stems/m^2/year whereas in Indian poly houses, yield is 150-200 stems/m^2/Year. Loose flower yield is 3-5 tonnes/ha/year.

Pests and Diseases

Insects

Many types of insects affect roses of modern era. Insect pests reduce decorative value of roses in the garden and decrease productivity of commercial rose cultivars. During severe infestation, these insects cause death of plants. The insect -pests of rose are described hereunder.

Aphids: Aphids are small and soft-bodied insects having piercing and sucking mouthparts. These grow early in the season on tender parts such as shoots, buds and flowers. They cause damage by sucking plant sap and infected flowers become malformed. Aphids secrete sweet honeydew on which fungi grow and produce black sooty mould. This reduces the ornamental value of rose. Aphids can be controlled by natural enemies such as lady bird beetles. Severe infestations can be controlled by spraying insecticides like monocrotophos (1.5 ml/lt), malathion (1.0 ml/lt) and dimethoate (1.0 ml/lt) at 10-15 days interval. It can be controlled by neem oil (10 ml/lts) spray.

Thrips: Flower thrips and western flower thrips are two most common species in roses. Thrips are tiny insects. Young thrips are of light yellow to lemon colour and are of spindle-shaped. These insects cause damage mainly by feeding on flowers. Heavy infestation reduces the beauty of the buds and flowers and prevents opening of buds. Thrips also infest plants by punching plant cells with their mandibles and sucking up plant sap of young and actively growing tissues. Silver or bleached damaged areas are developed on flower petals that ultimately turn brown. Petals and leaves become crinkled or distorted. Thrips can be effectively controlled by spraying insecticides like malathion (1.0 ml/lt), imidacloprid (0.3 ml/lt), Dimethoate (1.0 ml/lt) and spinosad (0.3 ml/lt). Spinosad is one of the most important insecticides that provides very good control of thrips

Leafhoppers: These are very small, light green or pale green insects which suck cell sap. Both adults and nymphs suck sap from the undersides of leaves and tender stems, resulting in spotted or yellow leaves. A white or yellow stippling of the leaves is the most common symptom in rose. Leaf hoppers can be efficiently controlled by spraying malathion (1.0 ml/lt), imidacloprid (0.3 ml/lt), and Phosalone (2 ml/lt).

Spider Mites: These are minute insects which are not visible to the naked eye. They feed on the underside of the leaves by sucking sap from plant cells. Heavy infestation by mites causes localized cell death which results in light coloured stippling. Leaves give a bleached or bronzed appearance on heavy population of mites. Severely injured leaves may curl and drop from the plant. During initial stage, mite infestation is only on the undersides of leaves whereas mites produce webbing on the top of leaves and other plant parts under heavy infestation. Mites are favoured by hot dry weather along with dusty conditions. Well watering of plants reduces the mite population. Mite infestation can be reduced by spraying water underside of leaves. These can be controlled by spraying miticides like dicofol (2.5ml/lt), soluble sulphur (3g/lt), vertimac (2ml/lt) and Omite (2ml/lt).

Scale: They can be easily noticed by the reddish encrustations on the shoots. Their bodies are covered with a hard scale-like covering which may be elliptical, tear-shaped, or oyster shaped, round. Scale damage plants by sucking plant sap with their thread-like sucking mouthparts. Heavy infestations reduce plant vigour and growth which may further cause the death of shoots or even whole plants. They can be controlled by spraying insecticides like dimethoate (2g/lt).

Chafer Beetles: These are beetles that feed on the growing points, making irregular holes and punches on the leaves. The grubs feed on the roots. These can be controlled by insecticides such as chlopyriphos (1.5ml/lt)

Termites: Termites cause damage to rose plants even before they are fully established. They destroy the underground parts. They are very difficult to control as they colonize under the soil. They can be managed by soil application of chlopyriphos (1.5ml/lt) and Imidachloprid (0.2 ml/lt) to the extent.

Mealy bugs: These are small insects with white cottony filamentous growth on the exterior. They damage the plants by sucking juices from the shoots. They can be effectively controlled by spraying monocrotophos (1.0ml/lt) and dimethoate (2ml/lt).

Diseases

Die Back (*Diplodia rosarum*): It causes death of the plant from top downwards. In the initial symptoms, blackening of the stem from the pruned end occurs. Heavy infestation causes death of the plant. This pathogen enters into the plant through wounds, injuries and pruning cuts. To control this disease, affected parts should be cut away and dump in the soil. The pruning cut should be slanting to prevent accumulation water. The pruned end should be sealed with fungicide mix such as Bourdeaux paste (4 part copper carbonate+ 4 part Red Lead + 5 part Linseed oil). The disease can also be controlled by spray of Carbendazim (2 g/lt).

Powdery Mildew (*Sphaerotheca pannosa*): It is a major disease of roses grown in greenhouses. It also attacks field grown roses. Symptoms include powdery growth on the lower side of the leaves. The disease appears during warm days and cool nights. Infested flower buds fail to open. The disease can be effectively controlled by sprays of Carbendazim (1 g/lt) and wettable sulphur (2g/lt).

Black Spot (*Diplocarpon rosae*): The disease is recognized by dark brown, circular spots with fringed borders. The spots expand and collapse, leaves turn brown and fall on heavy infestation. Rose species such as *R. wichurana, R. rugosa, R. laevigata, R. multiflora, R. virginiana* and *R. roxburghii* are resistant to black spot. This disease can be controlled with chlorothalonil (2g/lt), benlate (1ml/lt), Carbendazim @ 1 g/l twice at fortnightly intervals or Azoxystrobin @ 1g/l.

Downy Mildew (*Peronospora sparsa*): Downy mildew is common in greenhouse grown roses. Infections occur on young plants during extended periods of cool, humid weather in early to mid spring. Purplish red to dark irregular spots are developed on infected leaves. Downy mildew is effectively controlled by spraying chlorothalonil (2g/lt) and mancozeb (2g/lt).

Rust (*Phragmidium spp.*): Reddish orange pustules on leaflets and petioles describe this disease. In severe infection, defoliation occurs. It can be controlled by Dithane Z-78 (2g/lt).

References

Barbosa, Jose Geraldo, Joao Miranda dos Santos, Jose Antonio Saraiva Grossi, Paulo Roberto Cecon, Helenita P. Rodrigues and Fernando Luiz Finger. 2007. Quality and commercial grade of rose yields as affected by potassium applications through drip irrigation. Acta Hort., 751: 83-87

Bhattacharjee, S. K. 1994. Influence of secondary nutrients on Rosa hybrida cv. "Raktagandha". Floriculture- Technology, Trades and Trends (Prakash, J. and Bhandary, K. R. ed.) Oxford & IBH Publishing Co. ltd. pp.100-102.

Bhattacharjee, S. K. and Damke, M. M. 1994. Response of Super Star rose to nitrogen, phosphorus and potash fertilization. Indian J. Hort.,51(2):207-213.

Cairns, T. 2003. Horticultural classification schemes. In : A. V. Roberts, T. Debener, S. Gudin. Eds. Encyclopedia of Rose Science, Volume I. Elsevier pp.117-124.

De Vries D. P., and Dubois L. A.M. 1984. Inheritance of the recurrent flowering and moss characters in F_1 and F_2 Hybrid Tea x Rosa centifolia muscosa (Aiton) Seringe populations. Gartenbauwissenschaft. 49: 97–100.

Debener, T. 1999. Genetic analysis of horticulturally important morphological and physiological characters in diploid roses. Gartenbauwissenschaft. 64:14–20.

Debener, Th. and Linde, M. 2009. 'Exploring Complex Ornamental Genomes: The Rose as a Model Plant', Critical Reviews in Plant Sciences, 28: 4, 267 — 280

Gudin, S. 2000. Rose: Genetics and Breeding. Chapter 17. In : Plant Breeding Reviews, Volume 17, John Wiley & Sons, Inc. pp. 159-189.

Gudin, S. 2003. Breeding Overview. In : A. V. Roberts, T. Debener, S. Gudin. Eds. Encyclopedia of Rose Science, Volume I. Elsevier pp.25-33.

Hand, D.W. and K.E. Cockshull. 1975. Roses I: The effects of CO_2 enrichment on winter bloom production. J. Hort. Sci. 50:183-192.

Hareing, P.A. 1986. Modern roses 9. American Rose Society, Shreveport. LA.

Higson, H. 2007. The History and Legacy of China Rose, Quarryhill Botanoical Gardens, Glen Ellen, California (www.quarryhillbg.org/page14.html)

Hurst, C.C. 1941. Notes on the origin and evolution of our garden roses. J. Royal Hort. Soc. 66:73-82.

Maia, N., and Vénard, P. 1976. Cytotaxonomie du genre Rosa et origine des rosiers cultivés.p.7-20. In: Travaux sur rosiers de serre. FNPHP, Antibes.

Marriott, M. 2003. Modern (Post 1800). In : A. V. Roberts, T. Debener, S. Gudin. Eds. Encyclopedia of Rose Science, Volume I. Elsevier pp.402-409.

Mortensen, L.M. and R. Moe. 1983. Growth responses of some greenhouse plants to environment. VII. The effect of CO2 on photosynthesis and growth of roses. Meldinger fra Norges Landbrukshogskole 62:1-11.

Nybom, H., Werlemark, G., Esselink, D. G. and Vosman, B. 2005. Sexual preferences linked to rose taxonomy and cytology. Acta Hort., 690: 21-27.

Quest-Ritson, C. and Quest Ritson-B.2003. The American Rose Society Encyclopedia of Roses, DK Publishing Inc., New York, 448pp.

Rajamani, K., Thamburaj, S., Thangaraj, T. and Murugesan, S. 1992. Studies on the effect of certain herbicides in Rose cv. "Happiness". South Indian Horticulture, 40(2):121-122

Rathore, D. S. and Srivastava, U. C. 1992. Rosa Species (A bulletin), N. B. P.G. R. Publ., New Delhi, Pp:113.

Shrivastava, H. P. and Chandra, V. 1985. Roses for perfumery industry- R. bourboniana. Indian Rose Annual, 4: 198-213.

Viraraghavan, M. S. 1991. Rose breeding for the tropics. Indian Rose Annual. 9:45-60.

Wylie, A. P. 1955. Master memorial lecture on history of garden roses II. Jour. Roy. Hort. Soc. 89:8-24.

13

Carnation

S.R. Dhiman, Puja Sharma, Bharati Kashyap Arshi Sultanpuri, Kalkame Ch. Momin and Pratibha Chauhan

Carnation (*Dianthus caryophyllus* L.) is one amongst the commercially important cut flowers traded worldwide. The genus *Dianthus* has about 300 species of annual, biennial and perennial herbs of which *Dianthus caryophyllus* is the most important species commercially. The genus *Dianthus* in Greek means 'Flower of Zeus' or 'Divine Flower'. Linnaeus chose the species name, *caryophyllus,* after the genus of Clove, as the fragrance from carnation is reminiscent of clove. It is genetically a quantitative long day plant based on its photoperiod requirement (Blake, 1955). The *Dianthus* species are adapted to the cooler Alpine regions of Europe and Asia and are also found in Mediterranean coastal regions. Carnations were the first flowers to be cultivated on commercial scale in high altitude areas of tropics for export to Europe and North America.

It is cultivated throughout the world. The largest growing areas of carnation are in Bogota and Colombia. Carnation is also grown in Italy, Spain, Kenya, Sri Lanka, Canary Islands, France, Holland, Germany, Israel and USA. In India, main production areas are located around Pune and Bangalore. It is also being grown in Solan, Shimla, Mandi, Kullu, Chamba, Sirmour, Bilaspur and Hamirpur districts of Himachal Pradesh. As per the State Department of Horticulture - Himachal Pradesh, carnation occupied 44.07 hectares out of 705.77 hectare total area under floriculture in the state during 2018-19. The acreage of carnation is still expected to increase because of the favourable climatic conditions for its cultivation in different parts of Himachal Pradesh and other hilly states of the country.

In India, depending upon the regions, there is a wide difference in temperature, light intensity and humidity which not only affect the yield and quality of the flowers but also limit their availability for a particular period of the year. To produce quality flowers, carnation needs to be grown under cover, that is, in greenhouse which provides the plants with the optimum condition of light, temperature, humidity and carbon dioxide for proper growth and to achieve maximum yield of best quality flowers. Though, there are different types of the greenhouses, naturally ventilated polyhouses are preferred in mild climate in which temperature is reduced by ventilation (Ryagi *et al.,* 2007).

Origin and Description

Carnation belongs to the family of Caryophyllaceae and is believed to be the native of Mediterranean region. The centre of origin extends from Eurasia to South Africa. It is a half hardy perennial with branching stems and tumid joints. Leaf blades are simple, entire and usually narrow. Leaves are linear, glaucous, in opposite or decussate pairs. Stipules are scarious, entire or lacearate. Each stem forms a terminal flower and hence inflorescence is generally a terminal cyme, sometimes racemiform, bracteates. Flowers are bisexual, occasionally unisexual. Calyx free or united below, cylindrical with scaly bracts at base. Most carnations are diploid (2n=30). Tetraploid carnations are also available which produce larger flowers, but less productive. The hybrids continue flowering for longer periods which produce blooms continuously in mild weather.

Importance and Uses

Carnation is well suited for cut flowers, due to its excellent keeping quality, wide range of colours, ability to withstand long distance transportation and to rehydrate quickly after continuous shipping. It can also be used for bedding purposes, pots, borders, edging and rock gardens. Though cut carnations are traded in the world market year round, they are in particular demand for the Valentine's Day, Easter, Mother's Day and Christmas. Miniature carnations are now gaining popularity for their potential use in floral arrangement. In India, places having cool climate like; Kalimpong, Kodaikanal, Bangalore, Pune, Nasik, Himachal Pradesh etc. are most suitable areas for the production of carnation cut flowers, which may also be exported to Europe.

Classification

Based on the availability of large number of varieties and diversified cultural requirements, carnations are classified as:

a) **Chabaud or Marguerite Carnations:** These are annual carnations developed by crossing of *D. chinensis* with *D. caryophyllus*. Flowers are single or double and are propagated by seeds. Flowers are large with fringed petals and have shorter post harvest life. The various kinds of Chabaud are Giant Chabaud, Compact Dwarf Chabaud, Entant de Nice, Fleur de Camelia and Margarita.

b) **Border and Picotee Carnations:** The flowers of border type carnations are symmetrical and are the easiest to grow. The flower colour varies from single to blended colour with irregular markings. The petals are broad and smooth edged. The flowers are generally frilled with open centers. In Picotee type ground colour is without spot or bars. The edges are regular and of bright colour.

c) **Malmaison Carnations:** These are strong, sturdy and stiff plants with broad leaves. Flowers are large, double with well filled centres and are mainly pink coloured with good fragrance.

d) **Perpetual Carnations:** These are hybrids of different *Dianthus* species that flowers round the year. The plants are not hardy and are generally treated as cool greenhouse plants. They bear flowers round the year with long stalks which make them suitable for cut flowers. Perpetual flowering and greenhouse carnations are of modern origin.

The commercially important perpetual carnations can be grouped into four major groups: Standard, Spray, Dianthini and Mignon or Micro types. Out of these, the large flowering Standard types are the most commercially important types.

i) **Standard types:** The standard type was the first group of carnation used for large scale production. In this type, flower buds formed on short lateral shoots arising from the axis of the upper leaves are removed to leave one large, terminal flower on a long leafy stem, therefore, it has one large flower on an individual stem. Flowers of standard types are available in various colour combinations.

ii) **Spray types:** In spray or miniature type, terminal flower bud is removed at an early stage to encourage more even development of the lateral flowers, which then produce a multiple flowered stem. The spray types produce many flowers of smaller size and are better adapted to higher temperatures than the standard types.

iii) **Dianthini type:** These varieties are more or less closely resemble the spray types in appearance. However, these plants have longer stems, smaller flowers with higher production. The flowers of these types of carnation are yet to become popular in the market.

iv) **Mignon or Micro type:** Mignon or Micro types also have much similarity with the normal spray varieties. The difference is that these plants have shorter stems and yield is higher than spray varieties. However, demand for these types in the market is less.

Varieties

A. Standard Varieties	
Red	Master, Gaudina, Sangria, Aicardi, Turbo, Guapo, Cantalupo, Merlino, Peter, Sergio, Domingo, Decio, Monte Lisa, Don Pedro, Joker, Britannia, Tanga
White	Baltico, Madame Collette, Angelica, White Dona, Cool, Nordika, Dover, Snow Storm, Memory, White Giant, White Perfection, Icecap, Asso, Emotion
Yellow	Liberty, Hermes, Diana Yellow, Kiro, Soto, Exotica, Yellow Dot Com, Pinto, Esty, Tahiti, Diana, Laser
Pink	Pink Diamante, Pink Sim, Shocking Pink, Berta, Navona, Dumas, Star Ginevra
Light Pink	Varna, Rubesco, Kleos, Charmant, Cipro, Pink Dover, Bacio
Orange	Helix, Malaga, Tamarind, Orange Firato, Orange Prestige, Orange Pinto
Purple	Marathon, Golem, Lavender Lace, Sibilla, Farida, Margaret, Imperial Clove
Green	Lady Green, Lady Spring, Prado Fresh, Prado Refit, Prado Pino, Prado Mint, Prado Nova, Java
Bicoloured/ Stripes	Madras, RendezVous, Happy Golem, Bright RendezVous, Dark RendezVous, Tempo, Falicon, Forever, Harmony, Super Star, Raggio-di-Sole
B. Spray Varieties	
Red	Red Eye, Red Fuego, Red Vital, Aliser, Darling, Etna, New Elsy, Red Arrow, Karma, Electra

Yellow	Stella, Prestige, City, Imagine, Mila, Sonia, Abril, Eureka, Limon, Lior, Castillo, Annelies
White	White Prestige, Milky Way, Elvis, Coconut, Bounty, Vanilla, White Smart, Excel, Royalette, Tibet, Amalia, Virgo, West Crystal, Opale, Hermon
Pink	Rosa Bebe, Spur, Suprema, Celebration, Osiris, Donora, Fantasia, Nuria, Rossini, Medley, Barbara, Silvery Pink
Orange	Sunshine, Autumn, Fancy Fuego, Disney, Eilat, Tip Top, Target, Niky, Macarena, Amaya
Purple	Purple Chopin, Aida, Boreal, Duraongo, Roxette
Variegates/ Striped	Challenger, Martina, Saffora, Garfield, Naomi, Jacobine, Picaro, Sintonia
C. Pot Carnations	
Red	Charmtop, Pinky, Arosa,
Pink	Davinci, Maldeves, Briljant, Pink Surprise
White	Sunny, Cosmos, Cristal, White Surprise
Fancy	Cerratop

Growing Structures

Most of the cut carnations are commercially grown under protected conditions to ensure quality and yield. Design and orientation of greenhouse are of great importance so as to ensure sufficient amount of light and proper ventilation. A greenhouse having ridges in the north - south direction is ideal for proper ventilation and light intensity. The plants are grown in beds of 1-1.2 m wide running in the same direction and the path of 60 cm for movement in between the beds and towards the side wall is desirable. Poly greenhouse provided with fan and pad system can bring down the temperature by 8-10°C. However, top ridge and side ventilation also give fresh air exchange and thereby lowers the temperature inside the greenhouse.

Soil

Site selection and proper soil preparation are very important for the long term production of carnations. A well-drained sandy loam or loam soil rich in organic matter content with pH of 6.0-7.0 is most ideal for carnation cultivation. Soils with high amount of clay or silt can be improved by incorporating organic matter or compost. A soil EC of 0.8 - 1.2 dSm-1 during the vegetative stage and 1.2 - 1.5 dSm-1 during the generative stage is best for carnations. The soil must have sufficient aeration and drainage because the roots of carnation are highly susceptible to poor drainage conditions resulting in many fungal diseases like; *Fusarium* wilt. The selected site should be free from weeds, nematodes or soil-borne pathogens. Pre plant soil sterilization with steam, solarisation or 5 % formalin is beneficial in making soil free from weeds, nematodes and soil-borne pathogens.

Climate

For commercial cut flower production, carnations are grown in greenhouses by maintaining optimum growing environment. Growth and flowering of carnations are influenced by the environmental factors like; temperature, light, carbon dioxide, humidity etc.

i) Temperature

Temperature considerably influences the growth rate, flower production and quality of the carnation flowers. The areas having high light intensity during winters and mild temperatures in summers produces the best quality carnations. The standard carnation prefers a cooler climate than spray varieties. For quality flower production in carnation, the optimum night temperature is 10-11°C during winters and 13-15.5°C in summer months, whereas, the optimum day temperature range is 18-24°C. The rate of development of floral buds is primarily controlled by temperature and high temperature is needed for the development of flower bud to anthesis. High day and night temperatures, especially during flowering of carnation can result in abnormal flower opening and calyx splitting. Calyx splitting of flower can be minimized by regulating temperature of the polyhouse to 15.6°C on cool days and 18.3°C on sunny days. With increased temperature, stem length, mechanical strength of stem, flower size, number of petals per flower and keeping quality of carnation decrease considerably. On the other hand, very low temperature delays flower bud development markedly.

ii) Light

Growth behaviour of carnation plants is influenced by duration and intensity of light. Genetically carnation is a quantitative long day plant, i.e., flowering as such is not inhibited by short days but is definitely hastened by long days (Blake, 1957). Long photoperiod usually promotes flowering while short days tend to delay it. Flower quality can be improved by providing long days only for a short period (4-6 weeks) when the shoots have 4-7 pairs of leaves. Many of the carnation cultivars respond to 15-21 days of lighting. Depending on the cultivars there are differences in requirement of photoperiod. The critical photoperiod for most standard and spray carnations is about 13 hours. Long day photoperiods restricts the growth of lateral shoots of perpetual flowering carnations and accelerates early flowering, whereas, short day photoperiods promotes development of lateral shoots and delays the flowering during the vegetative phase. After flower initiation, the short day photoperiods have no effect on the flowering time of the terminal shoots and the vegetative growth of the lateral shoots. For low light intensity during the growth period, supplementary lighting throughout the night is very effective in promoting flowering. However, continuous long photoperiods produce long and thin shoots of inferior quality and hence, results in the poor flower quality.

During flower initiation, low light intensity encourages the production of more leaves before the harvest of flowers and retards the pace of flower initiation. Whereas, high light intensity accelerates the rate of flower initiation and the stems have lesser number of leaves at the time of harvesting. If the carnation shoots have at least 4-6 pairs of leaves, the increase in light intensity reduces the number of days to flower from 120 days to 90 days (Mukherjee, 2008).

iii) Humidity

The plants must be protected from rain and dew during flowering. Wet buds and flowers are susceptible to fungal diseases. For commercial cultivation, the humidity of greenhouse should be maintained at 80-85% during beginning of vegetative growth and 60-65% during full growth stage. Hot and humid climate is not suitable for carnation cultivation.

iv) Carbon dioxide

Carbon dioxide level affects both plant growth and flower quality in greenhouse carnation. The effectiveness of the carbon dioxide treatment greatly depends on the temperature and light conditions. High light intensity along with high levels of CO_2 increases photosynthetic activity, which in turn enhances floral induction (Bunt *et al.*, 1981). Carbon dioxide level influences both the plant growth and flower quality in carnation. Goldsberry (1963) suggested that enrichment of the atmosphere to an extent of 550 ppm CO_2 increased the number and weight of carnation flowers. The best quality and the quantity of flowers can be obtained when CO_2 concentration in the greenhouse is maintained at 300 - 500 ppm on cloudy days and 750-1500 ppm during sunny days under high light intensity and in a temperature range of 14-15°C. Under favourable environmental conditions, additional carbon dioxide increases the flower production by 10-30%.

v) Ventilation

The carnation being a cool-temperature plant, abundant fresh air circulation and ventilation should be provided for proper growth and flowering of carnation. In greenhouses having cooling systems, the exhaust fans and the cooling pads will take care of proper air circulation. Greenhouses without any cooling facilities should have vents on the sides or at the roof for proper ventilation. A ventilation of 25-30% of the polyhouse ground area is ideal. Very low ventilation results in the withering of plants due to drought.

Propagation

Carnation can be propagated both by seeds and vegetative methods. Seed propagation is normally practiced to raise marguerite type carnation that breed true from seeds and for the crop improvement through hybridization. However, perpetual flowering carnations are vegetatively multiplied by means of terminal stem cuttings. Micropropagation techniques have also been standardized for large scale multiplication of disease free plants.

a) Terminal stem cuttings

Commercially, perpetual carnations are multiplied through terminal stem cuttings obtained from healthy, vigorous and disease free mother plants maintained in a certified nucleus block under protected conditions.

Maintaining healthy mother plants in good vegetative condition is very essential because the quality and quantity of carnation cuttings and subsequent flower production greatly depends on the quality of mother plants. A diseased and weak mother block cannot produce healthy crop. Hence, proper raising of mother stock and the subsequent maintenance of mother plants is necessary throughout the growing period. These mother plants are grown with nitrogen to promote vegetative growth. Foliar feeding of plants with micronutrients @ 2.5 g/l and 15 ppm of ethrel after 3 - 4 days of every harvest of cuttings enhances the shoot production for cutting in the next lot (Gupta and Bhattacharjee, 2002). Spray of Dithane M-45 (0.25 %) and Bavistin (0.1 %) at fortnightly intervals reduces incidence of diseases. At the time of planting, application of NPK at the rate of 30g, 20g and 10g/m^2 along with 5 kg/m^2 well rotten FYM will give better results. Mother block should be grown in places where earlier no carnation plants were grown or after one cropping. The beds should be thoroughly sterilized to minimise the disease infection. The soil should be well drained in order to avoid water logging of plants and free from soil borne diseases.

Disease free and healthy cuttings should be collected from the mother plants as soon as they reached the right length and vigour. The cuttings are to be made from high quality mother plants which were never allowed to flower and none of the shoots had ever attained more than six nodes. Terminal cuttings of about 10-15 cm with 4-5 pairs of leaves for standard carnations and 8-10 cm long for spray carnations are harvested for commercial flower production in carnation. In the commercial production of rooted carnation plants, it is also important to store the cuttings which would allow cutting producers to regulate market supply during surplus production or peak demand to help accommodate propagation and production schedules. Low temperatures are commonly used for postharvest storage of *in vitro* propagules, transplants, and rooted cuttings (Zencirkiran, 2010). For long term storage, cuttings are placed in polyethylene bags or polyethylene lined boxes to prevent them from dessication. Storage of cuttings can be done in polythene bags at 0-1°C for several weeks before rooting or can be directly taken for rooting in rooting chambers having intermittent mist systems for maintaining temperature and humidity.

A study was conducted to find out the storage life of carnation cuttings at Dr. Y S Parmar University of Horticulture and Forestry, Nauni, Solan (Himachal Pradesh) where it was observed that un-rooted cuttings of carnation packed in cellophane lined boxes can be stored at 2°C temperature and 75 % relative humidity for 35 days without significant reduction in the quality and quantity of rooted cuttings (Momin, 2012). Harvesting of all the cuttings from a plant at one time should be avoided to prevent the plant losing too much of the assimilating leaves. Treatment of unrooted cuttings with a combination of fungicides Dithane M-45 (0.2%) and Bavistin (0.1%) for half an hour before planting reduces the spread of fungal diseases during rooting. Quick dip treatment of the basal end of cuttings with NAA @ 500 ppm before planting improves rooting of cuttings. The cuttings are planted in the plug trays filled in with rooting media, normally a mixture of sand and cocopeat, for better rooting of cuttings. More frequent misting is required during summer months. There are various types of media used for the rooting of carnation cuttings such as sand, perlite, vermiculite, cocopeat (or coconut fibre dust), sphagnum moss, vermiculite, etc. Mixtures of two or more materials can also be used as rooting media. However, Singh *et al*., (2002) observed the perlite as the best medium for rooting. In general, rooting medium having 20-35%, moisture content, 60% porosity, 30-40% aeration and a pH ranging between 6.0 and 6.8 are conisdered. The medium should be sterilized with formalin before planting of cuttings.

Recently a new propagation methodology for rooting of carnation cuttings has been introduced in which a trench is dug out up to a depth of 30 cm and the width in accordance with the width of the plug trays. Here, the trench is filled with a thin layer of sand and the plug trays containing growing media and carnation cuttings are placed over the sand layer. Just after planting, trenches are irrigated thoroughly. Then the trench is covered tightly with a polythene sheet and low cost portable propagation tunnel covered with plastic cover is placed over the trench. It has been observed that this method of propagation is better as compared to the conventional method in terms of more percentage of rooted cuttings with better quality roots and labour and water saving cost. Cuttings normally develop sufficient root system within 25-30 days of planting.

b) Micropropagation

This is a commercial method of multiplication of carnations on large scale through tissue culture. The main aim of *in vitro* propagation is to produce disease free planting material on a large scale. Shoot tips or apical meristems are mainly used for micropropagation to produce virus free plants. Surface sterilization of explants with mercuric chloride (0.1%) for four minutes has been found ideal for shoot tips and nodal segments. Murashige and Skoog basic medium containing NAA (1 mM), kinetin (20 mM), sucrose (15 g/l) and agar (6 g/l) is ideal for establishment and shoot multiplication and medium supplemented with NAA (1 mM), BA (3 mM) and IBA (5 mM) is ideal for rooting. The rooted plantlets are hardened on potting medium containing equal proportions of soil, perlite and organic manure.

Polyhouse Cultivation

Propagation of Carnation by Cuttings

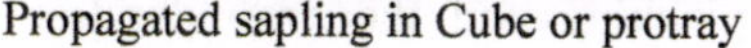

Propagated sapling in Cube or protray

Carnation crop in Agro net shed

Bed Preparation

For growing quality carnations, soil should be ploughed up to a depth of 60 to 70 cm so that it become loose and retain sufficient air and water as it will stimulate the rooting. To improve the aeration and soil fertility, organic matter should be added. Generally, well decomposed farm yard manure @ 25 kg, leaf mould @ 25 kg, neem cake @ 500 g and bone meal @ 200 g per m^2 is added. Adding coco peat (3 kg/m^2), humic acid granules (5 g/m^2), seaweed granules (5 g/m^2) and micronutrients (3 g/m^2) can improve the texture and nutrient status of the growing medium. The organic manures should be mixed thoroughly with soil followed by pre-planting sterilization of soil either with steam or chemical fumigants so as to eradicate pathogens and pests. Steam sterilization involves passing of aerated steam at 60-72°C through the soil covered with good quality plastic sheets. Of late, either Dazomet (30 - 40 g/m^2) or H_2O_2 (300 ml/m^2) is commonly used for soil sterilization. The soil is moistened first and then drenched with the fumigant and immediately covered with plastic. After 48 hrs the plastic cover should be removed and the soil should be thoroughly turned for proper aeration. The fumigant should be removed by leaching with plain water 3-4 times with turning of soil each time.

Before the planting, the basal fertilizer dose of single super phosphate @ 200 g/m^2, potassium sulphate @ 150 g/m^2, magnesium sulphate @ 50 g/m^2 and borax @ 2 g/m^2 along with 3-5 kg per m^2 of well rotten FYM should be evenly spread and thoroughly mixed with the media in the beds. Manures and fertilizers should be added to the soil two weeks before the planting of rooted cuttings. Generally, the beds having width of 1-1.25 m of desirable length (not more than 25 m) should be raised (15-20 cm) above the ground level for proper drainage. A path in between two beds should be sufficiently wide i.e., about 45 - 50 cm to carry out all cultural operations, efficiently. However, balanced development of the crop occurs when the beds are formed in the North – South direction.

Planting

Planting should be done preferably in the evening hours. Planting schedule is very important to regulate the flower production to get the maximum number of quality flowers to meet the consumer's demand and to obtain more profit. Under controlled greenhouse conditions, carnation can be planted round the year. In India, planting of carnation during mid September-November is ideal for northern plains to obtain flush of carnation flowers from February to April, whereas October-February planting is suitable for hilly areas to get flowers during mid-April to mid-July. In areas having snowfall planting should be done during March-April to get flowers during August to October. Mid-hills of Himachal Pradesh are practically suitable for year round flower production of carnation.

Spacing and Planting Density

Spacing and planting density are interrelated and usually depends on the planting pattern, cultivar, light intensity, planting time, pinching and the crop duration. Preferably, standard carnations are planted at the spacing of 20 x 20 cm from plant to plant and row to row, respectively and spray types are planted at the spacing of 30 x 30 cm from plant to plant and row to row, respectively. Spacing of 20 x 20 cm can accommodate 25 plants/m^2. Air movement between the plants and spray penetration is improved with wide spacing 30 cm x 30 cm under ordinary conditions. High planting intensifies the first flush of flowers but the quality of flowers declines gradually after the initial season while wider spacing spreads the yield and produces the quality flowers. Planting should be done at shallow depth to encourage rapid root growth. Deep planting will result into root rotting and stem rot. Therefore, cuttings should be planted at the same depth as they were in the rooting medium (Besemer, 1980).

Support System

Both standard and spray carnations require proper support system in order to get straight stems as bend and crooked stems decrease the flower quality and its market value. Poorly supported branches easily break, making them susceptible to diseases. Various kinds of material like; plastic mesh, string, bamboo, canes, etc. are used for plant support. In recent times, staking of plants with 4-5 layers of zinc galvanised metal wires or plastic (nylon) fixed to the frames is an important cultural practice in carnation growing. Before planting, the first layer of netting should be laid out. The frames can be made from angle iron or pipes, whereas, netting can be done with GI wire of 16 gauge for length wise fixing and nylon threads for width. Generally, the bottom nets having squares of 7.5 x 7.5 cm are placed 12 cm above the soil medium in the first planting and then the subsequent nets with squares of 15 x 15 cm are placed at the distance of 15 cm from one another.

Special Operations

a) Training

Training is a very important and continuous operation in carnation cultivation. This operation helps in keeping the plants within the specified area in the net to grow straight without bending at the bottom.

b) Pinching

Pinching or stopping is an important operation in the successful production of quality carnations and useful in flower regulation. Pinching is the removal of apical portion of the main stem at an early stage to encourage the lateral shoots to develop. Pinching should be done approximately 4-5 weeks after planting at or below the sixth node of the carnation plants. It generally increases the number of flowering stems but delays flowering. Three types of pinching methods are being followed in carnation:

i) **Single pinch:** It is the removal of the apical portion of main shoots below sixth node to give about 5-6 lateral shoots which produce flowers. This is done for early crop. This method is particularly applicable to cultivars which produce higher proportion of quality blooms like the hybrid standard carnations.

ii) **Pinch and a half**: This method is followed to regulate the supply of flowers throughout the year. It involves the single pinching of main stem tip to induce 4-5 lateral shoots. When these lateral shoots develop 5-6 pairs of leaves, only half the number of the lateral shoots are pinched. Two to three flowers are obtained in first flush and 6-8 in the later flush. This system reduces the amount of first crop and provides steady production of flowers without any peak time but reduces the quantity of the first crop.

iii) **Double pinch:** It is the single pinching of the main shoot below sixth node and pinching of all laterals again when they are 8-10 cm in length at 2-3 nodes approximately 5-6 weeks after first pinch. Generally 8-10 shoots are retained. This pinching produces larger number of flower bearing shoots and delays the flowering but produce weak shoots and poor quality and hence, is not commonly followed. Generally single pinch, pinch-and-a-half methods produce excellent quality flowers

c) Disbudding

Disbudding is an important operation in standard and spray carnations. In standard carnations, all axillary or lateral buds are removed and apical bud is retained on the shoots. These should be removed as soon as they can be handled between forefinger and thumb without damaging the leaves and stems to get good quality flowers. In spray or miniature carnations, central flower bud (terminal bud) is removed to encourage the lateral flower buds to develop a perfect spray formation.

d) Deshooting

Removal of unwanted shoots which generally arise in the leaf axil on the flowering stems is called deshooting. This operation is done manually when such shoots are about 2-3 cm long.

e) Weeding

Weeding is the removal of unwanted plants from the cultivation area. Generally, manual weeding is followed in carnation by using small hand hoes or forks. Weeding is combined with soil loosening for good aeration as well as water and nutrient percolation into the root zone.

f) Flower regulation

The main objective of a grower is to produce maximum number of excellent quality flowers at the time when the prices are high. Carnation shoots change from a vegetative to a reproductive condition when the plants have 6 pair of leaves. Carnations produce flowers in flushes as each set of lateral buds extends and terminates in a flower. A heavy flush of flower over a shorter period is not desirable unless it is intended for a specific date such as Diwali, New Year and Valentine Day. Time of flowering, duration of flush and subsequent flower production are directly affected by various factors such as, planting time, pinching methods, supplementary lighting and application of growth regulators. Several of these factors can be manipulated to control time of flowering and the amount and quality of flower production. Under mid hill conditions of Himachal Pradesh, flowers can be produced throughout the year as these factors can be controlled under naturally ventilated polyhouse conditions easily and flowers can be produced practically round the year. An experiment was conducted to evaluate the economics and profitability of carnation production influenced by planting dates at Dr. Y S Parmar University of Horticulture and Forestry, Nauni, Solan (Himachal Pradesh) and March planting was identified as the best planting time over other planting dates with highest cumulative returns to produce desired quantity and quality flowers to meet the growing domestic as well as international demand (Sultanpuri, 2018).

Nutrition

Carnation requires a high level of nutrition for optimum plant growth and better yield with high quality flowers. The deficiency of one or more of the major nutrients limits plant growth and flowering. Nutrient application in small quantities, but more frequently, favours better growth and flower production. At the time of land preparation sufficient quantity of organic manure must be added. Generally, recommended dose for carnation is FYM (5-10 kg/m^2), N (30 g/m^2), P (20 g/m^2) and K 10 g/m^2 for standard carnations, whereas, in case of spray carnation it is the same except that N is 40 g/m^2 instead of 30 g/m^2. The dose of nitrogen is given in two splits, i.e., one dose before planting and the remaining half after one month of planting.

Apart from the above fertilizers, bio-fertilizers and bio-control agents for the control of pest and diseases can be incorporated in the soil at the time of bed preparation. Biofertilizers like; *Azospirillum*, *Phosphobacteria*, *Trichodermaviridi, Pseudomonas fluorescens*, VAM etc. provide an alternative to agricultural chemicals as being more sustainable and more ecologically sound practices to increase crop productivity. It has been found that application of the mixture of biofertilizers containing vesicular arbuscular mycorrhiza species (VAM), *Azospirillum* and phosphate solubilizing bacteria (PSB) at the rate of 5-10 g in the root zone of the plant at the time of planting improve the crop growth and the quality of carnation flowers (Anonymous, 2007).

Fertigation control involves the determination of both timing and quantity of fertilizer and water application. Inadequate plant nutrition causes serious disorders in carnation cultivation and may eventually lead to decline of plant vigour and ultimately reduction of yield. Fertigation combines two main inputs required for plant growth and development i.e. water and nutrients. The right combination of water and nutrients is the key for high yield and quality. Fertigation is done with N at 100 ppm and K at 140 ppm twice in a week along with other nutrients like; Ca, Mg, Fe, Zn, Cu, Mn, Mo and B. Excess of K may cause Mg and B deficiency. Carnations are very sensitive to boron deficiency which causes excessive calyx splitting and abnormal opening of flower buds. Less than 20 to 25 ppm foliar boron levels cause shortened internodes, clubbiness, distorted flower buds and 'witch's broom' symptoms. On the other hand, excess boron (3000-5000 ppm) has been found toxic to the plants which expresses as leaf tip burn symptoms.

Irrigation

Irrigation should be done immediately after planting. Drip irrigation system is being followed in carnation cultivation after well establishment of rooted cuttings, generally after 3 weeks of planting. Laterals are laid out every alternate row and emitters are spaced at 30 cm apart. Overhead sprinkler irrigation is usually not used in carnation cultivation as it encourages the fungal diseases. The amount of water to be given per irrigation depends upon factors like air temperature, light intensity, humidity, soil type and soil temperature. Fully grown carnation plants require 4-5 liter of water/m^2/day. During summer months, the crop requires 2-3 irrigations in a week, whereas, in winters frequency of irrigation may be at the intervals of 7-10 days. It is advisable to check the irrigation water quality in terms of EC and pH. Stagnation of water should be avoided to minimize the incidence of diseases.

Harvesting

a) **Stage of harvesting:** The harvesting stage is fixed depending on the type of carnation and market demand. Stage, time and the methods of harvesting plays an important role in deciding quality of flowers.

Harvesting stages based on type of carnation

Type	Harvesting stage
Standard carnations	Paint brush stage, when petals have started to elongate outside calyx
Spray carnations	With two flowers open and rest of them showing colour

Harvesting stages based on market demand:

S.N.	Harvest stage	Target market
1.	Tight bud stage	This is for long distance markets. However, it is not practiced in commercial cultivation, since some of the flowers may not open at all after harvest.
2.	Paint brush stage	This stage is ideal for long distance markets or for use after a couple of days.
3.	Semi-open stage	This is ideal for short distance market. The flowers can be used in a day or two days after harvest.
4.	Open stage	This is ready-to-use harvest stage and not suitable for travel.

b) Method of harvesting

Carnation flowers are preferably harvested in morning hours when they are turgid. The flowers are harvested by either snapping the stem off at a node or cutting off with a sharp knife or small shears. The cut given on the stem should be smooth to avoid injury to flower stem or to the mother plant. As soon as the flowers are harvested, the cut ends should be kept dipped in water or preservative solution (sodium hypochloride at 1 ml/10 litre of water). Cutting the flower stems leaving 3-4 nodes above the ground is required for growth of the ratoon crop in the following season.

Post Harvest Handling

a) Precooling

The harvested flowers are kept at a temperature of 4°C to 6°C soon after harvesting to remove field heat..Rapid precooling of flowers maintains quality as well as increases the longevity of flowers.

b) Grading

The flowers are graded to different classes according to their quality. Before grading the foliage at the bottom half of the stem should be removed (stripping). Standard carnations are graded based on uniform size and length of stem, free from pests, diseases and physiological abnormalities.

The Society of American Florists suggests the following parameters to ensure quality:

1. Flower and leaves should be bright, clean and firm
2. Flowers should have fairly tight petals at the centre
3. Flowers should be symmetrical in shape and size representative of the cultivar
4. Flowers should be free from calyx splitting
5. Flowers should be free of decay and mechanical damage
6. Flower stems should be free of lateral buds and suckers
7. Flower stems should be straight and have normal growth.

Based on minimum flower diameter and stem length, the Society of American Florists has developed the following grades:

S.No	Parameters	Grades		
		Blue or Fancy	Red or Standard	Green or Short
1.	Bud diameter (mm)			
a)	Tight	50	44	None
b)	Fairly tight	62	56	None
c)	Open	75	60	None
2.	Stem length (cm)	55	43	30

The European Economic Community has developed the following grades of carnation based on stem length:

Description code	Minimum and maximum stem length (cm)
0	Less than 5 cm or flowers without stem
5	5-10
10	10-15
15	15-20
20	20-30
30	30-40
40	40-50
50	50-60
60	60-80
80	80-100
100	100-120
120	>120

c) Conditioning of flowers

Carnation flowers are very sensitive to ethylene. Ethylene accelerates senescence of flowers and sleepiness of petals. As flowers are sensitive to ethylene, the cut flowers should be treated with anti-ethylene preservatives such as silver nitrate, silver thiosulphate (STS), 1-methylcyclopropane (1-MCP), 6-methylpurine etc. as these chemicals inhibit the ethylene production in carnation cut flowers. However, STS (1mM) is not recommended these days as silver being a heavy metal is hazardous to the environment and its disposal after use becomes difficult. An external application of sucrose has been found to suppress the ethylene binding and ultimately reduce the ethylene sensitivity. Sucrose treatment also decreases the ACC oxidase (precursor of ethylene) and ACC synthase activities. Therefore, the flowers after harvesting and grading should be pulsed with 10% sucrose + 8-HQC + 25 ppm silver nitrate + 75 ppm citric acid 8-10 hours before transportation to increase storability and vase-life of cut carnations. After pulsing, the flowers should be stored at 2-4°C temperature and 95% relative humidity in water with a nutritive solution. This is necessary to prevent sleepiness in cut carnations.

d) Bunching

After grading, the flowers are bunched in two layers with 20 flowers per bunch for standard carnation types and 10 stems per bunch (consisting of 35 opened or partially opened flowers) in each stem for spray carnation types and kept at 0-2°C for 12-14 hours.

e) Packing

After conditioning, the flowers are wrapped with polythene or polypropylene sleeves to protect them from mechanical damage and to improve the appearance. Normally perforated sleeves are used as wrapping material. The basal ends of the stems can be placed in absorbent cotton saturated with water and enclosed in waxed paper or aluminium foils. Different types of corrugated fibre board boxes are used for packaging of carnation flowers. The boxes must be strong enough to withstand the weight of at least eight full boxes placed on top of one another under conditions of high humidity. For long distance transport, telescopic style boxes made up of corrugated fibre board are recommended. Normally a box size of 122cm, 50cm and 30cm length, width and height respectively is used. Standard carnations are packed with 24, 28 or 32 bunches per box according to the grade. Spray carnations are packed with 100 bunches per telescopic corrugated carton. All gaps inside the boxes should be filled with shredded paper. Sides of box should have vent holes with flap. Total vent size should be equal to 4-5% of the area of the end wall of the box.

Transportation should be done in a refrigerated van at 2-4°C temperature to maintain the cool chain up to cargo. However, for local markets it should be done by trains, buses and trucks during night hours.

Yield

Yield of carnations depends on type and cultivar, growing region and environment, planting time, plant density, pinching methods and also the period of harvest. In general, spray carnations produce more flowers than standard carnations. The number of flowers in standard carnations ranges between 8 and 12 flowers/ plant/year. In general 200-350 flowers/m^2 can be obtained from standard carnation, while 250 flowers/ m^2 can be obtain from spray carnation.

Insect Pests and Diseases

Red Spider Mite (*Tetranychus urticae*): This is the most serious pest in carnations. The mites are minute red insects which feed on the undersides of the leaves, suck the sap and eventually the leaves turn pale, withered, bronze and show severe webbing. Plant growth, crop quality, yield and vase life of carnation flowers decreased. In severe cases, plants become stunted with off-colour and smaller flowers.This pest is best controlled with miticide aerosols. Spraying of Malathion or Metacid (0.1 %) or Dicofol (0.2 %) at fortnight interval is reported to be effective.

Aphids *(Myzus persicae)*: Aphids suck the sap from the leaves and disfigure the young growth. In case of severe attacks, they leave sticky deposits on the leaves and flower buds. Aphids also act as vectors for ring spot and carnation mosaic viruses. It can be controlled by aphids, foliar spray of Rogor or Basudin (2 ml/l) is very effective.

Thrips (*Thrips tabaci*): Thrips also suck the sap from the leaves, causing them to turn yellow and patchy often with black specks and slight wrinkling. They also cause streaks in the flowers and the stems making them unmarketable. Red and pink cultivars are more susceptible. It can controlled by Malathion (2 ml/l) or Rogor (1 ml/l) or Sumithion (3.5 ml/l) can be used for effective control of thrips.

Bud Borer: Caterpillars are mostly a problem of the carnation bud. The eggs are laid in the buds and the larvae eat into the bud that is completely damaged. The larvae feed on leaves, flower buds and flowers. They make characteristic round holes in buds and flower heads. The infested buds fail to open. The attack by this pest is more during warm conditions. Caterpillars can be handpicked and manually destroyed. Alternatively, spraying of Novaluron @ 3 ml/l or Indoxacarp 14.5 SL @ 1 ml/l or Fenitrothion 50 EC at 1.5ml/l can also control the bud borer.

Root Knot Nematode (*Meloidogyne* spp): Symptoms appears as stunted growth of plants with galls or knot like swellings of various sizes on roots. Leaves are yellow in severe infestations. Nematodes can be eliminated by growing plants in fumigated soil. Application of Furadon, Aldicarp or Nemaphos @ 10g/m^2 controls nematodes in carnation.

Diseases: The important diseases of carnation and their possible control measures are:

Fusarium wilt: The disease caused by *Fusarium oxysporum* var. *dianthi*, a soil borne fungus, is one of the most serious diseases as it is causes severe damage to the carnation crop. This pathogen is very virulent when the temperature is high and the damage in soil is poor. It results is wilting of foliage, often only on a few branches followed by death. In the later stages, rotting of the stem is seen below ground level with internal brown streaking extending up to stem. If pulled the plant breaks off easily while the firm roots remain in the soil. Infected cuttings wilt and die rapidly.The best control measures are soil sterilization or chemical fumigation of the soil, use of pathogen free plants and proper sanitation in the greenhouse. Diseased plants should be rogued and destroyed to reduce the source of infection. Drenching the soil with Bavistin (0.1 %) and Dithane M-45 (0.25 %) at fortnightly intervals is quite effective. Benomyl or Rhidomil @ 2g/lit of water can also control the spread of the disease.

Stem rot: It is caused by *Rhizoctonia solani*. It affects the plants at soil level and the plant dies within a week or 10 days. The plants start wilting ollowed by yellowing of foliage and ultimately the death of plants. Stems sometimes show a brown discoloration and cracking just below soil level. The brown rot can extend upto the stem. Fluffy, light brown fungal hyphae can sometimes be observed on the surface of the rotting tissue. Early stages of the disease can be confused with *fusarium* wilt, but differ in that no internal brown streaking is observed.

Control

- Plant material derived from pathogen tested stock into fumigated soil.
- Relative resistance of carnation to *R. solani* can be increased by good air circulation, good drainage, shallow planting of cuttings and a low or medium fertility level.
- Incidence of the disease can be reduced by drenching with fungicide (Bavistin or Benomyl @ 2g/litre) before planting.
- Copper oxychloride (0.4 %) is also very effective.
- If disease develops infected plants should be removed and apply Rhidomil or Benomyl @ 2g/lit as a soil drenching.

Rust: It is caused by *Uromyces dianthi or U. caryophyllinus*. Early infections appear as pale green blister like swellings, which erupt releasing reddish to dark brown powdery masses of spores. Postules can be up to 10 cm in length and occur on stems, leaves and calyces. Severely infected leaves may turn yellow and die. The disease is common under warm humid conditions. It reduces plant vigour and quality of cut flower. To control wet foliage should be avoided and infected plants should be rogued. Maintain a regular preventive spray program using mancozeb @ 1.5g/lit, zineb @ 1g/lit and sulphur @ 1g/litre or Bavistin (0.1 %) / Dithane M-45 (0.25 %) to keep the disease under check.

Grey mould: It is caused by *Botrytis cinerea.* Initially a wet tan coloured blotch develops on petal tips and spreads rapidly through the petals to produce a fluffy gray mould. This disease can develop on cut flowers while in transit. The disease is favoured by high humidity. It can be controlled by reducing the humidity, maintain good ventilation and hygiene practices. Rovral @ 0.5g/litre or Benlate @ 1.5g/litre is registered for control of *Botrytis* on carnations.

Stem and root rot: It is caused by *Phytophthora* spp. Symptoms are seen as withering and yellowing of foliage, leaf death, external browning of stems and internal browning at nodes. Stem and root rot may be present. Wet conditions, over watering and badly drained soils favor developments of the disease. For effective control, over watering and poorly drained soils should be avoided. Drenching with Benomyl @ 2g/litre or Aliette @ 2.5 g/litre can also reduce the disease incidence.

Alternaria leaf spots: This disease is caused by *Alternaria dianthi.* Small purple spots appear on the leaves, stems and occasionally on the flowers. These develop in to spots upto 5mm in diameter with brown center surrounded by broad purple margin. Spores resembling black specks develop randomly in the center of spots. Heavily infected leaves may die. Infected branches may be girdled, particularly at the nodes. The temperature beyond 23.8°C is known to increase the disease incidence. Avoid excessive moisture in crops. Disease can also be controlled by spraying with Zineb @ 1g/litre, Mancozeb @ 1.5g/litre, Hexaconazole @ 1ml/litre or Propiconazole @ 1ml/litre

Fairy ring spots: This is caused by fungus *Cladosporium echinulatum.* Whitish tan coloured spots upto 5mm in diameter surrounded by a narrow red purple margin. Black pin-head sized spore masses occur in concentric rings on the surface of the spots. The disease occurs on the leaves, stems and calyces. Spray of Bavistin (0.1 %) or Dithane M-45 (0.25 %) or Mancozeb @ 1.5g/litre or sulphur @ 1g/litre regularly at 10 day intervals reduces the disease incidence.

Carnation Mottel Virus (CMV): CMV is the most common virus detected in carnation. Usually infected plants are symptom less, however some cultivars can show a mottling pattern on the leaves. It reduces flower quality and production and may cause inconsistent colour integrity in the petals. CMV has no known rector and is highly infectious. Prevention is the only option, as viral disease cannot be controlled by chemicals. Use plants derived from pathogen tested stock and maintain strict hygiene practices can minimize the disease incidence.

Physiological Disorders

Calyx splitting: Calyx splitting is an important disorder in carnation, which has been associated with many factors like genetic, environmental, nutritional and other cultural practices. The cultivars with short and broad calyx are more susceptible than the ones with long and narrow calyx. Irregular or fluctuating temperature during flowering also induces calyx splitting. Low temperature below 10°C leads to the development of an extra whorl of petals inside the calyx. The calyx unable to hold these extra growing petals later splits up. Nutritional make up of plants also influence calyx splitting. Low nitrogen, high ammonical nitrogen or low boron levels enhance calyx splitting. Closer spacing has also been reported to encourage calyx splitting.

Control

- Selection of cultivars such as, Espana, Carboret, Red Corso, Pamir, Raggio-di-Sole etc. that are less prone to calyx splitting.
- Regulation of day (20-25°C) and night (12.5-15.5°C) temperatures and maintenance of optimal levels of nitrogen (25-40 ppm) and boron (20-25 ppm) in the growing medium can minimize this disorder.
- Spraying of borax @ 0.1% at fortnightly intervals will reduce the disorder.
- Calyx splitting can be reduced by placing a rubber band around the calyx of the flower which has started opening.

Sleepiness: Sleepiness causes huge post harvest losses in cut carnation. It occurs due to exposure of flowers to ethylene or water stress. Also, the incidence of sleepiness has been found to be higher when the flowers are stored for a longer period or when they are exposed to higher temperature. Spraying of STS 0.4 mM before harvesting the flowers will correct this disorder.

Grassiness: Grassiness refers to failure of plants to produce flowers. This is a genetic disorder which varies from variety to variety. It can Removal and destruction of affected plants is the only way of correcting this disorder.

Slabside: This disorder refers to uneven opening of flower buds resulting in the petals protruding on one side only, giving an asymmetrical and lopsided shape to the flower. It is common during cooler periods. It can be overcome by gradually increasing the temperature to optimum level.

Calyx tip die back: Potassium deficiency and water stress cause tip die back. This disorder commences with browning of the calyx tip and it progresses downwards damaging a major part ofthe calyx. It is often followed by occurrence of secondary fungal infection which makes the flower unmarketable. It is controlled by spraying of potassium chloride @ 5g/l two times at 10 days intervals and providing adequate water @ 4.5 l/m^2 can minimize this malady.

Internode splitting: Splitting of internodes affects the quality of cut flowers. Splitting is due to boron deficiency. It can control by application of borax @ $2g/m^2$ will correct internode splitting.

References

Anonymous. 2007. Package of practices for floriculture and landscaping, Directorate of Extension Education, Dr. Y S Parmar University of Horticulture and Forestry, Nauni, Solan (H.P.).

Besmer S T. 1980. Carnation.In: Introduction to Floriculture. Larson R A, (ed.). New York, USA:Academic Press. pp. 49-79.

Blake J. 1957. Photoperiodism in the perpetual flowering carnation.Rep. 14th Int. Hortic.Congr. Netherlands, 1955.H Veemnan and ZonenWangeningen, Netherlands.331-336.

Gupta Y C and Bhattacharjee S K. 2002. Carnation. Tech. Bull. No. 15, AICRP on Floriculture, ICAR, New Delhi.

Momin Kalkame Ch. 2012. Effect of nutrient management on cutting production, storage and rooting of carnation (*Dianthus Caryophyllus* L.). PhD thesis, Dr. Y S Parmar University of Horticulture and Forestry, Nauni, Solan (H.P.).

Mukherjee D. 2008. Speciality cut flower production technologies. Kolkata: NayaUdyog. pp. 151-216.

Ryagi V Y, Mantur S M and Reddy B S. 2007. Effect of pinching on growth, yield and quality of flower of carnation genotypes grown under polyhouse. Karnataka Journal of Agricultural Science, 20(4): 816-818.

Singh K P, Suchitra, Raghava S P S and Misra R L. 2002. Effect of media on rooting of carnation cuttings. Journal of Ornamental Horticulture 5(2): 53.

Sultanpuri Arshi. 2018. Effect of staggered planting on growth, flowering and profitability of carnation (*Dianthus caryophyllus* L.) for extended flower availability. PhD thesis, Dr. Y S Parmar University of Horticulture and Forestry, Nauni, Solan (H.P.).

Zencirkiran M. 2010. Cold storage of rooted and non-rooted carnation cuttings. African Journal of Biotechnology 9(24): 3603-3606.

14

Chrysanthemum

L.N. Mahawer, A.L. Regar and Sadam Hussain

Chrysanthemum takes its name derived from the Greek word chryos meaning gold and anthos meaning a flower, as many species have yellow flower heads. The genus Chrysanthemum belongs to family Asteraceae. It is herbaceous perennial plant. The genus comprises some 200 species of annuals, perennials and shrubs from northern temperate zone (chiefly Europe and Asia), whereas some researchers claim it was originated only from China. Though botanists have retained only the annual species under the genus Chrysanthemum (Beckett, 1983; Dole and Wilkins, 1999).

Annual chrysanthemums which are propagated through seeds. Annual chrysanthemum comprise of three species *viz*., *Chrysanthemum segtum* (corn marigold), *Chrysanthemum carinatum* (tricoloured chrysanthemum) and *Chrysanthemum coronarium* (crown daisy or garland chrysanthemum). The crown daisy or garland chrysanthemum (*C. coronarium*) is a native from Southern Europe, branching annual with finely cut foliage reaching a height up to a metre, size of flowers varies from 2.5 to 4 cm and colour is usually in shades of yellow and white with cream zone at the center (Vishnu Swarup, 1967).

Importance and Uses

Chrysanthemum is the number one dollar earning flower in the United States and cultivars suitable for growing throughout the year have been evolved there. The members of National Chrysanthemum Society run into thousands in several countries. It ranks second (next to rose) in the international cut flower trade. In India, it occupies a place of pride both as a commercial flower, garden plant, pot plant and as a popular exhibition flower.

Large number of chrysanthemum cultivars which exhibit wide variation for characters like growth habit, size, colour and shape of blooms makes chrysanthemum suitable for varied purpose. Its erect and tall growing cultivars are suitable for background planting in borders and as cut- flowers. The dwarf and compact growing ones are suitable for front row planting or pot culture. The decorative and fluffy bloomed small flowered cultivars are ideal for preparation of garland and hair decoration. The extra large- bloomed cultivars are prized for their exhibition value. Certain species like *Chrysanthemum cinerariifolium* and *Chrysanthemum coccineum* are cultivated as sources of pyrethrum, a natural insecticide.

Origin and Distribution

It is native from Asia and Northern Europe. Most of the species originated from East Asia and center of diversity which is exist in China. Florist's chrysanthemum is a complex hybrid origin and has derived from a number of species such as *Chrysanthemum tweak* [(Kher, 1989; Arumugam *et al.*, 2006), *C. indicum* (L.) Desmoul. (Thistlethwaite, 1960; Beckett, 1983; Anderson, 1987)] bearing single yellow flowers, *C. japonicum* (Mak.) Kitam (Anderson, 1987), *C. ornatum* (Kher, 1989), *C. sinense* bearing single white ray flowers (Thistlethwaite, 1960), *C. satsumense* (Kher, 1989) and *C. veslitum* syn. *C. morifolium* (Beckett, 1983) and certain other species native of eastern Asia, viz. China and Japan (Hay and Beckett, 1971).

In the international cut flower trade rose is first and chrysanthemum ranks next to it. In India also, it is highly prized crop and has occupied a place of highly esteemed commercial flower with the name of `guldoudi, derived from 'gul-e-doud' meaning the 'flower of Doud' and a cultivar from India named 'Gul-doodi is reported to be growing in Holland since 1690. Commercially it is being cultivated in Maharashtra, Tamil Nadu, Andhra Pradesh, West Bengal, Rajasthan, Madhya Pradesh and Bihar. For local consumption it is also being grown in all the temple cities of the country. Its cultivation history is very old as evidently China had been cultivating it since 500 B.C. and Confucius, the Chinese philosopher has also described its yellow glory as early as 550 B.C. and the cultivated chrysanthemums after over 1,000 years of development were exported from China to Japan about A.D. 750 where Japanese further developed the Chinese cultivated varieties through cross- fertilization with wild Japanese varieties (Thistlewaite, 1960).

Taxonomy

Florist Chrysanthemum (*Dendranthema x morifolium*) a perennial group of plants, are half-hardy border perennials and hardy greenhouse plants which are usually heavily scented. They have alternate, lanceolate to obovate leaves, from nearly or quite entire, often lobed or more deeply dissected and terminal. Flower head is known as capitulum, which have two types of flowers. In a single chrysanthemum the outer rows of petals, known as ray florets, which are unisexual with pistillate flower, whereas yellow flowers appear in the centre, which is comprised from stamens and stigmas collectively called as disc florets. Various classes of chrysanthemum are based on the forms of the flowers angles of the petals of both types of florets and the size of the capitulum exist in flower.

Nazeer and Khoshoo (1982) stated that though basic chromosome number for the genus is n = 9, a wide range of ploidy level (hexaploidy, aneuploidy, polyaneuploidy, etc., i.e. 2n = 36, 45, 47, 51-75) is found in different cultivars. Self-incompatibility found in chrysanthemum is either protandry or sporophytic and involves more than one locus Fryxwell (1957). Drewlow *et al.* (1973) also reported that nine sibling clones selected from a highly compatible cross were all self-incompatible. Zagarski *et al.* (1983) when crossed 11 sibling lines obtained through 2-generation single plant self-pollination, in a complete diallel selfing and crossing with completely unrelated individuals, found that at least three genes were involved which governed self-incompatibility in chrysanthemum. While studying thc somatic-genetic analysis of the apical layers of 16 chimeral sports of cultivar. Stewart and Derman (1970) reported that a cultivar can be genetically of one colour in L and may be another colour in L_2 and as sex cells arise from L_2, a L_1 white but L_2 pink cultivar which appears of white flower colour will breed pink and vice versa on crossing. It is also recorded that for flower colour inheritance, the presence of one gene A ensures anthocyanin production, gene I inhibits carotene production, absence of A and I both give yellow colouration, absence of only A expresses white flower colours, A without I results into bronze and brownish red flowers, while combination of A and I results into pink, carmine and bluish-red flowers (Kher, 1989). Singleness is partially dominant over doubleness, in which semi-doubles are comparatively much more in the progeny, followed by singles including anemone types, and then doubles (Kher, 1989)' Chrysanthemums are required to be bred for flowering at low temperature, rapid growth following a short period of long days, longer stems (up to 80 cm), compact and short growth habits, good sucker production, less and smaller leaves to save planting space and for pollen-less cultivars. Kher (1986) states that a large number of outstanding spray and loose type chrysanthemum cultivars in India have been evolved through selection method such as Apsara, Birbal Sahni, CO1, CO2, Jaya, Jayanti, Sharad Singar etc.

In chrysanthemums, sports are of frequent occurrence so the better ones are isolated and multiplied. Artificially also, through irradiation with X-rays or ^{60}CO gamma rays the mutations in chrysanthemums are induced. Broertjes (1966) reported optimum X-ray close as 1500 R for producing chrysanthemum mutants though mutation frequency was low. Gupta (1971) reported three groups of cultivars with respect to optimum dose of gamma rays for induction of mutation limitations in chrysanthemum, viz. those highly sensitive to gamma rays where I.1)-50 appears below 1.5 KR dose, those cultivars where 1.1)-50 appears between 1.5 - 2.5 KR and those cultivars where 1.1) 50 appears above 2.5 KR dose. Induction of significant increase in the ray floret length through gamma irradiation has also been possible. Pink cultivars have been recorded producing more mutants in the direction of obtaining bronze, yellow and whites, though the mutants of reverse colours are difficult to obtain. Likewise, whites generally produce yellow and bronze red-yellow mutants. Broertjes *et al.* (1976) produced non-chimeral solid mutants by growing in vitro, the irradiated single cells of the epidermis, stating that mutation of L, layer produces flower colour mutations while L, layer mutation causes effect on plant height, vigour and flower size.

Classification

I. Classification based on National Chrysanthemum Society, England.

a. **Single:** Have one or more outer pistillate flower (ray florets) with disc florets at the centre.

b. **Anemones**: Similar to the class single except the disc florets are elongated and tubular forming a cushion. Disc flowers may be the same or a different colour from the ray florets flowers.

c. **Pompons**: A globular head formed by short uniform ray florets flowers, the shape is considered formal, disc florets flowers are not apparent, further classified into three distinct size,

(1) Small buttons (4 cm or less in diameter)

(2) Intermediate (4-6 cm in diameter)

(3) Large (6-10 cm in diameter)

d. **Decorative**: Have a floret arrangement similar to pompons since they are composed mainly of ray florets flowers but the outer rows are longer than central flower giving the flowers a flatter appearance or informal shape, sizes are mostly intermediate and large.

e. **Large flowered**: Blooms are greater than 10 cm and classified in many shapes. The disc florets are not apparent in most of these forms.

(i) ***Incurved double*:** Globose and formal with ray florets flowers similar in size to the disc flowers and that curve inward and towards the top.

(ii) ***Reflexed double*:** Less formal and globose than the incurved double with over lapping ray flowers curved downward except for the ray flowers.

(iii) ***Tubular ray florets*:** They are classified as

(a) **Spider**: Ray flowers tubular and elongated into the outer rows but short in center. The dropping outer row ray flowers are sometimes hooked on the ends.

(b) **Fuji:** Similar to the spider except the ray flowers may be shorter, droopless and hooks on the ends.

(c) **Quill:** Tubular ray florets flowers, long on the outside and short near the center, resembling feather quills. Ends of flowers are open and not flattened.

(d) **Spoon:** Similar to the quill except the outer row flowers are open and flattened, resembling a spoon.

II. Classification based on temperature requirement for flowering

i **Thermo zero cultivar**: Those cultivars in which flowering occurs at any temperature between 10-27°C but most consistently at 16°C night temperature. e.g. Shasta

ii. **Thermo positive cultivars**: The cultivars continues exposed to low temperature between 10° and 13°C inhibited and delayed floral bud initiation, minimum of 16°C required for bud initiation and at high temperature 27°C there will be rapid flower bud initiation but flowering was delayed.e.g. Cameo

iii. **Thermo negative cultivars**: Flower bud initiations occurred at low or high temperature between 10°C and 27°C but continuous exposure to high temperature delayed flower bud development cv. Defiance. Arumugam *et al.* (2006) suggested that chrysantheum requires two critical day lengths (photoperiods), one for floral bud initiation, other for floral bud development and these varies due to temperature and variety

III. Classification depending upon duration

Early	Medium	Late
90 days to bloom from final transplanting	100-110 days to bloom from final transplanting	110 and more days to bloom from final transplanting

Species and Cultivars

Species

The genus *Chrysanthemum* comprises from 250 species, out of which some of the important species are:

1. ***C. boreale*:** Believed to have been involved in the evolution of florists' chrysanthemum.
2. ***C. cinerariifolium and C. coccineum*** : Grown in temperate regions for making an insecticide called 'Pyrethrum'
3. ***C. carinatum*(Tricolour chrysanthemum):** Tri colour chrysanthemum, flowers 5 cm in diameter and winter season annual
4. ***C. coronarium*(Garland chrysanthemum):** Garland chrysanthemum, winter season annual. Flowers are yellow or white.
5. ***C. frutescens*(Paris daisy/ Pot chrysanthemum):** Small shrub, 60-90 cm tall, bearing white and soft yellow flower. Popularly grown as pot plants.
6. ***C. maximum* (White and Yelllow Cut Flower):** It is most widely grown for use of cut flower. Perennial in growth habit and which bears white and yellow flowers.
7. ***C. morifolium* (Florists' chrysanthemum):** Hybrid species resulted from repeated interspecific crosses among the elemental species extending over a period of more than 2500 years. Perennial, bear blooms as a composite inflorescence with two types of florets arranged on a flattened axis called capitulum. The outer florets are called ray florets(with only female parts) and the inner ones are disc florets(bear male and female florets) and a tubular corolla is made up of five united petals.
8. ***C. indicum*:** It is native from China and Japan and named inappropriately as *indicum.* Which contain yellow flower and involved in the evolution of florists' chrysanthemum.
9. ***C. japonicum*:** It is grown in Pacific coastal region of Japan and widely used as an ornamental plant.
10. ***C. sinense*:** Native of china and bears blooms of white ray florets. This is supposed to be source of today florists chrysanthemum.
11. ***C.sibiricum*:** It is an one of the parents of Korean hybrids evolved in early 1930s

Cultivars

Large no. of cultivars is present in chrysanthemum. In Japan > 50,000 cultivars, in India > 500 cultivars and in Britain > 6000 cultivars are reported.

I. Indian cultivars

a) Large flowered (Exhibit)

- White: Beauty, Snow ball, Innocence, Premier
- Yellow: Chandrama, Mountaineer, Super Giaint, Evening star, Pusa Centenary
- Purple and Pink: Pink cloud, Peter May, Royal Purple, Classic Beauty
- Red and Bronze: Bravo, Alfred Wilson, Distinction, Diamond

b) Small flowered

i) **For pot cultivation**: Dwarf, compactness, profuse branching, uniform spreading, simultaneous blooming habit, attractive novel colour, good colour retention quality and healthy leaves.

White Group: Excelsior, Honey Comb, Mercury,Niharika,Sharad Shoba, Sharad Mala

Yellow Group: Archana, Aparajita, Topaz, Indira, Liliput, Mayur, Peet Singar, Sharad Kanti

Red and Bronze Group: Arun Singar, Arun Kumar, Flirt, Jaya, Red Gold, Rakhee

Mauve Group: Ace, Charm,Fantasy, Mohini, Hemant Singar

ii) **For cut flowers:** The cultivar should possess healthy leaves, long erect stem, bloom uniformity, tough floret and long vase life etc.

White: Apsara, Himani, Birbal Sahni, Illini Cascade, Baggi

Yellow: Basanti, Freedom, Jayanti, Kundun, Nanako, Sujatha

Mauve: Ajay, Alison, Gaity, Nilima, Sharad Prabha

Red: Blaze, Dainty, Flirt, Jaya, Jubilee, Arka Ravi, Ravi Kiran

iii) For garland making: Cultivars should possess bloom diameter about 5.0 cm, disc absent or invisible, fluffy bloom, high yielding, long flower duration, good colour retention, yellow and white colour etc.

- White: Carol, Lilith, Sharad shoba
- Yellow: Vasantika, Hosur Yellow

II. Spray chrysanthemums for international flower trade

- White: Reagan White, Spider White, Cassa, Stallion
- Yellow: Reagan Yellow, Spider Yellow, Cassa Sunny
- Salmon: Reagan Salmon
- Pink: Reagan Improved, Glance Improved

- Orange: Tiger, Reagan Orange
- Red: Reagan Red

Indian Varieties

Red Gold –Flirt x Valentine : High yielder

IIHR 10 (Ravi Kiran) : Hybrid suitable for cutflower

IIHR 11 (Akash) : Hybrid suitable for loose flower

IIHR mutant 9 (Yellow Gold) suitable for loose flowers

IIHR Sl.5 & Sel-6 Seeding Selection: High yielders and suitable for loose flowers.

IIHR Sel.7, IIHR Sel.8, and IIHR Sel.9 released from IIHR.

NBRI-Agnisikha, Navneet .

IARI- Pusa Anmol, Pusa Guldusta for cutflower, Pusa Keshari, Pusa Centenary for standard, Pusa Chitraksha, Pusa Sona, Pusa Aditi for Pot mum purpose.

Growth requirements

Day length

Chrysanthemum crop requires long days for vegetative growth phase and short day length light for floral bud initiation and development phase (photosensitivity). There are varieties which are not photosensitive and flower at a certain age. This segment is very limited in both varieties and market share.

At present professional growers make use of the day neutral varieties programming them year round by manipulating the day length. This article only highlights aspects of this type of chrysanthemum cultivation. The production cycle of the chrysanthemum can be split into two *viz.*, vegetative phase (to get the correct stem length) and the generative phase (formation of the flowers).

Vegetative growth will occur when illumination is 12 hours or more i.e. long day (LD). Flower induction, the generative phase, requires at least 13 hours of night i.e. short day (SD). The plants built up florigen hormones needs for flower induction after 5 to 6 hours of short day. The time between the start of the short day period and flowering is called Reaction Time and which is expressed either in weeks or days. Each variety has its own specific Reaction Time. The reaction will be influenced by the different climatic conditions throughout the years, mainly temperature: the Reaction Time increases at both extremely high and low temperatures.

One can intervene in the process of flower formation by not letting the period of darkness exceeding 5 hours. This can be achieved through artificial lighting. Lighting can be done continuously or in cycles. The bulbs may be preferably by 150 Watts. The bulbs should give disperse light. Spot light types (mirrored bulbs) are not good since they give focused light. The number of bulbs required per square meter is 9, this way the Lux at plant level in all spots should come to 90, in those periods where nights are short to obtain flower induction one should go for black closing to arrive at the correct number of hours of darkness at least for 13 hours. Black closing is done by closing out all light with black plastic. Since errors in lighting are irreversible one should check the electrical system regularly. With the unreliable power supply in India generator is a must to ensure the availability of power to run the lighting system.

The artificial lighting or black closing, depending on the day length which is correlated to a particular time of the year. Thus it can occur that in some periods black closing is not required or both black closing and lighting are required. Understanding the mechanism of photosensitivity it is obvious that portions of shade created, for example the rolled up black plastic, gutters and trees have a negative effect on the uniformity of flowering.

Climate

Light is the source of plant activity. Light is however, the source of heat is also. During summer when the temperature starts rise, blocking out light with limestone and / or screens make sure that the Lux (the unit of measurement for light intensity) does not exceed 70,000. The dark period day length required for flower bud initiation 14-14.5 hours per day and for floral bud development 13-13.5 hours per day is essential for flower process. The minimum dark period day length required for flower bud initiation is 9-9.5 hr per day in chrysanthemum. For vegetative growth under long day condition requires 13 hours light and 11 hrs dark period. Whereas for floral phase under short day condition 10 hours light and 13 hours dark period is required.

The optimal temperature range for cultivation of chrysanthemum is 16-25°C. Temperatures below 16 °C but not lower than 12 °C which allows plant growth but most varieties need 16-18°C for flower induction, otherwise the plant will remain vegetative.

Temperatures from 25-35°C can relatively easy to be managed by cultivation techniques like applying limestone on plastic / glass, frequent overhead misting, different techniques and / or timings of closing the black plastic, installing a protective screen (shade net, alumunium net screen) and installing a pad and fan system. When the temperature exceeds 35°C especially the "darker colours" like red, purple and orange will tend to fade, the number flowers per stem will come lower down and also reduce flower size.

Soil

Chrysanthemum can grow on different types of soil however rich organic soils are favourable. The pH should be approximately 6.5, when the soil has a pH which is too high, use acid fertilizers (phosphates), when it is too low use Calcium Carbonate. Chrysanthemums prefer a well-drained slightly acid soil with a pH value of 6.5 (Brickell, 1992). Dole and Wilkins (1999) suggests a soil pH ranging from 5.7 to 6.2 and Kher (1989) from 6.2 to 6.7. The electric conductivity level (EC) indicates the salinity of the soil, it should be checked regularly. A soil – EC of 1 – 1.5 is good. When the EC is too high besides adjusting the fertilizer mixture, one can wash out the soil by excessive irrigation (not on heavy soil types).

Soil preparation

The soil should be leveled before planting to obtain a homogenous crop. The soil has to be made free from soil born diseases by summer ploughing of soil also known as soil solarization process for 30 days in May-June month and weeds once a year. This can also be done with methyl bromide or by steam. To improve the structure and water retaining capacity of soil can be mixed with cattle manure or coco peat. Since chrysanthemum has a shallow root system it is only required to mix it in the top layer (the first 25 cm).

Soil disinfection

If Chrysanthemum is to be grown for the first time, it is advisable to disinfect the soil. This can be done by fumigation under protected condition.

Irrigation

The availability of the correct amount of water is extremely important for a uniform growth. Quantity of water as well as distribution of water is important. This requires a good control and a correct irrigation system. Dry spots leads to retarded growth of plants and wet spots to a fungal disease called Pythium or Septoria leaf spot

Overhead sprinkling system with rotating sprinklers is the method. This not only gives uniform growth but also provides the required relative humidity in the greenhouse (75%). At the stage when flower bud starts opening (the last 2 weeks) over head irrigation should be stopped and watering should be done by hose pipes. Drip irrigation with 4 litre per hour emitters can be run for 30 minutes at three days interval except rainy days

Drainage

Good soil drainage is desirable for a number of reasons. The most important of these are:

a. To neutralize errors in watering
b. A constant ground – water level can be maintained
c. Maintenance of soil structure

Manuring and Fertilization

Chrysanthemum is a heavy feeder and requires large amounts of both Nitrogen and Potassium. Nitrogen is required at early stage and the plants need P throughout the growth period. As the buds appear, the proportion of K should be increased and be reduced. Sujatha and Mohandas (2004) concluded that *Glomus moschata*+ *Azotobactor* + 75% N + 75% P_2O_5 promoted yield of flower, *Glomus fascialuatum* + PSB (phosphorus solubilising bacteria) + 75% P_2O_5 promoted growth yield and which can save 25 percent application of phosphate fertilizer in chrysanthemum cv. Chandrika.

Chawla *et al.*(2007) reported that 300 kg nitrogen and 150 kg per ha phosphorus in chrysanthemum cv. Neelima resulted in higher yield of flower with the full dose of phosphorus and half dose of nitrogen should be applied as basal application and remaining half dose of nitrogen 30 days after transplanting under Udaipur . A basic fertilization of 4 kg NPK 10-10-10 and 5 kg triple phosphate per 100 square meter can be applied. In chrysanthemum , application of 25 kg N in the form of sheep manure and 75 kg N as urea per hectare should be given.

A fertilizer dose of 200 kg N, P, K/ha along with 50 t of FYM is optimum for increased flowering and yield (open field condition). Fertigation has to practiced for green house cultivation.

Nutrient requirement of chrysanthemum crop (ppm)

Plant age in weeks	N	P	K	Ca	Mg	B
0-3	120	30	220	78	20	0.33
4-7	130	30	230	78	20	0.33
8-12	150	30	250	78	20	0.33

At early stages establishment of overhead sprinklers are recommended. After 30 days drip/overhead fertigation starts. Five cubic meters of water with 60:60:60 ppm of N: P: K should be applied. At the vigorous growth stages 5-6 litre water/m^2 with 100:10:100 of NPK should be given.

Intercultural Operations

Weeding and hoeings are generally done manually as and when needed, normally 2-3 times. Besides control of weeds, the soil is made loose porous to provide aeration. Treatment with trifluralin , EPTC , nitrofen , DCPA, diphenamid, Oxyflourfen @0.75 kg a.i., Pendimethalin 0.75kg active ingradient per ha three days prior planting on moist soil or the day after transplanting of rooted cuttings markedly reduce the weed population. Reflective film and black film mulch can be effectively used to control weeds and conservation of moisture.

Cultural Practices

Special practices like staking, pinching and disbudding are important for proper growth and flower yield

Staking

Staking is a necessary operation to provide proper support and to maintain flower quality with straight stalk. There are only a few flowering varieties which neither require pinching nor staking and hence known as 'no stake no pinch' varieties. For standard varieties, the number of stakes will depend upon the number of main branches which have been allowed to produce the bloom. In spray types, 3-4 stakes are inserted on the bank of the pot and are tied with a string from the bottom to the top which gives good support to the developing branches.

Pinching

The objective of pinching is to encourage the side branches and it should be repeated to encourage more number of branches depending upon the number of blooms to be retained.

i. **Soft pinching** involves removal of the tip of shoot along with 2-3 open leaves. It is most commonly practiced because buds remaining on the stem break rapidly.

ii. **Hard pinching:** Hard pinch is generally used to reduce plant size especially in pots. Here buds do not break as rapidly as soft pinch.

iii. **Roll out pinching** is used only on short plants.It refers to simply pinching of buds just by rolling them with a gentle twist.

***Standard Chrysanthemum*:** In this case only single bloom on a branch is usually allowed to produce. The pinching is not done, if only one central bloom is desired on the main branch. Single pinching is done, if two flowers are desired, whereas double pinching is done for four flowers.

Spray chrysanthemum: These types produce numerous small to medium sized flowers. In such types two pinching are required to encourage lateral growth. First pinching is done at 4 weeks after planting and second after 7 Weeks of planting. As a general rule, rooted cutting are pinched 2 weeks after planting.

Disbudding

Disbudding is done to remove the side branches which arises from axillary buds, so that the number of flowers is limited and blooms of better size obtained. In standard chrysanthemum regular disbudding is done to produce single flowers on single stems.

Plant Growth Regulators

Crop growth regulation and flowering can be modified or controlled by use of growth regulators. Flower quality is also obtained by use of regulators. Spray of GA_3 at 50 ppm at 30, 45 and 60 days after planting is effective to increase flower yield. Spray of B-nine (5000 ppm) is beneficial to obtain the increased number of flowers per plant. Sidhu and Singh (2002) tried NAA and IBA at 250, 500 and 1000 ppm as quick dip treatment of chrysanthemum cuttings before inserting into river sand for rooting and found 250 ppm in both the cases as optimum for inducing rooting. Cuttings taken from young stock plants produce better rooting. Growth retardants, especially IBA and SADH promote rooting though CCC inhibits it (Arumugam *et al.*, 2006).

Micronutrient Disorders in Chrysanthemum

Iron Deficiency: Iron deficiency results into young leaves showing interveinal chiorosis in the early stages and are yellow in colour the severe stage with stunted growth and aeration. The use of undecomposed FYM is very important to prevent the Fe deficiency. Composts of pH 6 to 6.5 only to be used to avoid chlorosis. Foliar spray of 0.5 % Ferrous sulphate at pH 4.5 or Fe EDTA at 0.1% twice once at 30 days and another at 45 days after planting can be adopted correct this Fe deficiency.

Copper Deficiency: Copper deficiency is more widely reported in Chrysanthemums. It causes few flower buds opening, petals curving upwards. Also, flower bud initiation is retarded, terminal leaf base becomes chlorotic initially. Even veins are chlorotic in later stage. Copper suplhate application at 10kg/ha can correct copper deficiency.

Boron Deficiency: Boron deficiency causes peripherals flowers to lose turgidity before opening of the central flowers. Severe deficiency causes petal to twist and cup shaped. Soil application of Boron at 2kg/ha (15 – 20 kg borax/ha) and foliar spray of 0.2% Boric acid after 3 weeks of planting and again at 25% flower bud emergence can correct the disorder.

Boron toxicity: Excess boron application inadvertently or use of irrigation water high in boron creates boron toxicity (>125ppm) with reduced flower diameter and shortened plant height.

Harvesting

The correct stage of harvesting depends on the cultivars, marketing and other facilities available to a grower. Single cultivars are harvested when the maximum number of flowers opens before shading of pollens from the outer row of the disc florets. Anemone cultivars are harvested before the central cushion in the topmost flower is fully developed, whereas decorative types are harvested when the petals in the centre of the topmost flower is almost fully developed. In standards, harvesting is generally done when outer ray florets ceases to further develop. Pot-mums are sent to the market with half to fully opened flowers.

Standard chrysanthemums are harvested at unopened stage when only a few outer ray florets unfurl and the cut flower store equally well in dry condition to be handled conveniently with the least damage and equal or superior post-storage quality. Bud opening solution of 200 ppm 8–HQC +2 per cent sucrose can be used for stored flowers.

Post harvest

Depending on the variety plant start yielding flowers after 3-4 months of transplanting. For cut flower purpose stem is cut about 10 cm above the soil to avoid cutting into woody tissues. The lower 1/3 rd of stems are stripped of leaves at harvest and as soon as possible

After harvested the stem have to be cut at equal length (90 cm is the standard), bunched in five putting a rubber band at the base and sliding them into a plastic sleeve and putting the bunches in plastic buckets filled with 20cm of water. On the day of shipment the bunches can be packed in boxes.

Grading

Chrysanthemums are graded based on the stem length, flower appearance, number of flowers, stem straightness, colour and freshness of flowers. Standard chrysanthemum are graded into Blue, Red, Green and Yellow, whereas spray types are graded into Gold, Silver, Bronze and Make Up based on the quality parameters. In Dutch market, spray chrysanthemums are graded into extra grade and shorter grade. The lower leaves are stripped off upto 15-20 cm and bundled in units of 5 stems and secured with a rubber band.

Export Standards for Chrysanthemum

Parameters	Standard	Spray	Dwarf
Stem length	88-100cm.	77-88cm.	25-38cm.
Weight	30g/stem of 90cm.	30g/stem of 85 cm.	30g/stem of 30cm.
No. of flowers	Only 1flower with 5 buds.	10 flowers.	10-12 flowers.
Diameter	60-80mm.	35mm for half bloom. 45 mm for full bloom.	30mm.

Packaging

Most of the Standard chrysanthemums are placed in sleeves and packed in card board display boxes measuring 91 x 43 x 15cm. They are placed in the boxes according to the grades. For bulk packing of the spray chrysanthemums, 10, 15 or 20 stems are placed in sleeves according to the grades. Six sleeves, three at each end, are generally packed in each box, measuring 80 x 50 x 23cm.

Cold Storage

The stems in the buckets (after grading) are given a cut using sharp blade and precooled at 1°C minimum of 2 hours before packing. Chrysanthemum can be stored for 3-6 weeks period at 0-3°C.

Yield

The Yield of flowers in chrysanthemum varies depending on the types and the cultivars, the growing region and the environment, plant density, pinching, disbudding and other management practices. The average yield of loose flowers suitable for making garland and veni ranged from 7.5 to 15 t/ha while about 1,00,000 good quality sprays can be obtained from one hectare. Small flowered types of chrysanthemum produced 28.6 to 41.4 lakh cut flowers per hectare depending on the plant density.

Diseases and pests

Diseases

Root rot, (*Pythium, Phytophthora*) Phoma root rot, Foot rot, (*Rhizoctonia solanai*) Stem rot and wilt, (*Fusarium oxysporium*) Grey mould, (*Botrytis cineraria*) Ascochyta blight, (*A.chrysanthemii*) Septoria leaf spot, (*Septoria chrysanthemii*) Powdery mildew, (*Septoria chrysanthemii*) White rust, Verticillium wilt, Ray speck, Bacterial blight, Bacterial leaf spot, etc. are the disease infestations observed in chrysanthemum. Major disease problems are due to fungal infestation which can be controlled using carbendazim @2 g per litre of water or mancozeb @ 2 g per litre water. Root rot or wilt can be controlled by Captan, Mancozeb or Metalaxyl-MZ and Fosetyl-Al. Powdery mildew can be controlled using wettable suphur, Dinocap @ 1 ml/litre water or Calixin

@ 7ml/10 l water. Bacterial blight and leaf spots can be controlled by Streptocycline or Copper Oxychloride.

Fusarium Wilt (*Fusarium oxysporum pv Chrysanthemii*): Symptoms vary with the cultivar infected like yellowing of leaves, wilting and discoloration of the vascular tissue develops up one side of the plant. **Management**: Purchase culture-indexed cuttings free of the pathogen. Plant in pasteurized soil or soilless mix free of the pathogen. Maintain soil pH between 6.5 and 7.0. Use nitrate rather than ammonium forms of fertilizer. Apply a fungicide to protect healthy plants.

Aschochyta Blight: Flower development is retarded on one side of the bud. Petals exhibit a brown discoloration. Browning and blackening extends down the stem, causing the flower to droop. Brown to black irregularly shaped spots develop on leaves. Mangement: Avoid overhead irrigation. Apply a fungicide to protect healthy plants.

Powdery Mildew (*Golovinomyces cichoracearum formerly Erysiphe)***:** White, dry powdery fungal growth on upper surface of leaves.

Pythium Root and Stem Rot: Stems turn dark brown to black at the soil line. Plants are stunted, wilt, and die. Management: Plant in pasteurized soil or soilless mix free of the pathogen.

Verticillium Wilt (*Verticilllium alboatrum*)**:** The margins of lower leaves wilt and die. Or, the entire leaf dies. Symptoms proceed up one side of the plant.

White Rust(*Puccinia horiana*): Small, yellow to tan spots are observed on the upper surface of leaves. On the underside of the leaf below the spots, raised, pinkish to white to cream-tan areas develop in which spores of the fungus are produced.

Pests

Aphids, Hairy caterpillars, mites, Grubs, Thrips, Leaf miners, White fly,etc. These sucking pests showing symptoms of crinkling, curling and distorted leaves and buds can be controlled using systemic insecticides like Dimethoate @2 ml/L water or Imidachlorpid @3-5 ml/10 L water. Thiamethoxim @3-4 ml/L water is higly effective against mites. Hairy eating caterpillars found feeding on leaves making holes on leaves and eating buds. These can be controlled using contact insecticides like Monocrotophos 2ml/L water or Indoxocarb @ 3-4 ml/15 L water or Dicofol @ 2 ml/L water.

References

Anderson, N.O. 1987. Reclassification of the genus *Chrysanthemum* L. Hort Sci., 22: 313.

Arumugam, T., Jawaharlal, M. and Vijayakumar, M. 2006. Chrysanthemum. In: Advances in Ornamental Horticulture. Vol. 2. Herbaceous Perennials and Shade Loving Foliage Plants (ed. Bhattacharjce, S.K.), pp. 81-108.

Beckett, K.A. 1983. The Concise Encyclopaedia of Garden Plants, pp. 80-82. Orbis Pub. Ltd., Great Britain.

Brickell, C. (Ed.-in-Chief), 1992. The Royal Horticultural Society Encyclopaedia of Gardening, pp. 152-154. Dorling Kindersley, London.

Broertjes, C. 1966. Mutation breeding of chrysanthemum. *Euphytica*, 15: 156-162.

Broertjes,C.,Roest, S. and Bokelmann, G.S. 1976. Mutation breeding of *Chrysanthemum morifolium* using *in vivo* and *in vitro* adventitious bud techniques. *Euphytica*, 25:11-19.

Chawla, S.L., Mohammed, S., Mahawer, L.N. and Jain, M.C. 2007. Effect of nitrogen and phosphorus on vegetative growth and flower yield of chrysanthemum (*Chrysanthemum morifolium*)cv. Neelima. Annals of Agricultural Research New Series, 28(1): 25-28.

Dole, J.M. and Wilkins, H.F. 1999. Floriculture Principles and Species, pp. 292-304. Prentice Hall, New Jersey.

Drewlow, L.W., Asacher, P.D. and Widmer, R.E. 1973. Genetic studies of self incompatibility in the garden chrysanthemum *C. morifolium* Ramat. Theor. Appl. Genet., 43: 1.

Fryxwell, P.A. 1957. Mode of reproduction in higher plant. *Bot. Rev*., 23: 135-233.

Gupta, M.N. 1971. Mutation breeding of chrysanthemum. I. Production of new cultivars by gamma ray induced somatic mutations in vM1. Proc. Intern. Synip. on Use of Isotopes and Radiation in Agriculture. New Delhi.

Hay, R. and Beckett, K.A. 1971. Readrer's Digest Encyclopaedia of Garden Plants and Flowers, pp. 141-149.

Kher, M.A. 1986. Chrysanthemum. In: Ornamental Horticulture in India (eds: Chadha, K.L. and Choudhury, B.). ICAR New Delhi.

Kher, M.A.1989.Chrysanthemum.In:Commercial Flower.pp:453-486. Naya Prokash Kolkatta.-

Nazeer, M.A. and Khoshoo, T.N. 1982.Cytological evaluation of garden chrysanthemum. Current Science, 51:583-585.

Sidhu, G.S. and Singh, P. 2002. Effect of auxins on propagation in *Chrysanthemum morifolium*. In: Floriculture Research Trend in India (eds Misra, R.L. and Sanyat Misra), pp. 285-286. Indian Society of Ornamental Horticulture, Division of Floriculture and Landscaping, 1.A.R.I., New Delhi.

Stewart, R.N. and Derman, H.1970. Somatic genetic analysis of the apical layers of chimera sports in chrysanthemum by experimental production of adventitious shoots. American Journal of Botany, 57:1061

Sujatha, K. and Mohandas, S. 2004. Effect of biofertilizer on growth and flower production of chrysanthemum cv.Chandrika. National Symposium on Recent Trends and Future Strategies in Ornamental Horticulture, Dec. 1-4-2004(Eds) Reddy, B.S., Janakiram, T., Kulkarni, B.S. and Narayanaswamy P., UAS, Dharwad, p. 41.

Thistlethwaite, E.T. 1960. *Chrysanthemum*. Penguin Books, Middlesex, England.

Vince, D. 1960. Low temperature effects on flowering of *Chrysanthemum morifolium* Ramat. I. Hort. Sci., 35: 161-175.

Vishnu Swarup, 1967, Garden flowers. National Book Trust, India. New Delhi.

Wilkins, 1999. Floriculture Principles and Species, pp. 292-304. Prentice Hall, New Jersey.

Zagarski, J.S., Ascher, P.D. and Widmer, R.E. 1983. Self incompatibility in hexaploid chrysanthemum. *Euphytica*, 32: 1-7.

Zareer, M.A. and Khosoo,T. N. 1982. Cytological evaluation of garden chrysanthemum. Curr. Sci., 51:583-585.

15

Cultivation of Gerbera

Tanya Thakur and K. K. Dhatt

Gerbera (*Gerbera jamesonii Bolus*), commonly known as Transvaal Daisy, Barberton Daisy or African Daisy, belongs to the family Asteraceae is one of the most important cut flowers in domestic as well as global market. It ranks sixth among the top ten cut flowers. Gerbera can successfully be grown in the plains as well as in the hills. In India, its cultivation is confined mostly to Kalimpong, Bangalore and Pune. The flowers are known for long vase life (Chauhan, 2005) and good rehydration capacity suitable for long distant transportation. It is a diploid species with somatic chromosome number 50 (n=25).

Importance and Uses

Gerbera is very fashionable and widely used as a decorative garden flower and cut flower (Kanwar and Kumar, 2008). It is ideal for garden decoration as a bedding plant, pot plant and suitable for growing in rock gardens. It is also used in interior decoration in vases, flower arrangements and in bouquets.

Gerbera viridifolia is referred as 'the herb of milk' in the eastern Cape and is probably used to stimulate milk production in cattle (Dold and Cocks, 2002).

Origin

Gerbera is native to tropical Asia and South Africa. The species Gerbera jamesonii is native to Natal and Transvaal region, whereas Gerbera viridifolia is native to Cape region of South Africa. The modern cultivars (*Gerbera hybrida*) are evolved by hybridizing Gerbera jamesonii with Gerbera viridifolia (Leffring, 1973). The species of Indian origin are Gerbera anandria, G. kunzeana, G.languinosa, G. macrophylla, G. nivea, G. ovalifolia, G. maxima and G. piloselloides. In India, about 7 species are distributed in the temperate Himalayas from Kashmir to Nepal at an altitude of 1300 to 3200 meters.

Taxonomy

The genus Gerbera was named by pre-Linnean botanist, Jan Gronovius in 1737 in the honor of Traugott Gerber, a German naturalist and the species jamesonii in honor of Captain Jameson (Das and Singh, 1989).

Gerbera plant is stemless, tender perennial herb (Barreto, 2000) with radical, lanceolate, deeply lobed leaves sometimes leathery, narrower at the base and wider at tip and are arranged in a rosette at the base. The inflorescence is capitulum or head with inner disc florets (bisexual) and outer ray florets (pistillate). The daisy-like flowers grow in a wide range of colors including yellow, orange, cream white, pink, brick red, scarlet, maroon, terracotta and various other intermediate shades. The flower stalks are long, thin and leafless with single, terminal flower head.

Species and cultivars

The genus Gerbera consists of about 45 species which are half hardy and perennial in nature. Some of the species are described below-

Gerbera jamesonii (Barberton daisy) is the only cultivated species.

Gerbera asplenifolia - flower head pure white above and red beneath.

Gerbera aurantiaca - flower head orange and anther yellow

Gerbera kunzeana - flower scarcely open, chestnut brown.

Gerbera viridifolia - flower head dirty white above and red beneath, used as parent in hybridization.

Gerbera cantabrigensis (G. hybrida)- hybrid between Gerbera jamesonii x Gerbera viridifolia

Gerbera maxima - rediscovered from Pauri Garhwal (Uttrakhand, India)

Gerbera hintonii - endemic to Mexico

Depending on the arrangement of ray florets, the cultivars are divided into three types-

1. Single types consist of one or two rows of ray florets on the periphery of the disc florets.
2. Semi-double types consist of 2-3 rows of ray florets.
3. Double types consist of more than three rows of ray florets having bicolored flowers. Depending on the size and shape of the flowers, they are further divided into standard (>8 cm flower diameter), spider and mini (< 8 cm flower diameter).

Some of the cultivars are as follow

Dusty, Flemingo, Fradaisy, Fredeking, Fredorella, Maron Clementine, Nadja, Terraqueen, Uranus, Valentine, Vesta, Alexis, Amber, Anke, Apple blossom, Belitis, Caprice, Hildegard, Ibiza, Joyee, Kabada, Marleen, Pascal, Romilda, Rozamunde, Salmorosa, Sympathic, Easter Star, Eastelle, Nena, Terra Parade, Terra Son, Aruba, Asheley, Flora, Goldspot, Nevada. Rosabella, White Sun. Vino, Venturi, Sunset, Sundance, Sangriana, Siby, Sissy, North Star Ornella, Tropical, Cocktail, Panorama, Mirage, Pink Sensation, Tara, Thallasa, Diablo, Lyonella, Twiggy, etc.

Under Indian conditions, Jaffa, Sangria, Salina, Aida, Rosalina, Pink Elegance, Aalasmeer, Starlight, Rosula and Priyadarshi are commonly grown.

Some hybrids released by IHBT, Palampur are Him Gaurav, Him Saumya, Him Apoorva, Him Peace, Him Glow, Him Abha and Him Keerti.

Cultivar released by IIHR, Bangalore is Arka Krishika.

Depending on stem length-

Tall cultivars (>50 cm) are Funda, Sunanda, Dark Red.

Medium cultivars (40-50 cm) are Sunset, Ornella, Kolika, Pink Elegance, Ruby Red

Dwarf cultivars (30-40 cm) are JS Lal, Yellow Queen, Indu Kumari, Sangria, Red Monarch depending on flower shape and size-

Standard cultivars	Spider cultivars	Mini cultivars
Cabana, Blue Boy, California, Golden Gate, Jaska, Gloria, Zembla, Dalma, Shirley, Stardust, Ruby Red, Rosalin, Fantine	Chili, Punky, Spanky, Spiky, Vicky	Goldy, Red Hawaii, Jimmy, Lily, Nikky, Siby, Snoopy, Tiffany, Victory, Tweety, Winny

Climate

Gerbera can not tolerate extreme of weather conditions. It is grown under open as well as polyhouse condition in tropical and subtropical areas, whereas under greenhouse in temperate areas. The mild weather available in Bangalore and Pune allows its cultivation under open conditions but under extremes of North India, it is cultivated under polyhouses. The flower production under polyhouse is twice in 2nd and 3rd year that of 1st year. For good quality flowers, grow gerbera in polyhouse with ventilation, evaporative cooling and shading facilities. During summer, lowering of temperature is achieved by shading (green shade nets, aluminum coated nets) and evaporative cooling system. During winters, temperature is raised by heating or covering with single or double polythene layer. The height of the greenhouse/shade house should be minimum 5-6.5 m; so, there is proper air circulation. The two polyhouses should be atleast 4-5 m apart with polythene thickness of 200 micron or 800 gauge. To protect the plants in the monsoons provision for covering the top with polythene or plastic sheet is advised. The site should be 30 m away from buildings or windbreaks.

The optimum day and night temperature requirement are 27°C (16-20°C) and 12-14°C, respectively. The temperature requirement for flower initiation is 23°C and for leaf unfolding is 25 - 27°C. The flowering is harmed at below 12°C and above 35°C. The highest yield of flowers was obtained when plants were grown at 21 °C (Pettersen and Gislerod, 2003). The period of October-November and February-April appeared to be highly favorable for the production of flowers in Punjab region when minimum and maximum temperatures were around 10°C and 26°C, respectively (Kaur, et al 1996). Sunny or semi-shady locations are good for gerbera cultivation. Gerbera is short day plant with approximately 400w/m^2 light intensity affects crop production, thus artificial requirement. Plant requires sunshine in cool weather and shade during summers. For quality flower production 25% shade is required (Kaur *et al.*, 1996), thus, white shade net (50-70%) are used. In temperate areas, poor light conditions in winter affects crop and artificial lightening is done. Hence both light and temperature markedly influence the growth and flower production (Cockshull, 1985). The optimum CO_2 requirement inside polyhouse should be 1000-1500 ppm, whereas relative humidity should be 70-75%.

Soil

A well-drained, rich, light, highly porous soil with pH of 5.5 to 6.5 is most suitable for gerbera production. The salinity level of soil should not be more than 1 ms/cm. Gerbera are deep rooted plants and the roots go as deep as 50 to 70cm. The soil should be, light and well drained to have better root growth and better penetration of roots. Adjusting pH and better drainage are must for better crop production.

Propagation

Gerbera is propagated by seeds, vegetatively and micro-propagation (Kanwar and Kumar, 2008).

The seeds are set only if the crop is cross-pollinated. The seeds are sown immediately after harvest in any season showing 90% germination otherwise will lose viability. The seeds germinate at 15-20°C within two weeks; otherwise it may take up to 30 days and are ready for transplanting in 5-6 weeks. The plants produced from seeds will bloom in the second year and produce good flowers from the third year onwards.

Gerbera is propagated vegetatively through cuttings, side shoots, division of clumps and suckers. The side shoots do not emerge easily, thus, the production is enhanced by application of liquid manure.

The division of large clumps into smaller units is commercial and easiest method of propagation in gerbera. It is practiced in early September when the plants are to be set out in the field. The division of mother plant at the end of February produces an average of six plants and these starts flowering at the end of May. Before transplanting, the roots and leaves are trimmed keeping the central shoot intact. While planting, care must be taken that soil does not cover the central growing point. The division should be practiced after 2-3 years of planting.

Micropropagation has been successful for rapid and large-scale multiplication of disease-free plantlets in gerbera (Aswath and Choudhary, 2002). Direct shoot regeneration using shoot tips as explant is most convenient method of mass propagation (Huang and Chu, 1985); however, indirect shoot regeneration has been achieved from calli from explant like leaves, petal and flower buds (Posada et al, 1999). Murashige and Skoog (MS) media with modification is successfully used as culture media.

Planting

Before plantation of Gerbera, soil disinfection is absolutely necessary. In particular, the fungus Phytophthora is a menace to Gerbera. The solar soil sterilization is done by covering the soil with plastic sheet for 6-8 weeks during summers. The chemical sterilization is done by spraying or drenching the beds with formalin @ 7.5-10 lit/100 m^2 and covering the beds with plastic sheets for 7 days. Then flush the soil with water to drain the traces of formalin (Mukherjee, 2008). The other chemicals that can be used are Methyl Bromide @ 25 – 30g/m^2 and Basamid (Dazomet) @ 30 – 40g/ m^2.

Plough the field twice or thrice before planting. Gerbera is grown on raised beds to assist in easier movement and better drainage. The dimensions of the bed are height of 1.5 ft (45cm), width of 2 ft (60 cm) and pathways between beds of 1 ft (30 cm).

Gravel/sand can be added at the bottom for better drainage. While bed preparation, add 20 tones FYM, 100 kg Single Super Phosphate and 60 kg Muriate of Potash per m^2 as basal dose for better root establishment. Apply Magnesium Sulphate @ 0.5 kg per 100 ft^2 and $FeSO_4$ @ 10g/m^2 to avoid deficiency of magnesium and iron.

While planting, the crown of plants should be 1-2 cm above soil level. As the root system establishes the plants are pulled down. Therefore, the crown must be above the ground level at planting and also throughout the life cycle. The planting is done during spring and early rainy season. The best planting time in the plains is September-October and February-March. Early rainy season planting is suitable for 1, 1.5 and 2 year culture and spring season for 1.5 year culture.

Generally, two- four rows may be planted on one bed at row to row distance of 37.5-40 cm and plant to plant distance of 30.0 cm. This would generally accommodate 6 to 7 plants per m^2 and for large flowered varieties, 8-10 plants per m^2. After planting, rake the soil surrounding the plant every fortnight for proper aeration and maintain the humidity at 80 – 90% for 4 -6 weeks to avoid desiccation of plants.

Gerbera can also be cultivated in pots as bench system of planting.

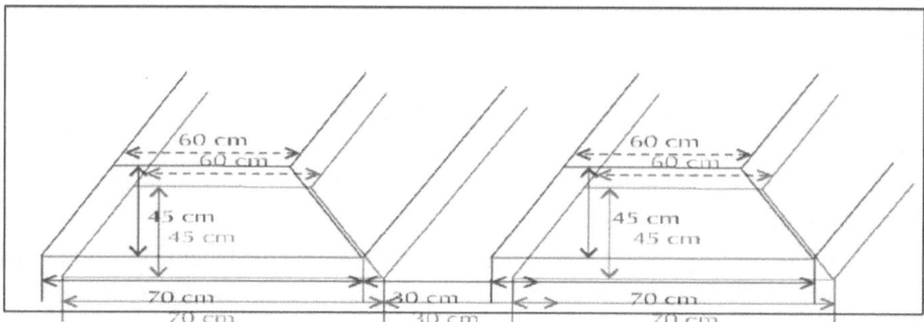

Manure and Fertilizers

Always analyze the soil once in 2-3 months to decide specific nutrient schedule.

Add 20 tonnes well-rotten Farm yard manure along with Single Super Phosphate 100 kg and Muriate of Potash 60 kg per m^2 as basal dose at the time of preparation of beds. After 4-weeks of planting i.e. when the plants are well set, CAN (88 kg per acre) should be applied at monthly interval during active growth period of the plants. After three weeks of planting apply NPK @ 1:1:1 @ 0.4g/plant/day for 3 months. Once flowering commences apply N:P:K @ 2:1:4 at the rate of 0.4 g/plant/alternate day (Bhattacharjee, 2006). Micronutrients should be given weekly or fortnightly as per the deficiency symptoms. Under high pH, deficiency of Fe and Mn is common. Boron deficiency causes base of young leaves to turn black coloured. Zinc deficiency symptoms can be identified with the C-shaped leaf structure caused by chlorosis on one half of the leaf blade which ceases to expand, while the other half of the leaf is normal.

Irrigation

Gerbera is moisture loving plant. The irrigation water should have pH of 6.5-7.0 and electrical conductivity of 0.5-1 ms/cm. To lower the pH of water, add acids in the water tank and then irrigate the plants. Immediately after plantation, irrigate the plant with overhead irrigation for four weeks to enable uniform root development. Thereafter, gradually change to drip irrigation. Drip irrigation is mainly for correct doses and fertilizer application. Generally, one dripper per plant is required. The aim is to provide sufficient irrigation in the 2nd year for extra foliage. The water requirement of gerbera plant may be approximately 300-700 ml per plant per day. Always water the plants before 12 noon. The relative humidity of air should not exceed 90-92%, as it will lead to deformity of flowers.

Intercultural Operations

Gerbera is perennial and remains in field for 2-3 years. The weeds appear only in the early stage of growth and 2-3 manual weeding are optimum.

The dis-budding is the removal of inferior quality flower buds at the initial stage after plantation to increase terminal flower size and maintain health of the plant.

The de-leafing is the removal of extra leaves for bud initiation and sanitation.

Due to daily irrigation, surface of beds become hard hence raking of soil around the plants is done to lose the top soil for proper aeration. This should be done twice in a month.

The removal of old, dry and infested leaves from the plant helps to main sanitation and keeping pest infestation away.

The quality blooms in gerbera are produced from November to December and February to March when the beds are provided overhead cover of plastic from November to March. The scorching heat of May-June is harmful for growth and adversely influences flower production. To reduce light intensity the plants should be covered with shading nets from April to September. The flower yield would increase two-to-three-fold by following these practices.

Harvesting and Yield

Gerbera is perennial crop with duration of 2-3 years, producing its first flower after 6-8 weeks of planting when 14-16 leaves appear on the plant.

The flowers are harvested when the outer two rows of the disc florets are fully expanded and perpendicular to the stalk. Harvesting id done by plucking the flowers in the morning or late in the evening or during the day when temperature is low. Cut the heel of the stem by giving an angular cut. Immediately after harvesting, put the flowers in water and keep for four hours at 7 to 8°C. The frequent re-cutting of stem ends is suggested to avoid bacterial clogging of stem. The flowers are relatively insensitive to ethylene.

Under greenhouse conditions, plant yield around 200- 240 flowers per sqm/ year and 85% flowers are Grade I, whereas under open conditions, 120-150 flowers per sqm/ year and only 15-20% flowers are Grade I.

Postharvest Management

The individual flower head is provided with plastic cup (4.5x4.5 inches) which protects it during transport and bunched together. The large bloomed cultivars are bunched as 10 stem per bunch and mini cultivars as 20 stems per bunch (Pedapati, 2017). These bunches are packed in corrugated boxes of 90x40x12 cm size which accommodate 250 flowers per box.

The cut flowers can be stored dry at 4°C for two days and wet at 2 °C for 4-7 days (Singh et al, 2013). Gerbera flowers has maximum vase life of 7 to 8 days.

Gerbera are graded according to stem length, flower diameter and maturity. A good flower has stalk length of 45 – 55cm (not <40 cm) and flower diameter of 10 – 12cm (not <7 cm). Gerbera are the only cut flowers that are not damaged by chlorine added to vase water.

Grading standards in the world market requirements

Grade No.	Stem length (cm)	Flower diameter (cm)	Flower colour	Preferences (%)
1	>60	>12	Pink	40
2	50-60	10-11	Reddish orange	20
3	40-50	9-10	Orange	20
4	30-40	8-9	Red	15
5	<30	7-8	Yellow	5

Insect-pests and diseases

Insect-pests

The insects like aphids, leaf-miners, thrips and white flies commonly infest the gerbera crop.

Aphids suck the cell sap from leaves causing leaf curl, yellowing and distortion. They also secrete sticky substance honeydew that host the development of sooty mould and attract ants. These can be controlled by spraying Rogor 30 EC or Metasystox 25 EC each @ 0.1% at fortnightly intervals.

Leaf- miners larvae feed on leaf surface making white tunnels, while adult puncture the leaves and petals. These can be controlled by removing infested leaves and spraying Dimethoate @ 0.1%.

Thrips suck the cell sap from leaves and ray floret causing papery leaves and deformed flower head. These can be controlled by spraying Rogor 30 EC @ 0.1% at fortnightly intervals.

Whiteflies appear in cluster on underside of leaves feeding the sap, thus causing leaf yellowing and drop. These can be controlled by use of yellow sticky traps or spraying Metasystox 25 EC @ 0.1% at fortnightly intervals.

Root-knot nematode (*Meloidogyne incognita*) make nodules in roots and causes yellowing of leaves. It may be controlled by immersing bare-root plants in hot water (Arora, 2010).

Diseases

Among diseases, foot or crown rot, root rot, powdery mildew, blight and leaf spot are common in gerbera. Foot rot or crown rot (*Phytophthora cryptogea, Sclerotlinia sclerotiorum, Sclerotium rolfsii, Rhizoctonia solani*) is a very serious fungal disease of gerbera. It affects the basal parts of the plant causing root rotting, crown becomes black and ultimately plant wilt. It is favored by high humidity and waterlogged conditions and can be controlled by drenching of soil with Thiram (0.3%), Ridomil MZ (0.2%) or Dexon (0.2%).

Root rot (*Pythium irregularae* and *Rhizoctonia solani*) fungi causes rotting of roots and ultimately wilting and drying of plant. It can be controlled by soil sterilization or soil drenching with Ridomil MZ (0.2%) or Thiram (0.3%).

Powdery mildew (*Oidium erysiphoides* fspp. *gerbera*) appears as white floury patches on the leaves and tender shoots. It can be controlled very effectively by spraying Benlate (0.1%), Sulfex (0.2%) or Karathane (0.05%).

Gray mould/ Blight (*Botrytis cineraria*) appears as gray spots on flower petals and causes rotting in the heart of flower. The relative humidity of more than 92% for two hours is favorable. It can be controlled by spraying Bavistin 0.1%.

Alternaria leaf spot (*Alternaria* spp.) appears as light brown spots at first and later turns dark reddish-brown, causes speckling of ray florets. It can be controlled fairly effectively by Indofil M-45 (0.2%) or Bordeaux mixture (1.0%) sprays.

Cercospora leaf spot appear as olivaceous in the beginning but turn black ultimately. It can be controlled by spraying with Benlate (0.1%) or Indofil M-45 (0.2%) at suitable intervals.

Physiological disorders

Flower bent at the neck region is due to loss of cell turgidity and lack of calcium.

Pre-harvest stem break is due to high root pressure due to water stress followed by watering and high humidity in the air. The blockage of stem base with bacteria decreases water uptake.

Premature wilting of flower is due to cloudy weather followed by bright sun causing carbohydrate depletion.

Double-faced flower is caused by imbalance of nutrients, too much growth and less flower buds.

Non-uniform blooming is due to physical injury to flower stem or pest damage or phytotoxicity.

Short stem length is due to high salinity level, moisture stress, and low soil temperature.

Iron deficiency causes interveinal chlorosis in leaves. It can be corrected by spraying $FeSO_4$ @0.5%.

Excess nitrogen and ammonia cause more foliage growth and flowers are too short and hidden in foliage.

References

Arora J S (2010) Introductory Ornamental Horticulture. Kalyani Publishers, New Delhi, pp. 90-99.

Aswath C R and Choudhary M L (2002) Rapid plant regeneration from Gerbera jamesonii Bolus callus cultures. Acta Bot. Croat. 61:125-134.

Barreto M S (2000) Studies on the effect of different substrate media on growth, flower quality and vase life of Gerbera under polyhouse conditions. M.Sc. (Agri.) Thesis, MPKV, Rahuri.

Bhattacharjee S K (2006) Advances in Ornamental Horticulture. Pointer Publisher, Jaipur. Vol. 6: 201-211.

Cockshull K E (1985) Gerbera. In: CRC Handbook of Flowering, Halevy A H, Florida, USA, pp. 43-47.

Das P and Singh S P K (1989) Gerbera: Textbook of Commercial Flowers, pp. 601-622.

Dold A P and Cocks M L (2002) The trade in medicinal plants in the eastern Cape province, South Africa. South African J Sci. 98:589-597.

Huang M C and Chu C Y (1985) A scheme for commercial multiplication of gerbera (Gerbera hybrida Hort.) through shoot tip culture. Jpn. Soc. Horti. Sci. 54: 94-100.

Kanwar J and Kumar S (2008) *In vitro* propagation of Gerbera- A review. Hort. Sci. (Prague) 35:35-44.

Kaur K, Gill A P S and Kumar R (1996) Effect of modified environment on plant growth and flower a production of Gerbera. Madras Agric. J. 83 (11): 681-683.

Leffring L (1973) Flower production in gerbera; correlation between shoot, leaf and flower formation in seedlings. Scientia Horti. 1:221-229.

Mukherjee D (2008) Gerbera. In: Speciality Cut Flowers Production Technologies. Naya Udyog, Kolkata. pp.321-374.

Pedapati A (2017) Grading and Packaging of quality grade cut flowers for export. Biotech Articles (online).

Pettersen R I and Gislerod H R (2003) Effect of lightening period and temperature on growth, yield and keeping quality of Gerbera jamesonii Bolus. European J

Posada M, Ballesteros N, Obando W and Angarita A (1999) Micropropagation of gerbera from floral buds. Acta Hort. 482:329-332.

Singh Kushal, Kumar G, Saha T N and Kumar R (2013) Post Harvest Technology of Cut Flowers. Venus Printers and Publishers, New Delhi. pp. 26-28.

REFERENCES

Arora, R.K. [illegible] Germplasm of [illegible] Horticultural Crops. [illegible] Publication, [illegible] pp.

[illegible]

[illegible] (1980) Studies on the growth [illegible] under [illegible] Conditions [illegible]

Bhattacharjee, S.K. [illegible] Advances in Ornamental Horticulture. [illegible]

[illegible] (1990) [illegible] Handbook of [illegible] [illegible] pp. [illegible]

[illegible] Textbook on [illegible]

[illegible] and [illegible] (2002) [illegible] South Africa. [illegible]

[illegible]

[illegible]

[illegible] (1993) [illegible] Hort. 228

[illegible] (1989) [illegible] Kolkata. pp. 321-324.

[illegible]

[illegible] (2007) [illegible]

[illegible] Hort. [illegible]

Singh, [illegible] (2014) [illegible] Technology. [illegible] New Delhi. [illegible]

16

Orchids

N.K. Meena and Ram Pal

Orchid is a valuable flowering plant belongs to family Orchidaceae. It is considered that the family recognise for maximum number species ranged between17,000 and 35,000 worldwide (Garay 1960; Willis 1973). The word *orchid* is derived from a Greek word *orchis* for testicle because of the shape of the root tubers in some species of genus *Orchis*. India is a reservoir of 1300 orchid species from 184 genera, accounts for nearly 7 per cent of worlds orchids genetic diversity (Arora, 1983; Rampal and Nagrare, 2006) and 1150 species belonging to 164 genera (Satish Kumar and Manilal, 1994) of which about 800 species are found in the North Eastern region of the country (Nagaraju *et al.*, 2006). Apart from species, more than one lakh hybrids developed in orchids. It is found all over the world from tropical to temperate and even alpine region. People always think that orchids are epiphytic only but it is not. It can grow in soil as terrestrial, on tree trunk (epiphytic) and on dead organic matter (saprophytes) and on rocks (lithophytes). Orchids produce unique flowers, affluent for their uniqueness in shape, size and colour, exquisitely attractive; remain fresh condition for long period of time in comparison to other flowers. Vase life of its flowers varied among species and hybrids ranged up to 40-45 days. Orchids are cultivated for trade of orchid species, cut-flowers, potted plants, flower basket, bouquet, stage decoration in high profile meets and functions and also used in many other value-added products (Kuehnle 2007). Many orchid species are used as food like *Cymbidium hookerianum, Dendrobium chrysotoxum, D. kingianum, D. cathenatum* and many more (Singh *et al.*, 2016), spices (*Vanilla planifolia* and *V. pompona*) and medicinal uses (*Dactylorhiza hatagirea, Flickingeria macraei, Vanda tessellata* and *Dendrobium macraei* as reported by Ram Pal *et al.,* 2020). The most popular and commercially cultivated genera are: Cymbidium, Dendrobium, Oncidium, Phalaenopsis, Vanda, Paphiopedilum, Mokara, Arundina, Paphiopedilum, Ascocenda and Cattleya in the country and even in the world.

Origin and Distribution

The Orchidaceae are a diverse and widespread family of flowering plants currently accepted 28000 species distributed under 763 genera (Christenhusz and Byng 2016 and WCSP, 2010). Its origin is confined to particular region or continent even than it varied species to species. There is a general belief that the orchids are mostly found in the tropics. But it is not correct, they are found in all over the world from frozen areas of Alaska to, the snow-covered areas of the Himalayas and the sandy deserts of Australia and Africa. The majority of world's temperate orchids are produced in Netherlands, New Zealand and Australia, whereas South East Asian Countries like Thailand, Indonesia, Malaysia, Brunei, Cambodia, Singapore and Burma are the reservoir of tropical orchids. New Guinea has more orchids, native to the island, than any other part of the world. In India the majority of orchids are available in the eastern Himalayas, northeastern hills, the western India and the south Indian hills. Cymbidium orchids are found Sikkim, Darjeeling district of West Bengal, Arunachal Pradesh, and few in nilgiri hills. *Aerides* is found in southern and northern India, Burma, Malaya and Indonesia, *Vanda* spp are common in India, South East Asia, Indonesia and Ceylon. *Calanthe* spp are mostly available in the wide forest range of Australia, Vietnam, China, Japan and South India. *Cdelogyne* and Bulbophyllum are more common in northeast India including Sikkim. Netherlands is world's largest producer and exporter of temperate orchids while Thailand leads in production and export of tropical orchids.

Taxonomy

The taxonomy of the Orchidaceae (orchid family) has evolved slowly during the last 250 years, starting with Carl Linnaeus recognized eight genera in 1753. De Jussieu recognized the Orchidaceae as a separate family in his genera Plantarum in 1789. The Orchidaceae is currently placed in the order Asparagales by the APG system III (APG 2009). APG system III described the Orchidaceae under five subfamilies *i.e.* Apostasioideae (2 genera and 16 species), Cypripedioideae (5 genera and 130 species), Vanilloideae (15 genera and 180 specie), Epidendroideae (more than 500 genera and more or less 20,000 species) and Orchidoideae (208 genera and 3,630 species).

Species and hybrid

Orchids are the most potential group of plant in floriculture industry. They are considered among top ten ornamentals for cut-flower and potted plants. The most valuable orchids for national and international market are *Cymbidium, Dendrobium, Phalaenopsis, Cattleya, Vanda, Paphiopedilum, Oncidium* and *Zygopetalum* on breeding is undertaking for hybridization. The first man made orchid hybrid is *Calanthe dominiyi* which developed by Mr. John Dominy through crossing between *Calanthe masuca* and *Calanthe furcata.* This hybrid flowered in the year 1856 for first time. Cattleya is one of the genera in which artificial hybridization is practiced since early days of orchid breeding. The first and foremost thing in any breeding programme is to decide what to breed and what for breed. Bose *et al.* (1999) discussed the breeding objectives and steps in orchids. The species and hybrids of few important orchid genera are giving below.

Table 1: Important species and hybrid of orchids in India

Sl.No.	Genus	Species	Hybrids
1.	Cymbidium	*Cymbidium lowianum* (Rchb.f.) *Rchb.f., Cymbidium lancifolium Hook., Cymbidium devonianum Pazt., Cymbidium dayanum Rchb.f., Cymbidium tigrinum Parish ex. Hook F., Cymbidium elegans Lindl., Cymbidium tracyanum Rolfe., Cymbidium whiteae King & Pantl., Cymbidium mastersii Griff ex. Lindl.*	Cymbidium 'Winter Beach Sea Green', Cym. 'H C Aurora', Cym. 'Pine Clash Moon Venus'. Cym. 'Mint Ice Glacier', Cym.'Red Star', Cym. 'Show Girl', Cym. 'Sleeping Nymph' Cym. 'Bob Marlin Lucky', Cym. 'Fire Storm Blade', Cym. 'Fire Storm Ruby' Cym. 'White', Cym. 'Un green', Cym. 'Margaret Threcher' Cym. 'Burgandian Sydney, Cym. 'Cecil Park', Cym. 'W.W. Wounders' Cym. 'Levis Duke Bella Vista' Cym. 'Silver Kivi', Cym. 'Sparkle Late green' Cym. 'Hotescences'
2.	Dendrobium	*Dendrobium nobile Lindl., Dendrobium moschatum Sw., Dendrobium densiflorum Lindl., Dendrobium fimbriatum Hook., Dendrobium chrysotoxum Lindl., Dendrobium chrysanthum Lindl., Dendrobium aduncum Wall. ex Lindl., Dendrobium hookerianum Lindl.,Dendrobium primulinum Lindl., Dendrobium lituiflorum Lindl., Dendrobium aphyllum (Roxb.), Dendrobium falconeri Hook., Dendrobium lindleyi Steud., Dendrobium herbaceum Lindl., etc.*	Dendrobium 'Thongchai Gold', Dendrobium 'Bangkok Blue', Dendrobium 'A. Abraham', Dendrobium 'Erika' Dend. 'Triple Pink', Dend. 'Madam Pink', Dend. 'Snow White', Dend 'Sonia-16', Dend. 'Sonia-17', Dend. 'Emma White', Dend. 'Diamond Star', Dend. 'Big White', Dend. 'Fire Wall', Dend. 'Green Wonder', Dend. 'Winter Frost', Dend. 'Ice Storm', Dend. 'Miss Singapore', Dend. 'Australian Lemon Paper', Dend. 'Little Lolita' Dend. 'Medam Pempador'
3.	Phalaenopsis	*Phalaenopsis mannii* Rchb.f., *Phalaenopsis lobbi* (Rchb.f.) *Aver., Phalaenopsis amabilis, Phalaenopsis formosana, Phalaenopsis Aphrodite, Phalaenopsis antennifera, Phalaenopsis deliciosa, Phalaenopsis mysorensis*	Phalaenopsis 'White Dream', 'Florida Snow', 'Brother & Sister', 'Harman Sweet', 'Yellow Queen', 'Sogo Smith', 'Sogo Zebra', 'Minho Valentine', 'Minho King Buty', 'Brother Lowrance', 'Leopard Prince', 'Golden Sun', 'Golden Amboin',

Sl.No.	Genus	Species	Hybrids
4.	Vanda	*Ascocentrum ampullaceum* (Roxb.) *Schltr., Acampe rigida* (Buch.-Ham.) *Hunt, Renanthera imschootiana Rolfe, Rhynchostylis retusa* (Linn.) Bl., *Arachnis rubra Rolfe, Vanda coerulea* Griff. ex Lindl., *Vanda pumila* Hook.f., *Vanda tessellata* (Roxb.) Hook.ex G.Don,	Vanda 'Pine River Blue', Vanda 'Pakchong Blue', Vanda 'Dokter Anek', Vanda 'John Club', Vanda 'S.D. Black', Vanda 'Robert's Delight Blue', Vanda 'Sunsai Blue Kaufan', Vanda 'Pat Delight', Vanda 'Kultana Voilet', Vanda 'R.B.S.D. Pink', Vanda 'Motes Indigo'
5.	Cattleya	*Cattleya dowiana, Cattleya maxima, Cattleya tigrina, Cattleya hardyana, Cattelya kerrii, Cattleya lawrenceana, Cattleya sincorana, Cattleya pumila*	Cattelya 'Bobi', 'Snow Tiger', 'Leopard', 'Berlin beauty', 'Snow Field', 'Berlin Moon', 'Summer Pink', 'Berlin Star', 'Amazon Blue', 'Brazilian Ruby', 'Brazilian Star', 'Brazilian Fire', 'Black Wonder', 'July Sun', 'Spring Breake', 'Moonlight Glitter', 'Summer Bouquet', 'Blue Bird', 'Yellow Sky'
6.	Paphiopedilum	*Paphiopedilum rothschildianum, Paphiopedilum bellatulum, Paphiopedilum insigne, Paphiopedilum venustum, Paphiopedilum hirsutisimum, Paphiopedilu vellosum*	Paphiopedilum 'Double Moon', 'Winter Vacation', 'California Sprit', 'Red Delight', 'Big Tiger', 'Black Paper', 'Asian Sunsine', 'Red Glorry', 'Night Leopard', 'Black Bim', 'World Record', 'Elizabeth March', 'Prince Edward of York', 'Michel Koopwitz', 'Saint Swithin', 'Mount Toro', 'David Hensun'

Orchid species from Rajasthan: Orchid species i.e. *Aerides crispum, A. multiflora, A. maculosum, Epipactice veratrifolia, Eulophia ochreata, Habenaria marginata, H.digitata, H. longicorniculata, Nervilia aragoana, Vanda tessellata, V. testacea* and *Zeuxine strateumatica* reported from Mount Abu, Phulwari Wildlife Sanctuary and Wildlife Sanctuary located in Aravalli hill range of Rajasthan (Shetty and Singh, 1991; Sharma, 2001 & 2002).

Climate

Orchid is a vast group of a number of orchidaceous plants grown from tropical to temperate region and even alpine zone. So that it is very difficult to describe by common climate requirement needs for their growth and flowering. However the optimistic climate requisite for growing of important orchids can be described. *Cymbidium* survives in good health and bear regularly in the places where the summers are moderately warm and bright, however, autumn and winter nights are sufficiently cool. They are most suited in temperature regime of 10°C and 25°C. A temperature difference of about 10°C between day and night is desirable. Although they can withstand low temperature, it would be advisable to maintain a night temperature of 11-13°C during winter to avoid the risk of injury. Adequate light is the most important factor to grow and flower well. The plants grow best with 3000 to 4500 foot candles of light intensity. Commercial *Cymbidiums* grow best under partial shade-partial sun conditions. Shading of plant is an important component to regulate growth and flowering. The extent of shade requirement may vary from location to location or altitude and seasonal variation of light and temperature. In the hot summer particularly April to June, the intensity of light is very high. Higher light intensity coupled with higher temperature also prevails in rainy season. Usually 50% UV stabilized shade net is sufficient to protect the plant from scorching heat under mid hill situation (5000 ft. msl) of northeast region of the country. Adequate humidity is essential to protect the Cymbidium plant from the dehydration and stress, especially in summer months. Overhead misting can be used in protected as well as open house to lower the leaf temperature, at the same time, raise humidity levels. In general, 70-80% humidity needs to be maintained during the period of active growth. However, during winter the humidity requirement is less (40-50%). Dendrobium can be grown successfully under effective temperature range 24-30°C during vegetative growth period while 10-13°C temperature required during flowering. The growing season of Dendrobium is rainy season from July to September needs 50-70 per cent relative humidity and abundant light. Phalaenopsis orchids require summer daytime higher temperature of 27-30°C and low temperature 22°C with 60-70 per cent relative humidity.

Table 2: Climatic requirement of important orchid genera

Genus	Vegetative stage		Flowering stage		RH (%)
	Temperature (°C)	Light (μ mole $m^{-2}s^{-1}$)	(°C)	Photoperiod	
Dendrobium	24-30	300-600	10-13	Short day	50-70
Cymbidium	10- 25	2000-3000 f.c.	10-12	Short day in spring and autumn	70-80
Cattleya	23-32	300-600	12-16	Short day	50-80
Phalaenopsis	20-23	150-300	10-25	No response	60-70
Zygopetalum	22-26	300-500	10-14	Short day	50-60

Growing structure

Orchid plants should be protected from all sorts of weather damages for production of market worthy spikes. In commercial farming for export purposes, it is preferable to cultivate *Cymbidium, Dendrobium, Phalaenopsis, Cattleya, Mokara, Vanda and Zygopetalum* in pots placed on benches under the controlled environmental conditions of the modern greenhouse to achieve perfection though it is cost intensive. Nevertheless, greenhouse with all sides open is also suitable. In mid hill situation, bamboo/ wooden structure with UV stabilized polythene on the top is generally used with success. However, structure with steel pipe and top covered with double layered polycarbonate and encircled with 11gauge iron net is added advantage in durability of polyhouse.

Propagation

Propagation in orchids is an important process for multiplication. It can be propagated through seed, vegetative method of propagation, and tissue culture technique.

Seeds: Majority of the orchids are propagated from seeds but this technique is very difficult to produce orchids by the farmers as it has minute seeds require the presence of a suitable mycobiont and supporting organic carbon source (substrate). Several other biotic and abiotic factors also contribute to the germination niche for orchids.

Vegetative propagation: Vegetative propagation is an easier method to grow and multiplication of orchids, but it is a very slow process as a few plants can be multiplied from one plant in a year. But it is essential to maintain the uniformity in a particular type of orchid, as plants raised from seeds will not be similar to the mother plant. The seedlings may show variations by combining the characters of both the parents. It is moreover difficult for a common grower to raise plants from seeds. In vegetative method of propagation, the following technique can be used.

Division: Division method technique is most suitable for the propagation of sympodial orchids that produce multiple shoot growth from single rhizome. In this technique, full grown old plants (5-6 years) should be used for division through uproot the plants from pots and separates the shoots with pseudobulb in three to four parts depend on the growth of plant used for division. The separated shoots should be repotted in new pot and media and apply the water and nutrient as per recommended package for particular species or hybrid. The orchids *Cymbidium, Dendrobium, Epidendrum, Cattleya* and *Zygopetallum* can be propagated through this technique.

Off shoots and Kaikis: This technique is more suitable for propagation of monopodial orchids like Ascocenda, Phalaenopsis, Vanda and Angreacum orchids some sympodial orchids like *Dendrobium nobile* in which new shoots or kaikis produce from upper portion of backbulbs and internodes of stems naturally. However, it can be produced by breaking or cutting of top portion of monopodial orchids. When kaikis produce 3-4 roots, it can be separated from mother plant and put in new pot and after suitable growth it should transfer in a suitable media and pots.

Cutting: Cutting method of propagation is suitable for making new plants of monopodial orchids like Vanda and Aranda. These orchids grown up to more height due to indeterminate habit, hence it can be managed by cutting and pruning. Due to cutting of apical portion of stem, new aerial roots produce. The sprouted roots with stems separate from main plants and potted in new pots and suitable media to produces new plants but it is slow process of plant multiplication. Through this method untrained person or farmers can produce new plants.

Backbulb technique: In this method of propagation, leaves or shoots less pseudobulb which separate from mother plants during repotting are used for preparation of new orchid plants. It is more suitable for *Cymbidium, Cattleya, Dendrobium, Coelogyne, Zygopetalum, bulbophyllum* and many more orchids produce psuedobulbs. In the process of this method, carefully clean the psuedobulbs by removal of scales and leaf layers after separation of psuedobulbs from the mother plants. Later on, these pseudobulbs socked in normal ware for some time to clean all dirt and then treat pseudobulbs with nutrients and hormones for rapid new sprouting. Apply IBA 1000-2000 ppm for *Dendrobium aggregatum*, NAA 60-120 ppm and IBA 250g/litre in Cattleya hybrid and Cymbidium, respectively are found effective. The prepared pseudobulbs put straight in the raised beds containing media (wooden dust, sand, FYM and leaf mold) and then watering regularly to maintain moisture in the bed. Similarly the pseudobulbs can also be put in the jute bags filled with wooden dust or complete media (wooden dust, sand, FYM and leaf mold). After one or two months, lateral sprouting come out from pseudobulbs and when roots developed completely in news shoots, should separates from main pseudobulbs and separated shoots transplant in new beds or in a new pots. After removal of lateral shoots from pseudobulbs, they can be put in sawdust or media and repeat entire process again to get new plants. This is very effective method of multiplication of new orchid plants, can be utilized by farmers or orchid growers.

Tissue Culture: Tissue culture technique is more prominent in orchids for large scale multiplication of planting material in India and abroad. In this technique of propagation, people are generally using seed culture by which *Cymbidium, Dendrobium, Epidendrum, Zygopetalum, Oncidium* and many other orchids are commercially multiplied in the country. The second most important method of propagation is meristem culture, wherein apical tips of young leaf of matured plants are cultured either in Heller's medium or other suitable media. *Phalaenopsis*, a monopodial orchid, was the first orchid genus to be propagated in vitro while *Cymbidium* was the first to be propagated by shoot-tip, and a comprehensive analysis of conditions required for the tissue culture and micropropagation of various orchids has been reviewed (Arditti and Ernst 1993). The orchid planting materials produced through meristem culture are free from any kind of viral diseases.

Growing media

The main purpose of growing media is to hold the plant in place and supply sufficient amount of nutrients and water. Cymbidium prefers fairly open compost. While selecting potting media, one should be cautious to see whether the media is keeping the roots moist but not wet, even drying of the mix, keeping the roots cool and avoiding large air pockets in the mix. A mixture of peat and perlite is recommended for use in controlled green house condition. An even more open mix would be preferable for growing cymbidium in shade house. Whatever may be the mix, the thumb rule for any potting media is that it should be free from water logging. A healthy growth growing media containing Leaf mould, FYM, Charcoal, Coconut husk, Rotten logs (2:1:1:1:1) is found to be beneficial. The pH of the media should be acidic and it ranges from 5.5 to 6.5. Sterilization of media is an integral part as most of the diseases get entry through media itself. Potting media requires sterilization and mixing homogeneously before filling in the pots. The commonly used methods for sterilization are heating and chemical treatment. The media needs to be cut smaller sized pieces and mixed thoroughly before sterilization. Some media require soaking over night e.g. Coconut husk to leached out undesirable chemicals. If heat treatment is used all the media components should be sterilized separately as they require different period of heating.

In case of chemical treatment, the media components need to be stirred thoroughly, wet with formaldehyde @ 20ml/L and kept covered with polythene sheet for at least 72 hours. After opening, sun drying is required for few hours to evaporate the obnoxious gasses. Phalaenopsis and Vanda are epiphytic orchids and their roots are aerial type, hence it need thick pieces of stone chip or brick pieces, coco-chip, Charcoal, leaf mould in specified ratio. Cattleya and Dendrobium need potting mixture of tree bark, fibres of tree ferns, coco-chip, coco-pit and perlite. After proper sterilization and drying of growing media, potting should be done to avoid fungal disease infection.

Pots and potting

Type and size of pots depend on the growth and size of orchid planting material used for potting. The most commonly used are earthen or plastic pots. The pots should have sufficient number of holes for aeration to root zone and drainage. The plastic pots are economic, long lasting, and convenient and above all require less frequent watering. A year old plant (6 inches size) should be planted in 4 inches size pot. Thereafter, it needs to be transferred in to 6 inches size pot. Smaller plants of less than 6 inches size must be planted in the community pots to check the mortality. Pots should be sterilized in boiling water or Tri-sodium phosphate. The potting should be done during active growth phase i.e. during April to June. Crocks or bricks chips should be placed at the bottom of the pot. The plant then placed centrally and sterilized media need to be placed all around the plant. Care should be taken to avoid any large unfilled cavities while potting. It is always wise to keep the pots on the benches to provide proper aeration and check the soil borne diseases. The density of plant depends on the size of plant as well as pot. At initial stage of growth (1.5 yr old), around 30 plants (4") can accommodate in a square meter area. The spacing will be wider as a plant grows and nine plants of 4-5 year old can be accommodated (6" size pot) in a square meter area. However, fully grown plants (7-8 year old) need more space for sufficient aeration and hardly three plants can be spaced in a square meter area.

Watering

Immediate watering of newly potted or planted plant should be avoided, possibly up to two or even three weeks, which will encourage the new root to grow in the fresh compost. Only plain water is advocated for irrigation until new root grow in evident. The amount of water depends on the potting mixture. Since orchids all their vegetative growth during spring and summer months require more water and it should be reduced during winters. During hot days plants should watered through overhead misting or sprinkling to lower the temperature and increase the humidity level in the growing areas. Always irrigation should be done in the morning to wet down the media thoroughly which will supply water vapour for a long period.

Fertilizers

Fertilizer requirements vary to species and hybrid of orchids as well as age of plants. In *Cymbidium*, water-soluble fertilizer (NPK 19:19:19) at 1g/l should be sprayed weekly only after being assured of new root growth. Very strong feed also should be avoided to protect the immature root from burning. Once the new roots establish themselves a regular watering and feeding programme can be adopted depending on the growing condition and the potting mixture used. Generally fertilizer with higher nitrogen percentage is recommended during growing season, while higher potassium is lead to favour blooming. Dendrobium is an evergreen orchid is heavy feeder than that of deciduous type. A dilute solution of balanced fertilizer twice a month should be sprayed. Fertilizer mixture in 10:10:10 nitrogen, phosphorus and potassium is generally recommended. Fertilizer dosage needs to be sprayed in 19:19:19 or 20:20:20 NPK should be applied with each watering during summer in *Phalaenopsis* orchid. However keeping mix strength same, during winter the frequency of fertilization in adjusted every other watering. The mixture of NPK in the ratio of 10:14:12 once a week during growing season promotes growth and flowering of plants in Vanda.

Training of spike

Cymbidium, Dendrobium, Phalaenopsis, Vanda and *Zygopetalum* spikes need to be trained as soon as they appear. A cane or bamboo split should be inserted behind the spike. Care should be taken to avoid any root injury. Once the spike has reached about 20 to 30 cm in length as orchid species, a tie should be applied. Tying of the spike should be secure and firm with the cane. Great care must be taken, as the growing spikes are brittle and a jolt or rough handling will break them. When the buds have begun to form, a second tie should be given just below the first bud. An extra tie also will be needed below the tip to keep the spike straight.

Pest and Diseases

Orchids are more vulnerable to a number of insect-pests and diseases. They attain pests and diseases incidence from hardening to flowering of plants at different level of damage. Orchids particularly Cymbidium, Dendrobium, Phalaenopsis, Oncidium, Arundina, Mokra and Cattleya get infested with aphids (*Macrosiphum luteum* and *Toxoptera aurantii*), two-spotted spider mite, *Tetranychus urticae*, thrips, *Dichromothrips nakahari*), few scale species (*Pinnaspis buxi* and *Coccus hesperidum*) and slug and snails. These insects attack on foliage, flower buds and full bloom flowers are need to be controlled appropriately at right time to escape the qualitative and quantitative loss. Maintaining clean cultivation, proper air circulation, correct watering, weeding time to time and balanced compost would keep the pests and diseases away. However, with the sudden change in the temperature and increase in humidity with alternate rain and sun, encourage incidence of pest and disease. In such situation, depending upon symptoms application of recommended chemical will helpful to contain and control the same. Regular pest monitoring and cleanliness are two important practices to keep the orchid pests under check. Aphids can be managed by using azadirachtin 0.03%EC @ 3ml/litre of water as foliar spray followed by acephate 75%SP @ 0.05%. Two-spotted spider mite can be managed through cleanliness and proper planting distance with ventilation under controlled conditions. Enhancing the activity of anthocorid bugs, *Blaptostethus pallescens* is good for keeping mite infestation under check. Spray the plants with Econeem 1500 ppm @ 3.5ml/litre followed by spiromesifen 22.9% SC @ 1ml/litre or propergite 57EC@ 1.5 ml/litre (if required). Thrips can be controlled using aqueous Artemisia leaf extract @ 10ml/litre or neem oil 5ml/litre of water followed by acetamiprid 20 SP @ 0.3 g/litre of water. Scales insects can be managed by keeping cleanliness and regular care of plants, manual removing of infested plant parts and destruction them. One spray of acephate 75% SP @ 0.05% or carbaryl 50 WP@ 1g/ litre of water are advisable, if needed repeat the spray at 10 days interval.

In orchids, black rot, crown rot, orchid wilt, blossom blight, anthracnose, leaf spot, bacterial soft rot, bacterial brown rot and some viral infection like Cymbidium Mosaic Virus and Odontoglossum Ring Spot Virus are the important diseases. These diseases are species specific and cause severe loss in particular part of the year. The peculiar symptoms appear on particular plant parts are the best source of disease identification. Fungal diseases can be managed by making clean cultivation inside polyhouses. Procure disease free planting material from accredited nursery while set-up new orchid farm. Avoid unsterilized potting media and pots. Separate infested plants from healthy plants and remove and destroy infested parts to check the further spread of the diseases. For effective control, Matco MZ or metalaxyl@ 1g/lit. or mancozeb @2g/lit can be used as spray or soil drenching. Bacterial diseases can be manage by reducing prolonged wetness by increasing air circulation and also avoid overhead irrigation. Infected plants should be drenched or sprayed with 8 hydroxyquinoline a dilution of 1: 2000 in water. Viral disease can be prevent by sterilization of tools used for handling plants, procure virus free planting material and avoid overcrowding of plants in the polyhouses. If possible, utilize the plants produce through meristem culture.

Yield and harvesting

An understanding of the correct stage of harvesting becomes more important because it influences the keeping quality, appeal of the harvested spike and yield for the next year. The spikes should be harvested when all the flowers are open. While harvesting utmost care should be taken to see that pollen caps of the flowers remain intact. The plant will produce only one spike at the first harvest, which gradually increases up to three or more in the subsequent harvest. Number of flower per spike will vary from 10-15 depending upon the hybrid and variety. One flower spike 10 to 25 individual flowers depends hybrid categorized as miniature, intermediate and standard type. The spikes need to cut at the base of the stalk. The cut surface should be smooth and never be crushed. A sharp knife or secateur is required to sterilize time to time with antibiotic-fungicide solution to avoid transmission of viral diseases. The harvested spikes need to dip immediately in a bucket of water till they are packed. In order to improve the quality of flower spike and vase-life pulsing with sugar (2 - 4%) are given to the cut spike. A well-grown Cymbidium will have a vase life of 4 to 6 weeks at the consumer ends, provided proper handling was maintained after harvest. A number of floral preservatives are also in use to prolong the shelf-life of the flower. A combination of biocide, sugar and hormone remarkably enhance the post harvest life. "Purafil" a commercial product, HQC, HQS and several other can also increase the vase life. Dendrobium hybrids ('Thongchai Gold', 'Bangkok Blue', 'Erika', 'Triple Pink' and 'Madam Pink') produce 10 to 25 flower spike/plant/season.

Packaging and transportation

Flower should not be packed in green house to avoid risk of contamination. Usually the ambient temperature of packing sheds should be 12 to 15°C as such cool stores are not necessary. However, if cool store is available then temperature should not be below 10°C, otherwise flower injury may occur. The flowers should be checked before they sleeved. Cushioning material should be put in the back of the sleeve to avoid any injury during transport. The ideal export box would be a two-piece box. Each stem in the box should be put in the tube containing water, thus stem has water supply during transport. Instead of small water tube, cotton wrapping also served the purpose. In this case, a piece of cotton needs to soak in water. After squeezing the cotton is to be wrapped all around the spike. Then a piece of polythene is tied with rubber band. Before the lid is placed box are spread with an aerosol insecticide. This would ensure final precaution against any possible insects in the box. In order to check movement of spike within the box during transit, the base of the spikes may be tied to the base of cartoon by adhesive tape. The spikes are usually packed in cartons or box with sufficient number of hole for aeration. The size of cartoon depends on the length of flower spike. The standard size of boxes for this purpose are 76cm L x 29cm B x10cm H, 98cm L x 48cm B x 20cm H and 98cm L x 40cm B x 30cm H. Normally, the cut flower boxes are transported in air conditioned vans from the farms to the nearest airport. From these airports, the orchid's cut flowers are sent by air to various destinations in India and abroad.

References

Arditti, J. And Ernst, R. 1993. Micropropagation of orchids. John Wilay and Sons, Inc. New York, pp 696.

Arora, Y. K. 1983. Indian Horticulture. 28: 11-12, 16.

Bose, T.K., Bhattachacharjee, S.K., Das, P. and Basak, U.C. 1999. Orchids of India. Naya Prokash, Calcutta.

Carolus Linnaeus. 1753. Species Plantarum, 1st edition, vol. 2, pages 939-954. Holmiae: Impensis Laurentii Salvii (Lars Salvius). (A facsimile with an introduction by William T. Stearn was published by the Ray Society in 1957).

Christenhusz, M. J. M. and Byng, J. W. 2016. The number of known plants species in the world and its annual increase. Phytotaxa. 261 (3): 201–217.

Garay, L. A. (1960). On the origin of Orchidaceae. I. Botanical Museum Leaflets, 19, 57–96.

Kuehnle, A. R. (2007). Orchids, Dendrobium. In N. O. Anderson (Ed.), Flower breeding and genetics. Amsterdam, the Netherlands: Springer., pp. 539–560.

Nagaraju, V., Upadhyaya, R.C., Barman, D., Nagrare, V.S., Bag, T. K. and Sarkar, J. 2006. Orchids in northern eastern states, A Technical Bulletin. pp. 2-3.

Ram Pal, Meena, N. K., Dayamma, M. and Singh, D. R. 2020. Ethnobotany and recent advances in Indian medicinal orchids. J.M. Mérillon, H. Kodja (eds.), Springer Nature Switzerland AG. Orchids Phytochemistry, Biology and Horticulture, Reference Series in Phytochemistry, https://doi.org/10.1007/978-3-030-11257-8_26-1.

Rampal and Nagrare, V. S. 2006. Orchid diversity of India, its conservation and sustainable

Satish Kumar C. and Manilal, K. S. 1994. A cataloge of Indian Orchids. Bishen Singh and Mahendra Pal Singh, Deharadun, India

Sharma, S.K. 2001. New record of Nervilia aragoana in Rajasthan. Journal of the Bombay Natural History Society, 98 (3): 493.

Sharma, S.K. 2002. Occurrence of Habenaria longicorniculata Grah. Family Orchidaceae in Mount Abu Wildlife Sanctuary, Rajasthan. Journal of the Bombay Natural History Society, 99 (1): 156.

Shetty, B.V. and Singh, V. 1991. Flora of Rajasthan. Vol. II. Botanical Survey of India. Pp. 453-860.

WCSP World Checklist of Selected Plant Families. Retrieved 2 April 2010. http://data.kew.org/cgi-bin/vpfg1992/genlist.pl

Willis, J. C. 1973. A dictionary of the flowering plants and ferns. 8th edn., revised by Airy-Shaw, H. K. Cambridge, UK: Cambridge University Press.

17

Anthurium

Sanchita Ghosh, Ajit Kumar, Tripti Tewari and Syed Khudus

Anthurium is commonly known as anthurium/tail flower/oilcloth flower (*Anthurium andraeanum*) and flamingo flower/flame plant (*Anthurium scherzerianum*). The word anthurium is derived from two Greek words *viz*., *anthos* meaning flower and *aura* meaning tail, referring to the spadix. Anthurium plants were growing undisturbed for hundreds of centuries in the rainforests of South America. Samuel Damon, a businessman and politician in the Kingdom of Hawaii brought one species out of hundreds of species of anthurium to Hawaii in 1889 for cultivation in his garden. The plants were introduced by various people in their gardens from Samuel Damon's garden. It is a tropical ornamental plant valued for its flowers and unusually attractive foliage. In India, it is grown commercially in Assam, Kerala, Tamil Nadu, Karnataka, Sikkim, West Bengal, Meghalaya and Maharashtra. In India, total anthurium flower production is 3230 tonnes with Assam producing highest anthurium production of 2050 tonnes followed by Meghalaya 740 tonnes (Seemanthini and Chandrashekar, 2018). Major markets in the world for per capita consumption of anthurium flowers are Europe (47.80%), United States of America (21.80%) and Japan (16.80%) (Misra and Misra, 2017). In anthurium, various ploidy levels exist like diploid (*Anthurium andraeanum* and *A. hookeri*), triploid (*A. scandens*) and tetraploid (*A. digitatum* and *A. wallisii*). Basic chromosome number varies in different species of *Anthurium*. Most of the *Anthurium* species have 2n = 30 chromosomes while in some polyploids, it is 2n = 60. However, a few species have 2n = 20 to 124 chromosomes (Petersen, 1989). In *Anthurium andraeanum*, somatic chromosome number has been found as 2n = 30 + 2B and this species has high percentage of meiotic abnormalities and karyotype differences suggesting its hybrid origin. Anthurium plants need medium care and reach up to a height of 45-50 cm bearing pale white to yellow coloured flowers with orange to red coloured bracts.

Importance and Uses

Anthuriums belonging to family Araceae are one of the most important cut flowers grown throughout the world for their showy cut flowers and attractive foliage. Anthuriums are tropical, evergreen perennial but herbaceous aroids having creeping, climbing or arborescent stems. *Anthurium andraeanum* is commercially cultivated for cut flower production owing to its unusually attractive heart shaped, waxy red, reddish orange, scarlet or white spathes and yellow and white straight spadix while;

A. scherzerianum is a smaller plant easier to grow and sold as potted plant. Among cut flowers, anthurium ranks eleventh position in the international flower tarde. They are easy enough to grow and very popular because of showy flowers, which consist of colourful leathery, shiny spathes and has huge export potential if they are given the right greenhouse conditions. Anthurium flowers have a long vase life of about 13-15 days, so widely used in flower arrangements, bouquet preparation and indoor decoration. Meanwhile, anthurium as a potted plant has a colossal demand for interioscaping (Seemanthini and Chandrashekar, 2018). They have medicinal values too as steam of foliage of anthurium can be used to alleviate joint pain during rheumatism and arthritis.

Origin

Anthuriums are native to Central America, Columbia, Brazil, Guatemala, Peru and Venezuela and were introduced into Great Britain during early 19th century.

Taxonomy

Genus *Anthurium* comprises of about 825 species of arum family (Araceae) but only 50 species are in cultivation out of which only 10-15 species are in trade. They are grown either for their very showy flowers i.e. spathes and spadices or for their very showy velvety leaves. Hence, they can be categorized under two natural groups *viz.*

i) Flowering such as *Anthurium andraeanum, A. bakeri, A. brownii, A. x ferrierense, A. ornatum, A. regale, A. regnellianum, A. robustum* and *A. scherzerianum*

ii) Foliage such as *A. clarinervium, A. corrugatum, A.crystallinum, A. holtonianum, A. leuconerum, A. magnificum, A. panduratum, A. papilionensis, A. splendidum, A. veitchii* and *A. waroqueacum.* Though all anthuriums flower but those having showy and large flowers (spathe and spadices) are the flowering anthuriums such as *A. andraeanum, A. brownie, A. regnellianum, A.spathiphyllum, A. scherzerianum*. Anthuriums where flowers are insignificant and the foliage is large, handsome and velvety are foliage anthuriums such as *A. crystallinum, A. magnificum, A. regale, A. veitchii, A. warocqueanum.*

Apart from the 11 anthuriums (flowering and foliage anthuriums), *A. aemulum, A. digitatum, A. x ferrierense, A. pentaphyllum, A. radicans, A. scandens, A. subsignatum, A.undatum*, etc. are creeping or climbing anthuriums. *A. acutangulum* is an epiphytic while *A. acutifolium* is terrestrial, *A. amnicola* is epiphytic, *A. andicola* is either epiphytic or epilithic, *A. bakeri* is an epiphyte, *A. bogotense* is an epiphyte and terrestrial. *A. clarinervium* and *A. corrugatum* are terrestrial in nature. Recently, three new species recorded in *Anthurium* are *A. thompsonii* (Arias-Granda, 1996), *A. minarum* (Sakuragui and Mayo, 1999) and *A. faustomirandae* (Perez-Farrera and Croat, 2001). The species cross easily among themselves. The present day anthurium cultivars are mostly hybrids resulted through hybridization involving various species of *Anthurium andraeanum* and *A. scherzerianum*. The dwarf type cultivars are suitable as pot plant, large types suitable for flower arrangements and foliage types as pot plant for indoor decoration.

Varieties

Red colour

Hawaiian: Ozaki, Kozohara, Aumana, Tyam, Ayashii, Asahi and Mickey Mouse.

Holland: Cancan, Avo Rosette, Avo Serge, Avo Nette, Tropical, Scarlette, Avo Red, Avo Claudia, Fla Success, Fla Red, Fla King, Mirjam, Pronto, Ingrid, Inka, Rio, Violetta and Jacqueline.

Others: Mauritius Red, Tanaka, Nova-Aurora, Red Elf, Calypso, Madame Butterfly and Splish-Splash.

Orange colour

Hawaiian: Nitta, Sunburst Orange and Diamond Jubilee.

Holland: Fla Orange and Avo Gino.

Others: Mauritius Orange, Horning Orange and Horning Rubin.

White colour

Hawaiian: Manoa Mist, Uniwai, Chameleon, Trinidad, Hidden Treasure, De Weese, Mauna Kea, Suchiro, Jamaica and Myron Myron Moori.

Holland: Acropolis, Fla Exotic, Uranus, Ave Avo Margarette, Avo Jose, Cuba, Geisha and Lima.

Others: Mauritius White, Lima White and Haga White.

Pink colour

Hawaiian: Abe Pink, Blush, Marian Seefurth and Candy Stripe.

Holland: Avo Anneke, Hoenette, Surprise, Bettine, Lunette and Sarina.

Others: Agnihotri, Lady Jane (Miniature pink), Paradise Pink and Passion (salmon pink).

Obake Types

Fantasia (cream with pink veins), Madonna (cream obake), Medori (green), Fla Rose (peach), Anuenue (green), Rico (rose), Lambada (white-green), Farao (bright orange with green ears) and Caroline Simmon (purple).

Climate

Anthuriums are found in a wide range of climates varying from sea level to 3000 m from sea level. Anthurium being a tropical plant thrives best at a temperature between 18-28°C. Temperature influences the period of growth as well as flower development. The range of light intensity for plant growth is between 2000-6000 lux. Anthurium grows well in warm greenhouse with shading direct sunshine and a humid condition. A shade level of 75% is ideal for its growth. However, degree of shade varies with the cultivar, age and climate. Insufficient shading damages the leaves and eventually causes death of plants. Anthurium plants can be shaded with Saran or UV stabilized agro-shadenets for uniform shading and more flowers per unit area. Anthurium plants require RH of 80%. Commercial cultivation of anthurium can be done either in open field conditions under favourable climatic conditions or in protected structures like polyhouse/greenhouse. However, the quality of flowers is superior in protected structures as they give protection to the crop against adverse weather conditions especially temperature and rainfall and microclimate (light, temperature, CO_2, humidity, etc.) of the crop growing area can be manipulated to a greater extent. The protected structures should be equipped with fan and pad system/air conditioners (for cooling), drip system (for irrigation and fertigation) and foggers/humidifiers (for humidity control). Now-a-days, high-tech polyhouses in export oriented units (EOUs) are equipped with climate control computers with sensors to regulate microclimate i.e. light, temperature, CO_2, ventilation and humidity.

Media

Different media are in vogue among anthurium growers. One of the most important components of anthurium cultivation is the choice of growing substrate or medium. It is known to grow best in a well aerated medium with good water retention capacity and drainage. A good medium needs to be able to anchor the roots and stems so that the plant will not topple over as it grows larger, yet it should provide sufficient moisture, nutrient and aeration to the plant (Singh *et al.*, 2019). In Mauritius, sugarcane bagasse and in the Netherlands, polyphenol foam, mineral wool and peat are commonly used. In general, highly organic, well aerated media with good water retention capacity and drainage are used for growing anthurium. Media should anchor the roots firmly. Presence of salts in the media is harmful to the plants. The media generally used in India contain wood shavings, leaf mould, course sand, small brick pieces, neem cake, cow dung, coir pith compost, charcoal and coconut husk pieces having ideal pH of 6.0. Mineral deficiencies can arise as a result of high or low pH. Addition of organic matter maintains a desirable pH. Proper growth depends upon optimum temperature (25-28°C), relative humidity (80 per cent), light and shade regulation (70 per cent shade though varying with species and cultivars, age and climate), rainfall and water management, wind management and CO_2 concentration (900 ml/m^3) if planted in greenhouse environment. The media for its growing may be anything from coco-husk to oasis via coco-peat, rice husk, tree bark, fern root, sphagnum moss, bagasse, wood shavings, rock wool, lava, charcoal, sand, gravels, etc. with certain amendments. Medium should have qualities like offering sufficient support to the plants, should be highly porous for adequate root aeration, should be able to hold required water, fertilizers and other necessary nutrients and drain out easily. It should have pH range of 5.2-6.2 with EC 0.7-1.3 dS/m and capable to degenerate quite slowly (Misra and Misra, 2017). Rapid degenerating medium is not desired as it will become messy soon causing water-logging thereby resulting in poor aeration in the medium. This condition will not only impede the roots to function properly and hinder their growth and nutrients absorbing capacity from the media but will also make the plants susceptible to various pathogens. *A. andraeanum* thrives best in gravel than in peat or in a mixture of sphagnum moss and coniferous forest soil. However, 1:1 mixture of soil and wood shavings and cow manure + tree fern fibre, sugarcane bagasse, coffee leaf mould, tree bark, powdered bark, black cinder, chicken manure, pumice, peat moss are equally good for production of anthurium cut flowers. Singh *et al.* (2011) found that medium consisting of saw dust + brick pieces + wooden charcoal + soil + sand + FYM (2:1:1:1:1:1) was found best for leaf area (229.11 cm^2), petiole length (21.31 cm), minimum days to flowering (260.47 days), stalk length (36.16 cm), spathe length and width (8.50 and 8.08 cm, respectively) and number of flowers per plant for anthurium cv. Flame under protected conditions.

Propagation

Seed Propagation

Anthurium can be propagated by seeds. To grow anthurium plants from seeds is an extremely simple but lengthy process. The seeds germinate within 10 days and can be transplanted within 4-6 months. It may take 2 years for a plant to bloom. About 20-30% seed derived plants are discarded because of low seedling vigour and poor vegetative growth.

Vegetative Propagation

Anthuriums can be propagated through various means like stem cuttings, rootstock division, suckers and through leaf axillary buds. After heading back, the rootstock is removed from the containers in February-March and the adhering compost is cleaned off. The fibrous roots are then divided in such a way that each piece should contain a growing point i.e. node. They are dipped in captan (0.2%) and planted in fresh compost or organic matter and sand medium (1:3) and kept in humid conditions until they are fully established. However, the most common method of propagation is through stem cuttings. Terminal cuttings from older plants having 1, 2 or 3 nodes or leaves and treated with IBA or seradix 1 (500 ppm), provide good rooting if kept under intermittent mist which helps to accelerate rooting and more survival of the cuttings. It has been seen that cuttings taken with 2 or 3 nodes perform better than 1 node cutting during multiplication. Presence of intact aerial roots leads to 100% survival of cuttings under intermittent mist. Anthurium can also be propagated through axillary buds. A mature leaf along with the axillary bud and the root at its base is separated out, planted in moist sand and kept in a mist chamber till the bud develops into a new plant.

Tissue Culture

Methods have been standardized for quick multiplication of anthurium hybrids through tissue culture. Leaf segments, spadix segments, vegetative buds and stem sections have been successfully used as explants to get callus on MS medium. Formation of callus from leaf segments is highly dependent on genotype, leaf age and size. Just unfolding leaves show highest callus as compared to older leaves. The total duration of tissue culture cycle from leaf explant to complete plantlet is 11 months: 2 months for callus induction, 3 months for callus multiplication, 4 months for sprout induction and leaf development and 2 months for root formation. When apical meristem or the axillary bud meristems are used as explant, the resulting plantlets are uniform and true to type. When any other tissue is used, the plants are produced through callus and some amount of variability is inevitable. Deflasked rooted plants can be planted out in net pots held in egg trays. Washed fine sand or mixture of course sand and leaf mould can be used as medium. There is better establishment when a fertilizer mix consisting of urea, magnesium sulphate and super phosphate is used and sprayed with 1% Indofil M 45. Larger plants perform better than smaller plants. Plantlets grown in the medium soilrite exhibit had maximum growth rate. The plants treated with fungicide solution can also be planted in community pots. 10-12 plants can be planted in an earthen pot having a diameter of 4-5”. Bhavana et al. (2018) developed highly efficient, reproducible and cost effective protocol for multiplication of anthurium lines from *in vitro* shoots, through suspension culture (using liquid MS media supplemented with BAP). The complete plants along with roots were obtained in large quantity within 45 days of suspension culture. Shoot proliferation was obtained with agar-solidified MS medium as well as liquid MS medium, at various concentration of BAP. After 45 days of incubation at 80 rpm, both the media gave best results at 8.8 μM BAP, whereas higher number (29.333±0.342) of shoots were obtained with liquid medium compared to solid medium (7.667±0.342).

Planting

Planting is done either in the beds or in the pots. In the bed, 10 plants/m^2 are optimum. In small scale planting, anthuriums are generally planted in earthen pots. A pot size of 22-30 cm having two holes is usually preferred. Initially the potting mixture is filled to about one-fourth. Subsequently, with the growth of the plant, fresh medium is added, which not only encourages the growth of the plant but also good anchorage. Young plants require repotting every year while adult ones, in every 2-3 years. The new pots should not be larger than the previous ones. Anthuriums can be grown as ground plants too. For a large scale production, land having gentle slope is more ideal. Planting should be done on raised beds with the same potting media which has been used for pots. The width of bed is generally kept at 1.5 m to facilitate various intercultural operations while length can be kept up to 5.0 m depending upon the choice of the grower and length of the polyhouse. Deep planting should be avoided. The planting distance of 45 cm x 45 cm can accommodate 29,640 plants per ha. However, a closer planting distance of 30 cm x 30 cm can accommodate 61,750 plants per ha.

Manures and Fertilizers

It is better to apply the fertilizers in smaller doses and at frequent intervals than giving larger doses at longer intervals. Application of manures in soil is done at every alternate month. A combination of FYM with about 2 g of 17:17:17:2 of NPK and Mg per plant once or twice a month should be given. For plants growing in pots, 5 g complex fertilizer dissolved in 500 ml water per pot should be given in the medium once in two months. Foliar sprays of 0.5-1.0% of 17:17:17 complex fertilizer could also be given to the plants at bi weekly intervals. Plants should be watered shortly after application of the fertilizers. The N, K and Ca are the important nutrients required in anthurium nutrition. A deficiency of Ca can cause fading of the spathe colour. Apply Ca @5 g/plant/month to recover deficiency of Ca. Reduce the dose of N when plants switch over from vegetative to flowering phase. An overdose of fertilizers, applied shortly before the harvest of spikes reduces vase life of flowers.

Irrigation

The plants should be watered at least twice daily during summer. Mist or sprinkler irrigation is the best for anthurium. The last watering in a day should be timed in such a way so as to leave sufficient time for the water to evaporate so that the plants are not damp during the night hours. The quality of water is also important. It should be preferably free from dissolved salts. The optimum pH of irrigation water is between 5.5-6.0.

Special Horticultural Practices

Plant growth regulators like GA_3 (100 ppm) can be used in anthurium to increase growth, sucker production, flower yield, quality and also to reduce juvenile phase. Pruning of leaves is done as the old leaves of anthurium grow at the expense of the flowers. Too many leaves even do harm to the flowers. An anthurium plant has to be thus pruned to retain adequate number of leaves without any adverse effects on the flower production and quality. This also improves the health of the plants in view of better air circulation. Pruned leaves that are healthy can also be sold as cut foliage. On an averages 3-5 leaves are sufficient on a plant. Most anthuriums naturally produce small suckers at the base of the plant to rejuvenate themselves. However, this is highly undesirable for the quality of the flowers as sucker formation also leads to overcrowding which results in crooked stem and decrease in flower size. Moreover, young suckers grow at the expense of main plants and withdraw energy from the parent plant, which does not benefit flower production. It is a good idea to remove suckers at an early stage.

Harvesting and Post-harvest Management

The flowers are harvested after the unfolding of the spathe is complete. Flowers are ready for harvesting after 8-10 months of planting. They are harvested in the morning with 50-60 cm long stalks. Most anthurium blooms are harvested at about three-quarters maturity because at this time they have the longest shelf life as cut flowers. The basal portion of the stalk of the flowers should be kept in water immediately after harvesting to prevent wilting. If the flowers are taken to long distance markets, a piece of cotton soaked in water is kept at the end of the stem to prevent desiccation. The spathe along with the spadix is then inserted into a polythene cover of appropriate size. This prevents bruises on the spathe due to pressing of spadices of adjacent stems. The open ends of the polythene cover are stapled to prevent movement of spathe inside the cover. Salvi (1997) recommended right stage of flower harvesting when 1/3 of flowers open on the spadix. After harvesting, the cut ends of the flowers are placed in lukewarm water (38°C) overnight and then graded as per the colour, length of stems, and size of the spathe and spadices. The flowers are graded as per Mauritius system or as per Dutch system. Mauritius system is based on average length and width of the spathe while Dutch system is based as per diameter of the bract and stalk length as furnished in Table 1.

Table 1: Grading of anthurium flowers*

Mauritius system of grading			Dutch system of grading			
Grade	Average length and width (cm)	Length and width (cm)	Grade	Flower width (cm)	Minimum stalk length (cm)	Number per box
Extra super	>15	35.6	Peewee	<6.25	25	21
Super	15	30.4-35.6	Mini	6.25-7.5	25	18
Large	12.5-15	25.4-30.4	Small	7.5-10.0	30	15
Medium	10-12.5	20.4-25.4	Medium	10.0-12.5	35	11
Small	7.5-10	15.2-20.4	Large	12.5-13.75	40	16
Mini	7.5	14.6-15.2	Extra large	13.75-15.0	45	12
Peewee	<7.5	14.6	Premium	>15	50	10

*Misra and Misra, 2017

The flowers of anthurium can be stored successfully at 13-17°C for up to two weeks which may last for 2-4 more weeks in an arrangement. The storage temperature below 13°C may cause blueing of flowers, especially the red coloured varieties though some varieties can tolerate even as low as 5°C. These can also be stored at 2-10% O_2 and at ambient temperature of 24-25°C where cool chamber is not available. The most commonly used corrugated box sizes for packing anthurium flowers are 21.6 x 50.8 x 91.4 cm^3 or 27.9 x 43.2 x 101.6 cm^3 which can accommodate 120 flowers (Salunkhe *et al.,* 1990). A box with 60 x 30 x 22.5 cm^3 dimension will accommodate 70-100 flowers. Cartoons should be lined with polythene sheets and insulated with moist paper to maintain proper humidity during transport. The flowers having blemishes, black spots, short stalks, deformity and discolouration are discarded while grading. Anthurium cut flowers after grading are packed with cut ends inserted either in water filled rubber balloon or in water filled plastic tubes being securely tied to the stem above the cut ends so that there may not be water leakage. Cotton swabs or pads saturated with water and enclosed in wax paper are placed above cut ends and securely tied around the cut portion to prevent desiccation and then the flowers are placed into a polythene cover of proper size after putting some soft protective material in between the spathe and spadix to prevent bruishing. The open end of the polythene should be stapled so that flower inside it remains in a position. In the Netherlands, the flower stems are placed in flasks containing water and pack them in moist boxes after putting a soft protective material between the spathe and spadix to save the flowers from bruishing during transportation.

Effective post harvest management ensures longer post harvest life of cut flowers. Pulsing in 2.5% 7-Up (a carbonated beverage), 500 ppm benzoic acid, 7.3 ppm of sodium hypochlorite, BA at 50 ppm for 12 h, triadimefon at 25 ppm for 8 h or silver nitrate at 4Mm for 10-60 minutes dip (within 12 h of harvesting) before storage or shipping improves the life of cut flowers significantly but to these treatments varieties differ in their response. Holding solution results in delayed spadix necrosis, spathe bluing and increase vase life up to 27 days as per findings of Kerala Agricultural Vellanikkara, Kerala.

Yield: The anthurium plant produces flowers throughout the year. On an average, 6-8 flowers are obtained per plant per year from the axil of every leaf for initial two years and 10-12 flowers per plant per year from 3rd year onwards. Leaves of anthurium are also sold at a reasonable price in the market. Seemanthini and Chandrashekar (2018) reported production of 25,865 flowers and 11,655 suckers/500 m^2/year in *Anthurium andraeanum* var. Tropical under naturally ventilated polyhouse.

Physiological Disorders

Flower abortion, flower deformation and rosette formation: Flowers on spadix abort, growth of the spathe is stunted and stem of flower is short. This disorder is mainly genetic in nature but can be prevented by restricting root activity or pressure promoting plant activity, reduction in temperature, rapid ventilation, use of gibberellins cooling and humidification systems.

Cracks: This disorder often occurs at the margin of the spathe during active growth period with temporarily high relative humidity. Lower humidity during night is the best possible remedy to reduce cracks.

Folder ears: The basal lobe of the flowers is not fully unfolded during the early stage. This disorder is variety dependent. There is no known remedy.

Sticking: In this case, flower does not open because the spathe is stuck. Sticking is variety dependent and occurs especially during rapid growth period. Lower relative humidity is the best possible remedy to reduce sticking. Loosening of flowers by hand at early stage can minimize this disorder.

Jamming: Jamming is strong variety dependent characteristic that more frequently occurs under arid conditions. The flowers jam in the sheath, since the leaves are wound too tightly around the flower. Such disorder occurs more frequently in cultivars with long sheaths. Keeping the growing media relatively moist is the best remedy.

Diseases and Pests

There are three major diseases in anthurium *viz.*, bacterial blight, root rot and anthracnose.

Bacterial Blight

Bacterial blight is caused by *Xanthomonas axonopodis* (*campestris*) pv. *dieffenbachiae.* Initially it causes small and angular water soaked spots on the leaf margins where after killing the tissue, bright yellow halo around the spot will be formed and spadix starts rotting from tip to downwards. This disease has been found recording heavy loss to anthurium plantings to the growers of Kerala (Dhanya, 2000). Strict sanitization of the growing medium, removal of affected parts of the plants and spraying with streptomycin sulphate or oxytetracycline will keep the bacterial disease under check.

Anthracnose (black nose/spadix rot): It is caused by *Colletotrichum gloeosporioides.* It is highly prevalent in high rainfall areas initially with tiny dark spots on the individual flowers of the spadix which afterwards spread in angular or triangular shapes even on leaves and flower stalks and spikes and it can be effectively controlled through 0.2 percent maneb application.

Root rot: Root rot is caused by complex of causal organisms like *Phytophthora* sp., *Rhizoctonia* sp. and *Fusarium* sp. The symptoms appear as plant stunting, size reduction of leaves and flowers yellowing, hanging of leaves at the edges and the browning of roots and stems. It can be controlled by drenching the plants with 0.1% furalaxyl, mancozeb 0.25% or thiram 0.25% at 10 days interval.

Viral diseases: Mosaic and malformation of leaves and spathes occur mostly in white cultivars. They are transmitted through vectors like *Bemisia tabaci* (silverleaf whitefly) and through mechanical tools also. So controlling the vectors will control these problems and such plants should be uprooted and destroyed.

Insect-pests

Anthuriums are highly sensitive to certain pesticides, therefore, it is not an easy task to control all the pests effectively. Green peach aphid (*Myzus persicae* and *M. circumflexus*), *Aphis gossypii, Aphis nicotiana*, infest anthurium by sucking plant sap from various parts of plants, inject toxic substances to the plants, produce honey dew which attracts ants and certain fungal infections and transmit viruses. There should be proper sanitation and these should be controlled through 0.2 per cent spraying of some systemic insecticides such as dimethoate or malathion. Pyrethrum extract can completely control the pests under greenhouse conditions. Caterpillars of tomato looper and beet armyworm chew various parts of flowers and leaves which may be controlled through catching and killing through lannate 0.1% sprayings. Slugs and snails chew over the root tips, leaves, buds as well as flowers, especially during night and batter the leaves which should be effectively controlled through spray of metaldehyde baits. Nematodes (*Meloidogyne incognita*, Cadet *et al.,* 1993, *Aphelenchoides fragariae, Pratylenchus infestans, Radopholus citrophilus* (Goo and Sipes, 1997), and *R. similis* (Bala and Hosein, 1996) infest on anthurium. *Meloidogyne* produces galls on the roots which show brown lesions. *Aphelenchoides* feed in leaves causing necrotic areas between leaf veins. *Pratylenchus* attacks only the roots by which leaves become yellow and plants may die. These pests can be controlled by planting comparatively more tolerant species or varieties such as *Anthurium pittieri* and *A. ravenii* instead of *A. andraeanum* cv. Midori, by planting healthy stock in disinfected media through steam or methyl bromide, or by treating the cuttings for 10 min. at 50°C temperature, by separating, planting media from natural soil of bed and through precautionary measures.

References

Arias-Granda, I. 1996. A new species of Anthurium Schott (Araceae) from Cuba. Revista Jard. Bot. Nac., No. 17-18, pp. 5-7.

Bala, G. and F. Hosein, 1996. Plant parasitic nematodes associated with anthuriums and other tropical anthuriums and other tropical ornamentals. Nematropica, 26(1): 9-14.

Bhavana G.P., B.S. Kumudini and C. Aswath, 2018. Micropropagation of anthurium through suspension culture using in vitro shoots. J. Appl. Hortic., 20(3): 196-201.

Cadet, P. E. van den Berg, L. Nema and E. van den Berg. 1993. Parasitic nematodes of flower crops in Martinique. PHM, Rev., Hortic. No. 341, pp, 53-58.

Dhanya, M.K. 2000. Etiology and management of bacterial blight of anthurium (*Anthurium and raeanum* Linden.). M.Sc. Thesis submitted to College of Agriculture (KAU), Vellayani, Kerala.

Gao, M.Y.C. and B.S. Sipes, 1997. Host preference of Radopholus citrophilus from Hawaiian anthurium among selected tropical ornamentals. HortSci., 32(7): 1237-1238.

Misra, R.L. and S. Misra, 2017. Anthurium. In: Commercial Ornamental Crops: Cut Flowers. R.L. Misra and Sanyat Misra (Eds.). Kruger Brentt Publishers, Ltd., U.K., pp. 47-61.

Perez-Farrera, M.A. and T.B. Croat, 2001. A new species of Anthurium (Araceae) from Chiapas, Mexico. Novon., 11 (1): 88-91.

Petersen, G. 1989. Cytology and systematics of Araceae. Nord. J. Bot. 9: 119-158.

Sakuragui, C.M. and S.J. Mayo, 1999. A new species of Anthurium (Araceae) from south-eastern Brazil. Feddes Report, 110(7-8): 535-539.

Salunkhe, D.K., N.R. Bhatt and B.B. Desai, 1990. Minor cut flower crops Anthurium. In: Post Harvest Biotechnology of Flowers and Ornamental Plants. Salunkhe, D.K. (Ed.). Naya Prokash, Kolkata. pp. 242-252.

Salvi, B.R. 1997. Optimization of shades, nutrients and growth regulators for cut flower production in anthurium (Ph.D Thesis). KAU, Vellanikkara, Thrissur.

Seemanthini, N.S. and S.Y. Chandrashekar, 2018. A study on yield and economics of growth regulators application in Anthurium andraeanum var. Tropical under naturally ventilated polyhouse. Int. J. Pure App. Biosci. 6(5): 314-318.

Singh, P., A. Arha and A. Bhardwaj 2019. Role of growing medium in yield and quality of anthurium : A review. Agric. Rev., 40: 229-233.

Singh, P., B.K. Dhaduk and S.L. Chawla 2011. Standardization of growing medium for anthurium cv. Flame under protected conditions. Indian J. Hort., 68(1): 86-90.

18

Gladiolus

L.N. Mahawer, Saddam Hussain, A.L. Regar

Gladiolus, commonly known as 'Sword lily' which derives its name from latin word *Gladius* meaning sword shape foliage. The name gladiolus was originally given by Pliny the Elder (A.D.23-79). It is an important florist crop, cut flower in the domestic and international flower trade. It is relatively easy to grow, suitable for cut flower, pots, hanging basket, bedding and exhibition. An engrossing cut spikes bear a large number of florets on rachis, which exhibit various sizes, forms, with smooth, ruffled, deeply crinkled or laciniated tepals. As a genus it was first proposed by Tournefort and which was accepted in botanical litreture by Linnaeus through his description "Species Plantarum" in 1753 and "Genera Planterum" in 1754. The gladioli are said to be cultivated since the days of ancient Greece. The genus Gladiolus, at present comprises with 255 species (Mishra *et al*.,1995; Goldblatt and Manning,1998) with wide distribution occurs in Central Europe, Mediterranean region, Central and South Africa are primary centre of origin. Campbell and Bower (2003) mentions about one more species *Gladiolus scabridus*- a new species 1978 from the mountains of northern Kwa Zulu- Natal and southern Swaziland, restricted to well- drained rocky habitat and where it is endemic. This species has enormous commercial potential have large number of pink flowers. So now number of recorded gladiolus species are more than 260. The gladiolus was brought into cultivation first during Greek period. Gladiolus has been documented in literature since 1578, as evidenced from the record Lyte's Nievve herbal. This was first time introduced in France and soon after spread to England, Germany, Holland and North America. In India, its cultivation dates back to 19th century as mentioned in the 'Firminger's Manual of Gardening in India (1863)' that Mr. Charles Gray of Conoor grew some gladioli from corms and seeds at his garden. The European Union gladiolus importing countries are Germany, Belgium, Luxembourg, Denmark, Spain, France, United Kingdom, The Netherlands, Italy and Portugal.

The improvement of gladiolus began uniquely toward the start of eighteenth century when Willium Herbert (1806) in England delivered interspecific hybrids, however completely end up being sterile and were rarely marketed. Afterward, in 1837, Hermann Josef Bedinghaus created *Gladiolus gandavensis* cross breeds in Belgium which demonstrated accommodating for creation of current gladioli. In 1596 John Gerard, plant specialist to Loard Burleigh had two European species: *Gladiolus communis* L. from Mediterranean area and *G. segetum* Ker from South Europe, it is proved by his list of the plant he developed around 1030 types. 'General Historie of Plantes'(1597), the acclaimed home grown by Gerard has notice of these two species. By 1862, John Parkinson presented *G. byzantinus* Mill in to Europe, from Mediterranean locale and distributed in his 'Paradisi in sole Paradisus Terrestris.' *G. byzantinus* Mill was reared in 1893 with *G.cardinalis* Curt by Dammann and Cowwrorkers, *Gladiolus tristis* (fragrant) was presented from cape locale in Chelsea Physic garden in 1745. *Gladiolus grandis'* Thumb was naturalized in Spain from South Africa. By 1816 French, Dutch and Belgium began raising hybrids, the first being *Gladiolus gandavensis* Van Houtte by intersection *Gladiolus cardinalis* and *Gladiolus psittacinus* in 1838 by M. Beddinghaus. *Gladiolus primulinus* with its hooded quality of its focal petals was gathered from bank of Zambesi River and has been widely utilized plant breeders. In 1956 Mrs. John Wright in New Zealand endeavored the second bigeneric cross utilizing cv. 'Filigree' x *Acidenthera bicolor* and built up the well known fragrant cultivar "Lucky Star". Around 23 types of gladiolus appear to be contributed for the advancement of present day gladiolus (Mishra and Kaicker, 1986), the most recent one being *Gladiolus callianthus murielae* (once in the past known as *Acidanthera bicolor murielae*). Practically gladioli have all hues with exception of genuine green, blue and dark black, however close by hues are accessible. These are not appropriate for rock garden or for naturalizing impact as the corms require to be lifted each year.

Uses

Cut spikes are used in flower arrangements, bouquets, vase and for indoor decoration. The corms are roasted and used for culinary purposes, since they contain large amount of carbohydrates (68-70%). Spikes of gladiolus have good keeping quality and can be transported to long distances by air flight and ship cargo. Gladioli are rich in starch (65.4 to 78.61%) with 12.6 to 18.5% protein. Gladiolus flower and bulbs were part of the regular diet of the ancient Romans. Gladiolus, a bulbous ornamental, is gaining popularity due to unsurpassed beauty, good economic value, it has become important commercial cut flower in India.

In India around 11660 ha of area occupied under gladiolus cultivation with production of 106.0 crore (Agricoop.nic.in accessed on 21.06.2020) Gladiolus is major growing state are West Bengal, MadhyPradesh Maharastra, Chattisgarh, Assam, Uttarakhand, Karnataka, Haryana and HP in India. The gladiolus is commercially cultivated in various pockets like Lucknow, Meerut, Bulandsahar, Ghaziabad, Kullu, Shimla , Solan in H.P and Haldwani, Nainital in Uttaranchal for supply the cut spike to Delhi market.

Origin and Taxonomy

Gladiolus (*Gladiolus X grandiflorus*) is native to Tropical and South Africa, the genus consist from approximately one hundred and fifty known species belongs to family Iridaceae. It had basic chromosome number n=15. Most of the species under the genus are heteroploid, which have very small chromosome ranging from 2n=30 to 120 (Diploid, triploid, tetraploid, pentaploid, hexaploid, octoploid and hyperaneauploid). De Vilmorin and Simonet (1927) reported the chromosome number in *Gladiolus primulinus* (hooded central petal) var. 'La Meurthe' as n=15, which was native from banks of Zambesi river near Victoria Falls in the Mediterranean regions and South Africa. It is a herbaceous plant bearing underground storage stems known as corms from which arise sword shaped foliage bearing terminal inflorescences of flowers known as spikes. These attractive, perennial herbs are semi-hardy in temperate climate and grow from rounded, symmetrical corms that are enveloped in several layers of brownish, fibrous tunics. Their stems are generally unbranched, producing 1 to 9 narrow, sword-shaped, longitudinal grooved leaves, enclosed in a sheath. The lowest leaf is shortened to a cataphyll. The leaf blades can be plane or cruciform in cross section. The inflorescence of gladiolus is a one- sided spike where widely funnel-shaped individual florets are directly attached on rachis in alternate arrangement, progressing from base to tip. Each floret bud is encased separately within its own spathe, which consists of two green bracts. Bloom has a short curving tube and 6 petal-like lobes, 3 upper ones usually larger, 3 lower ones often blotched and these are collectively known as perianth or segments. The perianth is funnel-shaped, with the stamens attached to its base.

The fragrant flower spikes are large and one-sided, with bisexual flowers, each subtended by two leathery, green bracts. The sepals and the petals are almost identical in appearance, and are termed tepals. They are united at their base into a tube-shaped structure. The dorsal tepal is the largest, arching over the three stamens. The inner three petals are narrower. The flower structure and positional barrier between anthers and stigma suggest that gladiolus is essentially an out breeder means cross pollinated (Ohri and Khoshoo, 1983). The style has three filiform, spoon-shaped branches, each expanding towards the apex. The ovary is trilocular with oblong or globose capsules, contains many winged brown colour and longitudinally dehiscent seeds. An ovary is inferior and tri-locular, axile placentation and the fruit will be a capsule. Self incompatibility was found in diploid species *G. tristis* (2n=30) and two other tetraploid cultivars. In this situation no seeds are set when stigma is fertilized with pollen of the same parent or of a different plant of the same clone (Singh and Singh, 1985). These flowers are variously colored, pink to reddish, light purple with white, contrast marking, white to cream and orange to red. The process of flower differentiation is complete within in 7-10 days. Flower bud takes 16 days to reach the full bloom stage (Roychoudhary, 1980). The unfurling of the petal starts early in the morning and complete opening of florets takes 22 to 24 hours. The anthers dehiscence occurs between 8.00 am to 9.30 am followed by anthesis. Mahawer and Misra (1993[a]) reported that floret starts opening from 9.00 am to 2.00 pm of previous day and complete flower opening on next day at 9.00 am to 2.00 pm , along with anther dehiscence starts from 9.05 am to 2.43 pm in gladiolus under Delhi. As per anthocynin pattern , it was divided into 18 groups and from which it was malvidin glycosides together with petunidin 3,5-di-O-glucoside as a minor compound but delphinidin glycosides were not detectable, red colour of flowers was found due to pelargonidin glycosides, pink flowers due to various anthocyanins, pelargonidin, cyanidin, peonidin, petunidin and malvidin glycosides though were completely scares compared with those of purple and red flowers, however, in many of the yellow and white cultivars no anthocyanins were found but a few which had coloured spot or streak on perianth contained extremely small amount of anthocyanins (Takemura *et al.*,2005) Gladioli were originally pollinated by long-tongued anthrophorine bees, sunbirds, noctuid and sphingid moths, long-tongued flies and several others. The new corm continues to develop and old corms below it gradually degenerated. At the base of the new corm arise the stolons, the tip of which swells and develop into the cormels.

There are some 226 recorded species scattered in Republic of South Africa. Main species includes *Gladiolus natalensis, G. cardinalis, G. communis, G. callianthus, G. arneus, G. recurvus, G.tristis.* Its flowers open in acropetal succession, one by one and spike lasts for 1-3 weeks in ordinary vase water, depending upon the season and variety.

On the basis of their geographical origin *Gladiolus* sp. can be divided in to four grouped

1. Eurasian group
2. East African group
3. Natalensis group
4. South African Cape species

Classification

Modern types of gladioli are classified into 6 types on the basis of plant height, flower size and arrangement on the spike.

1. **Grandiflorus or Large flowered hybrids**: This gladiolus group belongs large and exhibition type, vigorous plant structure with large size florets. Florets diameter ranges from 10 –20 cm and spikes length varies from 90 – 150 cm.
2. **Primulinus hybrids:** Plant structure are less vigorous than the large flowered type, plant height ranges from 75.0-105 cm, upper most inner petal form hooded over anther / stigma, floret diameter varies from 5.0-9.0 cm along with spike length ranges from 40-45 cm.
3. **Butterfly hybrids:** Plants height from 75.0-120 cm, floret diameter varies from 7.5-10.0 cm with attractive throat blotches, florets arranged closely and symmetrically on shorter spikes length than 45.0 cm.
4. **Miniature hybrids:** These type hybrid are recent origin with plant height ranges from 75-105 cm, floret diameter ranges from 2.5-5.0 cm, florets arranged on 40.0 cm spike length with ruffled petals. These type hybrids are slow multiplier with small size corms.
5. **Face ups:** The face ups are charming, upward and pleasant. The dwarf nature plant height ranges up to 60-90 cm and floret diameter up to nearly 5.0-6.0 cm.
6. **Colvillei hybrids:** These hybrids are developed by a Chelsean Nurserymen, Colvillei by crossing *Gladiolus tristis* (pale yellow) x *Gladilous cardinalis*(red), early flowering hybrid, highly suitable for green house, , star shaped floret with 5.0-6.0cm in diameter and plant height hardly grows up to 60 cm.
7. **Orchideola hybrids:** It is new group under gladiolus developed by Israel. The spikes in this group are light in weight with smaller florets on shortest stem.

North American Gladiolus Council has recommended modern system of classification and which has been adopted all over the world. This was based on a three digit number. The first digit indicates floret size, second digit -basic colour and third the depth of colour.

On the basis of floret size, gladiolus has been classified into 5 classes,

Class	Designation	Floret Size (cm)
1	Miniature	< 6.4
2	Small	≥6.4 to <8.9
3	Medium	≥8.9 to <11.4
4	Large	≥11.4 to <14.0
5	Giant	>14.0

Species and Cultivars

There are 30000 modern gladioli cultivars developed at world level through the hybridization by utilizing various species like *Gladiolus arneus, G. callianthus murielae, G. cardinalis, G. communis, G. natalensis, G. recurvus, G. tristis.* The cultivars which have more number of florest are in demand and colour liking in domestic /international market are – Red, Pink, Yellow and White cultivars. Spikes with good demand in the market should be robust, tapering and thick with green leaves and sheaths, > 18 florets / spike and disease free.

Colour	Variety
Pink	America, Applause, Dawn Pink, Deciso, Friendship, My love
Orange	Autumn Gold, Coral Seas, Fiesta, Petar Pears,Setting Sun
Red	Black Prince, Hunting Song, Oscar, Red Beauty, Victoria
Yellow	Anglia, Aurora, Folk Song, Green Willow, Golden Peach
White	Amsterdam, Classic, White Prosperity, White Friendship
Purple, Violets	Blue Moon, Her Majesty, High Style, Mayur, Pusa Sarang, Pusa Shingarika

Climate

Gladiolus require full sunlight for better crop, otherwise blasting may occur or plants may remain blind. The long day conditions from 12 to 14 hours photoperiod increase number of florets, spike length and percentage of flowering. Low light intensity results failure in flowering. Whereas high light intensity without proper temperature control also affects growth adversely.

Gladiolus thrives best from temperate zone to subtropical climate zone. It is a summer crop in the temperate regions, while a winter crop in the subtropical conditions, However in tropical climate it can be grown throughout the year. But, at all, there is no comparison of the quality of the spike as well as of the corms and cormels, their health and multiplication of the hill-grown crop under temperate region because there is soil is normally acidic and sandy-loam and the summer temperature is normally 18-30^{0}C which is quite congenial for its growing. Solar radiation is the most important factor for growth and development, flowering and yield of corms, particularly in winter flowering gladiolus under sub-tropical parts.

Thermal fluctuations and illumination are found to be necessary for induction of contractile roots. The light stimulus is perceived by the leaves and not the corms. Under low light intensity in winter, exposure of 1-4^{0}C at low temperature increases the occurrence of flower blasting and frost injury. The light requirement of the gladiolus is usually too high for good vegetative growth of the plants under winter conditions and the plant fail to flower because of flower-bud blasting.

However number of days to flowering is highly dependent on temperature. High temperature persist at flowering time i.e 20-30 ^{0}C accelerate early flowering up to 55 days. Corm storage at 4 to 7^{0}C is good for better growth and flower production. The dormancy in gladiolus corms can be easily broken by low temperature storage for 90 day treatment of corms at 4 -7 ^{0}C lower down abscissic acid, improve gibbrellins and cytokinenin hormonal level helps in sprouting .

Soil

Gladiolus can be grown in a various types of soils but soil should not contains more clay content or waterlogged properties, otherwise which results in corm rot. The soil should have proper drainage facilities. It should also contain sufficient organic matter. The soil should be sandy-loam and slightly acidic in nature. It can be grown effectively at soil pH 5.0 to 7.0, if the pH of the soil is higher than brought it lower sides by addition of organic matter and gypsum is recommended. The land must be brought to fine tilth by deep ploughing and cross harrowing and the beds must be raised to a height of 20cm from ground level to allow proper drainage, working path 60 cm and bed width 1.20 meter for drip irrigation system . Well decomposed FYM @ 5 kg/m^2 should be incorporated in the soil.

Description of Internationally popular cultivars

1. **Friendship**: Cut spikes are ready for harvest at 86-90 days after planting. Spike length is ranges from 90-100 cm,18-20 florets per spike and floret diameter ranges from 9.5-10.5 cm. Florets are carmine rose(Red Group 52C) with yellow throat. Cormel multiplication ranges from 15-20 per corm.
2. **Peter pears**: It is a mid-season cultivar with spike length of 85-90 cm, 16-17 orange red floret (Greyed Orange Group 170 D) and the florets diameter of 10 – 10.5 cm. Produce 40-50 cormels per corm.
3. **Priscilla**: It is a mid-season cultivar, spike length ranges from 60-70 cm, with 14-15 florets/spike. The florets are Lilac purple in colour with 9.5 – 10 cm diameter.
4. **Jacksonville Gold** : Cut spikes ready for harvest at 110 days. Spike length ranges from 80.0-95.0 cm with 16-18 floret on spike. Florets are light yellow in colour belongs to RHS colour chart yellow group:10A with bright yellow throat and 9.5 to 10.0 cm diameter. Cormel production ranges from 15 to 20 per corm.
5. **Rose Supreme**: Cut spikes ready for harvest at 100- 110 days. Spike length is ranges from 90-100 cm with 17-18 florets on spike. Florets are light pink in colour belongs to RHS colour chart red group: 52D with dark pink lines

creamish throat and 10.5 to 11.0 cm floret diameter. Cormel production capacity ranges from 30 to 40 $corm^{-1}$.

6. **Trader Horn**: Late group takes 120-130 days for flowering. Spike length is ranges from 90-100 cm with 18-20 florets on spike. Florets are bright red in colour belongs to RHS colour chart red group: 44B with prominent streaks and 10 to 10.5 cm in diameter. Cormel multiplication capacity ranges from 25 to 30 $corm^{-1}$.
7. **White Prosperity**: Late group takes 120-125 days for flowering. Spike length is ranges from 85-90 cm with 18-20 florets on spike. Florets are white in colour belongs to RHS colour chart white group: 155B and 10 to 10.5 cm in floret diameter. Cormel multiplication capacity ranges from 70 to 80 $corm^{-1}$.
8. **Suchitra**: Early group takes 90-95 days for flowering .Spike length is ranges from 85-90 cm with 16-18 florets per spike. Florets are light pink colour belongs to RHS colour chart red group:49A with dark pink stripes and florets 9 to 10 cm in diameter. Cormel multiplication capacity ranges from 85 to 90 $corm^{-1}$.
9. **Nova Lux:** Spike length is ranges from 90-100 cm with 16-18 florets per spike. Florets are ruffled and yellow colour belongs to RHS colour yellow group:12B with deep yellow throat and 9 to 10 cm in floret diameter. Cormel multiplication capacity ranges from 35 to 40 $corm^{-1}$.
10. **Sancerre:** It is ready for flower at 80-85 days after planting belongs to mid season. Spike length is ranges from 85-90 cm with 16-17 florets per spike. Florets are snow white colour belongs to RHS colour white group:155B and floret diameter ranges from 10.0-11.0 cm . Cormel multiplication capacity ranges from 50 to 60 $corm^{-1}$.

Varieties developed by research institutes in India

1. **IARI, New Delhi :** The cultivar Agnirekha, Mayur, Pusa Suhagin are developed from Sylvia seedling, Pusa Bindiya from Ratna Butterfly Seedling, Pusa Sukanya from Salmon Queen Seedling, Pusa Swapnil from Viola Seedling, Mayur(Sylvia Seedling), Pusa Sarang(White Oak Seedling) by selection methods, Pusa Swarnima by spontaneous mutant from Dhanvantari and various hybrids like Anjali (Sancerre x Rose Spire), Pusa Archana(Creamy Green x American Beauty), Chandni (Green Woodpecker x White Butterfly),Pusa Chirag(Cygnet x Little Fawn), Pusa Dhanvantari(Jr. Prom x Lucky Star),Pusa Gunjan(White Oak xOscar), Pusa Kamini and Pusa Mohini (Ave x Chritian Jane), Pusa Lohit(Cremy Green x American Beauty), Neelam (Sylvia x Patricia), Pusa Rangmahal(Red Bantam x Flaura Belli), Pusa Shweta(Wind Song x Pink Forst Seedling), Suchitra (Sylvia x Jo Wagenaar), Pusa Sunayana(George Mazure x Melody), Pusa Urmil (Tinker Belle x Break O Dawn),Vandana (George Mazure x Eurovision),PusaKiran, Pusa Shubham,Pusa Sinduri, Pusa Valentine, Pusa Vidusi, Pusa Unnanti are developed by hybridization methods.

2. **IIHR,Bangalore:** Hybrids like Aarti (Shriley x Melody), Apsara(Black Jack x Friendship), Darshan(Watermelon Pink x Shriley), Kum Kum(Water Melon Pink x Lady John), Meera(G.P.1x Friendship), Nazarana(Black Jack x Friendship), Poonam(Gelliber Herald x R.N.121), Sagar(Melody x Wild Rose), Sapna(Green Wood Packer x Friendship) are developed by hybridization and selection methods, where as Shakti, Shoba by induced mutation, Sindur,Tilak (Watermelon Pink x Lady John), new promising hybrids Arka Aayush, Arka Amer(Watermelon Pink x Aarti), Arka Gold (Greenbay x Gold Medal-412), Arka Kesar (Winks Glory x Sagar), Arka Pratham, Arka Naveen(Hybrid74-39-1xTropic Seas),
3. **NBRI, Lucknow:** Through hybridization method various hybrids are developed like Archana (Sylvia x Friendship), Arun(Sylvia x Fancy), Dhiraj (Beauty Spot x Psittacinus hybrid), Kohra(Sylvia x King Lear), Hans, Indrani, Manhar, Manmohan, Manohar, Manisha, Mohini, Mridula,Mukta, Pitamber,Sanyukta and Triloki (Friendship x *G.tristis*). Some the cultivars developed by selection methods viz Basant Bahar (Unias challenge seedling), Gazel (White Friendship seedling), Jwala(Psittacinus seedling),Kalima (Sylvia seedling), Priyadarshini and Smita (Lavensque seedling), Sada Bahar (Sylvia seedlings)
4. **PAU, Ludhiana:** Hybrids like Punjab Dawn(Suchitra x Melody), Punjab Morning(Sancerre x White Prosperity), Shan-e-Punjab(Suchitra x Melody), Punjab Glance developed PAU-Ludhiana
5. **Horticulture Experiment and Training Centre, Chaubattia(Almora):** These hybrids viz,Chaubattia Ankur(Oscar x Friendship), Chaubattia Arunima(Oscar x Motherfisher), Chaubattia Shobhit(Meria Goretii xTropic Sea) and Chaubattia Tripiti(Sunny Boy x Oscar) are developed by HETC, Chaubattia.
6. **MPKV Pune:** Gladiolus hybrids like Phule Ganesh, Phule Prena and Phule Neel Rekha by NARP, Ganesh Khind, MPKV Pune

Propagation: This is propagated by various methods like corms, cormels, seeds and tissue culture.

(i) Corms

Gladiolus is a commercially propagated through corms. A single corm produces an average of 1 to 3 flower-grade daughter corms in a season depending upon its size and the variety. Single corm can produced 10-50 cormels during crop growth and development period. This is a slow process and is also disease prone.

Selection of corms: In gladiolus healthy, disease free, large and medium sized corms with at least 2.5 to 3.8 cm diameter are utilized for better cut spike production. The North American Gladiolus Council (NAGC) has categorized the corms into three groups based on their corm diameter:1. Large sized -Jumbo (> 5.1 cm), No.1(> 3.8 to <5.1 cm) for production of cut spike, 2.Medium sized: No.2(>3 .2 to < 3.8 cm), No.3(>2.5 to < 3.2cm) 3. Small sized corms commonly known as cormels group have three sub-groups viz No.4(>1.9 to <2.5 cm), No.5 (>1.3 to <1.9 cm) and No.6(>1.0 to <1.3 cm), which is used as planting stocks for multiplication of flower grade corms in forthcoming seasons.

The cultivars which are slow multipliers can be propagated by division of large size corms in to 7-10 small pieces and smaller corm in 3-4 pieces. Corms must be divided in to smaller parts depends on the number of buds present on corm surface. Every split corm division should contain visible bud, portion of basal plate root zone and treated with systemic fungicides Dithane M45(0.2%), Carbendazim or Saff for 30 minutes and dried under shade for 6 days to healing the wounds and seven days before planting.

(ii) Through Cormels

Cormels are healthier planting material than corms. The multiplication of gladiolus through cormels is an in-expensive and rapid method which enables build up of large stocks with minimum cost. The cormels also have tendency to escape virus diseases even if the parental corm is infected that virus-free corms were slightly late in sprouting but were more vigorous for better developmental and floral growth parameters than the commercial method of propagation by corm. Lopez Oliveras *et al.*(1984) reported that corm of cultivar Petar Pears divided into 4 or 8 splits and planted in 50 per cent peat and 50 per cent perlite substrate results in large number of grade I corms(4.8 cm) while soaking of corms for 24 hours in 900 ppm GA_3 solution results in increased cormel production.

Flat bed planting system can be utilized for sandy soils and raised bed system can be adopted for poor drained heavy and clay type soils. The cormels are planted closely 3-5 cm apart in rows or ridges about 15-20 cm apart and 2.5 – 3.5 cm deep. Before planting, dehusking of cormels followed by soaked in cold water (15-20°C) for at least 24 hours to helps in uniform germination.

Diseased cormels are float on water surface and hence, these discarded. Ramachandrudu & Thangam (2006) reported that Jumbo size corms cut into three pieces and planted along with various grade of whole corms in gladiolus. They concluded that whole corm results in more vegetative growth than cut corms. Whereas, cut corms of Jumbo grade (>5.1 cm) results in improve daughter corm diameter and weight than corms of Jumbo and No.1 grades.

(iii) Seeds

This method is utilized for development of the novel genotype by open pollinated, hybridization and to create variability among population by the breeders. Mahawer and Misra (1993[b]) suggested that pollen used from just dehisced anther is proved better over undehisced anther, however the pollination at bud stage or just anthesis of florets proved better over other stages. Mature capsules are collected, after few days when the capsule is completely dry, the seeds are extracted and stored in dessicator till forthcoming season. Seeds should be sown very closely in pots, troughs containing potting media, (3-5 cm apart) on flat and raised bed, in rows 15 cm apart and covered by leaf mould layer from 2-3 cm thickness in rainy season in plains. The bed should be kept moist by frequent irrigation to facilitate seed germination. After germination single seed produce single leaf above soil level, a small cormel at its base in side soil and thus produced cormels are harvested at maturity. This method normally takes three years to produce flower grade corms from seeds.

Cultivation

Planting time of corms: The planting time of corms varies with the geographical location. In North Indian plains, the best time for planting the corms is from September to November, optimum being first fortnight of October (Bagde *et al.*, 2009) and for hilly areas from February to April when of risk of frost is over (Suneetha,1994). Staggered planting can, however, be done at 10-15 day intervals from mid-August to mid-December to get continued supply of spikes over a longer period and to match the market demand.

The effects of planting date on yield and quality of gladiolus corms and cormels have significant effect delay in planting date reduce the number of daughter corms per mother corm, number of cormels per mother corm (Zubair *et al.*, 2006). In temperate climates, the corms are planted after the winter, i.e. in March and April when frosts are over and climate becomes warm. In Bangalore, the planting season is June and October-November.

Method of Planting of Corms

Only the non-dormant corms should be used for planting. The emergence of root buds at the base of the corms shows that the corms are ready for planting. The corms should be treated with fungicides like tabecunazole, carbenazim or saff (0.2%) for 30 minutes and treated corm should be dry in shade before planting. When planting, the lower portion of the corm should be placed on the soil such that the bud at the top lies straight above. It is done to make sure that the stem grows erect and does not show crooked growth. At planting time, the soil should contain sufficient moisture to facilitate uniform sprouting of corms.

Planting of bigger size corms (4.6 – 5.0 cm dia.) at 20 X 20 cm apart and at 8 cm depth gives better production of spikes as well as corms and cormels. Very close planting must be avoided to reduce the competition. Flat bed planting in sandy soils and raised beds in heavy soils is suitable. Widest spacing (30x30 cm) resulted in the greatest plant height, leaves per plant, spikes per plant, spike length, florets per spike, floret diameter, corms per plant and diameter of daughter corm and lowest days to first spike emergence and days to first floret opening (Anwer and Maurya, 2005). The optimum spacing has been standardized as 25 x 30 cm for gladiolus cultivation in bay Islands at Andmans (Nair and Singh, 2004).

Field preparation

Gladiolus is a shallow rooted crop therefore deep ploughing is not necessary for normal soils. Soil cultivated to 25 cm depth is sufficient the bed size should be 6 x 2 meters. Raised beds with double row planting is followed to protect from excess water.

Planting

Planting depth is depends on soil type, soil temperature, corm size and planting season. Medium size corms are planted at 10-15 cm deep. A planting depth of 10cm produced the highest plant height, number of leaves per plant, length of leaves and corm production in gladiolus cv. Candyman (Rana *et al.*, 2005). The spacing adopted double row system of planting is 40 X 20 cm . The requirement of the corms per hectare is roughly 1-1.5lacs. .

Irrigation

At planting time sufficient moisture in the soil is necessary so that the crop does not require water till sprouting. Gladiolus requires water in plenty but does not grow well under water-logged conditions. Frequency of irrigation depends upon the soil type, weather conditions and rainfall. Normally in sandy soils, the crop should be irrigated at 7-10 day intervals, whereas in heavy soils, at less frequent intervals. Irrigation should be withheld at least 4-6 weeks before lifting of corms. Two litre per hour emitter inline lateral must be utilized for 30 minute duration through drip irrigation system and raised bed width can be kept 1.20 meter.

Intercultural operation

Mulching

Mulching helps to conserve moisture and reduce the weeds which in turn will save the nutrients to help improving plant growth. Organic mulches and plastic sheets can be used as mulches. Organic mulches such as hay, dried grass, saw dust, peat, clippings and coconut fiber are cheaper. Mulching with organic materials further improves the chemical properties of soil (Barman *et al.*, 2005)

Weed control

The field is irrigated before planting and the weeds are allowed to germinate. Hand weeding is done along with 4-5 hoeing before planting. Pre-emergence herbicides reported for gladiolus are pendimethalin(0.75 kg/ ha active ingredient) + hand weeding at 60 days after planting, diuron @ 0.9 kg, linuron @ 3.0 kg active ingredient / ha found effective. Post-emergence herbicides, 2, 4-D @ 1.5-3.0 kg active ingredient / ha has been found to reduce weed population.

Staking

Large-flowered varieties of gladioli grown outdoors are susceptible to lodging, hence need staking. The stems should be tied with strings to thin but strong supports. Plants tied loosely around the stem to allow further growth. Stems should be secured in position with bamboo twig supports. Salix twigs make good supports if these are 75-100 cm in length . Butterfly and miniature types not required staking and which are suitable for garden decoration.

Nutrition management

Fertilizer requirement of gladiolus is vary with climatic conditions, irrigation methods and soil types. In sandy soils it is necessary to provide fertilizers frequently mainly during rainy season. In heavy loam soil light or no fertilizer is required for flower production four applications of fertilizers are recommended:

1. Preplant incorporation
2. Side dressing at 2nd and 3rd leaf stage
3. Side dressing at slipping stage when the inflorescence emerges from leaves
4. Side dressing about two weeks after flowering to develop new corms and cormels.

Nitrogen promotes vegetative growth of gladiolus due to the fact that N application increases more metabolic transport for growth and delays flowering (Chanda *et al.*, 2000). Out of phosphorus and potassium, potassium shows marked effect on flowering. Application of N_{40} P_{20} K_{20}/m^2 significantly increased length of spike, florets per spike, diameter of first floret, life of spike including durability of the whole spike under field conditions in gladiolus (Sharma and Singh, 2007).

Biofertilizers improve vegetative and floral characters of gladiolus (Srivastava and Govil, 2007). Vegetative growth was enhanced most effectively by Azotobacter treatment. However, for quality spike production, PSB was found more effective. It was found that treatment of corms with biofertilizers increased the total rhizospheric bacterial population. This indicates that the improvement in various characters of gladiolus is due to the activity of rhizospheric bacteria, which is enhanced by biofertilizer inoculation suggested by Dongardive *et al.,*(2007).

Application of 500:200:200 kg/ha NPK is optimum for growth and flowering of gladiolus cv. White Prosperity. Commonly a 12:12:18 N:P:K compound fertilizer is applied prior to planting at 1 ton per hectare. Gladiolus can be damaged by fluorine and phosphatic fertilizers containing fluorine should not be used.The crop can be top dressed as required with calcium nitrate at an application rate of 200-300 kg per hectare. In calcareous soils it should be avoided. Regar *et al.* (2015) reported 250 Kg nitrogen dose found better for spike quality parameter gladiolus under Junagadh-Gujarat.

Flowering harvesting of spikes

Gladiolus takes 100-120 days to produce spikes. It depends on the cultivar, temperature and corm size. In India year round production is possible because of climatic diversity. The spikes of gladiolus generally exhibit vase life of about 7-15 days.

The spikes should be harvested at 1-2florets starts opened for local market and 3- 4 floret bud show colour on spike for distant market, kept for pre-cooling in water to remove field heat at morning or evening hours of the day when temperatures are mild. Increase in temperature causes a depletion of stored food. Spike should be harvested just above second nodes. Spikes should preferably be cut with sharp knives or secateurs. Care should be taken not to crush the stem as it releases exudates that promote microbial growth. While harvesting, at least four basal leaves should be retained on the plant to ensure proper development of corms and cormels. The stage at which the spike is to be cut should depend upon the transportation distance, consumer requirement and prevailing temperature conditions.

Yield

The flower spike yield in gladiolus is vary according to the cultivar, corm size, planting density and management practices etc. Gladiolus cut spike yield would be around 1.5 – 1.8 lakhs and corm yield 1.5 – 1.60 lakhs per ha.

Post harvest management

Grading and packing

Grading is done chiefly on the basis of the spike length and number of florets present on spikes. Essential characters considered for grading are spikes should be straight and strong, florets not showing excess opening, florets symmetrically arranged and well spaced on the spike, spikes should free from pest and diseases.

North American Gladiolus Council Specifications:

Grade	Spike length (cm)	Florets/spike
Fancy	>107	16
Special	96 – 107	14
Standard	81 – 96	12
Utility	<81	10

*In India, these grades are mentioned as A, B, C and D.

Packing

The spikes are packed in bunches of 10/12 spikes per bunch. The bunches are tied loosely with a rubber band and tightly packed in card board boxes. PAU, Ludhiana standardized a box size of 127x22x22 cm (empty tube light boxes) which can hold 250-300 spikes for packing for domestic markets. For export, the spikes are to be pre-cooled. Pre-cooling at 4-5 °C for 72 hours results in good keeping quality. These packed boxes should be transported vertically to avoid geotropic bending transportation and handling of cut spikes.

Rehydration and conditioning of spikes

Spikes may losses fresh weight during transit and sometimes show signs of wilting. If so, the spikes should be immediately be put in warm water. This process is referred to as rehydration or conditioning, which helps improve the turgidity and facilitate opening of florets.

During transport of uncooled boxes, vents covering an area of 4% is made in side walls to ensure free circulation of air and avoid heating of spikes due to respiration.

Storage of spikes

Spikes do not store well for longer duration under refrigerated conditions. Under wet storage condition – spikes are put in a bucket of water and placed in cool chambers at 4-5°C by which they store up to 7-10 days. Under dry storage conditions – the spikes should be sealed in polyethene /polypropylene 25 micron thickness and stored in cold rooms at 4-5°C and they store for 6 days. Pre-storage pulsing with 20% sucrose +300 ppm Alumunium sulphate for 24 hours at 20 – 25°C enhanced vase life. Bairwa *et al.* (2014) concluded that 4 % sucrose + 300 ppm $Al_2(SO_4)_3$ is better holding solution with harvest stage 4-5 floret show colour. Mahawar *et al* (2015) suggested 7 days storage duration with 4-5 florets show colour on spike (S2) and packed in polypropylene -100 guage sleeves were found most suitable for post-harvest life under modified atmosphere storage to avoid glut cut spikes of gladiolus.

Geotrophic bending of spikes

Gladiolus spikes exhibit negative geotropism. This means that if the floral spikes are kept flat or horizontal position in storage or transport, the tip points will curve upward. This is an Important cause of quality loss in gladiolus and this is a response of the plant to a change in the distribution of auxin within the stem. This is primarily due to the lateral downward movement of auxin, IAA and its accumulation on the lower portion of the spike. IAA causes asymmetrical elongation of cells in this region, thereby causing the upward bending of the spikes. To prevent bending of tips, the spikes should be held vertically in storage as well as during transportation.

Care of spikes in vase

Spikes are pulsed with 20% sucrose + 8 -hydroxy quinoline citrate @ 300 ppm, alumunium sulphate @ 300 ppm and sodium hypochlorite 50 ppm. Holding solution: Sucrose 4% + 8 - hydroxy quinoline citrate @ 200 ppm, $Al_2(SO_4)_3$ @300 ppm and sodium hypochlorite @ 50 ppm utilized for gladiolus. The basal 2-3 cm portion of the spikes should be re-cut on alternate days to expose xylem vessels to improve absorption of water. The vase water must be changed on alternate days and the faded florets must be removed.

Lifting and storage of corms: Harvesting of corms is done when the leaves turn yellow and dry fully. It generally takes 6-8 weeks after harvesting of spikes for the corms development to become mature and ready for lifting. Plant growth development stops at this phase. Irrigations should normally be withheld at least for 3-4 weeks before lifting of corms. In India, lifting of corms is carried out manually with small garden forks or 'khurpi'. After lifting the corms from the soil, the upper leafy portions should be removed by twisting and breaking the stalk. The old withering mother corms attached to the bottom of the newly-formed corms should also be removed similarly with the thumb.

The cormels should also be separated simultaneously and handled separately. The corms usually get damaged or bruised during harvesting and cleaning operations. Therefore, lifted corms and cormels should be treated with fungicides like azoxystrobin (0.1 %) and bavistin (0.2%) dry under shade condition. The corms or cormels of different cultivars must be handled separately and labeling properly so that they do not get mixed up.

Storage of corms: Storage of corms at low temperature (4-7°C) for three months is an established commercial practice. Corms are stored in perforated wooden boxes, bamboo baskets or nylon mesh bags with free air movement. It serves three main purposes like

i) To break dormancy of corms raised under warmer climates; corm dormancy is mainly due to the accumulation of Abscissic acid, palmatic and lineolic acid. Cormels exhibit a longer dormancy period than corms.

ii) To overcome unfavorable warm and dry conditions during summer season that intervene between lifting of corms and their planting in the subsequent season.

iii) To overcome premature sprouting of corms under unfavorable summer season.

Plant protection

Diseases: Botrytis Soft Rot or Spongy rot (*Botrytis gladiolorum*): The disease occurs in the cold storage when the temperature stays below 13-15°C. The disease is characterized by the formation of white-molded, soft, spongy corms. At early stage, the infection may be confined to surface only, but later on, it travels to the core as well. At high temperature, corms seal off infection, escaping further rotting. In the field, infection is characterized by circular brown leaf spots. The smallest spots are visible only on the upper surface of leaf.

Control: Leaf infection can be controlled by spraying the plants with 0.25% Dithane M-45, twice a week and by dusting healthy corms with Dithane M-45, during storage.

Fusarium Corm Rot (*Fusarium oxysporum* f. sp. *gladioli*): The fungus is carried in corms /cormels from the soil. Latent infection in corms may develop into active rotting of corms / cormels during storage and also during plant growth. Extra losses occur when plant growth is checked by poor soil aeration and drainage, flooding or hot weather. Addition of excess nitrogen in the form of cow dung manure results in additional casualties. The disease is characterized by rot of corms, stunted plant growth, late blooming or blind plants, greener and smaller buds and poor opening. The most serious loss is the reduction in number and quality of flower spikes due to the infection.

Control: Fusarium corm rot is difficult to prevent but can be controlled to some extent by growing disease tolerant cultivars and by various control measures. Hot water treatment of corm /cormels at a temperature of 50°C for 30 minutes before planting, use of clean soil or fumigated soil and treatment of corms /cormels with fungicide prevent / control the disease. The most effective treatment is soaking the corms in Benlate, Bavistin 0.02% , Azoxystrobin (0.1%) solution for half an hour before planting, after treatment corm should be dry under shade and soil application of Furadan/ Thimet 10G @ 3g/square meter at the time of planting is very helpful in preventing the incidence of this dreaded disease. Poonam Shree *et al.*(2018) suggested that carbendazim (0.1%)+ Azadirachtin (0.2%)+*Trichoderma harzianum* @ 20 g / lit/ kg (c.f.u., 2.5 x 10^6) for 30 minutes corm dipping treatment lower down gladiolus wilt.

Penicillium Rot or Storage Rot (*Penicillium gladioli*): This is commonly known as green mold. The disease occurs when the corms are stored in heaps after digging in the cold weather. Large reddish brown lesions on the sides of the corm characterize the disease. Sometimes lesions may be covered with bluish green masses of fungal spores. Hot humid conditions favorable for this disease.

Control: The corms during storage should be examined periodically and disease corms should be sorted out and discarded. Storing of the healthy corms may prevent the occurrence of the disease. Dusting of corms with Dithane M-45 is helpful in preventing this disease during storage.

Curvularia blight or leaf spots: (*Curvularia trifolii f.sp. gladioli*) this is soil- borne disease surviving in the soil for three years or more. The disease is very serious for the plants grown from cormels or seeds. This infects all the plant parts, viz. leaves, stalks, flowers, corms and roots. The disease becomes very serious at 18.3-32.2°C temperature range and during humid weather. First appearance is noticed in the form of leaf yellowing with browning of leaf tips, initially with circular to oval spots which afterwards may become irregular and extend lengthwise along the leaf blades. These pale –yellow spots turn tan or dark brown.

Control: Use of Mancozeb and Blitox (Copper Oxychloride + Zineb) each at 0.2% fortnightly spraying will control this disease.

Insects and Pest

Thrips (*Taeniothrips simplex*): This is one of the most serious pests of gladiolus. Yellow coloured nymphs and black adults damage leaves and spikes. Affected leaves and spikes develop silver streaks, turn brown, get deformed and dry if damage is severe. Attack on young plants reduces flower production. This pest also attacks corms under storage. Infected corms become sticky, shrivel and produce weak plants when planted.

Control: Spraying of methyl demeton 25EC or dimethoate 30EC @ 2ml/litre of water at 10 days interval provides significant control of thrips. Storing of infested corms at 2°C for 6 weeks and later treating them with hot water at 46°C completely kills the thrips on gladiolus corms.

Cut Worms *(Agrotis segetum):* The pest attacks mainly the newly planted gladiolus plants. Female moth lays eggs near ground on plant parts. Hatched larvae feed during nights on emerging shoots. Grown up larvae, which are clay coloured, cut the plants at ground level. Plants are vulnerable to attack up to 3rd leaf stage. They also damage underground corms and developing spikes.

Control: Deep ploughing during summer months exposes the pupae to predators. Poison bait consisting of carbaryl or malathion @ 0.1%, wheat bran and molasses scattered in the field effectively controls the larvae. Sprays of methyl parathion @0.05% or quinalphos @0.05% provides protection to foliage from cutworms

Leaf Eating Caterpillar *(Spodoptera litura):* This caterpillar mainly damages foliage of gladiolus plants. Female moth lays ash coloured eggs in groups on lower side of leaves. The young larvae feed on lower surface of leaves by scraping while greenish-brown mature larvae feed voraciously during nights on these leaves.

Control: Collection and destruction of egg masses and leaves infested with young larvae reduce pest build up. Setting up of light traps attracts adult moths and helps in monitoring pest population. Deep ploughing in summer exposes pupae to predators.

Spraying of quinalphos @0.05% or carbaryl @0.1% or chlorpyriphos @0.05% gives protection to foliage from the leaf-eating caterpillar.

Mites *(Tetranychus equatorius)***:** Attack by the mite begins at an early stage when plants are young. Green coloured nymphs with lateral specks and reddish-brown coloured adults colonize on leaves and suck the sap causing discoloration and wilting. The affected leaves fall off.

Control: Flumite 20 EC @ 1.0 ml, spiromesifen 22.9 SC @ 1.0 ml, propargite 57 EC @ 2.0 ml per liters of water significant control the mites.

Mealy Bug *(Ferrisia virgata)***:** Damage by mealy bug begins in the field on underground corms during dry conditions and carries on to the storage. Nymphs and adults damage corms by sucking the sap causing shriveling and drying of affected corms.

Control: Prompt collection and destruction of infested parts reduces spread of the pest. Crawling of ants on plants is the sign of beginning of mealy bug infestation. Spraying should be taken up at this stage. Sprays of Methyl Parathion 0.04% or Dimethoate 0.04% or Acephate 0.1 % at 15 days interval effectively contains mealy bug infestation

Root-knot Nematode (*Meloidogyne* **sp):** The symptoms of the attack are in the form of stunted growth, yellowing of leaves and heavy galling on roots.

Control: Use nematode-free planting material is the best method to avoid the nematode infestation. Hot water treatment of corms at 57.8°C for 30 minutes helps to control the nematode population. Intercropping or crop rotation with marigold reduces the severity of attack in the field. Application of carbofuran , phorate @ 1g active ingredient per square metre is found effective.

Rodents: Rodents cause a considerable damage to gladiolus corms in the fields as well as during storage, by eating them away. The damage during storage can be prevented by dusting the corms with 5% cythion and storing them in perforated trays. Poison bait is quite helpful in checking rodent menace in the field as well as during storage.

Physiological disorders

Fluoride injury: Leaf scorch of gladioli due to the presence of fluorine compounds in the atmosphere which accumulated on the tips of leaves and heavy application of super phosphate.

Tip burning: (scorching)**:** This disease manifests itself in the form of necrosis of the leaf tips and is due to the damage done to the roots by excessive watering or by fluorides present in the soil at low pH or in air.

Spike topple: Spike topple or bud rot of spikes is caused due to low calcium content in stalk tissues when spikes are kept in vase solution. In this case $CaCO_3$ @ 0.2-0.3% spraying is recommended.

Blindness or floral blasting: Blindness in gladiolus is due to unfavorable climatic conditions prevailing during growing period, mainly due to low light intensity.

References

Anwer, S. and Maurya, K. R. 2005. Effect of spacing and corm size on growth, flowering and corm production in gladiolus. Indian J. Hort., 62 (4):419-421.

Bagde, M.S., Golliwar,V.J., Yadgirwar, B.M. and Wankhede, M.N. 2009. as par sir name first, initial as par indication

Bagde, M.S.,V.J. Golliwar, B.M. Yadgirwar and M.N.Wankhede, 2009. Effect of planting dates on flower quality and yield parameters of gladiolus. J. Soils and Crops, 19(2): 351-354.

Bairwa, S.K., Bairwa, H. L., Mahawer, L.N. and Meena, Ravi K.2014. Effect of stage of harvest and holding solutions on vase life of cut spikes of gladiolus (Gladiolus x hybridus Hort.). Journal of Ornamental Horticulture,17(3&4):97-104.

Barman, D., Rajani, K., Ram Pal and Upadhayaya, R.C. 2005. Effect of mulching on cut flower production and corm multiplication in gladiolus. J. Ornam. Hort.,8 (2) :152-154.

Campbell, T.B. and Bower, J.P. 2003. Gladiolus scabridus- the road map to conservation and commercialization. Acta Horticulture, 624 (4):67-72.

Chanda, S., Barma, G. and Roychoudhary, N. 2000. Influence of different levels of nitrogen, phosphorus and potassium on growth and flowering of gladiolus. Hort., J., 13(1):64-65.

De Vilmorin, R. and Simonet, M. 1927. Nombre des chromosomes dans les genres Lobilis, Linum et chez quelques antres especes vegetales. Compt. Rend. Soc. Biol., 96(1):166-168.

Dongardive, S.B.,Golliwar, V.J. and Bhongle, S.A. 2007. Effect of organic manure and biofertilizers on growth and flowering in gladiolus cv. White Prosperty. Plant Archives, 7(2):657-658.

Goldblatt, P. and Manning, J.C. 1998. Gladiolus in Southern Africa. Fernwood Press, Vlaeberg, South Africa.

Lopez Oliveras, A.M., D. Lopez Perez and M. Pages Pallares, 1984. Studies on corm splitting of gladiolus. Anales del Inst. Nacional de Investing Acions Agrarais, Agricola, 27: 29-45.

Mahawar, T.C., Mahawer, L.N. and Bairwa, H.L. 2015. Response of storage duration, harvest stages and polymeric packaging films on post harvest life of gladiolus cut spikes cv. White Prosperity. Indian J. Hort., 72(1):100-106.

Mahawer, L.N. and Misra, R.L.1993a.Studies on blossom biology of gladiolus under Delhi condition. Journal of Ornamental Horticulture, 1(2):16-20.

Mahawer, L.N. and Misra, R.L.1993b. Studies on gladiolus pollination with regards to stages of anther and stigma. Journal of Ornamental Horticulture, 1(2):51-54.

Mishra,S.D.,Ogale, V.K. and Rode, V.A. 1995. Role of corm size in gladiolus flowering and corm yield. Ind. J. Pl. Physiol., 38: 241-243.

Misra, R.L. and Kaicker, U.S. 1986. Geohistorical development of gladiolus. NAGC Bull., No. 165,pp: 43-45.

Nair, S.A. and Singh, D.R. 2004. Effect of varieties and spacing on growth and flowering of gladiolus in Andamans. Indian J. Hort., 61(3):253-255.

Ohri, D. and Khoshoo, T.N. 1983. Cytogenetics of garden gladiolus. Proc. Indian Natnl. Sci. Acad., B49 No.3 pp:279-294.

Poonam Shree, Mali, B.L., Mahawer,L.N., Bunker, R.N. and Tanwar, N.S. 2018. Integrated disease management of gladiolus wilt caused by Fusarium oxysporum f.sp.gladioli. Journal of Ornaml. Hort.,21(3&4):78-85.

Ramchandrudu, K and Thangam, M. 2006. Comparative performance of gladiolus cultivars under coconut garden and open condition. Journal of Ornamental Horticulture, 9(3): 200-2003.

Rana, P., Kumar, J. and Kumar, M. 2005. Response of GA3, plant spacing and planting depth on growth, flowering and corm production in gladiolus. J. Ornm. Hort., 8(1):41-44.

Regar, A.L., Thumar B.V., Mahawer, L.N., Chawla, S.L. and Meena, N.K.2015.Effect of spacing and nitrogen on floral and vase life parameter of gladiolus (G.grandiflorus L.). The Bio Scan, 10(1):1-4.

Roychoudhary, N. 1980. Floral biology of gladiolus. Lal-bagh Journal of Mysore Horticultural Society, 25: 36-38.

Sharma, G. and Singh, P. 2007. Response of N, P and K on vegetative growth, flowering and corm production in gladiolus under mango orchard. J. Ornm. Hort.,10(1): 52-54.

Singh, B. and Singh, M.1985. Studies on floral biology of gladiolus. Prog. Hort., 17: 135-136.

Srivastva, R. and Govil, M. 2007. Influence of biofertilizers on growth and flowering in gladiolus cv. American Beauty. Acta Hort.,742: 183-188.

Suneetha, S. 1994. Effect of planting dates and floral preservatives on spike qualities of gladiolus (Gladiolus grandiflorus). M.Sc. Thesis. K.A.U., Vellayani, Kerala.

Takemura, T., Takatsu, Y., Kasumi, M., Marabashi, W. and Lwashina, T. 2005. Flavonoids and their distribution patterns in the flowers of gladiolus cultivars. Acta Hort., 673 :487-493.

Zubair, M., F.K. Wazir, Gohar Ayub and Sohail Akhter, 2006. Planting dates effect on production and quality of gladiolus corms and cormels. Sarhad J. Agric., 22(2):249-258.

19

Aster

B. D. Bhuj and Yograj Kushwaha

Botanically aster belongs to the family Asteraceae and scientifically known as *Callistephus chinesnsis* which is basically native to China aster and annual aster. It is native to China and Japan and it is cultivated worldwide as cut flower, loose flower, bedding plant, for flower decoration and for making floral ornaments. It can easily cultivate in the open fields and green houses for the production of cut-flowers. The genus Callistephus derives from two Greek words that is 'Kalistos' and 'Stephos' meaning 'most beautiful' and 'a crown', respectively. The name Aster comes from the Ancient Greek word αστήρ (aster),, meaning "star", referring to the shape of the flower head (Janakiram, 2006).Among various flowering annuals, China aster is having an important position in Indian floriculture industry. It holds the position next to chrysanthemum and marigold among the traditional flowers. In fact, it has widely replaced an already grown annual Chrysanthemum up to greater extent in Indian floriculture industry.

This flower is most preferred for wide cultivation by virtue of its wide variation in attractive color and longer vase life. It is also very much used for beautification in landscape gardening which create mass effect. In India it is cultivated in about 1,020 ha with a production of flower 800 million tonnes. It is commercially cultivated in Karnataka, Tamil Nadu, West Bengal and Maharashtra. In Karnataka it account for 6 % of the total area and 8.8 % of the total flower production.

Importance and Uses

Among the entire garden annuals asteris one of the most popular annuals grown throughout the world. Its cut flower last long and are used in vases and decoration. Aster flowers are also an eye-catching addition to beds and borders. Mix colors, or plant drifts of a single color to suit to any personal style. Some dwarf varieties are used as pot plants but remember that containers dry out quicker than the ground, and be vigilant about watering. In the garden aster flowers have excellent nectar source for attracting pollinators and beneficial insects. For edging and window boxes dwarf varieties are very much suitable. Several countries are also producing seeds in large amount.

Taxonomy

Aster [*Callistephus chinensis* (L.)Nees.] is an annual type plant having unbranched straight stem reaching upto a height of 100 centimeters. The chromosome number of aster is 2n=18. The ovate or triangular ovate leaves are alternately arranged. The plant shed their lower leaves before flowering.The newer leaves appeared on the top of the plant have smooth edges type leaf blade while the older one has serrated type of edges. The large solitary flower head grows on the main stem and its branches. *Callistephus chinensis* provides showy, 3-5-inch diameter blooms. Species plants are infrequently grown because of the huge number of superior cultivars that have arisen over the years. Cultivar appearances vary from single daisy-like flower heads with yellow centers to fully double flower heads with shaggy to pompom form. Cultivar colors include shades of white to pink, red, blue, violet, purple and yellow.It is a self-pollinated crop; approximately have 10% natural crossing. Flower head consists of both pistillate ray florets and perfect disc florets. Normally, the stamens and pistils do not mature simultaneously in the individual flowers (Janakiram, 2006).

Climate

Aster is extensively grown in tropical and subtropical countries during the winter season and spring. Locations having full sun to partial shade it grows vigorously. Light and temperature play a major role in the growth and flowering of aster. Day length has a remarkable effect on vegetative growth and flowering of aster. Less than 14 hours day length was found to cause resetting of leaves. For the production of year round the aster seedlings are lighted during short days from the time of emergence to flowering.

Temperature is also another factor influencing the growth and flowering of aster. 20-30°C day temperature, 15-17°C night temperature and 50-60% relative humidity are preferred for its commercial cultivation. At higher temperatures the aster plants will grow reasonably well but will have weaker stem and lesser number of petals. It is also recommended that for the year round production of flowers the night temperature should be maintained around 10°C.

Soil

Aster can be grown in a well drained soil having ample amount of organic matter. Before adding any amendments, it is advisable that soil testing should be done.Aster prefers a pH range of 6.0 to 7.5. The field should be well prepared by deep ploughing with the help of plough and cultivator.Mix the well-rotted compost, if necessary, and finally field should be saturated with water.

Species and Cultivars

The species *C. chinensis* has a number of cultivars which are available in different form of colour such as blue, purple, yellow, pink and red with a bright yellow centre disc. The diameter of flower ranged from three to five inches. The whorls of petals are found in single, semi-double, double types and they appear on a separate stem. The species are wilt-resistant, with sturdy stems and alternating green leaves.

Classification of Aster

Asters are classified on the basis of their growth habits, flower shape, size and appearance of florets. Double-flowered asters are categorized into 3 group tall, medium-tall and dwarf (Bhattacharjee and Kumar, 2002).

A. Tall

The height ranges from 60 to 80 cm in this group. The types and colours in the following sub-categories are mentioned below:

American branching: Stems are long and flowering coming up with wide spectrum of colour in different cultivars- white, light blue, dark blue, pink, rosy, salmon, scarlet and violet.

Bouquet powder-puffs: It has medium sized flowers on the stem and colour range in different cultivars as described earlier.

Chikuma strain: It bears different shades of flower colours such as white, yellow, purple, red, rose, red with golden yellow center and salmon-rose. The flowers are fully double, ball-shaped, and ideal as cut- flower.

Giant of California: This is a late flowering type and having large sized bloom with wide range of colours.

Paeony-flowered: It has straight growth and produces chrysanthemum-like flowers of various attractive colours. It is also suitable for cut-flower.

Princess and giant princess: It is famous for cut flowers due to its large sized flowers produced on many branches. It bears blue, pink, deep yellow, cream-white, citron-yellow, rose, carmine, deep red, white, etc. of extra large double flowers on long stems.

B. Medium-tall

The height ranges from 40 to 60 cm in this group. The types and colours in the following sub-categories are mentioned below:

Early Burpeenna: It produces early mum-like flowers about 7 cm diameters with semi-incurved petals in white, rose, scarlet and blue.

Giant comet: These are much-branched, hardy plants, and flowers having curled petals.

Giant Crego: The plants are bushy and vigorous having curled petals of white, shell-pink, light and deep blue, scarlet and deep rose.

Lilliput: The plants are erect and produces fully doubled small blooms.

Ostrich feather: It produces medium-sized flowers having curved petals with wide range of colours.

Pompom: It has globular-shaped blooms with quilled petals, in wide range of colours – yellow, light blue, dark blue, lilac, white, scarlet, rose, and also with white centre.

Rubens: Early-blooming type with pompom-type flowers, 7 cm across. It is also suitable for cut-blooms.

Unicum: It produces quilled petals type large blooms.

C. Dwarf Double

The height ranges from 20 to 40 cm in this group. It also has different type with varying growth and flowering attributes:

Colour Carpet: It is characterized by dwarf and uniform globular plants producing chrysanthemum like blooms.

Dwarf Chrysanthemum: It has compact bushy growth and having wide range of flower colours.

Pinocchio: These are excellent for bedding, edging and window boxes due to large number of star-shaped flowers appear on dwarf and compact plants.

It is worth to mention here that the following important varieties are being commercially cultivated in India under different climatic conditions:

Arka Kamini: Flowers are semi-double type, deep pink in colour; suitable for garland, floral decoration and cut flower.

Arka Poornima: Flowers are powder puff type, pure white in colour, disc tubular; suitable for garland, floral decoration and cut flower.

Arka Shashank: Flowers are powde rpuff with tubular ray florets, creamy white in colour; suitable for garland, floral decoration and cut flower.

Violet Cushion: Its flowers are violet and are superior to the local violet variety. It produces pompon type of flowers having 4-5 rows of ray florets with tubular disc florets. It blooms are 4.5 cm in diameter and each flower weigh 2 g. It has a stalk length of 20 cm and a vase life of 8 days.It is a higher yielding variety and produces 70 flowers/plant.

Phule Ganesh White: Crop duration 4-5 months, Productivity 16 to 18 lakh, flowers. Characters / features Colour of flower White,: Early, Semi spreading.

Phule Ganesh Pink: Crop duration 4-5 months Productivity 16 lakh flowers. Characters / features Colour of flower Pink,: Early, Semi spreading.

Phule Ganesh Violet: Crop duration 4-5 months Productivity 20 to 22 lakh flowers. Characters / features Colour of flower Violet,: Early, Semi spreading.

Phule Ganesh Purple: Crop duration 4-5 months Productivity 16 to 18 lakh flowers. Characters / features Colour of flower Purple,: Early, Semi spreading.

Varieties Developed Outside India

Pot'n Patio Blue, Pot'n Patio Pink, Bouquet Powder Puff, Bouguet White, Asure Blue, Giants Of California, Ostrich Plume, Crego, Crego Azure, Comet, Totem Ple, Giant Massagno, Benihanabi, Cactus Flower, Blue Wonder, White Kurenai, AriakePumpson, Invineible, Early Bird, Giant Curved, Super Giant, Super Princess, Early Charm, Salmon rose, Ostrich Plume Blue,

Wilt Resistant Varieties (Arora, 1990).

Early Wonder, Giant Comet, Queen of Market, Star Dust, Unicene Record.

Planting Season

Generally seed sowing should be done during August to October and sometimes earlier in June or July in areas where rainfall is low. For obtaining better quality flower earlier is useful in Northern Indian plains.But in mild climatic condition like inBengaluru,it can be cultivated throughout the year (Rao et al.,1997).

Propagation

It is commercially propagated by seeds. Seeds of asters have no dormancy and freshly harvested seeds must be sown. Approximately 2.5 - 3.0 kg seeds are required for one hectare. In raised beds(10cm) of 120 x 60 cm size the seeds should be shown either by broad casting or line sowing. It would be wise to sow the seeds at frequent interval of 10 – 15 days, so that continuous flowers can be obtained and also avoid glut in the market.The soil moisture should be kept ideal and a temperature of 21-22°C should be maintained. Asters seed will germinate in a week. Aster seedlings are usually transplanted when they have developed about three to four leaves. 30-45 days old seedlings should be transplanted in the actual plot. Planting should be done preferably in the evening to avoid direct sunlight and the beds should be well-prepared by adding sufficient compost. After planting, the soil around the seedlings is to be firm and irrigated judiciously.

Alternatively *in-situ*, seeds can be sown in the field. Sow the seeds on top of the soil about three inches apart. Gently press each seed into the soil to a depth of approximately 1/3 cm. It is not necessary to cover them with soil. Keep the soil evenly moist, never letting it completely dry out, but don't allow it to become waterlogged. In 2 to 3 weeks seedlings will sprout. Continue to provide even moisture. Use floating row covers if desired to protect against pest and disease issues. When the seedlings have 2-3 sets of true leaves, start thinning them. Remove enough so that the remaining seedlings are spaced 25-30 cm apart, depending on the variety (Randhawa and Mukhopadhyay, 1986).

A field experiment conducted at Kalyani and Pune indicated that the seedlings of china aster spaced at 30x30 cm apart gave the best flower yield. A spacing of 30x20 cm and planting in the first week of October for Kalyani, and November planting for Pune regions have been recommended to obtained more flowers and seed yield in cultivar "Ostrich Plume Blue". Under Ludhiana condition 20x20 cm was found to be best for obtaining highest flower yield (Bhattacharjee and Kumar, 2002).

Manures and Fertilizer: For aster cultivation manures and fertilizers should be applied in a judicious manner. Deficiencies of nitrogen and phosphorus result in poor growth and flowering. Stunted growth and less numbers of flowers counted due to nitrogen deficiency. Retarded vegetative growth and late flowering were caused by phosphorus deficiency. At the time of field preparation of 10-15 tonnes FYM/ha along with NPK 180:60:60 kg/ha as basal should be applied. Another dose of nitrogen @

90 kg/ha should be top dressed at 40 days after transplanting for better growth.With the application of Zn, Cu, B and Mn in spray form may leads to increment in quality flower production.

A field trial conducted at Pune and Kalyani recommended that 400:100:200 kg/ha and 300:200:200 kg/ha NPK respectively should be beneficial for aster crop. In Punjab a fertilizer dose of 30 g each of NPK per m2 has been recommended for higher and quality production of aster flowers (Bhattacharjee and Kumar, 2002). An another study conducted at IARI reported that application of 75 kg nitrogen 60 kg phosphorus and 100 kg potash per hectare combined with vesicular arbuscularmichorrhizae (VAM) and phosphobacterium were proved to be most effective in increasing the quality and yield of flowers(Kumar et al., 2003).

Irrigation

During vegetative growth and flowering sufficient amount of moisture should be maintained. Moisture stress during growth and development may affect the normal growth and flower production. Frequency and quality of water mainly depend upon climate and soil conditions. Frequency of irrigation would be more in lighter soil and less in heavy soil. According to soil moisture condition the irrigation should be done once in a week. The optimum soil moisture can also be maintained by use of mulch such as woodchip, shredded bark, or leaf mould.

Pinching: The terminal portion of shoot is removed at 30 days after transplanting to induce lateral shoots for the production of more number of quality flowers.

Intercultural operations: Weeds are problematic in aster cultivation. After transplanting of aster cropweed competes with aster at its full efficiency and suppress the growth of plant. There is a need of care and attention to remove the weed from the field otherwise weeds grow in field and reduce the growth and quality of aster flower. Weeding is generally done manually. Three to four weeding are required for the entire growth period. Application of oxodiazon at 1.25 kg/ha sprayed at 3 days after planting shows best result in controlling the weeds. Control of grasses and broad leaf weedswith napropomide and oryzalin has also been reported in this crop. For continuous blooming it is advisable to de-head the wilted and dried flowers regularly. Staking of plants should also be followed in tall varieties to keep them upright. In this regard use of bamboo sticks and twine would be helpful (Bhattacharjee and Kumar, 2002).

Role of growth regulators: Growth regulators have significant role in development of vegetative and floral characters and their responses vary with concentration, method and frequency of applications. A field experiment was conducted at UHS, Bagalkot, Bangaluru on different growth regulators. The study indicated that GA3 (200 ppm) enhanced the duration of flowering and yield of flowers. It was also reported that the highest flower yield and flower quality were influenced by the foliar application of cycocel (1500 ppm).

In china aster different growth regulators viz., GA3, NAA, CCC, MH and Paclobutrazol are used either for inducing or retarding growth parameters which ultimately results in better quality, increased flower yield and also enhancing the vase life of cut flowers.

Spraying of GA3 at 200 ppm enhanced the duration of flowering (90.33 days), number of flowers per plant (68.54), diameter (4.86 cm), flower weight (3.26 g), flower yield per plant (111.2 g) and vase life (22.88 days). Among growth regulators foliar application of CCC at 1500 ppm recorded the highest flower yield and flower quality parameters. Soil drench of paclobutrazol at 200 ppm delayed flowering while early flowering was observed with 25 ppm. Application of MH at 1000 ppm resulted in minimum plant height, inter nodal length, number of leaves per plant and leaf area in aster (Munikrishnappa and Chandrashekar, 2014).

Harvesting and Postharvest Management

For Cut Flower: Aster plants produce blooms after 70-84 days after transplanting and it varies according to varieties, season and cultural practices. Flowers are harvested along with stalks and sometime whole plant. Harvesting is done by hand when the flowers developed their original colour. For distant markets they should be plucked earlier as compared to local markets. For harvesting of loose flowers individual flowers are harvested with short stems attached.

Flowers with stalks are immediately placed vertically in a container having clean water. To minimize transpiration activity lower leaves should be removed. According to stalk length and flower size,grading should be done. To improve attractiveness and freshness of flowers, a clear plastic film should be used for its packing.

Yield

It depends on the cultivars, planting seasons and cultivation practices. On an average a fresh flower yield of 20-22 tons/ha, 15.0 to 17.5 tons/ha and 10.0-12.0 tons/ha can be obtained in rainy, winter and summer seasons, respectively.

Vase-life

Aster flower shows comparatively longer period of vase life than the other flowering plants. The life of the cut flowers determines by the microbial population in vase. To reduce the bacterial population in vase, a slow release chlorine compounds such as sodium dichloroisocyanurate and 1, 3, dichloro 5, 5-dimethyldydantoin can be adopted. A study conducted at IARI New Delhi observed that for extending vase life of aster flower (15.88 days) holding solutions containing 8-HQC 200 ppm and sucrose 1% can be used(Singh et al., 2003).

In general, the following steps should be followed for increasing the vase-life of flowers:

1. Cut the stems in slanting manner by use of clean and sharp secateurs.
2. Remove the lower leave upto 6-12 inches depending upon length of stalk.
3. Place the stalks in a vase of water and ensured that there is no leaves below the water level.
4. Cut the stems in slanting manner and change the water daily.

With this regimen, the vase life of flowers can be extended upto two weeks.

Insect-pests and Diseases

Asters shows susceptibility to a number of diseases and pests. As for example- fusarium wilt, aster yellows, rust, botrytis blight, etc. may cause problems in aster cultivation. Leaf hopper, aphids, spider mite, thrips, white fly, mealy bug and caterpillar are important among the pests.

Diseases

Wilt: It is a very common disease and severe also generally spread via fungus called *Fusarium oxysporumf. callistephi*, *Acrostalagmusvil morini*, Verticilliumalbo-atrum with *Fusarium oxysporumf. callistephi* causes this disease. Sterilization of soil and drenched with copper-based fungicide (Carbendazim 1g/l) are quite effective for checking the malady.

Collar or Root Rot: It is caused by Phytophthoracryptogea and showing similar symptoms to those caused by wilt fungus plants appear water-soaked and black lesions. This disease can be controlled by restricted irrigation and avoiding planting near the alternative hosts. Avoid continuation of cropping in the recurrent field. For minimizing the disease infectionthe soil should be drenched with Copper oxychloride @ 2.5g/l.

Rust: This is one of the severe diseases which reduces the respirational and transpirational activity of the plant by covering stomata of the leaves with bright yellowish orange spores. Causing agent is scientifically known as *Coleosporium solidaginis*. Spots appear on the lower side of the leaves especially on the young plants. During growing season or vegetative growth use of Calixin or Benlate (0.1%) is quite effective to control the disease.

Viral Diseases: This is also a serious disease of china aster caused by Chlorogeniccallistephus. It can be identified by its pale-yellow tinge on the bud of leaves and a large number of adventitious shoots. Ultimately plants become dwarf. Flowers are off-colour and ray-florets become yellowish green colour. This malady can be managed by removing and destruction of affected plant and monitoring the vector, leaf hopper. The leaf hopper can be controlled by suitable insecticide.

Pests

Leaf Hopper (*Macrosteles fascifrons*): A very serious feeding pest on aster plants already mentioned, has the capability of transmitting virus. Use of dimecron (1 ml/l) or rogor (2 ml/l) or any other suitable insecticide will be effective to control the pest.

Black Blister Beetle (*Epicauta pennsylvanica*): It is also known as aster beetle, attack on flowers and foliage, sometimes if not monitor it can completely destroy the plants. This pest can be controlled by periodic spray of malathion (2 ml/l) at an interval of 5-7days during peak infestation.

Semilooper: Skeletinization of leaves and the plant becomes whitish. The larvae feed on leaf buds and flowers. It can be control by spraying Carbaryl@0.5ml/l. or Quinalphos @ 1.0 ml/l

Leaf Miner: Adult female of leaf miners puncture the leaves and sometimes petals to feed on exuding sap. Punctures turn white, giving foliage a stippled or speckled appearance. As the larva grows the mine becomes longer and wider. Mining on leaves usually has less impact on plant growth and rarely kills plants. Unusually heavy damage stunts the plant growth and may cause infested leaves to drop. Spraying of Monochrotophos @ 0.5 ml/l or Imidacloprid @ 0.5ml/l will control this pest.

Potato Aphid (*Macrosiphum solanifolii*): It is a sap feeding insect and attacks on the leaves, stems and flowers. Spraying of Imidachloprid can control the aphid population.

Spider Mite (*Tetranychus telarius*): These are minute insets frequently disturbing asters. Discoloured and distorted foliage indicate the infestation. Use of kelthane @ 2 ml/1 will be effective.

Nematodes: Nematodes are the serious problem in asters. Asters are highly susceptible to root-knot nematode, Meloidogyne incognita. The stunted growth of infested plants and shortened leaves are observed. Another nematode Aphelenchoidesritzema-bosi also disturb the plant growth. At 20°C the nematodes increased their populations in seedling stage very rapidly and causessevere leaf malformation. Soil solarization, deep ploughing or soil pelting and use of furadan or thimet can control the infestation of nematodes.

References

Arora, J.S. 1990. Introductory Ornamental Horticulture. Kalyani Publication. New Delhi.

Bhattacharjee, S. K. and Kumar, P. N. 2002. China Aster: Agro-Techniques for Flower Crop Production . AICRP technical bulletin no.16. New Delhi, India: pp49

Janakiram, T. 2006. China aster. In: Advances in Ornamental Horticulture (vol.- 1), pp 247-266. Bhattacharjee, S. K. (Ed.). Pointer Publishers, Jaipur, India.

Kumar, P.;Raghava, S.P.S. and Misra, R.L. 2003. Effect of biofertilizers and yield of china aster. *Journal of Ornamental Horticulture.* 6(2): 85-88

Munikrishnappa, P.M. and Chandrashekar, S.Y. 2014. Effect of growth regulators on growth and flowering of china aster {*Callistephus chinensis* (L.) Nees.}. Agriculture Reviews, 35(1):57-63.

Randhawa, G. S. and Mukhopadhyay, A.1986. Floriculture in India, pp 358-359. Allied Publishers Pvt. Ltd. New Delhi.

Rao, T.M.;Negi, S. S. andRaghava, S.P.S. 1997. Kamini: A new China aster cultivar from IIHR, Bangalore, Floriculture Today.5 (8): 40

Singh, K.P.; Kumar, V. and Suchitra. 2003. Vase life and quality of chinaaster (Callistephuschinensis) cut flowers as influence by holding solution. Journal of Ornamental Horticulture 6(4): 362-366

20

Tuberose

P. K. Yadav and Priyanka Kumawat

Tuberose is an important half-hardy, perennial bulbous flowering plant. It belongs to the Amaryllidaceae family and having basic chromosome number (n=30). It is one of the important cut flower among the top ten cut flowers due to its beauty, elegance and sweet pleasant fragrance (Sadhu and Bose, 1973). It is cultivated for its long lasting spike as cut flower as well as loose flower purpose.Tuberose also known as Rajanigandha in Bengali, Gulchari and Galshabbo in Hindi, Sukandaraji and Nelasanpengi in Telegu, Nilasompangi in Tamil and Sugandharaja in Kannada.It is suitable under protected cultivation as well as open field condition (Brundell and Steenstra 1985). It has a huge economic potential and more demand for cut-flower purpose (Usman, 2013) and essential oil industry (Singh, 1995). Due to its immense export potential, cultivation of tuberose is gaining momentum day by day in our country.

Origin and distribution

Tuberose is cultivated all over the world and it is originated from Mexico. Tuberose is commercially producing in various countries such as India, China, Bangladesh, Mexico, Kenya, Italy, France, Morocco, USA, Hawaii and South Africa. In India, it is commercial growing in West Bengal, Karnataka, Tamil Nadu, Maharashtra, Uttar Pradesh and North Eastern part of India. At present scenery tuberose is cultivation in India 7. 95 lakh hectare and cut flowers is estimated to be 27.71 '000 MT and loose flowers1560.70 lakh (Anonymous, 2013).

Importance and uses

It can be produce in pot as well as in beds. Loose flowers are used mainly in garland preparation. It is commercially used for aesthetic purpose, birthday ceremony, floral arrangement such as; bouquets, rangoli, boutonnieres, Potpouri. It has some chemical compound which is responsible for fragrance such as Geraniol, Nerol, Benzylalcohol, Eugenol, and Methyl anthranilate. Due to its fragrance, longer shelf life and long spike length it is used for table purpose.Its sweet fragrance able to open the heart refers your mind and clam effect the nerves. It is extremely used in perfume and essential oil industry (Qureshi *et al.*, 2018). In recent time tuberose also use for beverage industry and food industry. Its powdered bulbs used as a medicine for gonorrhoea. Powdered bulbs are diuretic and emetic. Bulb rubbed with turmeric and butter are apply on red pimples.

Climate

Tuberose is widely cultivated in tropical, subtropical and temperate climate conditions. It requires warm humid, open sunny condition. Average temperature for production of flowers ranges from 20°C to 35°C.The crop is reported to flower profusely throughout the year, if the climate is mild and free from extremes of high and low temperature. Flower length and quality is reduces above 40°C temperature. Low temperature less than 10°Cadversely affect the quality of flowers and frost damage the plants and flowers. It is not strictly photosensitive but long-day exposure promotes vegetative growth,early spike emergence and also increases flower spike length. Humidity and temperature is the main factors responsible for decrease or increase in yield.

Soil

Tuberose can be growing any type of soil but ideal for sandy-loam or loam soil. Soil should be well aeration and good drainage with pH (6.5-7.5) for quality flowers production (Sharga and Sharma, 1994). It is also grow in saline and alkaline affected soil by using good agronomical practices. Well drained, organic matter rich soils are better for plant growth and development. Garden soil, FYM and leaf mould in 1:1:1 ratio suitable for pot culture.

Verities

Shringar (Mexican Single × Pearl Double): This variety develop through hybridization at Indian Institute of Horticultural Research (IIHR), Bangalore. It has medium spike with single type, highly fragrant floret. Flower bud has pinkish tinge. A spike has large and number of florets. It contains 0.135% concrete. It is basically use as loose flower purpose and yield of these flower are 15,000-20,000 kg/ha.

Prajwal (Shringar × Mexican Single): It was also released from Indian Institute of Horticultural Research (IIHR), Bangalore. Colours of flower buds are slightly pinkish and flowers are white. It gives 20% more yield compare to the Shringar cultivar. Flowers are use as loose as well as cut flower purpose.

RajatRekha: It is a Gamma ray mutant cultivar released from National Botanical Research Institute (NBRI), Lucknow. It is a single type cultivar. Flowers are silvery white streaks along the middle of the leaf blade. It contains 0.089 % concrete. It is used for beautification.

SwarnaRekha: It is also gamma ray mutant cultivar released from National Botanical Research Institute (NBRI), Lucknow. The flowers have golden yellow steaks along the margins of leaf. It is a double type cultivar and content 0.062 % concrete.

Suvasini (Mexican Single × Pearl Double): It is a double type hybrid released from Indian Institute of Horticulture Research (IIHR), Bangalore. It bears numerous, fragrant, bold and uniform florets on spike. It gives 25% more yield thanPearl Double cultivar. It is use as a cut flower purpose. It is nematode tolerant cultivar.

Pearl Double: It produces double type of florets with reddish tinged. It gives higher yield with quality flowers. It is use in flower arrangement and oil extraction. Concrete recovery has been found to be 0.06%.

Vaibhav (Mexican Single × IIHR-2): It is a semi double type cultivar. It produce medium spike and greenish bud. It gives higher yield compare to Suvasini cultivar. It is good cultivar for cut flower as well as pot planting.

Culcutta Single: It is a white colour, single type cultivar. It is suitable for West Bengal condition.

Culcutta Double: It a double type cultivar. Flower bud has pinkish tinged before opening and completely white after opening.

ArkaNirantara: It is a white colour, single type, prolong blooming, fragrant cultivar of tuberose. It is evaluate at Indian Institute of Horticultural Research (IIHR), Bangalore.

Cultural Practices

Selection of site

Tuberose require sunny place to attain noble vegetative growth and supreme flower production. To get higher production, site should be selected where the plants will get adequately sunlight throughout the developing period. Moisture holding capacity of the soil should be good for tuberose cultivation. This place should be secure from strong wind, hot and cold waves, so that there is no damage of flower stalk. Shady situation adversely affect the flowering and plant become tall and lanky. Plants are highly vulnerable to water stagnation and poor drainage hence it require well drained and aerated soil.

Land preparation

It is a very essential operation for plant growth and development. Land should be ploughed 2-3 times to the depth of 30-45 cm. First ploughing is done in January month and field left exposed to sun for 15 days, it help in destroying insect pest and diseases. Second ploughing done after one month of first ploughing and at that time incorporate well rotten FYM @ 50-60 tones/ha in soil. After that soil brought to a fine tilth by removing the weeds and breaking the clods.

Propagation

Seed

In suitable climate conditions only single type of cultivars capable to produce seeds. Temperature and moisture affect the sprouting of plants; 25°C soil temperature is suitable for increase seed sprouting. Before raising seedlings bed should be prepared by digging and incorporate 10 kg/ m^2 FYM in soil. The bulb are sown in rows 10-12 cm apart and 5 cm deep in heavy soil and 2.0 cm in light soil. Seeds germinate in 10-15 days after sowing.

Vegetative propagation

Bulb

Commercially tuberose ispropagated through bulbs. In low temperature condition bulb goes in dormant stage in winter months. Bulb dormancy can be broken by treating the bulbs with 4%Thiourea solution for one hour. Dormancy of bulbs can also be broke by using ethylene chlorohydrins. Before sowing of bulbs scale should be remove from the bulb. Size of bulb affects the quality of flowers. Spindle shaped 2.0 cm to 3.0 cmbulbs free from diseases is suitable for planting. Smaller size bulb take 2-3 years to produce flowering according to size of bulb. For planting in one hectare land About 1.30 - 1.60 lakh bulbs/ 10-12 tons of bulbs are required.

Division

In this method bulbnormally cut into 2-3 vertical sections, each segment must be contain bud and a part of the basal plate. Sprouting is depends on the size of the bulbs and only the segments from large bulbs (2 cm or more in diameter) regenerate well (Hussain, 1999). To reduce the incidence of diseases every sections of bulb is treated with fungicide and planted vertically in a rooting medium with their tips just showing above the surface. Bulblets are suitable to transplant when roots develop from the basal plate.

Micro-propagation

Micro propagation is a hi-tech method of quality planting material production. These plants are free from the nematode and diseases.

Bulb treatment

To break the dormancy bulb dip in 4% solution of thiourea. Pre-plant storage of bulbs at 10°C for a period of 30 days will expand the plant growth, increase number of spike and flower yield. More number of floret, early and longer spike emergence enhance by pre-planting treatment of bulbs with GA_3, etherel or thiourea. The bulbs are first thoroughly cleaned and treated with Bavistin (0.2%) for 30 minutes. To avoid the disease bulb should be treated withEmisan (0.2%), Thiram (0.3%), Captan (0.2%) or Benlate (0.2%) solution for 20-30 minutes. Bulbs should be treated with systemic fungicide before planting and contact fungicide before storing. Drying the treated bulb in shade before planting or storing.

Planting

Planting density effect the quality of tuberose. For fetching high economic returns, bulbs are planted at spacing of 30 x 20 cm or 20 x 20 cm (Yadav *et al.*, 1983) or 30 x 30 cm (Nagaraja *et al.*, 1999) with 5.0 to 7.0 cm depth. Close spacing 20 x 10 can also used for planting.Depth of bulb depends on the size of bulb and type of soil. Planting is done in sandy soil deeper than the clay soil (Yadav *et al.*, 2002). For one acre land 45,000 - 55,000 bulbs are required. It is planted in January-March in the plains and in April-May in the hills. In southern parts of India it is planted in the month of July-

August. In West Bengal planting done in 1st week of April and in Bangalore planting can be done throughout the year. Highest yield of spikes and flowers of cv. Single recorded when bulbs are planted in the month of April.

Nutrient Management

Manure and fertilizers is require for fulfill the nutritional requirement of plant for their good yield and quality production. High dose of, N and P markedly promotes leaf and bulb formation but high dose of N make the plant soft and vulnerable to wind and the plants become more susceptible to diseases and pests. Ideal fertilizer dose are FYM (20 tonnes/ha), a fertilizer dose of 200 kg N, 150 kg P2O5 and 150 kg K2O per hectare is recommended for tuberose production (Singh and Sisodia, 2017). The half dose of N and full dose of P and K has to be applied at the time of planting and the remaining half of N is apply in three split dose such as 30, 60, 90 days after sowing.

Water Management

Water management has vital role quality production of flowers. For enhance sprouting irrigate the field before sowing and further irrigation should be skip till the bulb sprout. Drainage facility must for this crop because too much moisture results in the rotting of bulbs and similar condition during flowering adversely affects the development of spikes and flowers. Tuberose is irrigated at one week interval in summer and 10 days interval in winter months.

Earthling up and Staking

Earthling up should be done up to 10-15 cm high at the stage when plant height reaches upto 20-25 cm. To prevent the plant from lodging staking is done by bamboo stick or iron angles and string or rope may be tied in three rows along the plant.

Weed management

Weed infestation create problem in growth of plants and reduces the yield. Frequent manual weeding is effective but it is very laborious and increases the cost of cultivation. To pull the weeds and loosen the soil hoeing between plants is necessary practice. Pre-emergence application of Gramaxone at 3 lit/ha or Diaron at 3kg/ha and ATP+CIPC mixture 2.0+1.5 Kg/ha, effectively controlled weeds in tuberose field.

Plant growth regulators

Growth regulator has vital role for acceleration of vegetative and reproductive stage (Vendrame *et al.*, 2005). Foliar application of GA3 at 50 to 100 ppm thrice at 40, 55 and 60 days after planting is found to be beneficial. Application of both together CCC at 5000 ppm and GA3 at 1000 ppm results in early flowering, increased flower stalk, number of flower florets production and improves the quality of flowers (Mandal *et al.*, 2018).

Harvesting of flowers

Harvesting is a very important culture technique, it should be done at proper stage and time, otherwise economical loss increases.Flowering started within 3-4 month after planting in the month of July but peak flowering appear in August-September. It takes 25-30 days in opening of flower bud after emergence of flower.It produces flowers throughout the year in Calcutta and Bangalore. Peak period of flowering in Northern plains is June to December. Tuberose should be harvest when lower pair of flowers is fully open as a local market and as distant market lower flower buds has just burst (Prasad *et al.*, 2016). Harvesting should done in early morning or evening by a sharp knife above 4-6 cm from ground.

Ratoon cropping

For taking ratoon crop after harvesting the main crop,the flower stalks are cut to the base. About 3-4 ratoon crops can be taken from a single planting. Plots should be well-manured and irrigated for ratooning. To stimulate appropriate growth and development of plants, fertilizer dose as given in the main crop should be applied in two equal split doses in January-February and April. All agronomical practices should be follow as in case of main crop. Ratoon crop gives early flowering as compared to main crop. The ratoon crop results in more number of spikes but reduces number of florets, length of spikes and weight of flowers. Therefore, ratoon crop should be used only for loose flowers or oil extraction purpose.

Harvesting of bulb

Bulb takes 7-8 months for maturity and at this time flowering is over and plant growth ceases. Old leaves become dry before digging out the bulbs. Irrigation is stop before few days of digging the bulbs and soil is allowed to dry before uprooting. The leaves are cut at the ground level and bulbs are taken out from the soil by digging with spade without damage the bulb. Adhering soil should be cleaned and store them in suitable place.

Curing and storage of bulb

After harvesting bulbs are shorted and graded depends on the size and placed on shelves to cure. Curing process enhance by maintain temperature 27 to 35 °C. To avoid the fungal infection position of bulbs should be change in every few days. Hanging of the bulb on walls by tie dozen of bulb together in small cord. Storage of bulb should be done in shady, well aerated place. Storage of the bulb at 30 °C for 6 weeks promote the production of commercial size bulb. Longer storage duration deteriorate the quality of the spike and also reduce number of bulbs.

Yield

Yield is always depends upon variety, bulb size, planting time and density of plants. A well-managed crop produces 2-4 lakhs / ha spikes and 10-15 ton / ha loose flowers. In addition, 18-30 tonnes/ha of bulbs produced by tuberose (Singh and Sisodia, 2017). Harvesting of bulb after 3 year gives 21.3 ton bulb (Yadav *et al.*,1983).

The yield of tuberose single type variety is observed 100-140q/ha in the first two years in third year yield was recorded 90q / ha. In all three years approximate yield is observed 30q/ha. However cut flower spike are obtained 500 thousand in first year,

550 thousand in second year and 310 thousand in third year. After three years crop is again planted. The concrete yield is obtained 27.5kg.

Post-harvest handling and packaging

Loose flowers

Loose flowers graded according to the size and packed in bamboo baskets capacity of about 10-15 kg flowers. It is transported to the local market where they are sold by weight. Proper humidity should be maintained by spraying of water time to time if it is transported in distant market.

Cut flower

Field heat is remove by placing the spikes in water just after harvesting. Grading should be done for cut flowers purpose depend upon colour, number of floret, length of the spike, injury of the spike, weight of spikes. Length of single type spike is 90-120 cm and it contains 20-25 flowers. Double type varieties have 75-95 cm long spike and contain 40-50 flower per spike. After that round bundles made and each having about 100 sticks. The stem portion of the bundles is wrapped in wet newspaper sheets. The bundle of flower should be covered in soft, white tissue paper or polythene to avoid damage. Bundle place in two parallel rows, paper waste should be place in between the bundle to avoid bruising injury. For long distance transportation corrugated cardboard boxes are more suitable. Size of box is very according to length of spike. Length of the box should be about twice the width and its width about twice the height. For single type flower size of box is 130× 60 ×30cm and for double type spike size of box is 100× 50 × 30cm.

Vase life

Shelf life increases by dipping the spike in pulsing solution before packing, it containing 200 ppm silver nitrate ($AgNO_3$) and 4mM silver thiosulfate STS (Bakash *et al.*, 1999) or solution containing sucrose 2% + 8 HQC (200 ppm) + $AgNO_3$ (50 ppm) (Sudagar et al., 2010) and later bundles packed in the card-board boxes and shipped by air to reach the destination quickly.Cut spikes can be store at 7-10 °C temperature for 3-5 days (Bhattacharjee and De, 2010). The solution of Benzimidazole, SADH, Ascorbic acid and 8-HQC is very effective for opening of flower of tuberose spikes and enhancing the vase life of flowers (Pathak,1981).

Disease

Flower bud rot (*Erwinia* sp.): It is a bacterial disease, it cause dry rotting of the buds with brown necrotic discoloration of peduncles. The diseased plants should be uprooted and destroyed. The disease can be controlled by the spray of Streptomycin (0.01%).

Leaf blight (*Botryliselliptica*): This fungus seriously damages the leaves. Application of ammoniacal copper @ 0.2% or Greeno @ 0.5% for 15 days intervals help in control the disease.

Alternaria leaf spot (*Alternariapolyanthi*): It is a very common disease which adversely affect the yield. The symptom of the disease is appearance of brown spots with faint concentric rings on the mid-rib. The disease can be controlled by the spray of Bordeaux mixture @ 0.4% or Zineb 0.5%. Spraying of mancozeb or iprodione @ 0.2% also help in controlling this disease(Singh and Sisodia, 2017).

Stem rot (Sclerotiumrolfsii): The main symptom of the disease is prominent, coarse and mycelia masses appear on the leaf surface or near the soil level and later this sport enlarge and cover whole leaves. Leaves become yellow, droop and finally dried up and fall down. Application of brassicol@ 1 % or Zineb 0.20% thiram @ 0.2% help in controlling this fungus.

Insect pests

Grasshopper: This insects feed on young leaves and flower buds. Application of insecticide such as 0.1 % Roger or 0.1 % Malathion for fortnightly interval reduces the incidence of grasshopper.

Aphids: It is very small in size, soft bodied, green, deep purple or black colour insect which is feed on the flower buds and growing points. These pests may be controlled by spraying 0.1 % Malathion or dimethoate at 10-day interval.

Nematodes: Nematode is the very noxious pest among the others pest. Extremely damage cause by root-knot nematodes, reniform nematode and greasy streak nematode. It cause stunted plants growth and extensive yield losses. It reduces the leaf size, flowers look sickly andfinally the roots rot. Application of Thimet or Furadan at 25 kg/ha or carbofuran 3-5 kg/ha, neem cake 1.5 tonne/ha controls nematode infestation.

Red Spider Mites: It is mainly appear in hot and dry condition and can be seen at lower surface of the leaves. Mites are usually red or brown in colour and multiply very fast. It is suck plant sap, result formation of yellow stripes and streaks on the foliage and stunted the plant growth. Application of Kelthane @ 1.2% con. is effective to control the mites (Singh and Sisodia, 2017).

References

Anonymous. 2013. www.nhb.com.

Bakash, A., Khan, A.M., Ayub, A., Shah, M.A. and Afzal, M. 1999.Effect of various chemicals on the vase life and quality of cut tuberose flower. Pak. J Bot., 2:914-916.

Bhattacharjee, S.K. and De,L.C. 2010. Advanced Commercial Floriculture. Aavishkar Publishers, Distributors, Jaipur.

Hussain S. 1999. Effect of bulb size and depth of planting on growth and flowering of tuberose (*Polianthestuberosa* L.) cv. single. M.Sc. Thesis. Dept. Horticulture, Univ. Agriculture, Faislabad, Pakistan.

Mandal, M.,Maitra, S. and Mahata, D. 2018. Production technology of tuberose (*Polianthestuberosa* L.) cultivation.Journal of Pharmacognosy and Phytochemistry, 7(6): 2360-2364.

Nagaraja, G.S., Gowda, J.V., Narayan and Farooqui, A.A. 1999. Effect of planting densities on growth and flowering in tuberose (*Polianthestuberosa* L.) cultivar 'Single'. Mysore J of Agril. Sci., 33(2):206-209.

Pathak, S. 1981. Ph.D thesis ,University of Burdwan ,West Bengal.

Sadhu, M.K., and Bose, T.K. 1973. Tuberose for most artistic garlands. Ind. Hort. 4:159-164.

Singh, A.K. and Sisodia, A. 2017. Textbook of Floriculture and Landscaping, New India Publishing Agency, New Delhi. p. 418.

Singh, K.P. 1995. Improved production technologies for tuberose (*Polianthestuberosa* L.). Indian Institute of Horticultural Research, Hessargarhatta, Bangalore. Rev. Res. India., 7:1996-1998.

Sudagar, I.P., Sankarnarayannan, R. and Aruna, P. 2010. Effect of chemicals in increasing the vase life of tuberose cultivars. Asian J Hort., 4:421-423.

Usman 2013. Marketing of Cut-Flowers: A case study of district Kasur. M.Sc. (Hones.) Thesis (unpublished). Inst. Agric. Res. Econ. Univ. Agri., Faisalabad.

Vendrame, A.W., Maguire, I. and Moore, K.K. 2005. Growth of selected bedding plants as effected by different by different compost percentages. Florida State Hort. Soc., 18:368-371.

Yadav, L.P., Bose, T.K. andMaiti, R.G. 1983. Response of tuberose (*Polianthestuberosa* L.) to nitrogen and phosphorus fertilization. Indian Agric., 28(1):53-62.

Yadav, L.P.,Maiti, R.G. andBose, T.K. 2002. Tuberose. In Bose, T.K., Yadav, L.P., Pal, P., Das, P. and Parthasarathy, V.A. (ed.) Commercial flowers Vol 1stNayaProkash, Culcutta, p. 624.

21

Bird of Paradise and Heliconia

K. Nihad

Zingiberales is an important order in Kingdom Plantae with eight families in which most of them are important as far as commercial horticulture is concerned. The order includes fruits (bananas and plantains), spices (ginger, turmeric, cardamom), medicinal (arrowroot, galangal) and ornamental (Heliconias, Bird of Paradise, Marantas, Canna, Costus, etc) plants. Among the ornamental zingiberales, plants of Strelitziaceae and Heliconiaceae are gaining popularity for its aesthetic preferences and marketable cut flowers. The unique colour and enhanced vase life of inflorescences makes them as flowers of choice. The plants of other families such as Marantaceae, Cannaceae, Costaceae, Zingiberaceae, Lowiaceae and Musaceae are also gaining momentum in floriculture business for their cut foliages and inflorescences. Some plants such as calathias are admired as potted plants and for indoor gardening.

Heliconia Society International (HSI), USA is the official International Cultivar Registration Authority (ICRA) for the genera Heliconia, and related ornamental Zingiberales. HSI sponsors students' grant program supporting research projects on aspects of the botany and horticulture of ornamental Zingiberales conducted by graduate and post-doctoral students (https://www.heliconia.org/).

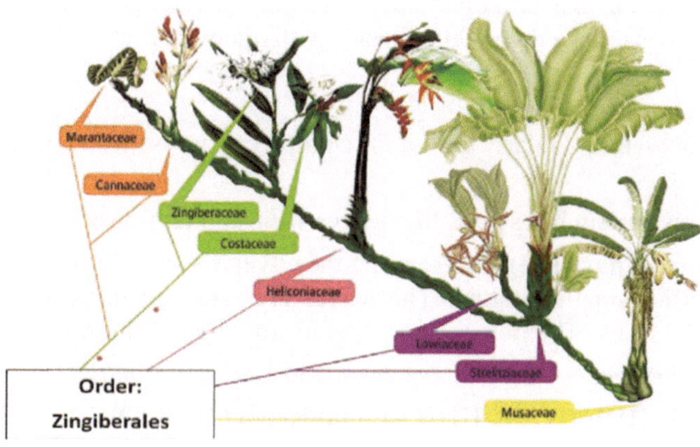

Phylogenic tree of order Zingiberales

Bird of Paradise

Introduction

Bird of paradise (BOP) belongs to Strelitziaceae family of Order Zingiberales. The plant gets its name from the beautiful inflorescences resembling the head of brightly coloured bird. The genus *Sterlitzia* includes around five species of monocot plants. Among the species, *S.reginae* is a commercially grown ornamental plant for its cut flower and foliage. The inflorescence resembles the head of a crane-like tropical bird with orange wings and deep-blue beak. In India, the plant is grown in sub-temperate to sub-tropical regions like Himachal Pradesh, Darjeeling, and Western Ghats etc.

Importance and uses

The crop is widely grown in gardens for aesthetic purpose. The cut flower and foliage of *S. reginae* are much preferred items in tropical flower arrangements due to its longer vase life and unique brilliant speck of colours.

Origin: It is a native of subtropical coasts of South Africa.

Taxonomy

Sterlitzia sp. are evergreen rhizomatous monocot perennials. It has an underground stem (rhizomes) with clump-forming pseudo stems. Mature stumps attain 120- 150 cm height with a plant spread of 90 -150 cm. The leaves arise spirally from the base of clump in a fan-like pattern. The leaves are thick, fleshy, concave and oblong with bluish green midrib. The leaf lamina is about 15 cm wide and 45 cm long. The leaf lamina are grayish green in colour with smooth waxy texture. The inflorescences are produced above the canopy level. The flowers are enclosed in large boat-shaped spathe which opens and emits bright coloured sepals and petals. Each inflorescence stalk has six or more florets with orange sepals and blue petals enclosing stamen and style. The petals of the flower are arranged in dissimilar patterns which are erect and pointed like a bird's beak. It has five stamens and a three-celled ovary with many seeds.

Species and cultivars

***Sterlitzia reginae* (Crane flower):** popularly grown for cut flower purpose with bright orange and blue flower heads.

***S. reginea* var. Juncea:** This variety has sword-like leaves with smaller flowers

***Strelitzia nicolai* (Giant bird of paradise):** It resembles the traveler's palm (*Ravenala madagascariensis)* but with a smaller stature. The leaves are arranged in fan-like pattern with leaf scars on the trunk. The flowers are white in colour with blue beaks.

Other species are *Sterlitzia nugusta*, *Sterlitzia kewensis*, and *Sterlitzia candida.*

Studies conducted at ICAR-IIHR (ICAR-IIHR Annual Report 2008-09)

Forty-four genotypes of Bird of Paradise (*Strelitzia reginae*) are evaluated for growth, yield and floral characters. The mean values were 128 cm for plant height with 140 leaves and 22 dichotomous branches. Inflorescence-length was 83.09 cm with 4.31

cm stem diameter. Month wise inflorescence yield was recorded and the yield was maximum during September month. Number of florets per inflorescence ranged from 6.2 to 11.8.

Climate

Bird of Paradise is a sub-tropical ornamental plant. Its flowering is directly related to atmospheric temperature. It prefers a temperature range of 17-27^{0}C. At higher temperature it promotes leaf production in the expense of flowering. It comes up well under open field conditions .Flower production is adversely affected when grown under shade. Too much exposure to sunlight, especially during summer months, may result in leaf scorching.

Studies conducted at ICAR-CPCRI during 2014-19 showed that under tropical humid condition it takes around three years for flower initiation. The flowering is triggered only when the day temperature is below 30^{0}C. Flowering season in this agro climatic condition is from July to November (Nihad *et al.*, 2019).

Soil

It can be grown in all types of soil but the plant prefers moist organic rich soils with good drainage. If grown in soils with low organic content, care should be taken for supplying additional dose of organic manures. Soil temperature is one of the key factors affecting the growth of Bird of Paradise. Being a subtropical ornamental, the soil temperature should be kept below 30 to 35 ^{0}C. This can be maintained through addition of coir pith compost @ 2kg/plant during October- November. Drip irrigation can be given for ensuring the soil moisture especially during summer seasons.

Propagation

Mature rhizomes are commonly used as planting material. Seedlings take three to five years for flowering where as plants grown from divisions will flower in one or two years.

Planting

For commercial cultivation, Bird of Paradise can be planted at 60 cm x 60cm spacing. Rooted rhizomes are planted in pits of size 60 cm x 60 cm x 60cm filled with top soil, sand and organic manure (1:1:1). It can be planted throughout the year except during heavy rains.

Manures and fertilizers

Bird of Paradise is a heavy feeder plant which require nitrogen, phosphorous and potash in equal proportion. Commercially available complex fertilizer mixtures like 17:17:17 or 18:18:18 can be used for this purpose. The plants are supplied with NPK mixtures @ 5g/plant in fortnightly intervals from third month after planting. Application of slow release manures such as neem coated urea is recommended, as leaf scorching is observed in plants supplied with excess quantity of fertilizers. Mulching the plant basin with well rotten organic mulches like coir pith compost during January-February is ideal. It provides essential micronutrients, conserves moisture and reduces weed growth.

Irrigation

During summer months irrigation is applied twice a week. The soil should be moist and free from water logging. Drip irrigation is preferred during commercial cultivation.

Intercultural operation

The dried flowers will remain in the plant for longer time. For enhanced flowering, periodic pruning of senile flowers and leaves is recommended. In commercial plantations, staking is done by tying with two to three strings of rope along the rows of plants supported by GI pipes.

Harvesting and yield

Harvesting is done by gently pulling the flower stems. The inflorescences are harvested when the florets starts opening and bends at 90° angle with the stem. For longer transportation, harvesting can be done at tight bud stage prior to floret emergence from the spathe. A healthy plant produces more than 10-15 marketable flowers in a year.

Post harvest management

The harvested inflorescences are dipped in water for about an hour to remove field heat. The flower stem is trimmed (60 -80 cm length) using a sharp knife after submerging its bottom in water for preventing the clogging of conducting vessels. The longer the flower stem more will be the vase life. The cut end of the stem is wrapped with moist cotton swab, packed in paper boxes (120 x 30 cm) and stored at 8°C temperature. The vase life can be enhanced by dipping the cut flowers in solution of 10 % sucrose + 250 ppm 8-HQC + 150 ppm citric acid for 48 hours.

Pests

The major pests of Bird of Paradise are mealybugs, scale insect, aphid and white flies.

Control: Spraying with 5% neem oil is suggested.

Diseases

Root rot (*Thielaviopsis basicola, and Rhizoctonia* sp.) due to water stagnation is the major problem in Bird of Paradise.

Control: Drenching with Hexaconazol 1ml/l is recommended.

Heliconia

Introduction

Heliconias is a large genus of ornamental monocots belonging to Heliconiacea family of Order Zingiberales. Heliconia takes its name from the Greek word *helikonios* representing Mount Helicon, the home of ancient Greek gods. The genus is made up of about 200 species including a variety of hybrids and cultivars. Flower production in Heliconia is related to the rate of vegetative growth as well as plant density. The marketable flower yield varies with nutrient content of the growing medium and light intensity.

Importance and uses

Heliconia plants are grown mainly for commercial cut flower purpose. The inflorescences have 10-12 days of vase life and can be used in stage decorations, bouquet making and long flower arrangements etc. Among the tropical flowers, Heliconia is outstanding for its diversity in form, colour, size and particularly, its durability (Rodrigues, 2008). Apart from the use as cut flowers many species are suitable for landscaping as well.

The main criteria used by the farmers for Heliconia inflorescence are the number of open bracts and the length of the inflorescence stem (Costa *et al*, 2006b). According to Powell (1991), Heliconia flowers could be cut when two or three bracts are open, but tighter flowers could also be used effectively in floral arrangements.

The harvesting stage of inflorescence is very important for fetching higher market price as it will not open further once it is removed from the mother plant. The flower characters of *H. stricta* is classified based on the stages of opening (Nihad, 2013). Inflorescences can be marketed from the flower opening Stage I to III which will take 15 to 20 days. Hence, these flowers are lesser subjected to vagaries in market as the harvesting can be delayed up to three weeks.

Just Emerged to Just Opened (Stage I to II)

The days taken for the inflorescence from just emerged from terminal end of pseudostem (Stage I) to just opened stage (Stage II). The inflorescence is deep red in colour.

Just Opened to Fully Opened (Stage II to III)

The days taken for the inflorescence peduncle to emerge and elongate resulting in complete opening of inflorescence (Stage III).

Longevity of Flower (Stage III to IV)

The days taken for fully opened flower to senescence (Distinct loss of color, with bracts going from dark red to blackish).It is taken as the

The developmental stages of *Heliconia hirstutella* inflorescence has also been documented (Ballah and Starr, 2010). The opening characters were classified into six stages. Each of these can be further subdivided into two sub-stages, which varied in their distinctness among the stages.

Stage 1. Emergence: Inflorescence emerged from terminal end of pseudostem. Peduncle was not visible, bracts still closed without nectar secretion. Inflorescence was reddish-orange.

1A. Tip of bract one was visible.

1B. Bract one fully emerged.

Stage 2. Elongation: Peduncle emerged and elongated. Inflorescence rotated relative to axis of pseudostem. Bracts remained closed and were bright reddish-orange. Nectar droplets appeared on cheek of basal bracts.

2A. Peduncle began to appear and elongate, elevating inflorescence away from pseudostem. Inflorescence orientation was parallel to axis of peduncle.

2B. Inflorescence rotated to angle of 45-90° from axis of peduncle. Bract two starts to emerge.

Stage 3. Opening of first bracts: Nectar droplets appeared on all bracts. Flowers and fruits appeared. Inflorescence had bright reddish-orange with some yellow.

3A. Bracts began to open, bracts one and two unfurled, each with droplets of nectar. At least one flower was present in either or both of these bracts. Immature fruit was present in bract one only.

3B. Bracts three and four unfurled, with nectar droplets. Flowers were present in all bracts with immature fruit. Mature fruit were seen only in bracts one and two.

Stage 4. Inflorescence opened completely: Flowers and fruits were visible in all bracts, with dried pedicels in bracts one to four. No nectar secretion was seen on bracts one and two. Bracts were brick red-orange.

4A. Bracts five to six unfurled. Flowers were visible in all bracts, with fruits at all stages of maturity. Pedicels were seen only in bracts one to four. Nectar droplets were seen in bracts three to six.

4B. All bracts fully unfurled.

Stage 5. Early senescence: New-flower production reduced, darkening of bracts and cessation of nectar secretion occurred.

5A. Bracts darken, immature and mature fruit were present in all bracts together with dried pedicels.

5B. Further darkening of bracts was observed which led to development of dull red to pale orange in parts. The ratio of dried pedicels to flowers and fruits increased.

Stage 6. Full senescence: Flower production stopped entirely, with clear decay in all parts of inflorescence. There was distinct loss of color, with bracts going from dark red to blackish.

6A. Only mature fruit was present in some bracts.

6B. Inflorescence dried and became lifeless and its parts disintegrated

Origin

The crop is native to Central and South America and some islands of the South.

Taxonomy

Heliconia is the only genus of the family Heliconiaceae, which was formerly included in Musaceae family. In 2003, based on the APG (Angiosperm Phylogeny Group) system of plant classification, heliconia plants were identified as distinct from other plant species and grouped into a separate family *viz.* Heliconiaceae. Heliconias produce underground stems (rhizomes) and erect shoots. The rhizomes branch off and produce new shoots. New branches and shoots develop from buds or eyes on the rhizomes. The shoots are composed of overlapping leaf sheaths which make up the above ground pseudostem and support the large leaf blades and inflorescences.

The inflorescences are modified leaves called bracts, which may be erect, pendulous or spiraling in the shapes of bird's beaks, lobster claws or fan shaped and colors of reds, pinks, gold, oranges and splashes of a mixture of colours (Castro and Gonclaves, 2007). The flowers are mostly hidden by large, colourful bracts. The floral tubes are made up of six variably fused tepals with inferior ovaries. Within the flowers are the pollen-producing stamens and the stigma on a long style.

Species and cultivars

Berry and Kress (1991) has described a large number of heliconias with distinct floral characters. New varieties of Helcionias are being registered as per the standards of International Society for Horticultural Science (ISHS) and added to the gene bank through HSI.

Heliconia is classified into different groups based on:

1. Inflorescence character
2. Growth habit
3. Phyllotaxy

Heliconia plants are categorized into two based on the inflorescence character:

a). Erect types (e.g. *H. stricta*),

b). Pendent or hanging types (e.g. *H. chartacea*).

Based on growth habit they are classified into two:

a). Spreading types

b). Clumping types.

The spreading types fill and colonize quickly and are capable of covering a substantial area of land in a few years. These types are mainly used for landscaping purpose. (e.g. *H. psittacorum*)

The clumping types grow more slowly and new pseudo stems develop on the edge of the clump and the centre of the clump hollows out. Most of the commercial cut flower varieties comes under this type (e.g. *H. stricta, H. chartacea*).

Based on the phyllotaxy heliconias are again grouped into three:

a). Musoid: Erect leaves with long petioles like bananas

b). Zingiberoid : Horizontal leaves with short petioles resembling gingers

c). Cannoid: Intermediate with short to medium petioles supporting leaves at about a forty degree angle, like Cannas

For cut flower industry, the species characteristics of interest are:

- Production of inflorescences during the whole year,
- Short flowering cycle,
- Light flowering stems for lower transportation costs,
- Inflorescence stems longer than 80 cm,

- Inflorescence stems with diameter thick enough for better resistance to handling and for lighter total weight of inflorescence,
- Inflorescences with no wax and no hair; and bracts arranged in one plane for easier handling and packing,
- Inflorescence width and length of more than one meter,
- Firmness of bract,
- Bract with few or no flowers inside,
- Post harvest durability or vase life of more than seven days.

Heliconia species of commercial interest with strong seasonal flowering periods were noted by Criley (1999). They were *H.angusta, H.bihai, H.caribaea, H.caribaea x H.bihai, H.collinsiana, H.lingulata, H.rostrata, H.stricta, H.wagneriana* whereas the *H.psittacorum* cultivars and hybrids were observed as species with longer periods of bloom. In another study by Costa *et al.* (2006b), the number of days between the shoot emission and the inflorescence emission (DBSI) was observed in 10 genotypes of Heliconia. The shortest was 105 and 45 days and the longest was 126 and 93 days respectively.

The list of some popular commercial cutflower varieties of *Heliconia* in India:

Type	Variety	Scientific name	Market value (2014-15) (Rs. per inflorescence)
Erect type	Iris**	Heliconia stricta	20-30
	She*	H. orthotricha	20-25
	Sunrise*	Heliconia stricta	25-30
	Jaquinii	H.caribaea x H.bihai	45-50
	Kawauchi*	H.bihai X H.caribaea	40-50
	Caribbean	H.caribaea	60-70
Pendant type	Sexy Pink	Heliconia chartacea	50-60

Nihad *et al.*, 2016 ,* Suitable for commercial cultivation in coconut plantations (Nihad et al., 2018), ** Variety suitable for intercropping in highly shaded (more than 60%) plantations (Nihad *et al.*, 2019)

Loges *et al.* (2006) studied in detail the floral traits of 18 Heliconia genotypes from the Heliconia Germplasm Collection of Federal Rural University of Pernambuco State. The genotypes were *H. psittacorum* L.f. *X H. spathocircinata* Aristeguieta cv. Golden Torch Adrian; *H. psittacorum* L.f. x *H.spathocircinata* Aristeguieta cv. Alan Carle; *H. psittacorum* L.f. cv. Strawberries & Cream; *H. psittacorum* L.f. cv. Suriname Sassy; *H. psittacorum* Red Opol; *H.pseudoaemygdiana* L. Em. & Em; *H. psittacorum* Red Gold; *Heliconia* x *nickeriensis* Maas & de Rooij; *H. latispatha* Benth (orange); *H. latispatha* Bentham cv Yellow Gyro; *H. rauliniana*; *H. latispatha* Benth cv. Distans; *H.rostrata.*; *H. rostrata*; *H. wagneriana* Peters; *H. bihai* (L.) L. cv. Kamehameha; *H. psittacorumx H. spathocircinata* Aristeguieta cv. Golden Torch and *H. bihai.* Number of days from inflorescence emission to harvesting (DIH), fresh weight of stem (FWS), number of leaves in the stem at inflorescence emission (NLI), diameter of stem at 20 cm under the inflorescence (DI), stem length (SL) and inflorescence length (IL) are some important traits to be considered for genotype selection. The inflorescences were harvested when they had two to four open bracts. Significant (5%) difference was observed for all the traits among genotypes. The DIH ranged from 14 to 23 days. This information was important to help the growers on planning the harvesting program according to the species cultivated. The FWS varied from 40 to 360 g and the DI 18 to 34mm. The SL ranged from 56 to 125cm, and the IL from 15 to 87 cm.

According to Syvertsen *et al.,* (1995), the amount of light absorbed by a leaf, and the diffusion pathway of CO_2 through its tissues depended, at least partially, on its thickness. Negative relationships between leaf thickness and growth rates (Poorter, 1990) and photosynthetic rates (Enriquez *et al*., 1996) have been observed and thicker leaves were associated with increased longevity (Mediavilla *et al*., 2001). White and Montes (2005) reported that leaf thickness was often used as a tool to screen species and/or cultivars for productivity.

Climate

Heliconias are well adapted to all major agro climatic zones of India. The genus is made up of about 100 species along with a large number of hybrids and cultivars. At optimal temperatures of 21-35°C, a flower can be harvested 8-9 weeks after emergence of the shoot. At temperatures below 21°C, growth rate is reduced proportionally until 10°C is reached. At this temperature all growth stops and cold damage symptoms appear, first as small black spots on the floral rachis near the point of bract attachment, and eventually as blackening of the entire inflorescence and necrosis of the foliage.

Broschat *et al., (*1984a) reported that the growth and flowering of some Heliconias such as *H. psittacorum* were known to be limited by light-intensity. Yet some other species appear better adapted to slightly shaded conditions and were easily injured under full sunlight in the tropics (Criley,1995). *Heliconia stricta* can be grown from open to 40% shade (Berry and Kress, 1991). Under 50% shade, *H. stricta* plants produced fewer number of flowers (Broschat and Svenson, 1994) in temperate condition where as the variety Iris grows well under 60-65% shade under coastal humd tropical condition (Nihad *et al*, 2019). Eventhough, *Heliconia psittacorum* plants grown under 30% shade cloth had a lower total yield of flowers they produced longer stems, bigger flowers of better quality and a higher marketable yield (Powell, 1991). *H. angusta* was found mainly in shaded places, forming smaller shoots, about 0.70 m in height inside the forest, while those found at the border of the forest reach about 1.75 m (Simao and Scatena,2001). Plant height, leaf number, leaf area, number of suckers, number of spikes, no.of flowers/ bract and vase life were the highest in 25% shade followed by 50% shade in many shade loving Heliconias (Sheela, 2008).

Flower production in *Heliconia psittacorum* increased as fertilizer rate increased and was substantially greater under full sunlight than 63% shade (Broschat and Donselman, 1983b;Manarangi *et al*., 1988).According to Kress (1990), Heliconia species found in open areas achieve a larger size than those species restricted to shaded areas. This fact probably was due to the difference in the light intensity and to the degree of vegetative competition that occured among the different species in these habitats (Stiles, 1975). According to Broschat and Svenson (1994), plants grown in open condition had more inflorescences than those in shaded condition, possibly due to abortion of flower buds under suboptimal light intensity.

The mean maximum height of species growing within 15° of the equator (7.8 m) was 29 times greater than the height of species between 60 and 75 °N (27 cm), and 31 times greater than the height of species between 45and 60 °S (25 cm). There was no evidence that the latitudinal gradient in plant height was different in the northern hemisphere than in the southern hemisphere (P =0.29). A 2.4 fold drop in plant height at the edge of the tropics (P =0.006) supported the idea that there might be a switch in plant strategy between temperate and tropical zones (Moles *et al.*, 2009).

More shoots, however, were produced at the higher photosynthetic photon flux (PPF) level (10.1 compared with 8.3 shoots). The duration from emergence of shoot to the appearance of inflorescence was significantly less at 32°C day/20°C night than at 24°C day/20°C night (140 and 146 days, respectively) and was unaffected by PPF. This duration also was significantly affected by the interaction effects of order of shoot appearance and the number of leaves subtending the inflorescence. The second shoot to emerge had the shortest duration from shoot emergence to inflorescence emergence. The number of leaves subtending the inflorescence increased at higher temperature and decreased as rate of shoot production increased but was unaffected by PPF (Catley and Brooking, 1996).

Soil

Heliconia, being an under storey forest crop, requires soils rich in organic matter with good drainage. But through adequate nutrition, it can be grown commercially even in soils low in organic matter. Studies conducted at ICAR-CPCRI, Regional Station, Kayamkulam where the soils are sandy loam of the order Entisol with very low organic matter (less than 0.2%), reveals that addition of organic matter through vermicompost can enhance the growth and performance of the crop (Nihad,2013). Mulching with coir pith compost or farm waste is recommended during summer season for maintaining the soil moisture status. The plants cannot tolerate water logging continuously for more than five days. Care should be taken for providing proper drainage during rainy season. Most of the varieties prefer soils with less clay content, but according to Heliconia growers of India, psittachorum varieties come up even in clayey type soils.

Propagation

For clumping types, rhizomes taken from seven months old healthy vegetative sucker are used for planting. Before planting, the leaves are removed and the rhizomes are trimmed (around 1m length) with a slanting cut using a sharp knife. For spreading types, rooted rhizomes with two to three leaves are used for planting.

Planting

Planting can be done throughout the year except during winter and heavy monsoon seasons. However, ideal time for planting is from August to November. The rhizomes are planted in pits (30 cm x 30 cm x 30 cm) after filling with top soil mixed with dried cow dung (1 kg/pit) and bone meal (250 g/pit). Mulching with dried leaves or coir pith is done after planting. Depending on species, rhizomes start sprouting at 25 to 45 days after planting.

At least 250 plants are needed for starting commercial cultivation. When grown as intercrops, care should be taken for leaving the root zone of main crop so as to avoid competition for water and nutrients. If planted in coconut gardens, rhizomes are to be planted at 1 to 1.5 m apart, leaving a distance of two meter around the coconut basins. It can also be planted in the interspaces of rubber plantations or as a component crop in multi species cropping system.

Manures and fertilizers

Heliconia can be grown either as purely organic or integrating organic manures and chemical fertilizers. The manures and fertilizers are applied at quarterly intervals beginning from three months after planting. For organically grown Heliconia, 200 g vermicompost and 100 g neemcake are applied per plant at three months interval. Half the dose of vermicompost and neemcake (100 g and 50 g per plant) along with 13:5:13 NPK (5 g/plant) can be given at three months interval for integrated method of cultivation (Nihad *et al.*, 2013). For soils low in organic matter, drenching diluted cow dung slurry in the ratio of 1:10 at six monthly intervals enhances the production of quality inflorescence.

Ferreira and Pires (2005) studied in detail about the growth and flowering of Heliconia. Increased vegetative growth delayed flowering in Heliconia. Delay in floral initiation resulted in the development of additional leaves. Heliconias responded satisfactorily to application of fertilizers resulted in vigorous plants with increased size of inflorescences. Phosphorus and potassium positively influenced increase in the length of buds in *H.psittacorum* "St. Vincent Red." According to Clemens and Morton (1999), The number of shoots and flowers produced per rhizome were maximum at low N to K ratio (1:1).

Nutrient deficiency symptoms: Leaf yellowing with marginal necrosis is the common symptom noticed in Heliconia during nutrient and water stress (Nihad *et al.*, 2018). During initial growth phases, *H. stricta* is prone to K deficiency which can be managed through one time application of MOP @ 60g/plant.

The critical nutrient level in Heliconia leaves of spreading (var. Parrot) and clumping (var. Pink lobster claw) types was estimated from University of Hawaii (Uchida, 2000) is listed below;

Elements	N	P	K	Ca	Mg	S	B	Fe	Mn
Unit			(%)					(ppm)	
var.Parrot	1.67-1.79	0.27-0.38	1.72-2.13	0.75-0.81	0.33-0.38	0.36-0.39	10-15	30-40	26 -93
var. Pink lobster claw	2.34	0.27	1.77	1.34	0.75	0.3	09	3.8	142

Irrigation

Heliconia plants always require moist soil. It needs to be irrigated once in two days during summer. The frequency of irrigation can be reduced to once in four days by providing mulching with coir pith compost during February-March

Intercultural operation

Regular weeding is required till eight months after planting. Thinning of eight months old suckers with less than 7 cm diameter should be carried out monthly for promoting production of more number of quality inflorescences. Old and senile leaves and inflorescences are to be removed at periodic interval for promoting more flowering.

Harvesting and yield

Depending on species Heliconia starts flowering from four to eight months after planting. The harvesting is usually done after 4 pm by cutting the rhizome at ground level along with the inflorescence. After cutting, the outer leaves are to be stripped off and the top most leaf blades are cut leaving the petiole.

The average harvesting interval (interval between emission of inflorescence bud and the harvesting day) varied from 14 and 4 days (Hybrid Golden Torch) to 27 and 9 days (H. bihai cv. Nappi Yellow) (Costa *et al.*, 2009a). The number of leaves observed at the time of inflorescence emergence may be a useful indicator for producers to quantify the plants expected to bloom for market planning. The number of leaves on the pseudostem at inflorescence emergence ranged from 5.13 to 6.29 in *Heliconia psittacorum* (Rocha *et al.*, 2010) and four to five leaves per shoot in *Heliconia stricta* (Cabral and Benedetto, 2010).

With proper nutrition around 40 to 45 marketable inflorescences are produced in a year from a single clump. Inflorescences of around one meter length and nine centimeter stem girth with two or more open bracts are selected for sale. Smaller inflorescences can be marketed through value additions such as bouquets and table top arrangements.

Post harvest management

After harvest, the cut end of the inflorescence stem is dipped in tap water for about an hour to remove the field heat. These are then washed in water for removing soil and dust. The excess water is wiped off and inflorescences are graded based on their length. Inflorescence with fewer flowers inside the bracts is ideal for marketing. This character will reduce time and cost of cleaning, minimise occurrence of insects, unpleasant odours from water accumulation and organic matter deterioration.

Inflorescences are graded based on their length, stem girth and spike width.

Grades	Length of inflorescence (cm)	Stem girth (cm)	Spike width (cm)
Grade I	>100	>9	>25
Grade II	75-100	7-9	20-25
Grade III	<75	<7	<20

Nihad *et al.*, 2016

Inflorescence can either be packed individually or in bulk. Care must be taken to exclude ants, bees etc before packing. Bulk packing needs a minimum of 45 inflorescences. Uniform grade flowers are packed in a single box. Different materials such as aluminum foil, butter paper and news papers or recycled papers can be used for wrapping the inflorescence.

Pests

Leaf-Feeding Insects: Hispine beetle (Chrysomelidae), leafhopper (Cicadellidae), stinkbug (Pentatomidae), grasshopper (Orthoptera) and white flies (Aleyrodidae)

Insects infesting Pseudostem and Rhizome: Scale insects (Coccidae), mealybugs (Pseudococcidae), Spittlebugs (Cercopidae)

Insects infesting inflorescences: Heliconia inflorescence (bracts) are infested by a range of flower feeding Diptera, including species of Bibionidae, Drosophilidae, Tephritidae, Richardiidae,Simuliidae, Syrphidae, Chironomidae and Sciaridae. This is seen in plants grown in areas where there is a large source of decaying organic matter.

Management

Increased usage of nitrogenous fertilizers may result in increased pest attack. Jet spraying with water can wash off many of the insect pests such as hoppers and white flies. Spraying with 0.5% neem oil can be adopted in mild infested plants.

Diseases

A. Foliage diseases

1. Causal organism (*Calonectria spathiphylli*): Yellowing and marginal necrosis of leaves, sheath spots and petiole blight. Rots of the sheath and petiole interfere with water movement to the leaf, causing water stress and producing dry leaf edges. As rots of the sheath expand, less water moves up to the leaf blade, and leaves become yellow. Eventually the leaves die, resulting in premature loss of older leaves. The photosynthetic capacity of the diseased plants is reduced by multiple sheath and petiole infections affecting flower production.

Susceptible varieties: *H. angusta* cv. Holiday, *H. bihai* cv. Kamehameha and cv. Lobster Claw One, *H. caribaea* cv. Purpurea, *H. indica* cv. Spectabilis, *H. mutisiana, H. psittacorum* cv. Parakeet, *H. psittacorum* x *H. spathocircinata* cv. Golden Torch , *H. stricta* cv. Dwarf Jamaican and cv. Fire Bird, and *H. wagneriana*

2. Causal organism (*Bipolaris incurvata*): The disease begins as small, water-soaked flecks and spots. The fungus continues to grow in the leaf tissue, and the spots enlarge. After two weeks, many spots coalesce to form bigger (9.5 to 35 mm in diameter) spots with oval or irregular shape with a yellow hallow. Then it turns light brown with a darker edge and holes are formed on the leaf as diseased tissue falls out. In advanced stages of infection, the petiole, sheath, and floral bracts are also spotted with faint brown to purplish-red spots. Spotting of floral bracts makes flowers unmarketable.

Susceptible varieties: *Heliconia stricta* cv. Dwarf Jamaican, *H. orthotricha, H. chartacea,* and *H. mutisiana*

Other fungi causing leaf diseases are: *Exserohilum rostratum*, *Pyriculariopsis* sp., *Cercospora* and *Pseudocercospora* sp.

A. Rhizome and root diseases

1. **Causal organism (*Calonectria spathiphylli*) :**Severe root and rhizome rots kill plants or cause rapid plant decline. Root and rhizome rots of field heliconia start at the center of clumps with old diseased stalks, which are dry and collapsed, and develop outward. Root rots prevent proper anchorage, and taller diseased heliconia cultivars are prone to toppling.
2. **Causal organism (*Phytophthora nicotianae*):** Symptoms: Diseased stems have brown rots at the collar with rotten roots. This is a slow spreading disease. Healthy, vigorous plants gradually decline in one to three years affecting flower production.
3. **Causal organism (*Pythium* sp.):** It is a soil borne fungi causing rhizome and root rot. Moisture and poor drainage greatly favour diseases caused by *Pythium.* The pathogen is transported to new locations by the movement of contaminated soil and water or infected plants
4. **Causal organism (*Rhizoctonia* sp.):** Root rotting is noticed in Heliconia plants due to *R. solani* or other *Rhizoctonia* species

Colletotrichum spp., *Pestalotiopsis* sp., *Phyllosticta* sp., *Phomopsis* sp., *Acremonium* sp., and *Fusarium* spp. are the other fungi recovered from diseased plant parts of heliconia.

Disease management : Use disease free planting materials. Avoid water logging and ensure proper drainage and adequate soil moisture. Proper weeding and harrowing of soil for air circulation to be ensured. Field sanitation/Soil solarisation can prevent soil borne pathogens. Supply of adequate organic manure and plant nutrients helps in maintain the plant vigour and improved disease tolerance. Use of resistant varieties in disease prone areas.

B. Bacterial wilt

Causal organism (*Pseudomonas solanacearum*): Leaf rolling, marginal necrosis, wilting and shoot die back. Leaves curl initially due to water stress caused by vascular plugging following infection of roots and rhizomes. As the disease advances in the rhizome, drying and browning of leaf edges occurs, followed by formation of large patches of necrotic tissue towards the midrib. Usually, these symptoms are more pronounced on older leaves. Eventually, the entire leaf turns dark brown with an oily appearance, resulting in leaf loss. Within the rhizome, a dark brown discoloration of the vascular tissue runs longitudinally down the center. Often, milky ooze is associated with this brown vascular discoloration.

Management: Remove and destroy the infected plants is the best method for controlling the spread of disease. Proper weeding is necessary as *Pseudomonas* *solanacearum* survives for long periods in soil and in many weed hosts.

Prophylactic spraying with a mixture of turmeric powder and soidum bi carbonate in 10:1 (@0.15% ,twice at weekly interval) or Streptocyclin 100 ppm spray can be done before monsoon season.

C. Nematode infestation

Plants infected by nematodes exhibit symptoms similar to those caused by water stress and nutrient deficiency such as yellow leaves, excessive leaf curling and wilting. The major disease symptoms are brown, rotten and swollen roots or root knots, and root lesions. Nematode infections of roots may occur alone or in association with pathogenic fungi such as *Calonectria spathiphylli, Rhizoctonia* spp., and *Pythium* spp.

Management: Soil solarisation can be done for managing nematode infestation.

Photos of Different Heliconia Flowers

References

Annual Report 2008-09. Indian Council of Agricultural Research Indian Institute of Horticultural Reserch, Hessaraghatta, Bangalore. 135 p.

Ballah,S.T. and Starr,C.K. 2010. Composition and activity of ant visitors to inflorescences of *Heliconia hirsuta* (Heliconiaceae). Sociobiology, 6(3) : 585-599

Berry, F. and Kress, W. J. 1991. Heliconia - An Identification Guide. Smithsonian Institution Press, Washington.103p.

Broschat, T. K. and Donselman, H. M., 1983a. Production and post harvest culture of *Heliconia psittacorum* flowers in South Florida. Proc. Florida State Hort. Soc. 96: 272-273.

Broschat, T. K., Donselman. H.M., and Will, A.A., 1984a. 'Andromeda' and 'Golden Torch' Heliconias. *Hort. Science,* 19(5): 736-737.

Broschat, T.K. and Svenson,S.E. 1994. DCPTA enhances growth and flowering of Heliconias. Hort Science, 29(8): 891–892.

Cabral, A. and Benedetto, D. 2010. Preliminary studies on the production of cut flower Heliconias in the Colorado. Hort. Argentina. 29 (69): 33-40.

Castro C.E.F., May, A., and Gonçalves, C. 2007. Nomenclature review of species of genus Heliconia (Heliconiaceae). Rev. Bras. Hortic. Ornam. 13: 38-62.

Catley, J. and Brooking, I. 1996. Temperature and light influence growth and flower production in Heliconia Golden Torch. Hort. Science, 31:213-217

Clemens, J. and Morton, R.H. 1999. Optimizing mineral nutrition for flower production in Heliconia 'golden torch' using response surface methodology. J. Amer. Soc. Hort. Sci. 124(6): 713–718.

Costa, A.S., Loges, V., Castro, A.C.R., Guimaraes, W.N.R., and Nogueira, L.C.2009a. Heliconia genotypes under partial shade: I. Shooting and blooming. Acta Hort. 813: 609-614.

Costa, A.S., Loges, V., Castro, A.C.R., Nogueira, L.C., Guimaraes, W.N.R. and Bezerra, G.J.S.M. 2006b. Number of shoots and blooming of Heliconia cultivated under partial shade. In: Mercuri, A. and Schiva,T. (eds.), Proceedings of XXII EUCARPIA Symposium,11-15 September, 2006, Sanremo, Italy, pp. 20.

Criley, R.A. 1995. Culture profile of Heliconia psittacorum. Bul. Heliconia Soc. Intl. 8 :9-11.

14. Criley,R.A. 1999. Seasonal flowering patterns for Heliconia shown by grower records.In Maloupa,E.(ed.), Proceedings of the Fourth International Symposium on New Floricultural Crops; 22-27 May 1999; Chania,Crete, Greece.pp.372

Enriquez, S., Duarte, C.M., Sand-Jensen, K.,and Nielsen, S.L. 1996. Broad-scale comparison of photosynthetic rates across phototropic organisms. Oecologia 108: 197–206.

Ferreira, D.B. and Pires, L. 2005.Effect of NPK fertilization on the initial productivity of inflorescences Heliconia sp. Electronic Proceedings of the II Seminar of Research and Graduate Studies [CD-ROM]. Available: http:// www.conference.ifas.ufl.edu/gardener10/ Proceedings07.pdf. [21 Sep.2011].

K. Nihad, Mukesh K. Berwal., K. Balachandra Hebbar., Ravi Bhat., A. Abdul Haris and S. V. Ramesh.2019. Photochemical and biochemical responses of heliconia (Heliconia stricta 'Iris') to different light intensities in a humid coastal environment Hortic. Environ. Biotechnol.60:799-808 https://doi.org/10.1007/s13580-019-00173-1

Kress, W.J. 1990. The diversity and distribution of Heliconia (Heliconiaceae) in Brazil. Acta Botanica Brasilica 4: 159-167.

Loges, V., Costa A.S., Castro, A.C.R., Nogueira, L.C., Guimaraes, W.N.R., Motta, R.M., and Castro, M.F.A.2006. Inflorescence morphological characters of Heliconia genotypes in Brazil [abstract]. In: Abstracts, XXII EUCARPIA Symposium (Section Ornamentals) "Breeding For Beauty"; 11-15, September, 2006, Sanremo, Italy,p. 25. Abstract No.23

Manarangi, A., Sakai, W.S., Gerken, C., Crowell, M., Nielson, G., and Short, R. 1988. Growth and flowering of Heliconia psittacorum cv. Parrot in Hawaii. J. Hawaii Pacific Agr. 1: 1–3.

Mediavilla, S., Escudero, A., and Heilmeier, H. 2001. Internal leaf anatomy and photosynthetic resource-use efficiency: interspecific and intraspecific comparisons. Tree Physiol. 21: 251–252.

Moles, A.T., Warton, D.I., Warman, L., Swenson, N.G., Laffan, S.W., Zanne, A. E., Pitman, A., Hemmings, F. A., and Leishman, M. R. 2009. Global patterns in plant height. J. Ecol. 97(5): 923-932

Nihad K, Krishnakumar V and Ravi Bhat .2016.Floriculture for enhancing profitability of

Nihad K, Krishnakumar V and Sheela VL. 2013. *Heliconia stricta* as intercrop in coconut- Impact of cropping system and nutrition on floral traits. *J Plant Crops* 41: 251-254

Nihad,K. Abdul Haris,A and Krishnakumar,V. 2019. Performance of Alpinias and Bird of paradise in coconut gardens-A pilot study. Kalpa CPCRI Newsletter 38(1):4

Nihad,K., V. Krishnakumar, A. Abdul Haris, Ravi Bhat and P. Chowdappa.2018. Heliconias: A potential intercrop in coconut ecosystem. Bulletin Heliconia Society International-A journal of the zingiberales 24(4): 8-9.

Nihad,K., V.Krishnakumar and. Sheela. V.L 2018. Relationship between stress and flowering in Heliconia (*Heliconia stricta).* International Journal of Agriculture Sciences. 10 (3): 5137-513

Nihad, K. 2013. Nutrient management practices for Heliconia under open condition and as intercrop in coconut garden. PhD Hort.), Kerala Agricultural University,Thrissur. P.142

Poorter, H. 1990. Interspecific variation in relative growth rate: on ecological causes and physiological consequences. SPB Academic Publishing, Germany 168p.

Powell, J. 1991. Ornamental Flower Research: 1989-90, Tech.Bull. No. 170, Berrimah Agriculture Research Centre, Darwin, Australia.15p.

Rocha, F.H.A., Loges, V., Costa, A. S., Aragao, F.A.S., and Santos, V. F. 2010. Genetic study with *Heliconia psittacorum* and interspecific hybrids. Crop Breed. Appl. Biotech.10: 282-288.

Rodrigues, P.H.V.2008. Somaclonal variation in micropropagated *Heliconia bihai* cv. Lobster claw plantlets (Heliconiaceae). Sci. Agric. 65(6): 681-684.

Sheela, V.L.2008. *Flowers for trade*. In: Peter, K.V. (ed.), New India Publishing Agency, New Delhi, pp 193-210.

Simao, D.G. and Scatena, V.L. 2001 Morphology and anatomy in *Heliconia angusta* Vell. and *H. velloziana* L. Emygd. (*Zingiberales: Heliconiaceae*) from the Atlantic forest of southeastern Brazil. Rev. Bras. Bot. 24(4): 55-65.

Stiles, F.G.1975. Ecology, flowering phenology, and hummingbird pollination of some Costa Rican Heliconia species. Ecology 56: 285-301

Syvertsen, J. P., Lloyd, J., Conchie, M. C., Kriedemann, P.E., and Farquhar, G.D. 1995. On the relationship between leaf anatomy and CO_2 diffusion through the mesophyll of hypostomatous leaves. Pl. Cell Env. 18: 149-157.

Uchida, R. 2000.Recommended plant tissue nutrient levels from some vegetable, fruit and ornamental foliage and flowering plants in Hawaii. In: Silva,J.A. and Uchida,R.S. (eds.). Plant nutrient management in Hawaii's soils, approaches for tropical and subtropical agriculture, College of tropical agriculture and Human Resources, University of Hawaii, Manoa. Available:http://scholarspace.manoa.hawaii.edu/bitstream/handle/10125/1908pnm 0. pdf?sequence=1[25 Aug.2011]

White, J.W. and Montes, R.C. 2005. Variation in parameters related to leaf thickness in common bean (*Phaseolus vulgaris* L.). Field Crop Res., 91: 7–21.

22

Dahlia

Simrat Singh, R. K. Dubey and K. K. Dhatt

Dahlia (*Dahlia variabilis* Desf., family: Asteraceae, origin: Mexico) is regarded as a premium specialty cut flower besides its utilization for garden decoration and growing as a potted flowering plant. The dahlia flower symbolizes elegance, inner strength, creativity and dignity. It is herbaceous rooted perennial valued for their majestic and attractive display of colourful spectacular flowers. The generic name "Dahlia" was coined by Swedish botanist "Andreas Dahl". Dahlia was introduced in India during 1857 under the auspices of the "Agri-Horticultural Society of India", Calcutta.

Dahlia exhibit wide range of flower from and size ranging from miniatures (less than 2.5 cm across), to giant (more than 40 cm in diameter). Dahlia flowers profusely and its easy cultivation have made these very popular worldwide. The commercial value of this crop has been exploited in certain countries only. The Netherlands is claimed to be the largest producer of tuberous rooted dahlias. Other countries cultivating Dahlia as a specialty cut flower are Japan, France, South Africa, the U. K., Italy, Germany and the United States. The commercial cultivation of dahlia in Indian subcontinent is limited to hills and plains of Eastern India (Misra and Misra, 2017).

Holland takes the lead in dahlia bulb production cultivation neeary 400 ha, followed by France and Chile. There are 41 international associations concerned with the cultivation and spreading awareness about dahlia as a potential specialty cut flower and flowering potted perennial. Societies such as National Dahlia Society in the UK, the American Dahlia Society (ADS), other groups of New Zealand in European countries, France and Germany have been maintaining Dahlia germplasm for developing novel flower colour and breeding dahlias for their uniform and profuse flowering, longer flower duration, stem sturdiness and longer vase life.

Importance and uses

Different flower forms of Dahlia arc extensively used for beautification of residential homes, garden exhibitions, flower bed decoration and as a cut flower (Bhattacharjee et al, 2006). Dwarf types of Dahlia are excellent for raising as single or mixed borders. Giant or large flowered decorative dahlias are suitable for pot culture in terrace garden or for decorating the veranda. The long, sturdy stemmed Dahlia is utilized as specialty cut flower for flower arrangements. Pompon dahlias are also good for cut flowers and sometimes for making garlands.

The Dahlia flowers are a potential source of flavonoids (anthocyanins, butane and flavones) as petal pigments. Cultivars evolved from the *Dahlia pinnata* are believed to possess edible tuberous roots- a rich source of carbohydrates for human diet. The inulin found in the tubers (found in inner cortex and parenchyma tissue) is a rich source of polysaccharides that can be converted into fructose for further utilization in pharmaceutical industry. Other useful compounds that have been isolated from Dahlia tubers are polyacetines, tetrahydropyrans, suberine, ferulic aci and glycerol (Moldovan *et al,* 2017).

Origin

Dahlia is believed to be a native of Mexico and was discovered by the Aztecs during 1570's who named it "*Cocoxoehitl*". They used to grow it as a medicinal plant as well as a flower for worship. The plant was first mentioned by Francisco Hemandez, physician of Philip II in 1651 and thereafter, Vincente Cervantes, Director of the Mexico City Botanic Garden in 1789. The readers can refer to Shewell-Cooper, (1976) for gaining insights into the history of dahlia.

Taxonomy

(Sorensen 1969) proposed the most comprehensive document on the dahlia taxonomy. He recognized 27 species and 4 varieties of *Dahlia* through extensive field collections and observations followed by cytological studies. (Willis 1966) described about 20 species of dahlias.

Species and cultivars

The genus Dahlia comprised of 35 wild species that are endemic from Mexico, however only four of these species namely *D. coccinea* (single red flowers), *D. pinnata* (bluish red flowers), *D. merckii* (lilac and yellow flowers) and *D. imperialis* (white red tinged flowers) have constituted the genetic basis for the evolution and development of modern day Dahlia cultivars. The most widely known is *D. pinnata*, which is derived from a fertile hybrid between *D. coccinea* and *D. sorensenii* likely produced in Mexico about 500 years ago (Marina, 2015). Owing to wide variation in flower forms of Dahlia, there was need to classify the wide morphological variants based on flower size and form. The American Dahlia Society (ADS) has classified Dahlia into 20 "classes" (Forms) that generally represent the inflorescence type based on shape and arrangement of ray florets (Table 1)*.

Table 1: Classification of Dahlia based on inflorescence type

Code	Description	Code	Description
AN	Anemone-flowered	BA	Ball
C	Straight Cactus	CO	Collarette
FD	Formal Decorative	IC	Incurved Cactus
ID	Informal Decorative	LC	Laciniate
MB	Miniature Ball	MS	Mignon Single
N	Novelty	NO	Novelty Open
NX	Novelty Fully Double	O	Orchid-flowering
P	Pompon	PE	Peony-flowering
S	Single	SC	Semi-Cactus
ST	Stellar	WL	Water Lily

*Adapted from Washington State University (WSU), Master gardener program

Cultivars

According to Swami Vinayananda - a pioneer Dahlia breeder, around 20,000 cultivars of Dahlia that are recognized with International of dahlia at Royal Horticultural Society (RHS), England. Due to heterozygous and cross-pollinated nature of Dahlia, the development of new cultivars is in progress and the evolution of varieties with novel colour and other desirable traits is increasing every year. Around 57 DUS characteristics have been described under European UPOV species code of Dahlia. Some of the outstanding cultivars of dahlia based on inflorescence form that are prominent in trade and grown throughout the world are as follows:

Flower type	Cultivar name and colour
Anemone Flowered Dahlia	Guinea (yellow), Scarlet Comet (bright scarlet)
Ball Dahlia	Alltamy Cherry (bright red), Blossom (pale pink), Camano Candy (pink), Crusader (red), Dr. John Grainger (burnt orange), Risca Miner (purple)
Cactus Dahlia	Alvas Doris (bright red), Arthur Laslie (bright red), Athalie (deep pink)
Collerette	Christmas Star (red and white), Claire De Lune (yellow), Komeet (red and buff), Mars (red), Thais (purple and white).
Decorative	
Giant Decorative:	African Queen (purple to crimson), Alvas Supreme (yellow), Barbara Marshall (red), Basudev (light purple) Bhikkus Vivek (vermilion), Blue Monarch (bluish red), Camano Titian (pink blends)
Medium Decorative	Alloway Cottage (yellow blend), April Dawn (pink and white bleed), Bhikkus mother (orange to tan with white tips), Chitchor (bicolour, blackish red with white tips) King Soccer (orange blend), Lord Buddha (bicolour, scarlet with white tips)

Flower type	Cultivar name and colour
Large Decorative	Alden Galaxy (red), Arthur Hambly (rosy lavender), Black Out Sport (variegated, with blackish red stripes over petals), Islander (deep pink), Kenora sunburst (Yellow)
Miniature Decorative	Christine Hammett (apricot), Eastwin (purplish red), Elizabeth Hammet (lavender) Delight (white with lavender bluish), Robert Walker (purple and white blend)
Fimbriated Dahlia	Cabaret (bicolour, carmine and white), Cheyenne (flame), Phenomenon (bocolour white and pruple), Frilly Dilly (Pink)
Pompon Dahlia	Andrew Lockwood (lavender), Ezar Willo (purple), Diana Gregory (lavender and white),
Semi Cactus Dahlia	Amelisweerd (pale bronze), Bob Fitzjohn (light bronze) Carol changing (bronze yellow) Daleko Jupiter (red and yellow), Davenport Sunlight (yellow)
Water Lily Dahlia	Christopher Taylor (red), Erin Ann (pink), Island Dawn (Pink and yellow blend)
Miscellaneous	Golden Star (yellow), Jescot Julie (orange), Pink Giraffe (pink).

Dahlias are often grouped by the size of the bloom with a letter code. The following *table mentions the letter codes and their corresponding sizes:

Code	Name	Size
AA	Giant	over 10 inches in diameter
A	Large	over 8 to 10 inches in diameter
B	Medium	over 6 to 8 inches in diameter
BB	Small	over 4 to 6 inches in diameter
M	Miniature	up to 4 inches in diameter
BA	Ball	over 3.5 inches in diameter
MB	Miniature Ball	over 2 to 3.5 inches in diameter
P	Pompon	up to 2 inches in diameter
MS	Mignon Single	up to 2 inches in diameter

*Adapted from Romer and Nelson, (2008)

Climate

For open field cultivation of dahlia, the selection of site is important. The site should be protected from strong winds that may lodge the tender and succulent stems laden with big blooms. A cool atmosphere free from frost is needed for proper growth and flowering.

Dahlia plants response to varying day and night temperatures. The day length also influences the flowering of the shoot and tuberisation of the roots. Ideal temperature of 25° C day and 16° C night has been found desirable for the dahlia to produce healthy and vigorous vegetative growth with quality flowers. It has been reported that the cooler temperatures (less than 18°C) combined with short days promote tuber formation in Dahlia. It has also been reported that the number of ray and disc florets increased with decreasing night temperature.

Dahlia prefers to grow under high light intensities (4500-6000 f.c.). Dahlia roots began to from tubers in short days, therefore, rooted cuttings of Dahlia are subjected to long days (under protected conditions) by provising 14-hr day length through night interruption. It is always advisable to maintain 2000 f.c. of light intensity best growth of plants.

Soil

Dahlia plants prefer a sandy loam, well drained and fertile soil rich in organic matter with a pH range of 6 to7. Dahlia plants can be grown successfully both in pot and in the ground.

Propagation

Dahlia plants can either be propagated from seeds, tuberous roots and cuttings. Multiplication by grafting is possible but this is useful for preserving rare and weak growing types. Propagation through tissue culture utilizing apical meristems as explants may be used for producing virus-free plants.

Seed Propagation

Seed propagation is mainly followed to raise dwarf bedding single dahlia plants and for developing new varieties. Plants developed through seeds show segregation and a heterogeneous population. The seeds are sown after cessation of heavy rains during September-October in the plains, and during summer in the hills. Sowing of seeds should be done thinly on raised nursery beds or in seed pans. The seeds are covered with a thin layer of screened leaf mould and watered through a fine rose can. Seeds germinate in about 10 days. A temperature of 18 to 28° C has been found favourable for germination of dahlia seeds. High temperature and low humidity conditions are not favourable for seed germination. Seedlings grow very fast and become ready for transplanting within 20-25 days.

Vegetative Propagation

Division: The dahlia plants can be propagated by division of tuberous roots each containing a called 'heel' with 1-2 buds. As these buds are not located on the tubers but are found on the crown (swollen portion of stem near the ground level) to which tubers are attached, care should be taken to contain at least one bud while dividing the tubers. After harvesting, it is therefore suggested to allow curing of the buds by placing the harvested and cleaned roots devoid of soil in a warm, moist place for a short period before dividing the roots. An average tubererous root weight of 80 to 120 g is optimum for proper bud development.

Cutting: Dahlias are commercially propagated by terminal softwood cuttings. For this purpose, the selected tuberous roots with 1-2 buds began to sprout in August when placed in moist and well drained medium. New shoots rise from the buds at the crown portion. The cuttings of 7 to 8 cm length with two to four pair of leaves are taken from the healthy shoots in September-October. These cuttings are treated with IBA powder (2500 ppm) as a quick dip to encourage faster rooting after planting in washed coarse sand for rooting. It has been observed that the rooting characteristics such as percentage of rooting, root number per cutting and root quality vary with the cultivars and the climate of region.

In areas with mild climate, the cuttings can be taken from emerging shoots periodically from July to mid-September. Swami Vinayananda has also advocated late cutting for preservation and propagation of dahlias. In this method, the cuttings are taken from the flowering plants and the resultant plants are subjected to a regular pinching off of the tips to make the plants bushy and encourage many soft branches. The cutting taken from the emerging succulent branches make quality planting material.

Grafting: Grafting is practiced for preservation of rare and weak germplasm. The tops of tubers are given a slanting cut and desirable portion of the scion is excised to make it compatible union with the scion. Both the stock and the scion are placed and tied together with raffia or polythene strips. They are then carefully planted in a pot containing a well drained potting mix.

Miller (1982) suggested two methods (French and English method) of grafting, namely based on the technique of excising the bud from the tubers. The English method of grafting consisted of removing the bud from the tuberous roots and replacing it with a bud from that of another desirable cultivar. French method involves the removing of buds giving a wedge shaped cut on the root stock and scion. These scion buds are then placed carefully on the rootstock tubers. However, the English method of grafting dahlia tubers was found better.

Planting

A dahlia tuber should be planted 4 to 6 inches deep from the ground level. The tubers are planted on a slight slant taking care of the orientation of the emerging bud toward the surface. After planting the exposed surface of the tubers are covered with 2 inches of soil/organic matter mix. It is advisable to insert the stakes of desirable height 6-8" from the base of the crown to support the taller plants by tying with a suitable gunny thread.

Manures and fertilizers

For commercial cultivation of dahlia, the field should be dug deeply and manured with well rotten FYM a few weeks before planting. At the time of land preparation, organic matter comprising FYM (5 kg per square meter) should be thoroughly mixed. Dahlia plants are responsive to the application of organic matter. In temperate regions, poultry manure is mostly preferred, but it may reduce the keeping quality of tubers. It has been reported that for tuber production soil should not be too fertile but should be well drained.

According to Swami Vinayananda (1990), beds may be prepared by digging the ground deep to a depth of 45 cm after incorporation of 6" layer of well rotten FYM. Addition of 60 gm SSP, 30 g MOP and 150 g bone meal per square meter should be applied. Dahlias are also grown in pots for quality flowers for exhibition purpose. A potting mix containing equal parts of sandy loam soil, decomposed FYM and leaf mould are mixed thoroughly on volume basis. In addition, 100 gm of mustard oil cake, pieces of charcoal, 50 gm of steramcal and trace amount of garden lime should be mixed as a soil amendment to improve the physic-chemical characteristics of the growing media.

Irrigation

Dahlia plants respond to deep irrigation once tubers get sprouted. Dahlia tubers are susceptible to rot especially in damp soils. The tubers should be planted in soil at field capacity and soil should never get dried after the tubers are sprouted and plants began to grow. Once plants get established, a deep irrigation 2-3 times per week in the summer is sufficient or scheduling can be done as per the prevailing climate. The soil should never completely dry out. Drip irrigation system is also advisable to irrigate dahlia plants that may suitably be fertilized with water soluble fertilizers through fertigation.

Intercultural operations

Staking: Tall dahlia varieties are prone to lodging and need to be staked. Small varieties that only grow to heights of 24 inches or less may not need staking unless they are able to support themselves or possess strong and sturdy branches. Stakes need to be atleast 4 - 5 feet tall or of a desirable height, depending on the mature height of the dahlia. It is advisable to stake the dahlia right at the time of planting the tubers.

Disbudding: Dahlia plant tend to branch profusely from the lateral buds that sometime may cause the plant to appear bushy that may affect the flowering. To promote the uniform growth and interception of adequate light within the dahlia plant, it is advisable to disbud and / or disbranch the plants during their active growth prior to bud initiation.

At the onset of crown bud that is removed to give space and divert the translocation of nutrients to developing companion buds to make the plant appear uniform in flowering. Contrary to the above method, Disbudding may also be done to remove the companion side buds to enable the central crown bud grow over a long sturdy stems. The flower that develops from the crown bud will also be larger due to diversion of applied nutrients to the developing bud.

Disbranching: Disbranching is the practice of removing the lateral branches from a dahlia stalk to get larger and quality blooms spaced uniformly all over the plant. Depending upon the branching habit of a particular variety, 5-8 branches are kept to ensure uniformity and profuse flowering on the plant. The miniature dwarf varieties need no disbranching. The size and shape of the dahlia bush is also improved considerably if timely and need-based disbranching is done. In addition, better air circulation is promoted which helps development of powdery mildew and other fungal diseases. Disbranching should be done carefully to avoid bruising injury to the rest of the plant.

Pinching: Pinching operation is done to promote shorter, bushier plants with better stems for taking cuttings. By pinching off the main stem, the lateral buds began to grow from the leaf axils below the pinch or cut. When the dahlia plants attain 12 -18 inches in height, pinching above the 4th or 5th full set of leaves will create a stronger, bushy and uniform plants with more blooms. It will also make the plant less top heavy.

Deadheading: Periodic deadheading the dahlia blooms as and when they get dried encourages more buds that flower subsequently. Dahlias will bloom profusely and continue to give flowers from January till last week of April if periodic deadheading is done.

Harvesting of flowers

Dahlia Flowers are harvested early in the morning or late in the afternoon at the fully opened stage. It is advisable to cut the bloom with as long a stem as possible. After harvest, stems are immersed in warm water of 50° C in a bucket and are kept in a cool and dark place for conditioning.

Vase Life: The vase life of cut dahlia flowers varies between 3-5 days at 4-5 °C. Generally, a vase solution containing 10 % sucrose + Ag NO_3 (50 ppm) is effectively used to enhance vase life of cut dahlias. Bud opening solutions can be used for opening of immature flower buds. For long distance transportation of flowers, the stems should be placed in water after removing all leaves to avoid bacterial contamination in vase water that may interfere in water uptake.

Harvesting of tubers

A single dahlia tuber planted during onset of autumn (Oct) will multiply into a number of tubers by late summer. Generally, tubers in dahlia began to develop under short days. The tuberization in dahlia is responsive to day length. The critical day length for tuber formation is 11-12 hrs. Dahlia plants exposed below this critical photoperiod began to store food in the roots for tuber formation. Plants that are grown under natural photoperiods develop more fibrous roots with greater shoot growth and later emergence of flowers than those that are grown under short days.

After leaves of the plant tend to dry, the tops are given clean cut close to the ground level. The field is lightly irrigated to contain adequate soil moisture but not too damp that may cause rotting in tubers. The dahlia tubers are dig gently in clumps, cleaned to remove any soil. The tubers are carefully separated with a sharp cut using a knife. Care should be taken that each of the tubers must contain an eye or viable bud attached around the crown of the "mother" plants. If the tubers are pulled off without an eye, they tend to rot and are rendered worthless. Separated and cleaned tubers are placed in warm well ventilated damp-proof room to let the cuts heal-over for a day or two before these can be packed for storage. Dahlia tubers must be labeled properly with name of the cultivar, colour etc. to ensure safe and easy identity during and after storage.

Storage of tubers

Tubers of Dahlia can be stored in layers of dry sawdust sawdust, peat moss, or coconut coir. Tubers should not be allowed to suffocate and be placed in well ventilated room. Under hilly conditions susceptible to frost, the dahlia tubers can be stored in a box covered with sand or saw dust to prevent freezing injury or loss of vigour. Ideally, tubers should be stored in a cool environment maintaining a uniform temperature of 4-7°C.

Swami Vinayananda (1983) also reported that tubers of Dahlia can be preserved in the ground itself ensuring presence of required soil moisture after flowering is over. The tubers in the soil should be left undisturbed, and they may sprout during Aug-Sept. Afterwards, the tubers may be lifted for division or for taking terminal cuttings after sprouting. In plains areas, the potted dahlia plants are shifted to the shade under large trees after heading back the dried branches close to ground level. The tubers are left in the pot itself (pot tubers) are kept moist to prevent from drying.

Pot cultivation of Dahlia

The pot cultivation of Dahlia is gaining popularity due to the availability of wide range of flower forms in different colours. Dahlia being tuberous rooted, prefers porous, well drainage soil rich in organic matter. Selection of pots is also very important. Earthen pots of 20-30 cm diameter filled with growing media mix comprising loamy soil: FYM: leaf mould @ 3:3:2 is considered as ideal for optimum growth and flowering.

Dahlia being a heavy feeder requires enrichment of potting media with essential micro-nutrients. Addition of 30 g bone meal, 20 g horn meal, 15 g SSP, 5 g of SOP per 20-25 cm pots is advisable for healthy growth and development of Dahlia. It is desirable to add potassium (K), as dahlia has a high intake of this element, due to its large flowers that continue to bloom for several weeks. Dahlia is quite responsive to the application of liquid manures. A liquid mix comprising cowdung, mustard oil cake and rotten fish given bi-weekly by diluting in equal amount of water is beneficial for Dahlia. Foliar spray of urea may be given once a week after the establishment of plants, and the spray can be continued till emergence of the buds. The colours of the dahlia blooms intensity with the application of manganese sulphate, fish bile and bone meal at weekly intervals. Once the buds began to show colour, manuring should be stopped. Application of dihydrogen orthophosphate (0.3%) and potassium nitrate at 7 day intervals produces large-sized dahlia blooms with bright colours.

Insect pest and diseases

1. **Aphids:** Aphids generally act as a vector for transmitting viral diseases such as mosaic spotted wilt. The aphids also suck the sap and affect the general vigour of plants. Common species of aphids observed on Dahlia are *Myzus persicae, Brachycaudus helichrysi and Aphis fabae*. The aphids are controlled by spraying with 1 % metacid.
2. **European Corn Borer** (*Pyraussta nubilalis*): The young larvae bores into succulent shoots, flower buds and feed on leaves which get distorted and turn brown. The affected shoot tips die completely. The borers bore down within the stems causing wilting of plant. The pest can be controlled by spraying with 1% Dichlorovos.
3. **Leaf Hoppers** (*Empoasca fabae*): The nymphs and adults suck the sap and the leaves turns pale yellow initially, turning brown and brittle at later stage. Spray 1 % Malathion to control leaf hoppers.

4. **Beetles** (*Oxycetonia albopunctata*): The beetles feed on developing buds and flowers and the blooms began to show distorting. Spray with Carbaryl (0.1 %) is effective to control this pest.
5. **Thrips:** Several species of thrips (*Megaleurothrips distalis, Microcephalothrips abdominalis* and *Thrips hawaiiensis* have been reported to infest flowers. These thrips feed on succulent tissues and flowers at full bloom. Spraying with 1 % Malathion has been found effective to control thrips.
6. **Red Spider Mite** (*Tetranychus urticae*v) These mites are very serious pests of dahlia damaging the foliage by feeding on the undersides of leaves, causing a white speckling on the leaf surface. Spray with 1 % Endosulphan is effective to control the pest.
7. **Powdery Mildew**: Powdery mildew *(Erysiphe chicoracearum)* is one of the most important fungal diseases of dahlia. It is characterized by the development of white or grey powdery mass on the upper surface of the leaves. The disease is most prevalent in the early stages of plant growth and reappears at the end of the flowering season. The disease is mainly responsible for the rotting of seed heads.

Control: Sprays with Bavistin, Benlate (0.1 %) or wettable sulphur (0.2 %).

1. **Bud and Flower Blight:** This disease is caused by the fungus *Botrytis cinerea.* It occurs in cool and moist climates and produces water-soaked spots on buds and flowers. The infected areas get covered with grey-coloured mycelium and spore growth, which ultimately become faded and turn brown in colour. It can control by spraying with Zineb or Captan (0.2 %).
2. **Stem Rot**: The fungi *(Sclerotinia scleroliorum)* infests the main stem and branches near the base. The young seedlings also get infected with typical symptoms of light brown water soaked patches covered by white mycelia growth over the infected stems. Plants ultimately wither and die. To control stem rot, overcrowding of plants should be avoided. The potting mix may preferably be sun exposed to kill harmful fungi.
4. **Wilt:** Wilt is a serious disease caused by *Fusarium oxysporum* f. *dahliae* and *Verticillum dahlia.* The plants began to wilt and die due to blockage of xylem vessels as a result of toxins produced by the fungus. Wilted plants should be destroyed and only disease-free planting stock should be used for propagation.
5. **Leaf Spot**: Leaf spot caused by *Cercospora grandissima* and *Cercospora dahliicola.* The symptoms appear in the form of angular to irregular dark brown spots over leaves which later on coalesce, forming large patches. The disease is more severe in humid climates. Diseased leaves and plants should be removed and buried. Spraying with Bavistin (0.1%) or Zineb (0.2%) is effective.

Bacterial Diseases

1. **Crown Gall**: Crown gall caused by *Agrobacterium tumefaciens.* The symptoms are characterized by irregular growth of tissue near the crown portion. Infected plants become stunted and began to show decline. The roots of seedlings should be dipped in Streptomycin solution prior to transplanting. Uproot the infected plants and bury them.
2. **Bacterial Wilt**: The disease is caused by the bacterium *Pseudomonas solanacearum.* The infected plants began to droop and wilt suddenly. The tissue at the crown portion began to rot and start oozing yellow slimy exudates. Dip treatment of roots in Streptomycin solution is the only remedial measure.

Viral Diseases

1. **Mosaic**: This is the most serious virus disease of dahlia. The leaves show mottling and typical vein-banding symptoms. The virus is sap transmissible and infected plant exhibit a mosaic or pale yellow appearance. It is also transmitted by vector *Aphis gossypii and Myzus persicae*

References

Bhattacharjee S.K. Vinayananda S and De L.C. (2006). Dahlia. Advances in ornamental horticulture. Vol p: 181-198

Miller, H (1982). 'Dahlia of today'.Puget Sound Dahlia Association.

Marina LJ (2015). Cultivation of Dahlia: Review. Cultivos Tropicales. Vol 36 (1), p: 103-110.

Misra, R.L and Misra, S. (2017). Dahlia,Commercial Ornamental Crops-Traditional and Loose Flowers. Kruger Brentt Publishers,U.K.p:45-70.

Moldovan I, szekely-Varga Z and Cabtor M (2017). Dahlia an unforgettable flower-A new perpective for therapeutic medicine. Hop and Medicinal plants. No. 1-2. P:56-68.

Romer J and Nelson D (2008). Growing Dahlias. Iowa State University. P:1-4.

Sørensen PD (1969) Revision of genus Dahlia (Compositae, Heliantheae, Coreopsidina) Rhodora71:309-416.

Shewell-Cooper W.E. (1976). Basic book of dahlia growing. Barrie and Jenkins, London.

Swami Vinayananda (1993). Dahlia breeding. Dahlia society of India pp:79.

Swami Vinayananda (1990). 'Dahlias'. Flower Show Souvenir. Delhi, India: Dahlia Society.

Swami Vinayananda (1983). Dahlias of Today. Puget Sound Dahlia Association, pp. 24-25.

Washington State University (WSU) Master gardener program. Dahlias. C026. http://extension.wsu.edu/spokane/master-gardner-program/home-lawn-and-garden.

Willis, J C (1966). Flowering Plants and Ferns. U.K.: Cambridge University Press.

23

Lilium

M.R. Dhiman

Lily is one of the most beautiful and popular ornamental bulbous flowers. They belong to the genus Lilium of the family *Liliaceae*, which comprises of 110 species classified into seven taxonomic sections: Lilium, Pseudolirium, Martagon, Sinomartagon, Archelirion, Leucolirion and Oxypetalum. All species are distributed over the mountainous area in the Northern Hemisphere (10° to 60°), mainly in Asia, North America and Europe. Especially, China, Nepal, Korea and Japan are the major centres of lilium gene pool around the world. Being second amongst the bulbous flower crops in the commercial market, Lilium has 4th position among cut flowers in Flora Holland (Flora Holland Annual Report 2017). In the language of flowers lily is known as symbol of purity and innocence. Majority of cultivated lilies are either hybrids or selections. A large number of species and hybrids can be used in ornamental gardens even though the lily requires little more attention than any other bulbs. Many species of lily (*L. leichtlinii*, *L. lancifolium*, *Lilium candidum* and *L. auratum*) are also edible and having medicinal value.

Importance and uses

Lilies are one of the most important bulbous ornamental plants with varies uses, grown in border, beds, pots, landscape gardens and are excellent cut flowers of magnificent appearance and beautiful flowers. Roman naturalist and writer, Plinius, recorded that salves and oils were prepared from leaves and flowers of the Madonna lily, or *L. candidium*. Some species are also known for medicinal and food values, which increased its economic importance many folds. The mucilaginous substance coming out from the bulbs while cooking is used as an ointment for tumours, ulcers and inflammations. The plant was really more of a miniature household pharmacy. Different compounds have been reported in this genus, such as carbohydrates, proteins, fats, amino acids, vitamins, alkaloids, phenolics, flavonoids, carotenoids, sterols, steroidal saponins, and steroidal glycoalkaloids. Alkaloids, steroidal saponins, flavonoids and polysaccharides have been isolated from *L. lancifolium* and *L. brownii* var. *colchesteri* bulbs. Flowers of *Lilium candidium* contain an alkaloid. Extract of *Lilium polyphyllum* bulb has been used as refrigerant, expectorant, aphrodisiac, diuretic, antipyretic and tonic (Warrier *et al.*, 1997; Dhyani, 2007). In the customary system of medicine, this species reported to restore health immediately and works as antioxidant in the body. Lily oil is an aid in childbirth. The white lily was also renowned as a cosmetic.

Origin and distribution

Lilies are endemic to mountainous region of the Northern Hemisphere up to South Canada and Siberia and their Southern limit is Florida and the Nilgiri hills of India. Lilium, which is taxonomically and phylogenetically regarded as an important clade of the core Liliales, appears to have evolved in the Himalayas approximately 12 million years ago, despite the lack of fossil records. Approximately, 55 species occur naturally in the Southwest and north of China, which is considered to be the global centre of wild lily diversity. The East coast of Asia, the West Coast of North America and the Mediterranean region are the three most richly garnished places. Far East is the home of almost half of the world's lilies and majority of our finest garden lilies originated there. A few species have been found in Northeast India (*Lilium wallichianum, L. napalense* and *L. polyphyllum*) also. The distribution of the lily is mainly based on environmental conditions like altitude, temperature, humidity, light intensity, etc. A species may have a limited distribution, for example, *Lilium regale* is found only in one steep-sided Chinese valley; others can be wanderers like the champion globe-trotter like *Lilium martagon* known from Siberia across to Poland and down to the Balkans. In Taiwan, the trumpet lily (*Lilium formosanum*) called after the islands former name, grows from sea-level to some 3600 m.

Taxonomy

Lilium belongs to the subclass Monocotyledonae and family Liliaceae. It is an herbaceous perennial. The lily bulb is composed of firm, fleshy scales that store food for the following season's growth. The bulb consists of short stem, or axis, to which the scales are attached. The axis also called the basal plate is the most important part of the bulb which produces the roots, scales, and buds for new growth. The colour of the scales is one feature by which a lily species may be identified; this colour may change on exposure to light. The lily bulbs are concentric or rhizomatous. Most European and Asiatic lilies have concentric bulbs. In species such as *Lilium lankongense, L. nepalense*, and *L. wilsonii*, the stem travels underground for some distance before emerging, bearing bulblets away from the mother bulb. This habit has been termed stoloniferous. The rhizomatous type of bulb is best developing in the Eastern American species (*Lilium canadense, L. michiganense*, and *L. superbum*). A modification of the rhizomatous type of bulb is exemplified in the Western American species *Lilium pardalinum* and *L. parryi.*

A mature lilies' flowering stem may be as short as a few inches in *Lilium nanum* or other high alpine species, or as tall as 250 cm, as in *Lilium leucanthum* var. *centifolium* or *L. superbum*. Some stems rise straight from the bulb, as in *L. martagon* and *L. regale*, others travel horizontally underground before emerging, as in *L. lankongense* and *L. nepalense*. The leaves range from the narrow grass like foliage of *L. pumilum* to the broad, lanceolate leaves of *L. auratum* var. *platyphyllum*. Some species, such as *L. taliense*, produce a naked, asparagus-like stem that rises 30 cm or more before the leaves expand. *Lilium martagon, L. hansonii*, and their hybrids, as well as several North American lilies, bear their leaves in regular whorls around the stem, with gaps between the leaves. In most of other lilies, such as the trumpet and Oriental species and hybrids, the leaves are arranged alternately. This is also same in *L. candidum* and *L. x testaceum*, but their leaves diminish in size from the base upward. Several species and hybrids (*L. lancifolium*) bear small purple bulbils in the axils of their leaves.

The lily inflorescence may be a raceme, an umbel, or a single terminal flower. The flowers of lilies are quite diverse in form and colour, and this diversity contributes enormously to the charm and beauty of the genus. The basic forms are Turk's-cap (*Lilium martagon*), trumpet (*Lilium longiflorum* and *L. regale*) and bowl shaped (*Lilium auratum*). The carriage of the flower is another feature used to differentiate lilies. Flowers may be upright or up-facing, outfacing, or pendent. Within the flower are six stamens composed of slender filaments or stalks with the anthers or pollen-bearing organs at their tips. Lily pollen varies greatly in colour from species to species and among hybrids, ranging from soft yellow to dark brown. The centre of the flower contains the pistil composed of ovary at the base, a long style, and a three-lobed stigma at the tip. At the base of each tepal there is a narrow groove, the nectar furrow. Nectar is secreted here to attract pollinating insects and birds. Some species and hybrids, however, lack nectaries.

The form of the lily seed capsule differs among species and hybrids, ranging from the relatively short capsule of *Lilium candidum* to the long, slender capsule of *L. formosanum*. All lily capsules are divided into three two-part sections with papery dividing walls, inside which the numerous flat seeds are stacked like corns in a wrapper. All lily seeds are flat. Some, such as those of *Lilium auratum*, have large "wings" – the papery margins that aids in wind dispersal of the seeds. Others, such as those of *L. polyphyllum*, have very little wing tissue. In a fertile seed embryo appears as a line through the endosperm, the darker mass in the centre of the seed.

Species and cultivars

The genus Lilium comprised of around 110 species (GRIN 2016), which have been taxonomically divided into seven sections: Lilium, Pseudolirium, Martagon, Sinomartagon, Archelirion, Leucolirion and Oxypetalum (De Jong 1974). Some of the important species and their descriptions are given below:

***Lilium alexandrae* (syn. *Lilium nobilissimum*, *Lilium japonicum* var. *alexandrae*, *Lilium longiflorum* var. *alexandrae*, *Lilium makinoi* var. *alexandrae*, *Lilium ukeyuri*)**: It is native to the south of Japan. Flowers are trumpet shaped and pure white in colour. It is sometimes confused with *Lilium longiflorum*. It is almost similar to *Lilium Longiflorum* but unlike it, in *Lilium alexandrae* the tepals are free, the style has a distinct curve and the stigma is brownish.

Lilium amabile: It is native to Korea. Flowers are deep red with strong fragrance, which may be sometimes unpleasant. It is free flowering from even small sized bulbs.

Lilium auratum: This species is also called Gold Band lily due to the presence of gold colour. The flowers are fragrant, 15-25 cm in diameter, bowl shaped with tips of petals recurving. At the base of the tepals are papillae and the centre of the tepal has a channel which leads to the nectarines.

Lilium candidum: It is well named as *candidum*, which means dazzling white as the flowers are pure white in colour. Flowers are fragrant, 12-14 cm in diameter and are associated with religious feasts and symbolism, flower June – July.

Lilium columbianum: Commonly known as Columbia tiger or Oregon lily. It is native to northern California through Oregon and western Idaho and into Washington State and British Columbia. The flowers are dull orange and the petals recurve almost into a ball. The flowers are not so attractive but are produced in large numbers.

Lilium formosanum: It is native to the island of Formosa. An interesting point of this species is that with changing altitude the change in its morphology can be easily observed. The plants are at lower elevations are larger in size and produce flowers of pure white colour, trumpet shaped and of 25 cm length while with increasing elevation the size of the plants as well as flowers decreases. Thus the plants at highest elevation are much smaller and flowers have a red reserve in tepals.

Lilium henryi: This species is native to China. The shape of flowers is similar to *Lilium speciosum*; hence it is also called Orange *speciosum*. The bulbs are 15 cm long and also broad. The flowers are bright orange with numerous dark carmine spots, 7-9 cm in diameter and 8-10 in number in a raceme. The flowers are likely to bleach in full sun, so the species should be planted where it is shaded during the hottest part of the day.

***Lilium longiflorum* (Easter lily)**: It is native to Japan. Flowers are pure waxy white, funnel shaped, mild fragrant, 5-6 per plant and 10-15 cm long. The bulb is globular and large up to 10 cm diameter.

***Lilium Martagon* (Turk's - cap lily)**: This species is native to Europe. The flowers are pendant and when fully open the stamens are quite prominent, jutting out from the centre of the flower. Flowers are purple with many darker spots on the tepals and are 3 cm long.

Lilium nepalense: It is native to Nepal. Flowers are bowl-shaped, fragrant, pendulous, and green with red purple inside, 15 cm in diameter and 4-5 per plant. Bulbs are globular, diameter 8-18 cm with broad thick scales and purple in colour.

***Lilium pardalinum* (Panther Lily)**: It is native to the coastal ranges of California. Flowers are bright yellow, spotted brownish purple, 10 cm in diameter and pendulous. Bulbs are scaly rhizome.

***Lilium parryi* (Lemon lily of California)**: This species is native to southern California. The colour of flowers is lemon-yellow with maroon spots. Flowers are funnel shaped; the tips of the tepals are curling back and are 8 cm in diameter.

Lilium philadelphicum: It is native to Canada and Carolina. The flowers are cup shaped, upright and orange to deep red in colour.

Lilium polyphyllum: It is native to the western Himalayas. In the wild it grows up to 20 cm in height on the hills but in cultivation it grows only to 10 cm. Flowering takes place during June-July on the hills. Flowers are creamy yellow, tinged and dotted purple, fragrant and 4-5 per plant.

Lilium regale: It is native to the western China. This species flowers in July but the colour is not fade in the sun. Flowers are small, white, tinged yellow at the base with purple midribs, trumpet-shaped, 10-15 cm long and 3-4 in number per plant.

Lilium speciosum: Native to Japan. It is one of the most beautiful lilies. The species name 'speciosum' is quite suitable as it means good looking, showy and splendid. Flowers are delicately fragrant, white, suffused rose pink in the centre and spotted blood red with a green stripe at the base. The tepals are recurved and twisted in such a way that the centre of the flower is well exposed with quite prominent papillae and stamens.

Lilium tigrinum syn. ***Lilium lancifolium*** **(Tiger lily)**: It is native to Japan. Flowers are bright salmon red with purple-black spots. This pattern of colouration resembles with the skin of the tiger, which gives it its name of Tiger-lily. Flowers are 8-10 cm in diameter and 10-15 per plant. It produces deep purple bulbils in the upper leaf axils.

Lilium wallichianum: This species is native to western Himalayas regions. Bulb globular, 5-8 cm in diameter, edges slightly notched, broad, thick reddish. Stem 1.5 to 2.0 cm long, green smooth, leaves erect, 10-12 cm long. Flowers 2-3, 20 cm in diameter, fragrant, waxy, creamy white, flowers during July-August.

Cultivars

In recent years, due to interspecific hybridizations, a wide variety of hybrids are developed. These hybrids are divided morphologically and taxonomically into seven divisions each representing a separate group. Some of the outstanding cultivars of each group are classified below.

Division I Asiatic hybrids: Hybrids derived from the following species and interspecific hybrids: *amabile*, *bulbiferum*, *callosum*, *cernuum*, *concolor*, *dauricum*, *davidii*, *L.* × *hollandicum*, *lancifolium* (syn. *tigrinum*), *lankongense*, *leichtlinii*, *L.* × *maculatum*, *pumilum*, *L.* × *scottiae*, *wardii* and *wilsonii*. Flowers usually small to medium-sized, few to fairly numerous, bowl-shaped, flat or with tepals recurved, usually with an open centre, up-, out- or down-facing. Flower colour often uniform, or with contrasting tepal tips and/or throat, rarely with contrasting white tepal margins. Spots absent, or when present, well defined and often rounded. Papillae, if present, usually inconspicuous. Conspicuous brush marks sometimes present. Tepal margins usually smooth or slightly ruffled, with relatively inconspicuous nectaries. Flowers usually with little or no scent. Inflorescence sometimes with secondary buds. Buds sometimes hairy. Leaves scattered, narrow to medium width. **Examples**: Angela North, Avignon, Ariadned, Black Eyed Cindy, Citronella, Connecticut King, Enchantment, Mont Blanc, Willow Wood etc.

Division II Martagon hybrids: Hybrids of martagon type derived from the following species and interspecific hybrids: *L.* × *dalhansonii*, *hansonii*, *martagon*, *medeoloides* and *tsingtauense*. Flowers usually small, often numerous, mostly down facing, typically with rather thick, recurved tepals and often of Turk's cap form, and often borne in a narrow, cylindrical inflorescence; sometimes out-facing with tepals hardly recurved. Spots usually present, numerous, often on at least three-quarters of each tepal and sometimes also on the edges of the outside. Tepal-margins usually smooth. Flowers with little or an unpleasant scent. Buds are often hairy. Leaves are typically whorled, often rather broad. Bulbs are often mauve or orange-yellow. Early in flowering. **Examples**: Cadense, Claude Shride, Cranberry Dancer, Larissa, *L.* × *dalhansonii* 'Marhan', Orange Marmalade, and Rose Tattoo etc.

Division III Euro-Caucasian hybrids: Hybrids derived from the following species and interspecific hybrids: *candidum, chalcedonicum, kesselringianum, monadelphum, pomponium, pyrenaicum* and *L.* × *testaceum.* Flowers mostly small to medium-sized, few to fairly numerous, often bell-shaped to Turk's cap-shaped, often down-facing, in a relatively short inflorescence. Flower colour often in rather pale muted shades. Spots are absent to numerous. Brush marks are absent. Tepal-margins smooth, often gently reflexed. Flowers are often scented. Inflorescence lacking secondary buds. Leaves scattered. Many are lime-tolerant.

Examples: Ares, June Fragrance, Moonlight Madonna, *L.* × *testaceum,* Zeus etc.

Division IV American hybrids: Hybrids derived from the following American species and interspecific hybrids: *bolanderi, L.* × *burbankii, canadense, columbianum, grayi, humboldtii, kelleyanum, kelloggii,*

maritimum, michauxii, michiganense, occidentale, L. ×*pardaboldtii, pardalinum, parryi, parvum, philadelphicum, pitkinense, superbum, vollmeri, washingtonianum* and *wigginsii.*

Flowers small to medium-sized, often in a pyramidal inflorescence, mostly down-facing. Flower colour often strong yellow to orange or orange-red, often with contrasting centre and tepal-tips. Spots often very conspicuous, distributed over at least half of each tepal, usually rounded and often surrounded by a paler halo. Papillae absent or inconspicuous. Brush marks absent. Tepals rather narrow, margins smooth, usually gently to strongly reflexed. Flowers with little scent. Pedicels often long and slender. Buds usually without hairs. Leaves usually whorled (at least in part). **Examples**: Bullwood, Bellingham, Del Norte, Eric Mayell, Lake Tulare, Mexico, Oliver Wyatt, Shuksan etc.

Division V Longiflorum lilies: Hybrids or selections derived exclusively from *formosanum, longiflorum, philippinense* and *wallichianum.* Flowers are medium-sized to large, often few, trumpet-shaped. Flower colour typically uniform internally (white). Spots, papillae and brush marks absent. Tepal-margins smooth. Flowers usually scented. Inflorescence lacking in secondary buds. Leaves are scattered, narrow to medium width. **Examples**: Formobel, Formolongi, Kiyosato, *longiflorum* 'Slocum's Ace', Yatsugatake etc.

Division VI Trumpet and Aurelian hybrids: Hybrids derived from the following species and interspecific hybrids: *L.* × *aurelianense, brownii, L.* × *centigale, henryi, L.* × *imperiale, L.* × *kewense, leucanthum, regale, rosthornii, sargentiae, sulphureum* and *L.* × *sulphurgale* (but excluding hybrids of *henryi* with all species listed in Division VII). Aurelian hybrids are derived from a combination of *henryi* and trumpet lilies. Flowers medium-sized to large, trumpet-shaped, bowl shaped, flat or with tepals reflexed (the flatter or star-shaped flowers usually with a shorter funnel-shaped base, so centre of flower is closed). Flower colour white, cream, yellow to orange or pink, often with a contrasting star-shaped throat and/or strong bands of colour outside. Trumpets usually scented, without spots, papillae or brush marks; other types often with spotting as small streaks gathered in two bands at the base and sometimes with conspicuous papillae. Tepal-margins smooth or twisted with irregular ruffling, tepal-tips often reflexed. Inflorescence sometimes with secondary buds. Leaves scattered, narrow to medium width. **Examples:** Carnival Queen, Gold Eagle, Golden Splendor, Herald Angel, Louise, *L.* × *kewense* 'White Henryi' etc.

Division VII Oriental hybrids: Hybrids derived from the following species and interspecific hybrids: *auratum*, *japonicum*, *nobilissimum*, *L.* × *parkmanii*, *rubellum* and *speciosum* (but excluding all hybrids of these with *henryi*). Flowers usually medium-sized to very large, few to fairly numerous, usually bowl-shaped, flat or with recurved tepals. Inner tepals often very broad, margins often ruffled or twisted and usually overlapping at base so the flower has a more 'closed' centre. Flower colour mostly white to pink to purplish red, some golden yellow; ground colour often white, with a contrasting colour as a central ray, or suffused over the whole tepal, except (in many cases) at the margins. Spots absent to numerous and conspicuous, sometimes covering more than half of each tepal, variable in form but often with diffuse edges. Papillae often present and conspicuous, even when flowers are unspotted. Nectaries often large and conspicuous. Flowers usually scented. Leaves scattered, usually broad to very broad. Often late flowering. Examples: Casa Blanca, *L.* × *parkmanii* 'Imperial Silver', Journey's End, Star Gazer, Woodriff's Memory etc.

Division VIII Other hybrids: Hybrids not covered by any of the previous Divisions (I-VII). Includes all interdivisional hybrids, such as *longiflorum*/Asiatic (LA) hybrids, *longiflorum*/Oriental (LO) hybrids, Oriental/Asiatic (OA) hybrids and Oriental/Trumpet hybrids (Orienpets or OT hybrids). Hybrids of *henryi* with *auratum*, *japonicum*, *nobilissimum*, *L.* × *parkmanii*, *rubellum* and *speciosum* (that are excluded from Divs VI and VII) are placed here. Examples: Amber Star, Chinatown, Elegant Lady, Eureka, First Crown, Pink Jazz, Red Planet, Wild Treasure etc.

Division IX Species and cultivars of species: Includes all species and their subspecies, varieties and forms, and cultivars selected there from (excluding those derived exclusively from any one of the following: *formosanum*, *longiflorum*, *philippinense* and *wallichianum*, which are placed in Div. V). Examples: *auratum* var. *rubrovittatum* 'Crimson Beauty', *canadense* var. *editorum* 'Chocolate Chips', *lancifolium* 'Splendens', *leucanthum* var. *centifolium* 'Kristine's Charm', *nepalense*, *speciosum* 'Coral Queen' etc.

Sub-Divisions

Lilies in all Divisions may be further classified on the basis of flower aspect (up-, out- or down-facing) and flower form (trumpet-shaped, bowl-shaped, flat, with tepals recurved). Flower aspect is given before flower form and the two are separated by a forward slash (/). Flower aspect is encoded as:-

a. up-facing
b. out-facing
c. down-facing

Flower form is encoded as:-

a. trumpet-shaped
b. bowl-shaped
c. flat (or with only tepal-tips recurved)
d. with tepals recurved (i.e. with more than just the tip recurved or reflexed and with the Turk's cap form as the ultimate state)

Examples: **'Madison' I(a/b)** indicates an Asiatic hybrid with up facing, bowl-shaped flowers.

'Lightning Bug' II(b/d) indicates a martagon hybrid with out-facing flowers with recurved tepals. **'Milden Red Ball' IV(c/d)** indicates an American hybrid with down-facing flowers with recurved tepals. **'Snowdon' VII(a-b/-)** indicates an Oriental hybrid with flowers that vary from up- to out-facing, with no information about the flower form. ***speciosum*** **'Crimson Glory' IX(b-c/d)** indicates a selection from *speciosum* with out- to down-facing flowers with recurved tepals.

Some of the important varieties under different cultivated groups are mentioned in Table 1.

Table 1: Varieties recommended for cultivation

Asiatic hybrids				
Varietiy	Growing Period (days)	Plant height(cm)	Number of bud	Colour
Brunello	90 -100	90-100	6-7	Orange
Bariton	80-90	120-130	7	Orange
Gironde	90-100	110-120	7	Yellow
Navona	90-100	80-90	6	White
Nello	110-120	120-130	6	Orange/red
Tresser	100-110	100-110	6-7	Orange
Oriental hybrids				
Acapulco	90-100	120-130	4	Dark pink
Bernini	100-110	120-130	5	Dark pink
Laguna	100-110	100-110	5	White
Mero Star	90-100	120-130	6	Pink/Red
Mother Choice	100-110	120-130	6	White
Rialto	110-120	120-130	5-6	White
Siberia	110-120	100-110	5	White
Sorbonne	110-120	100-120	5	Pink, White Edge
Tiara	90-100	110-120	5	Pinkish White
Tiber	100-110	100-110	6	Reddish Pink

Contd.

LA-hybrids				
Bright Diamond	90-100	130-135	5	White
Ceb Dazzle	90-100	100-110	6-8	Yellow
Cilesta	80-90	120-130	6	Orange
Courier	80-90	100-110	6	White
Ercolano	90-100	90-100	5	White
Eyeliner	90-100	120-130	5	Yellow
Golden Tycoon	100-110	120-130	6	Yellow
Litowen	80-90	120-130	6	White
Menorca	90-100	130-135	5	Salmon
Serrada	90-100	110-120	6	Yellow
Pavia	90-100	100-110	6	Yellow
OT-hybrids				
Avocado	110-120	130-140	6	Salmon /yellow
Yelloween	100-110	130-140	5	Yellow
Torrango	100-110	110-120	4	Pink
Rexona	80-90	110-120	4	White
Belladonna	100-110	100-110	4	Yellow
Addison	100-120	120-130	7	White

Propagation

Generally, lilies are propagated by seeds, bulblets, bulbils, division, and scaling, although in vitro culture using explants from scales is commonly used (Aartrijk *et al*, 1980; Magnani, 1988).

Seed Propagation

It is the cheap and easy method of propagation and used only for evolution of new varieties. The only demerit in this method is that it involves a considerable time lapse. Seeds are sown during winter in pots and are transplanted outdoors in the spring. An ideal soil mixture should have loam, peat and sand in the ratio of 7:3:2. Moisture level should be maintained in the soil so that seeds do not suffer a dry spell. Germination of seeds in lilies may be of four different types: (1) *immediate hypogeal*, (2) *delayed hypogeal*, (3) *immediate epigeal,* and (4) *delayed epigeal*. In hypogeal germination, there is no cotyledon, which emerges above the ground. However, true seed leaves emerge above the ground. Before the emergence of true seed leaves, the food from the seed is transferred to a point which is underground and in between the root tip and true seed leaves. Due to transfer of food material, the point gets swollen and becomes a miniature bulb. If the growth cycle is continuous, it is called hypogeal immediate and if a incubation period is necessary after the initial germination, it is called hypogeal delayed.

In epigeal germination, a cotyledon emerges above the ground and the endosperm nourishes the developing seedling. Just above the root tip a small node is apparent, which then becomes the bulb. When it has sufficient strength, true leaves are produced.

Epigeal immediate types may be sown as soon as ripe, but this is not always practical. Epigeal delayed requires a definite cold period before germination occurs.

Vegetative propagation

Lilium is vegetatively propagated by bulblets, axillary bulbils, separation of scales, division of bulbs, leaf cuttings and through micro- propagation. Vegetative propagation allows plants to grow into true-to-type. Propagation through scales is a rapid means of multiplication.

Bulb scales and bulblets

Lilies are commonly multiplied by scales. Mature bulbs are required for successful forcing of lilies. Mature bulbs force rapidly and predictably after low temperature treatment. Maturity of bulbs could be determined by the weight, extent of scale filling, degree of summer dormancy remaining and responsiveness of flower inducing treatments. Bud removal is a common practice to obtain large sized bulbs. Raju and Misra (2002) reported highest bulblet formation with early disbudding. Singh *et al.* (2016) reported maximum increase in number and weight of bulblets per plant when flower shoots were harvesting at 40 cm above the ground surface. Highest bulblet formation was reported with 100 ppm NAA or IBA in FYM + soilrite. Bulb production suffers from problems such as double nose bulbs (i.e., bulbs having daughter meristems) and premature sprouting of the daughter bulbs due to low temperatures during the summer treated scales at 22^{0}C, followed by a cold treatment (0, 3 or 5^{0}C), for a total of 14 or 11 weeks. Treatment at 22^{0}C delayed emergence and promoted production of many small bulbs.

Propagation through scales depended largely on variety. Seradix B No.1 stimulated rooting but increased bulblet yield in certain varieties. Darkness inhibited leaf emergence from bulblets in vivo but this was reversed by chilling treatment (4±1^{0}C) for I week. The season and position of scales (outer, middle or inner scales) also influences the propagation. It was found that the number, weight and diameter of bulblets were higher on scales taken in 1st June than on those taken on 1st March or 1st September and the numbers of bulblets were promoted by IBA (500 or 1000ppm), IAA (500ppm) or NAA (up to 500 ppm); bulblet weight was promoted by IBA as well as by BA (1000ppm) and Kinetin (500 or 1000ppm). Incubate the scales in a well-ventilated room at 24±1^{0}C until the bulblets and roots are fully formed. The duration of incubation depends on the variety. Asiatic hybrid lilies require 10-12 weeks, trumpet species and hybrids 12 to 14 weeks, and Oriental hybrid lilies 16 to 18 weeks. After removing the trays of scales with bulblets from the incubator, store them at an intermediate temperature (4° to 10°C) for three to four weeks, and then place them in cold storage (2°C) to break dormancy for specified period. When soil conditions and temperatures are favorable, the scales are planted in outdoor rows or bed. The bulblets have now been conditioned to sprout readily, so it is important not to plant them too early and expose them to frost damage. Scale plantings are usually allowed to remain in the ground for two growing seasons. Many Asiatic varieties could produce flowering stems during the same season they are planted, which is an indication of superior vigor in the clone. In some lilies, new bulbs are produced from buds around the rim of the bulb's basal plate, which would develop into replacement bulbs for the following season. Each species or hybrid reproduces at its own rate -a very small bulb, a weak species, or a struggling bulb may produce just one new bulb; two new bulbs is more normal and, with really happy, robust bulbs, a three- to six fold increase is possible.

When we lift bulbs at the end of summer, we will find the new bulbs still joined at the basal plate, but they can easily be cut or snapped off. If bulbs are left for two seasons before lifting, the original connecting tissue will have perished, and of course last season's bulbs will have produced their own successors. The Asiatic and LA-hybrid lilies are most suitably propagated by these methods than the Orientals. These may be lifted in late summer, split and replanted in fresh spots or returned to their former positions, once these have been well dug over and envigorated with generous additions of humus. Bulblets may be formed on the underground or above ground part of the stem. The number and size depends on the species or cultivar and the strength of the individual bulb. The stem of a vigorous one can generate up to a dozen or more bulblets -perfectly formed small bulbs with scales and roots ready for independent life once the stem dies back. The number of these bulblets to be harvested from the buried part of the stem is positively correlated to the amount and vigor of the stem rooting. To increase the harvest by number and, even more significantly, by weight and volume, stem rooting needs to be encouraged. This can be done initially by providing an open, gritty, humus-rich soil and planting the bulbs with at least 10-15cm deep above their noses, kept moist but not over-wet. Stem rooting activity can be dramatically increased by humus mulches from late spring onwards.

Axillary bulbils

Small, dark bulbils appear in the leaf axils; they swell to become purple-black mini-bulbs, sometimes starting to produce leaves and roots before they are fully ripe, and fall to the ground. Bulbils formation can be enhanced by removing of flower buds. If bulbils are harvested and potted up before late summer a good proportion of reasonable sized ones can be got into such energetic growth as to bear one or more flowers the very next summer.

Micro propagation

Use of micropropagation techniques is increasing day by day. It is the best method to obtain true to type as well as disease free plants. Different explants used for this purpose may be bulb scales, stem apices, apical meristem, floral organs etc. Misra *et al.* (2001) used segments of bulb scales as explants in Asiatic lily cv. Orange Pixie. *In vitro* raised bulbs exhibited 100 percent sprouting in the field as compared to *in vivo* raised bulbs. An efficient system has been developed for *in vitro* plant regeneration of *Lilium longiflorum* cv. Nellie White by culturing receptacle sections from flower buds. A section size of 3-4 mm was found to be optimal (Duong *et al*., 2001). Anthers are used for raising haploids. Anthers which can give true to type diploids of greenhouse plants have a significantly higher capability to from callus and regenerate bulblets than the field grown plants (Niimi, 2001).

Climate

Lilies prefer mild climatic conditions. It is better to grow them in partial shade as direct sun may affect them adversely. Knowledge of climatic factors on growth and flowering of lilies is important for successful forcing. Temperature strongly affects the growth, stem elongation and bud development. Temperature above 30⁰C and photoperiod in the early stage prevent loss of flower buds. Lilies are susceptible to high temperatures (>30⁰C) but they tolerate and respond to high light intensity (2000 μmol $m^{-2}s^{-1}$). Average daily temperature (ADT) affects flowering rate in *lilium longiflorum*, which profoundly decreases as average daily temperature increases and flower abortion occurs when ADT is >24⁰C. The optimum day/night temperature combination that could shorten length without interfering with flowering was found to be 16/24⁰C day/night. High temperatures being detrimental to the crop growth, temperature-dropping methods were tried in lily. Of the five different methods tried, water sprinkling over the white shading net over roof gave best results. Shading in the early vegetative growth phase helps better growth. Effect of shading (25, 50 and 75 %) was tried on different lily cultivars. All lilies cultivars recorded favourable height, fresh and dry weight, number of flowers, stem diameter and leaf area by increasing shading percentage.

Soil

Lilium can be grown on a wide range of soil but the ideal soil should be well drained. Water logging particularly during dormant period can be lethal to the plant. Saline soils are not suitable for growing lilies. Most lilies prefer soil with pH 6.5 to 7.5. However, *Lilium philadelphicum* prefers acidic soil while *Lilium candidum* prefers alkaline soil. Easter lilies require non-acidic soil that is high in calcium. The critical electrical conductivity (EC) for growth was 2.8 dS/m. Lilies cannot survive in soils with an EC of >10.8 dS/m. application of NPK fertilizers to the soil would increase EC. Nutrient uptake decreased with increasing salinity, and significantly decreased when EC was >4.4dS/m.

Growth, development and flowering

Environmental factors affect the growth and development as well as flowering in lilies. Light and temperature are among the important environmental factors.

Light

There are three aspects through which light can influence the growth and development of any plant. These are light quality and quantity, photoperiod and intensity. Lilium shows different growth behaviours with the changing light quality. If inner daughter scales are treated with red or far-red light, shoot emergence is significantly increased (Roh and Wilkins, 1974). Long photoperiod promotes stem elongation (Heins *et al*., 1982). When lilies grown under reduced irradiance were increased plant height, elongated internodes and pedicels and poor flower quality (Miller and Langhans, 1989a). Long days are also used to substitute vernalization to some extent. The bulbs are responsive to photoperiod since bulbs exposed to long photoperiods during the last two weeks of vernalization have fewer leaves and accelerated flowering. High irradiance is often associated with accelerated plant development but this effect is probably due to increased plant temperature (Heins *et al.*, 1982). Reducing greenhouse irradiance decreased leaf and flower soluble carbohydrate concentrations and increased the amount of bulb reserve hydrolysis during forcing (Miller and Langhans, 1989b).

Temperature and forcing

Optimum temperature for the growth of most lilies is about 16-20°C. Plant height becomes short as vernalization duration increased. Once bulbs are harvested and shipped to growers, traditional bulb vernalization is given to accelerate flowering. Bulbs are packed in moistened peat moss and stored at 4.5-7.0°C for 6-8 weeks. Bulb vernalization induces rapid and uniform shoot emergence and flowering but the number of flower buds is reduced in Easter lily, but increased in Asiatic hybrid lily. When *Lilium longiflorum* bulbs received 3 weeks of bulb vernalization, shoots emerged early and the reduction in the number of flowers was minimal as compared to bulbs that received 6 weeks of vernalization (Roh and Wilkins, 1976).

Dormancy

Dormancy is the period of rest when most of the physiological activities are slowed down. In fact, the existence of rest periods is convenient for the geophytes since it permits easy handling, storage and transportation of the bulbs. Hence, dormancy can be called a necessary evil in the life cycle of geophytes. As *Lilium* is a summer flowering bulbs on the hills, its dormancy period generally occurs in winter when the temperature is low. The site of dormancy in Easter lily is considered to be the daughter scales (Roh and Wilkins, 1977). If daughter scales are not present, there will not be any dormancy. However, it is difficult to have bulbs lacking daughter scales. The endogenous level of GA-like substance is lower in dormant bulbs as compared to non-dormant bulbs (Kim *et al.*, 1999). Thus, gibberellins are used for breaking dormancy. It can also be broken by cold treatment. A cold treatment of 4°C for 6-8 weeks or GA at the rate of 100-150 ppm found to be optimum for breaking dormancy.

Planting

The need of planting clean, healthy and disease-free bulbs can hardly be overstressed. Time, depth and method of planting depends on the type of lily to be grown and size of the bulbs, although there are certain general rules for all members of the lily groups. The planting depth affects the stem, leaf and root growth of lilies in forced cultivation. Bulbs are planted in mounded rows, 15-25 cm below the top of the mound. With shallow planting, basal roots establish well and flower stems become shorter and stem diameter more compared with deeper planting. When bulbs are planted deeply, the growth of stem roots is promoted. However, number of days to shoot emergence and flowering, and pant height are reduced by longer periods of cold treatment. Percentage of winter sprouting decreased at later planting times and at deeper planting depths. Percentage of winter sprouting also decreased as bulb size increased. Generally, plants in the unheated plastic house produced more daughter bulbs than plants in the field (Kim *et al.*, 1994).

Spacing

Spacing of lilies is mostly depended on the size of the bulbs. Plant size varies considerably between the groups, cultivars and bulb size planted, all of which influence the optimum planting density. Planting density is also affected by the season and the soil type. Decreasing planting density increased soil temperature and improved cut flower quality. A high planting density is needed in the months of high temperatures when light intensity is high. In winter or under poor light conditions a lower planting density should be used. Planting density per group, type and size of bulb in cm per net m² of bed should be used is mentioned below:

Group	Bulb size (cm) 10-12, 12-14, 14-16, 16-18, 18-20
Asiatic hybrid	60-70, 55-65, 50-60, 40-50
Oriental hybrids	40-50, 35-45, 30-40, 25-35, 25-35
LA-hybrids	50-60, 40-50, 40-50,

Irrigation

The amount of water depends on type of soil, climate and the variety. Too much or too little watering will result in uneven, delayed emergence and growth; reduction in stem length; *Pythium* (by excessive watering); and even flower bud desiccation among certain susceptible cultivars. Water requirement in summer is 6 to 8 lit / m^2 / day and in other season are 4 to 5 lit / m^2 / day. First two weeks irrigation only by using water can or shower. Third week onwards it is recommended to use drip for irrigation. Lilies are sensitive to salt. High salt contents will produce roots that are hard, brittle and yellow to brown in colour. A high salt content will also reduce the roots capacity to absorb water, and this will lead to a reduction in the height of the crop.

Manuring and fertilization

The nutrient requirement depends on the soil type. There was a significant interaction between growing medium and fertilizer application on growth and development of lilium. For optimum growth, fertilization is a necessary operation. Fertilizer doses for most of the lilies are 140KgN, 280Kg P_2O_5 and 200Kg K_2O per hectare. Normally Indian soils are not deficient in K_2O, so it is applied only when it is deficient. Only a small portion of Nitrogen is applied at planting, the rest should be applied during the period of rapid growth. In Lilium longiflorum, primary fertilisation factors are described by Seeley (1950). These are as follows:

1. For minimising the incidence of leaf scorch from fluoride the pH of the planting medium should be maintained high (soil 6.5-7.0; soil-less media 6.1-6.5).
2. High levels of calcium (supplied from pre-plant incorporation of lime and from calcium nitrate applied during growth).
3. High levels of nitrate nitrogen for optimum leaf quality.
4. Freedom from elevated soluble salt levels that cause root injury.

The absorption and utilization of nitrogen increases during development of inflorescence. If dry weather prevails, it is better to go for liquid fertilization. But when cloudy weather prevails, it is apt to go far a top dressing of slow release fertilizer (Miller, 1991). Nitrogen and phosphorus both are linked to leaf senescence. Fluoride induced leaf scorch is one of common nutritional disorders. Boron toxicity also leads to leaf tip scorch.

When the balance of elements in the soil is disturbed by there being too much of one in proportion to another, this disturbs the absorption of certain elements by the plant. The table-1 lists the elements that can affect each other in this way.

Table.1: Elements that affect absorption of other elements

Too much of this element	Reduce the absorption of this element
NH (ammonium)	Ca (calcium), Mg (magnesium)
K (potassium)	Ca (calcium), Mg (magnesium)
Mg (Magnesium)	NH (ammonium)
Mn (manganese)	Fe(Iron)
High EC level	Ca (calcium)

Mulching

One way of conserving moisture in between watering is by mulching. Mulching is advantageous for lilies. It keeps the soil cool, loose, discourages weeds, control soil borne pathogens and control black body radiation. It provides a fluffy, nutritious medium for the hungry stem roots The mulch can consist of with rice hulls, rice straw, pine needles, upgraded black peat, etc. A certain amount of care should be taken when mulching due to the possible presence of the fungus *Rhizoctonia solani* in some mulching materials.

Staking

Depending on the time of the year and the cultivar, it should be necessary to stake plants during the growing periods. Chrysanthemum netting is frequently used to support lilies and during the growing period the height of this netting should increase parallel to the growth of the plant.

Post harvest handling

The flower should be harvested at the appropriate harvesting stage, that is, when first flower bud shows the colour. Treatments for extending the vase life include two separate steps. The first step is to pulse the flowers with chemicals such as STS, sucrose, (Han, 1992), or growth regulators immediately after harvest. Especially, GA_3 1000ppm and $MnCl_2$ when added in pulsing solutions were found to be very effective in prolonging the vase life and preventing the leaf yellowing in lilies (Nowak and Mynett, 1985). GA_3 and $MnCl_2$ might maintain a structural role in the chloroplast membrane system and stimulate photosynthesis (Salisbury and Ross, 1985). The second step is to hold the cut flowers in preservative solutions. Preservative solutions containing 150ppm HQS +2 percent sucrose +50ppm GA_3 or 30 percent soda water were reported to be effective in 'Corderia' (Song *et al.*, 1996).

Harvesting and storage of bulbs

When the aerial parts start yellowing and ultimately dry, bulbs are lifted. Lifting is generally done on the hills in November when the temperature starts falling. Bulbs should be dug carefully and should be properly cleaned, graded and treated with suitable fungicides. Most of the *Lilium* species bear bulblets. These bulblets are also separated, cleaned and treated with the fungicides. Lifting should always be done at the proper time and maturity stage. If it is delayed, the bulbs become maturer and less dormant. The outer new scales are the site of the maturity factor. Bulbs are graded according to their size. In the U.S.A., the commercial grades are 15-17.5, 17.5-20, 20-22.5 and 22.5 – 25 cm in size (Larson, 1980). The bulbs should be cooled to 0-2°C within 3 weeks of lifting and before storage. Storage temperature for most of the lilies is -2 to 2°C.

Diseases and pests

Lilies are attacked by a number of diseases and pests. Penicillium, Bulb/ Scale rot & Fusarium stem, Rhizoctonia, Phytophtora, Pythium and Botrytis are the main diseases of lilies. Common diseases of lilium and their control measures are listed below:

Botrytis Blight: It is caused by *Botrytis elliptica.* It is probably the most serious disease of field grown lilies. Grey-brown to dark brown speckles, sometimes with a dark green edge, that measure 1 to 2 mm across appear on the leaves. The infestation can begin in the middle of the leaf surface or at the edge where it will be crescent shaped; the result will be a stunted, malformed leaf. If the leaves are severely infested, the tissue can wither, turn yellow, shrivel and finally become papery. Botrytis can also infest the stems. The exterior layer of the stem turns grey-green to dark brown. The leaves will then yellow, wither and drop off. Buds infested during a very early stage of development display brown raised spots on the outer petals. As they develop, they become malformed and can rot altogether. Flowers that have opened are extremely vulnerable to Botrytis damage in the form of greyish, watery-looking, round spots known as "pox". The common name for Botrytis is "fire". Warm humid weather favors the disease. Infection occurs most frequently between 10°C and 18°C. This disease could be controlled by reducing planting density during a period of a high RH, watering the soil in the morning and not sprinkling during times of little wind or high RH and carefully remove crop residues after the completion of cultivation. Copper fungicides like Benlate or Bordeux mixture @ 2 % are effective in controlling the disease.

Bulb and Scale rot: Bulb rot is the most serious bulb disease of Asiatic and Oriental lilies as it destroys the bulb and results in total crop loss. This disease is caused by a fungus *Fusarium oxysporium* and *Cylindrocarpon* of lily. The disease generally occurs during storage and the site of infection is the base of scales. The symptom of disease including chocolate or dark brown rot that extends into the scales resulting in scale detachment from the basal plate. Above ground, this can be identified by the premature yellowing of the lowest leaves that will then turn brown, rot and fall off. This can be controlled by steam sterilization of soil is very effective. Use disease free planting material; avoid the excessive use of nitrogen fertilizer that will encourage the disease, over watering should be avoided and use resistant cultivars. Soil drenching with Bavistin is effective and also treating bulbs in Benomyl solution at 30°C for 30 minutes.

Penicillium: The infestation usually comes from *Penicillium* fungi commonly found in the environment and starts during storage when the spores enter wounds in the bulb tissue. An excessively high temperature and an excessively low RH during storage promote the problem. Damage to the bulbs increases the risk of a *Penicillium* infestation. *Penicillium* show up on the bulb scales as brown rotten spots covered with a white fungal weft that later turns blue-green and is accompanied by masses of spores. Once established, the rot slowly spreads throughout the storage period, even under low temperatures (-2°C). To avoid these diseases keep the bulbs from drying out during storage and processing, and store the bulbs at the lowest possible temperature. Do not plant bulbs with infested base plates. Plant infested lots as quickly as possible, preferably from December through March (when they can have a slow start).Keep the soil at the proper moisture level before and after planting.

Root rot: Root rot is also a major problem in lilium cultivation. It is caused by several pathogens including *Fusarium, Rhizoctoni* and *Pythium* species. The disease results into die back of foliage starting at the base of the plant and moving up the stem. The extent of root rotting may vary but the root decline contributes to non-uniform development of the plants. Hence to prevent the disease, steam sterilization or fertilization of soil/media should be done and maintain proper moisture/ drainage and avoid the over watering.

Foot rot: It is caused by the fungus *Phytophthora nicotianae*. The stem just below the soil surface is infected by the fungus which results in withering and ultimately death of the plants. The disease occurs mostly in over-watered soil. To prevent the disease, soil should have good drainage facility. Affected plants should be removed and destroyed.

Black scale and scale tip rot: Anthracnose or black scale is caused by *colletotrichum lilies* and appears as black and rotted outer scales that reduce salability of the bulbs. At harvest, brown lesions develop on the scales which enlarge in storage and the scales shrivel and turn black (Linderman, 1985). This pathogen may also cause a serious disease of petal tissues in *Lilium longiflorum*. Control involves the use of healthy planting stock. Scale tip rot is characterized by darkening of scale tips and the surface of outer scales that shrivel and rot.

Sclerotium: This disease is caused by a fungus: *Rhizoctonia solani.* It is transferred to the plants from the soil and develops most rapidly under moist conditions and at temperatures above 15°C. Leaves that have come into contact with the soil will wilt and start to rot. Brown spots appear on the foot of the stem, which then rots all the way through and collapses. Very characteristic of this fungus is the presence of white strands of hyphae, and later the formation of round sclerotia on the diseased tissue and surrounding soil. These sclerotia are first white and then younger leaves and the growth point are usually damaged. The underground stem parts can display brown stripes and spots that are usually elongated. The emergence of stem roots is suppressed, development is delayed, and flowering is poor or nonexistent because the flower buds have dried out at an early stage. Control involves the drenching of soil with bavistin is effective.

Pests

Aphids: Aphids are the most serious pest of lilies as they transmit various viral diseases. They suck the sap from the leaves and stem. They may also infest bulbs in store. Different broad spectrum insecticides can used to control the aphids.

Lily Beetle: Larvae of the lily beetle may sometimes cause severe damage on leaves, stem and flower buds. They foul the plants with black slimy excrement.

Bulb mites : *Rhizoglyphus echinopus* is usually a secondary pest attacking the bulbs. It enters through prior mechanical injury to the bulb. Damaged bulbs should be discarded. Hot water dip is also effective.

Thrips: These pests cause scarring of the foliage. They eat the surface of the leaves and often an attack can be severe before being noticed. The bulbs should be stored at a temperature below 10°C to control this pest.

Slugs and snails: They attack the young shoots as tender shoots are attractive food for these pests. Slug and snail bait should be used as a precautionary measure.

Nematodes: Foliar nematode *Aphelencoides fragaric* causes yellow blotching and necrosis in foliage. Hot water dip treatment is effective in controlling nematodes. Other methods of field control include crop rotation, rogueing, removal of plant debris from the field, etc. (Linderman, 1985).

Viral diseases : Viruses also attack lilies to a great extent. Lily symptomless virus (LSV), lily mottle virus (LiMV), cucumber mosaic virus (CMV), etc., attack the lilies. Affected plants become dwarf and crippled. LSV produces vein clearing and is transmitted by aphids in a non-persistent manner.

LiMV produces yellow mottle and mosaic on the leaves and the plants show stunted growth. Sometimes, flowers get malformed with colour break. Aphids are the non-persistent host for this virus. LiMV can be detected in hybridization assays since the coat protein gene has been sequenced and virus specific primers have been produced (Dekker *et al.*, 1993).

CMV also produces mosaic patterns on the foliage and also necrosis on the plants. *Myzus persicae* is a common vector of CMV.

Physiological disorders

Leaf scorch: Leaf scorch occurs at the time when flower buds are not yet visible. First of all the young leaves are slightly inward and then a few days later greenish-yellow to whitish spots appear on the scorched leaves. In severe cases all leaves and the tender young buds will be lost. Leaf scorch occurs when there is a disturbance in the balance between the absorption and evaporation of water which causes a calcium deficiency in the cells of the youngest leaves. This disorder can be managed effectively by keeping soil moistened before planting, plant bulb with good root system, plants to an adequate depth i.e., allow 6cm-10cm of soil on top of the bulb, prevent large difference in green house temperature and air humidity and maintain the R.H of approximately 75%.

Bud drop and bud desiccation/Abscission: Bud drop can occur when the flower buds reach a length of 1 to 2 cm. Buds turn light green in color when plants receive insufficient light. Bud desiccation occurs if rooting conditions are poor. i.e. too dry soil. Controlled by providing proper sun light (2000-3000 lux) and bulbs should not be allowed to dry out during planting.

Papery leaves: The occurrence of papery leaves is related to weather conditions outside previous to harvest, and the RH in the greenhouse. Under bright sunlight, lily plants being produced under a high RH (exceeding 80%) will transpire slowly and accumulate excessive moisture. After harvest, the stomata on the leaves of these lilies will no longer be able to close. Areas of the leaves will then desiccate and take on a papery appearance. A high RH combined with low light conditions will also increase the risk of papery leaves. Maintain RH below 80% by watering beneath the crop, shade during excessive sunlight and by providing sufficient light in the greenhouse effectively control the papery leaves.

Nutrient deficiency

Iron deficiency: The leaf tissue between the veins of the young leaves is yellowish green. This occurs particularly among rapidly growing plants. The greater the iron deficiency the more yellow the plant will become. This abnormality occurs especially in calcium rich soils (high pH) and on light, sludge soils or on sites with excess water. It may also occur if soil temperature is too low. Iron deficiency can be control by ensuring that the soil is well drained with a sufficiently low pH and good root system. Iron chelate should be added to soils with a pH higher than 6.5 before planting @ 2-3 g/m^2. To prevent leaf scorch, the chelate should be applied in the early evening to a moist crop.

Nitrogen deficiency: It results in the whole leaf becoming lighter in color and is often more noticeable when plants are about to bloom. It should be controlled by applying sufficient quantities of nitrogen based on the soil samples.

Other deficiency problems: One or more other deficiency problems can occur during lily cultivation. Only some of these can be identified by a discoloration of the leaves. By applying the required element in time, the symptoms can be prevented or eliminated. In addition to the previous two deficiency diseases described, here are a few more that may occur.

Calcium (Ca): Plants remain small and the colour of the leaves is light to pale green. Leaf tips bend downward and sometimes turn brown at the very end. The leaves are sometimes very pale green with white spots. Root development is poor. A calcium deficiency can be prevented by liming the soil before planting with calcium carbonate ($CaCO_3$). Other fertilisers can also contribute to eliminating a calcium deficiency. These include magnesium carbonate ($MgCO_3$), magnesium oxide (MgO) and magnesium hydroxide ($Mg(OH)_2$).

Phosphate (P): Plants remain somewhat smaller. Colour is dull pale green. The more mature leaves develop brownish-red leaf tips. A deficiency of phosphate can be difficult to solve during cultivation since the phosphate level should be sufficient previous to cultivation. Increasing the phosphate level can be done by fertilising with dicalcium phosphate ($CaHPO_4$). This fertiliser does not contain any fluoride. Scatter phosphate previous to tilling the soil.

Potassium (K): Plants remain small and are somewhat stocky, poor growth rate; younger leaves are dull yellow-green with a brown discoloration except for the leaf tips. Small white necrotic spots appear over the entire surface of the leaves. Ultimately, the leaf tips wither. Potassium deficiency can be eliminated by fertilising with such fertilisers as nitrate of potash (K_2O). This fertiliser can be added to the sprinkling water.

Magnesium (Mg): Plants remain small. Leaves are light green and bent downward. Brownish-white spots sometimes occur along the length of the leaves. The most mature leaves show the worst symptoms. A magnesium deficiency can be eliminated by applying magnesium sulphate with a sprinkler circuit or scattering it among the plants.

References

Aartrijk, J. Van and Blom-Barnhoorn. G.T. 1980. Effects of sucrose, mineral salts and some organic substances and the adventitious regeneration in vitro of plantlets from bulb-scale of Lilium speciosum 'Rubrun'. Acta Hort. 109:297-302.

De Jong, P.C. 1974. Some notes on the evolution of lilies. Lily Yearbook North Am. Lily Soc. 27: 23–28.

Dhyani, A.2007. Exploring Lilium polyphyllum in Uttarakhand, India. The lily yearbook of North American lily Society, pp: 79-82.

Duong, T.N., Bui Van Le. And Tran Thbh Van, K. 2001. Manipulation of morphogenetic pathways of L. longiflorum transverse thin layer explants by auxin and cytokinin. In vitro cellular and developmental boil.' Plants, 37(1):44-49.

GRIN. 2016. Germplasm Resources Information Network. Available at http://www.ars-grin.gov. Accessed, 29 Feb 2016.

Han, S.S.1992. Role of sucrose in bud development and vase life of cut Liatris spicata (L.) Willd. Hort Science., 27:1198-2000.

Heins, R.D., Pemberton, H.B. and Wilkins, H.F.1982. The influence of light in lily (Lilium longiflorum Thunb.).1. Influence of light intensity on plant development. J.Amer. Soc. Hort. Sci., 107: 330-335.

Kim, K.W., Kim,K.A., Byun, M., Kim.K.W., Kim,K.A. and Byun, M.S.1999. physiological changes before and after breaking dormancy of L. honsonii bulblets regeneration in vitro. J. Korean Soc. Hort. Sci., 40(6):751-754.

Larson, R.A. 1980.Introduction to Floriculture. New York, U.S.A.: Academic Press Inc.,

Linderman, R.G.1985. Easter lilies. In: Diseases of floral crops. Vol.2. (Ed.). Strider, D.L.L.New York, U.S.A.: Praeger Publishers,pp. 9-40.

Magnani, G.1988. The effect of temperature levels during scale multiplication on the yield of bulblets of Lilium. Colture Protette, 17(11):69-74.

Miller, R.O.1991. Lilies. In: Ball Red Book. (Ed.) Ball, V. 15th edition. West Chicago, III, U.S.A.: George J. Ball Publisher, 625-651.

Miller, W.B. and Langhans, R.W.1989a. Carbohydrate changes of Easter lilies during growth in normal and reduced irradiance environments. J.Amer. Soc. Hort. Sci., 114:310-315.

Miller, W.B. and Langhans, R.W.1989a. Reduced irradiance affects dry weight partitioning in Easter lily. J.Amer. Soc. Hort.Sci., 114: 306-309.

Misra, P. and Datta, S.K.2001. Acclimatisation of Asiatic hybrid lilies under stress condition after propagation through tissue culture. Current Science, 81(12):1530-1533.

Niimi, Y., Dong Shang, H., Fujisaki, M. and Han, D.S. 2001. Production of virus free plantlets by anther culture of Lilium x Enchantment. Scientia Hort., 90(3-4):325-334.

Raju, V.S. and Misra, R.L.2002. Response of Asiatic hybrid lily to flower bud removal. J. Ornamental Horti., 5(2): 74-75.

Roh, S.M. and Wilkins, H.F. 1974. Red and far-red treatment accelerate shoot emergence from the bulbs of L. longiflorum cv. 'Nellie White'. Hort. Science., 9:38-39.

Roh, S.M. and Wilkins, H.F. 1976. The relationships of bulb vernalization to shoot growth, flowering and scale bulblet formation in L. longiflorum Thunb. "Nellie White". Thesis Collection. Yenngnam University, 10: 295-305.

Roh, S.M. and Wilkins, H.F. 1977. The physiology of dormancy and maturity of L.longiflorum Thunb cv. Nellie White bulb. III. Maturity scale removal and vernalization treatment. J. Korean Soc. Hort. Sci., 18(2):187-193.

Salisbury, F.B. and Ross, C.W. Plant Physiology. California, U.S.A.: Wardsworth Publishing Company, 1985.

Seeley, J.G. 1950. Mineral nutrient deficiencies and leaf burn of croft Easter lilies. Proc. Amer. Soc. Hort. Sci., 56:439-445.

Singh, M.K., Sindhu, S.S. and Dhama, A.S. 2016. Evaluation of performance of varieties and effect of different levels of harvesting of flower shoot on bulb production of LA hybrids in northern plain conditions. Indian J. Agricultural Sciences, 86(8): 1082-1085.

Song, C.Y., Bang, C.S., Lee, J.S. and Lee, D.C. 1996. Effects of post-harvest pre-treatments and preservative solutions on vase life and flower quality of Asiatic hybrid lily. Acta Hort., 414:277-285.

Warrier, P.K., Nambiar, V.P.K. and RamanKutyy, C. 1997. Indian Medicinal Plants. A compendium of 500 medicinal plants. Arya Vaidya Sala. Orient Longman.

24

Statice

Ranjit Singh and Parminder Singh

Statice, botanically known as *Limonium*is very highly demanded ornamental plant. It is also known as 'Sea Lavender' owing to its color and place of growth. The genus name *Limonium* is from Greek name 'leimon' which translates to origin 'a meadow', most suitable as many species are found growing near marshy lands or salty areas of sea shore. The statice plants have very wide adaptability and can be seen growing under a wide range soil and climatic conditions like sea shores, marshy lands, deserts and plains etc. Statice figures in top 20 cut flowers sold worldwide but information on its trade, volume, export-import and number of stems sold etc. is not available. The reason for this may be attributed to classification systems, data collection categories and frequently changing nature of the assortments in the international flower market. Comprehensive trade statistics are difficult to locate for Statice. However, Statice figured in top 10 imported products in the Netherlands way back during 2013, 2014 and 2015. Though it is presently not among the top 10 flowers category, still it enjoys the immense popularity among the growers owing to its ease in cultivation and other characteristics. It is cultivated as annual as well as perennial crop, as different species have been domesticated for this purpose.

Importance and uses

Statice is used in flower bouquets as low-cost filler. It replaces gypsophylla in preparation of flower arrangements and bouquets. Unique characteristics of its branches make it very special for bouquets, corsages and other flower arrangements. As a filler it is very much used in making mixed bouquets. As per Royal Flora Holland figures during 2018, *Limonium* recorded 21 % increase in volume compared to 2017. Its bright purple, pink, yellow or white flowers complement the bouquets (Hamrick, 2003). Statice is also grown for garden decoration as it is suitable for borders, rock gardens, pots and for alpine houses. The calyx of the flowers of statice retains its original colour even after dying. So, it is also used as dry flower, as solely or in making bouquets and arrangements. *Limonium profusum* species used for extraction of medicines from its roots (Arora and Pal ,2002). It is grown worldwide like Mediterranean region, Europe, Japan, Korea, China, Israel, Kenya, Zimbabwe, USA, Colombia and India. In India it is cultivated in almost all the states. However, in Punjab it is cultivated as winter annual for seed production and also as filler for making flower arrangements.

Origin, Morphology and Taxonomy

Statice is native to Europe; the Mediterranean regions, Asia and Canary Islands. Caucasus mountains are reported to be the place of origin for *L. tataricum* species. Many species are found growing near or around the coastal marshes which are characterized by salty or saline soils.

The sea lavender or statice, *Limonium sp.* belongs to family Plumbaginaceae which falls in caryophyllale order in kingdom plantae. The *Limonium sp* may be annual, biennial or perennials with deciduous and ever green nature. Plants are 60-65 cm tall. Leaves are simple to entire or sometime pinnatifid or lobed in different species. Leaves are stalkless, rosettes from which plant grows upright. Inflorescence is panicle or corymb. The stems which bear inflorescences only carry scale leaves also known as winged stem. The individual flowers are small 4-10 mm long with 5-lobed calyx and corolla and five stamens. The color of the flower may be pink, or violet to purple or yellow or white. Many species are apomictic. Fruit is capsule which is very small containing a single seed, partly enclosed by the calyx which is persistent and retains its color even after drying.

Climate

Statice is cultivated in open and sunny situation. Night temperature range between 13-15° C and day temperature range 23-27° C is preferred temperature regime for production of flowers. Temperature is reported to influence the flowering of Statice when employed through seed treatment. The seed treatment to low temperature is reported to reduce the time period from sowing to flowering and also increased the number of cut flowers produced (Yoo *et al.,* 1998). Vernalization of seedlings at 2-3° C for 30 days also have been reported to cause early flowering (Azuma *et al* 1983). Farina *et al*(2000) also concluded that low temperature exposure of the plants at suitable growth stage is necessary for blooming of statice. Five weeks at 6° C applied to plants with more than 16 leaves are sufficient for induction of flowering. Devernalization was also not observed after the exposure of the vernalized adult plants to constant 32° C for 4 weeks. Chin Dan *et al* (2009) also have reported beneficial effect of low temperature treatment of plants of *Limonium* cv. Fantasia as low temperature treatment for six weeks has resulted in early flowering, also improved the quality of cut flowers, yield and also increased fresh weight of stems. *L. perezii* is free flowering species, does not require any vernalization.

Limonium sinuatum is classified as facultative long day plant with critical day length of 13 hr. Long days also promote flowering in many species (Chen *et al*, 2010). Low light intensities are detrimental to flowering in statice as the flower stocks become corky and weak, thus reduce the quality of the produce. Seedling illumination combined with low temperature treatment at 2-3° C for 50 days resulted in increased number of cut flowers produced (Katsutani *et al* 1998).

Soil

Statice is grown successfully in wide range of soils, however it prefers soils which are rich in lime content with soil pH 6.5-7.5 and EC 0.7 mmhos/cm. It is resistant to drought hence we can think of sandy soils also. Sandy loam textured soils with ample organic matter and good drainage are most suitable for many species, except *L.altaica* which prefers clayey soils.

Species and cultivars

The basic chromosome numbers for genus *Limonium* is reported as 6, 7, 8, 9 (Darlington and Wylie, 1969). Polyploidy is well documented in this genus. Laguna *et al* (2016) has also reported aneuploid and polyploid species which are occurring in this genus through apomictic seed production. There are around 150 species under genus *Limonium. Limonium* species are diverse in growth habit and cycles which include annual (*L. lobatum, L. sinuatom ssp, L. beaumieranum*), perennial herbs (*L. sinuatum ssp sinuatum*) and even woody shrubs up to 2m (*L. arborescens*) (Morgan and Funnel, 2018). Some species are listed below:

***L. album* (Coincy) Sennen:** Perennial species, found near region of Murcia, Spain at an altitude of 300-500 m. on rocky slopes.

***L. alutaceum* (Steven) Kuntze:** Native to west, Ukraine

***L. amoenum* (CH wright) R.A. Dyer:** Native to S.W.Cape Province

***L. aureum* (L.) Hill:** Erect 30 cm tall perennial but grown as annual. Calyces are yellow to orange good for cut flower.

***Limonium acuminatum* L. Bolus:** Native range is S.W. Cape.

***L. bellidifolium* (Gouan) Dumort.:** Commonly known as matted sea lavender and is found in salt marshes of Europe and South West Asia. It is compact evergreen woody-based perennial and grows from rhizome.

***L. gmelinii* (Willd.) Kuntze:** Commonly known as Siberian statice. Perennial, very attractive with smoky blue flowers. Very attractive to butterflies and bees.

***L. otolepis* (Schrenk) Kuntze:** Perennial, rhizomatous and herbaceous in nature. Flower color pale blue purple.

***L. perezii* (Stapf.) F.T. Hubb.:** Commonly known as Perez sea lavender and Sea foam statice. It is native of the Canary Islands and is widely used genus. It is perennial with woody rhizome which hold its basal rosette with triangular shaped leaves.

***L. peregrinum* (P.J. Bergius) R.A. Dyer:** It is endemic to Cape west coast. Grows to 100 cm. Leaves are long and oblanceolate. Petals are magenta with calyx dull pink in color.

***L. purpuratum* F.T. Hubb. ex L.H. Bailey:** Its native range is South West cape province.

***Limonium platyphyllum* Lincz:** Perennial, branched 60-62 cm tall, long leaves. Six mm long deep Lavender to Blue flowers, calyces white. Also known as broad leaved statice.

***L. sinuatum* (L.) Mill.:** It is common statice or wavy leaf sea lavender or annual statice. Its plants are erect. It is herbaceous perennial. But in India grown as winter annual. Plant is branched and 50-60 cm tall. Leaves are long to lance shaped deeply lobed with wavy margins, basal rosettes. Flowers are small funnel shaped pink, white or blue in color. Violet blue calyx and white corolla. It requires very low maintenance and used as cut flower.

***L. sinense* (Giard) Kuntze:** Herbaceous perennial with up to 60 cm height. The underground part i.e. roots of this species are reported to have an antiviral compound gallic acid which inhibits the hepatitis C virus infection by blocking early viral entry into the cells (Hsu *et al.*, 2015). This species occurs near China sea shore, Ryukyu islands,Taiwan and Vietnam

***L. tataricum* (L.) Mill** syn *Goniolimon tartaricum* commonly known as Tartarian Sea Lavender or Tartarian Statice or German Statice. It has pink, violet or white colored flowers. It is perennial in nature. It has more compact shape with small flowers.

***L. vulgare* Mill.** It is perennial plant that can grow up to 30 cm height. The root of this species has medicinal value. They are decongestant and depurative (Freethy, 1985). Plant is also known to repel moths. (Freethy, 1985) and root is source of tannin (Uphoff,1959).

Cultivars: Morgan and Funnell (2018) stated that comparatively narrow range of closely related species is currently being used in the trade. They have also reported that the cultivars are categorized horticulturally as Altaica, Latifolium, Sinensis and Limonium or statice.

I. Altaica: Hybrids of *L. gmelinii, L. altaica L. latifolium and L. serotinum*. They include Emille cultivars.

II. Latifolium: Hybrids between *L. latifolium and L. otolepsis*

Eg. Misty Blue, Beltlaard

III. Sinensis: Includes hybrids between *L. Sinensis*and, *L. fortunei*, *L. tetragonum* and *L. aureum*

IV. Statice comprises of *L. sinuatum*

Some of the selected cultivars are given (Mukherjee, 2008)

Blue Colour: Blue Bonnet, Blue Moon, Turbo Blue, Misty Blue, Forever Blue, Excellent Blue, Sea Sun Birds, Aqua Sun Birds.

Lavender Colour: Lavender Emille, Lavender Queen, Sky Sun Birds.

Pink Colour: Excellent Pink, Forever Pink, Pink Emile, Atlanta Sun Birds, Caldo Sun Birds.

Purple Colour: Turbo Purple, Forever Purple, Purple Monarch, Excellent purple.

Rose Colour: American Beauty, Early Rose, Sunset Rose, Rose Light, Sophia, Rose Strike

Yellow Colour: Forever Gold, Forever Yellow, Gold Coast, Sunset Yellow, Turbo Yellow,Shiny Sun Birds

White Colour: Forever White, Iceberg,Turbo White, Snow Birds, Silver Sun Birds

China Flamingo, China Red, Misty Blue, Misty White are perennial cultivars of statice.

Yellow cultivars are more prone to root fungi than blue cultivars, hence need more care while cultivation.

Propagation

Statice can be propagated through seeds, vegetative means and micropropagation. Most of the perennial cultivars grown under protected conditions are propagated through micropropagation.

Seed propagation

Seed propagation is easy and best method and is common for annual types. Seeds of statice are planted as winter annual. One-gram weight consists of around 350 seeds. Seeds are sown during September-October in plains when the temperature falls down. Best temperature for germination of seeds is 14-18°C. Seed germinate within 5-9 days. Seedlings when acquire 3-4 cm length, are transplanted in the main field. In hills transplanting time is March-April. Avoid early sowing of seeds as it is reported that seeds require vernalization. Vernalization of seeds certainly improves germination, plant growth, stem length and weight of fresh cut stems. Vernalization of statice seeds at 2-3°C for 30 days has been reported beneficial for early flowering and also increased the number of flowers per plant (Azuma *et al* 1983).

Vegetative propagation: Perennial statice cultivars are propagated through division of clumps as in case of some grasses. Each separated portion from the clumps must contain some roots.

Micropropagation: Perennial *Limonium* species are multiplied through micropropagation for commercial cultivation in countries like Israel, Holland, Korea, Japan etc. Micropropagated plants have advantages in many ways. They are uniform in growth and color, disease free and bear flowers at the top of the stem with long stems. Shoot tips and axillary buds at flowering stage have been successfully used as explant for micropropagation. MS medium has been used with various supplements of growth regulators depending upon the need i.e. for shoot and root regeneration. Plant regeneration from cell suspension culture has also been reported in statice. Lateral buds from flowering stock were used and cultured in vitro in half strength MS medium. Repeated sub-culturing was done along with low temperature treatment at 2-4°C to the shoots thus regenerated. It was found that low temperature treatment of shoots with less sub-culturing results in more and uniform flowering (Miyama, 2009).

Planting

Three to four weeks old seedlings are transplanted or when they have attained 4-5 cm height. Since statice is chiefly cultivated from seeds as winter annual so planting time is Mid-September to Mid-October for open field conditions. Plant spacing of 60×30 cm has been reported optimum for November transplanting for various yield characteristics. While planting of statice for landscaping and as border or rock garden generally 25-30 cm spacing is followed. Plants generally come to flower after 3-4 months. Approximately 14000-15000 plants are required for planting one-acre. Very close plants viz. 30×30 cm will result in more number of cut stems but quality may be compromised. Continuous cropping can be harvested, if the climatic conditions are mild provided with 2-4 weeks interval in seed sowing as well as transplanting. The perennial *Limonium* varieties can also be grown outdoor but the attack of Botrytis blight may be a challenging factor. Hence, they can be cultivated in low cost protected structures like bamboo made structures or even naturally ventilated polyhouse. The perennial varieties will flower many times a year and are not in much demand of cool or chilling hours as in case of annual varieties. For greenhouse cultivation 30×30 cm spacing is followed. The bed width is 1m with 40 cm wide path between the beds is recommended. Two rows of plants are planted per bed. This will accommodate 6 plants per net meter square area. Cultivar Emille is suitable for high density planting, it can be planted 8 plants per net m^2 area. The plant spacing may be followed 40×40 cm or 30×30 cm depending upon the weather i.e. winter planting or summer planting scheme. The perennial crop may last for 3-4 years.

Manuring

Field should be well prepared by incorporating well rottened Farm Yard Manure @ 20 Tonnes per acre at least 15-20 days before actual planting. Statice plants do not require much nutrition. The basal dose for statice has been recommended by the Netherlands grower companies that consists of potassium sulphate, magnesium sulphate and triple super phosphate each @ 30-40 gm per meter square area. In another recommendation application of 100 kg/hectare each of N, P, K is emphasized for proper crop growth. Half of this N and full P and K should be incorporated in soil before transplanting as basal dose. Remaining half N is applied one month after first application. Lot of work have been reported on nutritional aspects but again it is based on soil characteristics. The liquid fertilization is also practiced in statice. The target figures for main element in liquid fertilization in perennial statice is as (values in mmol/liter)- NH_4^+ 0.1, Nitrate(NO_3^-) 2.5, Potassium(K^+) 1.3, Chlorine(Cl^-) 0.4, Sodium(Na^+) 0.4, Sulphate(SO_4^-) 1.5, Calcium(Ca^{2+}) 1.5, Bicarbonates(HCO_3^-) 0.5, Magnesium(Mg^{2+}) 1.0, Phosphate($H_2PO_4^-$) 0.1. The EC of the water through which fertilization, if applied through drip is very important. It must not increase 0.5 m mohs/s.

Irrigation

As the Statice grows from basal rosette leaves, in 3-4 months the crop will be coming to harvesting. It grows very fast at early stages. Soil moisture stress at any stage may lead to economic losses. So, adequate amount of moisture in the soil is essential. The water supply at flowering stage may reduce but not stopped. At early stages of plant growth overhead irrigation can be employed. After rosette formation care should be taken that leaves do not rot due to accumulation of water on them. Flower bud must not come in contact with water, it may lead to development of botrytis blight. The statice plants absorb enough water from ground during the night to compensate water requirement of the plant. In terms of salinity tolerance *L. sinuatum* is reported to be moderately tolerant while *L. Perezii* is sensitive (Grieve *et al* 2005).

Intercultural operations

Support

The plants particularly the perennial types are to be supported by wire mesh netting. A single layer of 17×20 cm mesh wire netting is used. Supporting poles are placed oneither side of the beds at 3.5 m to 5 m intervals (Anonymous, 1997).The support is required both for indoor and outdoor cultivation as the flowering stems has the tendency to lodge downwards.

Weed control

Many weedicides have been reported for statice but one must be careful about the label claims as well as adequate dosages. As the improper utilization may lead to development of resistance among the weeds. Innova (1999) has reported that six herbicides applied have resulted in increased cut flower yield. Pre emergence herbicides applied before transplanting and a second application over the top before flowering with Chlorothal-dimethyl, oxadiazon and oxyfluorfen proved safe for control of weeds. Alachlor, Chlorothal-dimethyl, Napropamide and Ozyzalin have been reported effective even against grasses also during winter.

Extending flowering period

Some work has been reported that flowering can be extended in statice from over and above its natural flowering time. Low temperature treatment has been tried. Blue and White cultivars of *L. sinuatum* were tried by vernalization (6° C for 5 weeks) of seedling to produce early (February) crop. For production of flowering during mid-October to end November white cultivars were unable to produce flowers in response to vernalization.

Harvesting and Yield

As mentioned earlier harvesting can be expected in 3-4 months after planting depending upon variety and climatic conditions. Flower stems should be cut when 90% of the flowers show color and are open. Flowers should not be picked too early or at immature stage as they will not open in vases. Harvesting stage can be ascertained when most of the calyices have been opened and start showing color. The stems are cut with sharp secateur and are put in water immediately after harvesting.

The yield of cut stems also varies with variety. Generally depending upon the cultivar, it varies from 12-20 stems per plant per year (100-125 stems/m^2/year). Blue colored varieties are reported to be low yielder as compared to yellow cultivars.

Post-Harvest Handling

After the flowers have reached at post-harvest handling facility, the lower 2-3 cm portion is recut and is placed in plain water. They will remain fresh for 2 weeks in plain water. Gibberellic acid @ 300 ppm has been reported to enhance the flower bud opening when used along with certain biocides like silver nitrate or aluminum sulphate. The flowers of statice can be stored at 5° C.

Stems are graded based on uniform stem length, free from disease/ insects and freshness of the produce. The stems are made into bunches of 5 each and then 5 bunches (25 stems) are packed in one package especially polythene sleeves which are supplied by auction centers in the Netherlands. In local market 20 stems per bunch are sold without any packing.

Drying of flowers

Dried flowers of statice are also prevalent in the market for flower arrangements and bouquets, due to their special ability to retain color even after drying. The flowers dried properly will remain in good condition for 2-4 years. For drying of statice flowering stage of harvesting is very important. The proper stage is when half of the flower are open in on the stems. The leaves are removed from the stems. They are made into bunches of 15-20 each. They are tied from the base lightly. Then these bundles are hanged upside down in well ventilated dark room till they dry up completely. Alternatively stems of statice can also be dried by soaking in glycerol and water (1:2 ratio) solution followed by microwave treatment at 34° C for one minute.

Insects Pests and Diseases

Insects: Most of the statice varieties come to flowering during February onwards in Northern plains. At this time, it may be attacked by various sucking insect pests and also, by insects which damage foliage by cutting and eating the leaves thus reduce the quality. Various sucking pests like Aphids, thrips can be controlled by spraying rogor (2ml/L) or Imidacloprid (40 ml/100 L of water) or thiamethoxym(2ml/L) one or two sprays are enough. Try to use green labelled insecticides. Mealy bugs are not big problem but sometimes may attack the statice crop. For mealy bugs Profenofos 2-2.5 ml/L or clothianidin (15 ml/100L water) can be used effectively. For other types of insects like larvae etc. which cut and chew the foliage, one can go for chloropyriphos(2ml/L) or chlorantraniliprole (1-2 ml/L) or spinetoram (100g/100L of water) as broad-spectrum insecticides.

Diseases

Damping off: This disease occurs in nursey stage. The seedlings are infected with this disease. The collar region is affected first, then the seedling dies. The causal organism may *Pythium spp, and Phytophthora spp.*\. The disease can be suppressed by drenching the nursery beds with some fungicide like carbendazim @ 2g/liter or try other suitable fungicide.

Powdery mildew (*Oidium sp*): Powdery mildew colonies will be visible on the infected portion. This is more likely to occur in warm winters. It can be controlled by spraying Tridemorpz @ 0.5 ml/L or Carbendazim @ 2g/L or Zineb 300 g dusty powder.

Downy mildew (*Peronospora statices*): This fungus produces spores at temperature range of 15-18° C which is easily available during day time during winter in Northern Indian plains. The fungal spores penetrate through the stomatal pores on lower side. Chlorotic or reddish colored spots on young leaves appear which may be followed by brown oval or round spots on the edges of the leaves. To control this disease spray application of azoxystrobin @2ml/liter or forsetyl or mancozeb @ 2g/L is recommended.

Grey mould: (*Botrytis cinerea*) This fungus infects the leaves and flowers causing irreversible damage. Very low temperature along with high humidity favors the spread of this disease. Soft water soaked, small brown spots appear on the leaves and flower buds or petals. In dark colored petals the spots are brown white and in light colored petals they are white and circular. These spots develop into fuzzy grey mould, the spores are air borne thus infecting more plants. It can be prevented by

During cultivation; - Do not allow flowers to come into direct contact with water. During processing; - Never place damp flowers in the cold store, Do not bunch flowers too tightly so that moisture from breathing can evaporate, Remove damaged or loose leaves, as they can be easily affected by Botrytis, Keep buckets clean.

We have to manage growing environment (if growing indoor) or we can go for certain fungicides like carbendazim(2g/L) or benomyl(2g/L) or chlorothanonil(2g/L)

Leaf spots: Various fungi are reported to cause leaf spots in statice. They may be *Cercospora insulana, Alternaria sp* and *Phyllostricta sp* etc Spray application of chlorothanonil(2g/L) or mancozeb(2g/L) will be effective enough to control these leaf spots.

Anthracnose (*Colletotrichum gloeosporiodes*): This fungus causes crown rot and lesions on the leaves, stems and young flower buds. Spray the crop with carbendazim(2g/L) or mancozeb(2g/L) at weekly interval can control the disease very effectively.

Bacterial leaf spot and Bacterial wilt: Bacterial wilt (*Pseudomonas caryophylli*) are characterized by unilateral chlorosis or entire wilting. Where as in bacterial leaf spot (*P. andropogonis*) leaf and flower scape lesions are seen. For bacterial diseases we can spray antibiotics.

Viral Diseases: Number of viruses are reported to infect Statice. Some of them are Clover Yellow Vein Virus, Cucumber Mosaic cucumovirus, Tomato Spotted Wilt tospovirus etc. We have to control the vectors which cause their spread. This may be any insect or sometimes our cutting implements may also help in spread of viral diseases.

Phytoplasma diseases: Some workers have reported aster Yellow phytoplasma disease in some varieties of Statice. They will cause degeneration of chloroplast, flower phyllody. Witches broom, leaf reddening, stunting growth, shoot dieback, malformation, flower discoloration and seed sterility. We must remove and destroy such plants and control the vector (insect pests) responsible for the transmission of phytoplasma.

References

Anonymous 1997. Production Manual Perennial Statice (Free flowering varieties) APEDA, Ministry of Commerce Govt. of India pp 9

Arora, J S and Pal P. 2002. *Statice*In Commercial Flowers Vol 1&2 (Eds. Bose T K and Yadav LP) Nayaprokash Calcutta. PP 576-584

Azuma A, Shimasaki Jand Inubushi S .1983. Acceleration of flowering of statice (*Limonium sinuatum* Mill) by seed vernalization. *J.Jap.Soc.Hort.Sci.*51(4):466-474

Chen J, Funnell K A and Morgan Ed .2010 A model for scheduling flowering of a *Limonium sinuatum* × *L. perezii* hybrid. *Horti Science* 45(10) 1441-1446

Chin Young Don, Dong Chun An, Ju Chean Hwang, Ju Yeon Song, Byoung Ryong Jeong. 2009. Effect of Low Temperature on Flowering and Cut Flower Yield of *Limonium* spp. 'Fantasia' Korean J Hortic Science and Tech 27(3): 380-385

Darlington CD, Wylie AP .1969. Chromosome atlas of flowering plants, 2nd edn. McMillan, New York

Farina E, Dalla Guda and Scordo E. 2000. Effect of low temperature and photoperiod on flowering of Limonium gmelinii Acta Hortic,. 541 pp:93-199

Freethy, R.1985. From Agar to Zenery: The Crossword Press, Marlborough U.K. Grieve C M, Poss J A, Grattan S R, Shouse P J, Leith J H and Zeng L. 2005. Productivity and mineral nutrition of *Limonium* species irrigated with saline waste waters. Hort. Science., 40(3): 654-658

Hamrick D. 2003. Ball Red Book Crop Production Volume -2, 17th Edition, Ball Publishing Batavia, Illinois USA Pp 506-511

Hsu Wen-Chan, Chang Shun-Pong, Lin Lie-Chwen, Li Chia-Lin, Richardson C D, Lin Chung-Ching and Lin Liang-Tzung. 2015. *Limonium sinense* and gallic acid suppress hepatitis C virus infection by blocking early viral entry. Antiviral Research 118 pp139-14

Innova I (1999) Effect of some herbicides systems on flower production of statice (*Limonium sinuatum*) Rastenievdni-Nauki.36(4) Pp 212-215

Katsutani N, Kajihara S, Kawamoto Y and Hara H. 1998. Forcing of Statice (*Limonium sinuatum* Mill.) by low temperature treatment of seedlings with illumination to prevent their rot *Bull Hiroshima Perfectural Agri. Res.* Centre No. 66 53-59

Laguna E, Navarro A, Perez-Rovira P, Ferrando I, Pablo Ferrer-Gallego, P. 2016. Translocation of *Limonium perplexum* (Plumbaginaceae), a threatened coastal endemic. Plant Ecol 217(10):1183–1194.

Morgan and Fennell K. 2018. Limonium In Handbook of Plant Breeding- Ornamental Crops (Eds. Huylenbroeck, Van J.) Springer International Publishing AG. Switzerland PP 520-27

Mukherjee, D. 2008. Specialty Cut Flowers Production Technologies, Naya Udyog, Calcutta pp 447-473 Sub culturing *in vitro* on subsequent flowering of *Limonium sinuatum* Mill. 67(4) pp 632-634

Uphof, J C T.1959. Dictionary of Economic Plants Weinheim, Engelmann. 400pp

Yoo, D L, Ryu S Y, Kim W B, Nam C W Choi Y E Kim B Hand Kim J K .1998. Effects of high land raising and seed low temperature treatment on yield of statice (*Limonium sinuatum*) cut flowers under structures during winter season in low land RDA J.Hort Sci.40(1150-154).

25

Jasmines

Safeena S.A.

Jasmines constitute a group of traditional flower crops which are commercially grown for their fragrant flowers and essential oil production. There are trailing, climbing, and erect growing species and cultivars in the genus *Jasminum*. The genus *Jasminum* comprises of more than 200 species which are known to occur in different tropical and subtropical regions of the world. The countries growing Jasmines on a commercial scale are India, France, Italy, Morocco, Algeria, North Africa, Egypt and Spain. More than 80 *Jasminum* species are found in India, of which only four species viz., *Jasminum sambac* (Mogra / Gundumalli), *Jasminum auriculatum* (Jui / Mullai), *Jasminum multiflorum* syn *Jasminum pubescens (*Kunda / Kakda) and *Jasminum grandiflorum* (Jai / Pitchi) are widely used for commercial cultivation. Jasmines are commercially grown in many parts of the country as a dry land crop. The leading states in jasmine production are South Indian states like Tamil Nadu, Karnataka and Andhra Pradesh. In India, the commercial cultivation of jasmines are popular in Athoor, Coimbatore, Dharampuri, Dindigul, Erode, Krishnagiri, Madurai, Nilakottai, Periakulam, Salem, Thiruchirapalli, Thirunelveli, Thiruvallur, Tirumanagalam, Vellore, Virudhunagar etc. in Tamil Nadu; Bangalore, Bellary, Kolar, Mysore, Shimoga, Tumkur etc. in Karnataka ; Hyderabad, Khammam, Mahbubnagar, Mancherial, Ranga Reddy, Sangareddy, Vikarabad, Warangal Rural, Yadadri etc. in Telangana; Anantapur, Chittoor, Cuddapah, East Godavari, Guntur, Krishna, Kurnool, Nellore, Prakasam, West Godavari etc. in Andhra Pradesh; Ahmednagar, Aurangabad, Pune, Sangli etc. in Maharashtra; Ahmedabad, Anand, Bharuch, Bhavnagar, Kutch, Vadodara etc. in Gujarat; Gazipur, Jaunpur, Kannauj etc. in Uttar Pradesh; Ajmer, Jaipur, Kota, Udaipur etc. in Rajasthan; Panskura Ranaghat etc. in West Bengal; Guwahati, Hoja, Jorhat etc.in Assam ; and parts of Punjab and Delhi. Jasmines contributes markedly to the national economy and it is estimated that annually more than 20 crores worth of jasmine flowers are produced and traded in India and are also exported to countries like Malaysia, Singapore, Sri Lanka, UAE and United States (Safeena *et al.*, 2017).

Importance and Uses

The flowers of jasmines are used for various purposes viz., for making special type of flower strings called veni, garlands, decorating hair of women, worshipping / religious offerings, floral decorations etc. The flowers of jasmine are mostly worn by women in their hair in south India. This crop is extensively cultivated for the unique fragrance of their flowers. The flowers of jasmine are used for extraction of essential oil which is used in preparing high grade perfumes, perfumed hair oils and attar, soap and cosmetic industry, colognes and flavouring the beverages etc. The main beauty and uniqueness of Jasmine is its odour, which cannot be imitated by any known synthetic aromatic chemical, still has a unique status in the perfume world (Safeena *et al.*,2016). Jasmine oil blends with every floral scent and hence widely used as an important perfumery item throughout the world. It is in good demand in the world market and used widely in cosmetic and perfumery industries and have a great export potential. Jasmines are also enjoyed in the garden, as a house plant. *Jasminum species* are suitable for planting in shrubbery, arches and pergolas. The flowers of Arabian jasmine (*Jasminum sambac*) are reported to be used for flavouring tea. They also have medicinal uses which has growing demand in India as well as in many developed countries of the world.

Origin and distribution

The genus *Jasminum* contains more than 200 species native to tropical, subtropical and warm temperate regions of the Eurasia, Australasia and Oceania. The center of diversity of Jasmine is in South Asia and Southeast Asia. The pattern of distribution of some important *Jasminum* species is given below: -

J. auriculatum : India

J. grandiflorum : Subtropical Himalayan Region

J. multiflorum : India, China, Myanmar, Malaysia, U.S.A

J. favreri : Myanmar

J. humile : Tropical Asia

J. flexile : India

J. officinale : Iran, India, China

Taxonomy

Taxonomic name *Jasminum* is a genus of shrubs and vines belonging to the family Oleaceae. The basic chromosome number of the genus *Jasminum* is 13, and most of the species are diploid (2n=26). But, natural polyploidy also exists in *Jasminum species*, mostly in *Jasminum sambac* (2n=39), *Jasminum mesnyi* (2n=39), *Jasminum angustifolium* (2n=52) and *Jasminum flexile* (2n=52). Jasmines are either erect, spreading, trailing or climbing shrubs and vines. Leaves of jasmine are opposite or alternate in arrangement. They can be simple, trifoliate, or pinnate. The flowers are usually very fragrant, white, cream or yellow in colour, borne in cymose clusters with a minimum of three flowers, sometimes solitary on the ends of branchlets. Each jasmine flower consists of about four to nine petals, two locules, and one to four ovules. They have two stamens with very short filaments. The bracts are ovate or linear. The calyx is bell-shaped. The fruit of jasmine is a berry that turns to black colour when ripe.

Species and Cultivars

Some selected species of *Jasminum* include *Jasminum humile* L. – Italian jasmine, Italian Yellow Jasmine, Peeli Chameli, Semmaligai; *Jasminum adenophyllum* Wall. – Blue Grape Jasmine, Pinwheel Jasmine, Princess Jasmine; *Jasminum mesnyi* Hance – Japanese Jasmine, Prim Rose Jasmine, Yellow Jasmine; *Jasminum angustifolium* (L.) Willd.; *Jasminum auriculatum* Vahl – Indian jasmine, Needle-flower Jasmine; *Jasminum sambac* (L.) Aiton – Arabian jasmine, Sambac jasmine; *Jasminum arborescens* – Tree jasmine, Muta, Bela, Nava Mallige; *Jasminum calophyllum* – Pandal Malli; *Jasminum flexile* and *Jasminum nittidum*.

The four commercially important *Jasminum* species which are cultivated widely in India are *Jasminum sambac, Jasminum auriculatum, Jasminum multiflorum* and *Jasminum grandiflorum*. Amongst these, the first three species are mainly cultivated for use as fresh flowers whereas the last one is cultivated mainly for concrete extraction. The brief description of these species with their important cultivars are given below.

Jasminum sambac: It is popularly known as Mogra / Bela / Gundumalli / Arabian Jasmine / Tuscan Jasmine / Malligai. *Jasminum sambac* is one of the most common cultivated species in India. The flower buds of this species are white with single or multi-whorled petals. The leaves are ovate, simple, opposite and smooth (glabrous). The important varieties of *Jasminum sambac* are : Adukkumalli, Arka Aradhana (Flower yield is 8t/ha ,Concrete yield is 14.95 kg/ha), Butt Mogra, Bela Japani, Double Mogra, Elusuthu malligai, Gundumalli (Average yield of flowers is 7-8t/ha , Estimated concrete yield is 15kg/ha), Iruvatchi, Kasthurimalli, Khoya , Madanaban, Motia, Mysore Mallige, Nilakottai Mallige, Oosimalli, Ramaban, Ramanathapuram gundumalli (Flower yield is 7 to 8 t/ha.), Rameswaram Mallige, Single Mogra, Sujimallige, Virupakshi etc.

***Jasminum auriculatum*:** It is commonly known as Jui / Juhi / Mullai / Malle / Indian Jasmine / Needle flower jasmine. It produces star shaped white scented blooms. Leaves are simple and occasionally trifoliate. The important varieties of *Jasminum auriculatum* are: Co-1 Mullai, Co-2 Mullai, Long Point, Long Round, Manamadhurai Mullai, Medium Point, Nilakottai Mullai, Periakulam Mullai, Short Point, Short Round, Parimullai (Concrete recovery is 0.29 per cent ,Flower bud yield is 8 t/ha), Co-1 Mullai (Concrete recovery is 0.34 per cent, yield 8.8 t/ha), Co-2 Mullai (yield - 11 t/ha) etc.

Jasminum multiflorum

It is commonly known as Kundphul / Kunda / Kakada / Kundo / Tundam / Kundamu / Star jasmine / Winter Jasmine. It is a highly resistant species of jasmine with high ornamental value but it is not scented. It bears complete white flowers in clusters which are widely sold as loose flowers. The important varieties of *Jasminum multiflorum* are: Arka Arpan (Flower yield is 6t/ha), Mysore Kakada etc.

Jasminum grandiflorum

It is popularly known as Chameli / Jai / Jahi / Pitchi / Jajimalli / Jathimalli / Royal Jasmine / Spanish Jasmine / French or Catalonian Jasmine. The leaves are opposite, pinnate, compound with 7-9 leaflets of equal sizes. Flowers are axillary or terminal, five petalled, white and often tinged with purple outside. The flowers possess unique and sweet fragrance. The important varieties / clones of *Jasminum grandiflorum* are: Arka Surabhi, Bangalore Pitchi (J.G.1), Co-1 Pitchi, Co-2 Pitchi, Coimbatore Pitchi (J.G.2), Lucknow Pitchi (J.G.3), Nilakkottai Pitchi, Periakulam Pitchi, Tenkasi White Pitchi (J.G.4), Thimmapuram Pitchi (J.G.5), Thrum type and Triploid (J.G.6) etc.

Salient features of some *Jasminum grandiflorum* varieties

Co-1 Pitchi: Clonal selection from germplasm. It is suitable for use as both loose flowers and extraction of essential oil. Flower buds are pink tinged with long corolla tube, pink streaks are found on external surface of the flower petal, concrete recovery is 0.29 per cent, concrete yield is 29.42 kg/ hectare. Average flower yield is 10 t/ha.

Co-2 Pitchi: It is an induced mutant evolved by treating the vegetative cuttings of Co-1 Pitchi with gamma rays @ 1.5 kR , Buds are bolder, white with pink tinge and have long corolla tube, The flower bud is 4.14 cm in length as against 4.00 cm in Co-1 Pitchi and weight of 100 flower buds is 10 g. It is best suitable for earlier and quicker tying of buds in garland making, flower yield is 11.68 t/ha.

Arka Surabhi: It is a clonal selection earlier known as Pink Pin. It is remarkably tolerant to drought, it is a high yielder of essential oils. Flowers are also sold as fresh loose flowers, concrete recovery is 0.35 per cent, per hectare yield of concrete is 35 kg and average flower yield is 10 t/ha.

Climate

Jasmines can be cultivated in a variety of climate. They can be grown in the open field for commercial flower production. Mild tropical climate is considered to be ideal for the optimum growth and flowering of Jasmines. It ideally prefers warm summer, sunny days, moderate rainfall and mild winter for their successful cultivation. They are sensitive to frost.

Soil

Jasmines can be planted on a wide range of soils. Well-drained sandy loams and red loams under tropical conditions are best suitable for its cultivation. In clayey type of soils, increased vegetative growth and reduced flowering is observed. Clayey soils can be improved by addition of more organic manures and lime. Jasmines can be grown in sandy soils also, with liberal application of manures and assured supply of water.

Propagation

Most of the cultivated *Jasminum species* generally do not bear seeds and the plant is reproduced solely by different methods of asexual propagation. Jasmines can be propagated by cuttings, layering, sucker, grafting, budding and tissue culture (Safeena *et al.,* 2016). Layering and cuttings are the main methods of propagation. Easiest method of vegetative propagation is by stem cuttings. Most successful method of layering is ground layering. Commercially *Jasminum auriculatum* is propagated by Semi hard wood cuttings, *Jasminum grandiflorum* by Terminal cuttings and *Jasminum sambac* by Terminal and Semi hardwood cuttings (Safeena., 2018). Higher rooting percentage and increased number of roots are obtained when cuttings are taken during April – September. Better rooting of cuttings can be obtained by planting in a medium of coarse sand under mist and also by using any of the rooting hormones like IBA (1000 ppm), IAA (1000 ppm) and NAA (1000 ppm) for terminal cuttings. 2500 ppm of rooting hormones are used for semi hard wood cuttings. Simple and compound methods of layering can be practiced during June-July to October-November. Layers will be ready for planting within 90-120 days.

Planting

Jasmines are perennial in nature and once planted, it remains in the same spot for 10-15 years. Hence for planting of jasmines, land with proper drainage, irrigation facilities and sunny location are essential. Jasmines are generally planted during the rainy season. Pits of 45 cm x 45 cm x 45 cm have to be dug at least one month before planting and exposed to sunlight. Pits should be filled with 2 parts of well rotten farmyard manure and one part each of fresh earth and coarse sand, a few days before planting. Healthy and strong well-rooted plants are planted in 45 cm^3 pits according to the spacing recommended for each species during June - November.

The distance of planting plays a key role in flower yield. The recommended spacing for planting of commercially important Jasminum species are given below. *Jasminum sambac* are planted at a spacing of 1.25 m x 1.25 m. *Jasminum auriculatum* are planted in pits dug at a spacing of 1.8 m x 1.8 m. *Jasminum grandiflorum* are planted at a spacing of 2.0 m x 1.5 m. *Jasminum pubescens. Syn. Jasminum multiflorum* are planted at a spacing of 1.2 m x 1.2 m.

Manures and Fertilizers

Jasmines responds well to application of manures and fertilizers. Too much of manuring encourages vegetative growth in Jasmines and hampers quality and quantity of blooms. In *Jasminum sambac,* FYM or compost 10 kg and NPK at 60:120:120 g per plant is applied in two split doses. *i.e.* once after annual pruning (November) and again during June - July. In *Jasminum auriculatum,* FYM or compost 10 kg and NPK at 120:240:120 g per plant is applied in six split doses at bimonthly intervals. The first dose is given immediately after annual pruning (January). In *Jasminum grandiflorum,* FYM or compost 10 kg, NPK at 60: 120: 120 g per plant is applied in 2 split doses. *i.e.* once in December after annual pruning and again in June - July. In *Jasminum multiflorum,* FYM or compost 10 kg, NPK at 120:240:120 g per plant is applied in 2 split doses in December after pruning and again in June - July. An increase in flower yield may be obtained if micronutrients like Mg (40 kg/ha), Boron (5 kg/ha) and Zinc (10 kg/ha) are applied along with NPK fertilizers.

Irrigation

Irrigation should be given to jasmines immediately after planting followed by weekly irrigation depending upon the weather conditions. It prefers moderate watering. Constant and adequate water supply during peak flowering season is essential for high yield of flowers. During flowering season, water should be applied twice a week if there are no rains. After flowering is over, the water supply can be cut off. During summer, irrigate twice a week and once a week during rest of the months.

Intercultural operations

Weeds compete with the nutrients applied to jasmine and causes obstruction in movement during harvesting of flower buds. Manual weeding is effective but expensive. The first weeding in Jasmine plots should be done 20-25 days after planting and subsequent weeding have to be done once in 2-3 months. Mulching is another effective method to reduce the weed population. After the first rains, the basins of jasmine plants should be mulched thoroughly which keeps most of the weeds under control. Mulching should be followed by 2-3 hand weedings a year. Chemical weed control is effective and economical. Spraying of post emergent weedicides like paraquat and glyphosate (2.0 kg/ha) can be practised to control both grassy and broad-leaved weeds. Strengthening of irrigation channels and bunds are the other intercultural operations that should be followed for jasmine cultivation.

Pruning

In jasmines, flowering habit is terminal and axillary. So, increasing the number of shoots or branches would increase the yield in jasmines, for which pruning is essential. Pruning is generally done to get the desired crop. Pruning influences the growth of new healthy shoots, flower bud initiation, differentiation and ultimately the flower production in Jasmines. Normally, irrigation is withheld prior to pruning and the plants are pruned by removing all the past season shoots including dead and diseased branches. *Jasminum sambac* bushes are pruned to 50 cm height from the ground level during last week of November. The bushes of *Jasminum auriculatum* are pruned to 45 cm height from ground level during the last week of January. Pruning is done in *Jasminum grandiflorum* during the last week of December to 45 cm height from ground level. Pruning is done in *Jasminum multiflorum* during last week of October to 45 cm height from ground level (Safeena *et al.*, 2020).

Harvesting and Yield

Flowering commences in Jasmines after 6 months of planting. Jasmines gives economic yield only from the third year of planting and produces maximum flowers up to 12-15 years and then the yield starts declining. The stage of harvesting in Jasmines depends on the purpose of flowers to be harvested. For fresh flowers, flower buds of jasmines are plucked either in the evening or early morning hours when the buds are unopened and fully developed. For the purpose of jasmine concrete extraction, only freshly picked fully opened flowers are required. The flowers of *Jasminum grandiflorum* usually open in the evening hours between 5-7 p.m. Fully opened flowers should be harvested in the early morning hours, preferably before 10.00 a.m. There will be gradual reduction in the yield of concrete if flower picking is delayed. The yield and quality of the concrete will be considerably reduced if the flowers are picked during hot hours. Hence, always care should be taken to harvest the fully opened flowers in the early morning hours before weather becomes hot. Avoid damage to flowers during harvest as it will affect the shelf life of fresh flowers and concrete recovery.

Yield in Jasmines depends upon the cultural practices followed by the farmer. In *Jasminum sambac*, flowering commences in March - April. Unopened flower buds which are fully developed should be picked in the morning hours. Yield of about 5t / ha can be obtained from *Jasminum sambac*. In *Jasminum auriculatum*, flowering extends from May to November. Fully matured unopened flower buds should be picked in the morning hours. Yield of about 5,000 kg of flower buds/ha can be obtained from *Jasminum auriculatum*. In *Jasminum grandiflorum*, the flowering season starts from May to October. Unopened flower buds which are fully matured should be picked in the morning for fresh flower trade. For oil extraction fully opened flowers are to be picked before 10 a.m. The flower yield is 6 t/ha of flower buds in *Jasminum grandiflorum*. *Jasminum multiflorum* is the species which flowers profusely for longest duration. The flowering season in *Jasminum multiflorum* starts from November to December and bears flowers upto August. The flower buds which are fully developed and unopened should be picked in the morning for fresh flower trade. The flower yield in *Jasminum multiflorum* is about 3-4 t/ha of flower buds.

Post-harvest management

Grading: There are no standard grades available for jasmines. The flowers may be graded according to the corolla tube length, size of the flower bud, shape of the flower bud and freshness.

Packing: Harvested flowers should be provided with cold treatment before packing. Packing should be functional, economical and attractive for better acceptance in the markets. Corrugated cardboard boxes are good for distant market. Wholesalers pack jasmine flowers in bamboo baskets. Packing should be done in such a way that there is maintenance of some moisture and air circulation in the baskets. It is advisable to sprinkle water on the newspapers covering the inside of the basket. Cover the top portion with paper again and close with a bamboo basket cover or a gunny sack.

Marketing, Distribution and Transportation

Transporting of jasmine flowers is done through trucks, ships (Refrigerated) etc. Before long distance transportation it is better to keep jasmine flowers in bamboo basket which should be covered under moist muslin cloth. Jasmine flowers should not be handled badly or damaged during transportation. Damage to flowers during transit will affect shelf life of fresh flowers and concrete recovery.

Export Packaging Technology for Jasmine

Major mode of transport of Jasmine flowers for export is by air. Jasmine flowers reach the Dubai and Singapore markets within 24 hours after harvesting, while it takes around 36 – 48 hours to reach the New York market. High levels of post-harvest losses (35 to 40%) was reported during transit. Jasmine flowers show signs of wilting with an abrupt loss of fragrance within 24 -36 hours after harvesting, since they are delicate in nature. One of the major problems faced by the exporters is lack of a proper packaging technology for export. Whenever there is delay in sending the flowers due to flight cancellations, the exporters face heavy losses, because the jasmine flowers cannot be preserved beyond a day. Jasmine flowers retain their freshness only for a day in ordinary packaging. Hence, any packaging technology that would extend the post-harvest life for even few hours would benefit the exporters (*Source*:TNAU).

An export packaging technology has been standardized for jasmine flowers by Tamil Nadu Agricultural University (TNAU), Coimbatore under the ICAR-NAIP sponsored project 'Value Chain on Flowers for Domestic and Export Markets', in collaboration with a successful flower exporter, M/S. Vanguard Exports (TNAU Agritech Portal., 2014).

Export packaging technology for Jasmine for Dubai, Market

The flowers of jasmines were collected from the auction centres and then strings were made with different shape and size based on the requirement of the buyer. These strings are further treated with floral preservatives (Boric acid @ 4%) to preserve its freshness and improve the shelf life. Treated strings of jasmine flowers are packed in well ventilated corrugated fibre board (CFB) boxes with butter paper lining. After proper packaging the CFB boxes are airlifted to respective destination (*Source*: TNAU).

Export packaging technology for Jasmine for USA, Market

Jasmine flowers were collected from auction centres and loose flowers are made into strings and related products. These jasmine strings are further treated with 4% boric acid and packed in aluminium foil lined light weight cardboard boxes. These boxes are then packaged in large thermocole boxes with intermittent ice gel packs to maintain the temperature and humidity inside the box. After proper packaging, these thermocole boxes are transported in refrigerated vans to airport and then despatched to the respective destination (*Source*: TNAU).

Impact of the advanced packaging technology

- Shelf life of flowers increased to 72 hours against 36 hours in normal packaging
- Post-harvest losses in jasmine were reduced from 40% to 10%
- The export volume increased to 1000 kg/ day from 500 kg/day

Extraction of jasmine concrete (*Source*: http://ecoursesonline.iasri.res.in/):

The concrete obtained from *Jasminum grandiflorum* is a wax like substance which contains the natural perfume of the flower together with plant waxes, albumin and colouring matter. Natural perfume is available in jasmine flowers in very meagre quantity (0.25%) in the form of volatile oil. Steam distillation process for the extraction of the volatile oil could not be followed in jasmines as it does not yield the perfume oil on steam distillation. Hence, the method of Solvent extraction is practiced for extraction of jasmine concrete. The principle behind this method is that the odoriferous substances of the flowers are allowed to be absorbed by a highly volatile solvent and then the solvent is evaporated leaving the odoriferous principles (Hemla Naik and Kannan., 2012).

The fully opened flowers of *Jasminum grandiflorum* should be plucked and transported in clean baskets or cloth bags. While harvesting care should be taken not to damage the flowers and sweat, dirt etc., of the person who pick the flowers do not contaminate them. The contamination if any during picking and transporting will reflect in the concrete quality. The flowers which are damaged yield concrete with more percentage of unwanted wax and other plant materials. The harvested flowers should be stored in a cool location and processed within a maximum period of two hours.

Two steps are involved in the processing of jasmine flowers.

Step 1 : Dissolving the perfume material by soaking the flowers with the solvent

Step 2: Removal of solvent from the perfume material by evaporation

Step 1 : Treating the flowers with the solvent

The fully opened flowers of *Jasminum grandiflorum* are treated with the solvent so that all the odoriferous material in the flowers gets dissolved in it. The most crucial factor in solvent extraction process is the quality of solvent used. Commercial food grade hexane is one of the good solvents used in large scale for the extraction of floral concrete across the globe. Food grade hexane available in the market should be purified before using it as solvent. Food grade hexane is treated with liquid paraffin at 5% and distilled at 70°C. The residue which is not evaporating at 70°C is rejected as impurities. The purified hexane is a colourless liquid with a boiling point ranging from 60°C to 80°C. The jasmine flowers are soaked in purified hexane at 2 litres per kg of flower for about 30 minutes. For ensuring the complete extraction of perfume from the flowers, rotate the container having flowers and solvent slowly for 20 minutes in the rotary type of extractor. In this process, the solvent penetrates the flowers and dissolves perfume substances along with waxes, pigments, albumin and colouring matter also, which have to be removed at later stage to get the pure perfume oil.

Step 2 : Evaporation

After the complete extraction of the perfume from the jasmine flowers, the perfume laden solvent is led into the evaporator. The solvent is filtered and concentrated by evaporation at a constant temperature of 75°C. At 75°C, the solvent evaporates leaving the perfume and other plant waxes. Vapour of the solvent is condensed to liquid again for recycling. Concentrated liquid which contains perfume, wax, pigments and solvent is distilled in a vacuum distillation unit for complete solvent removal in the still. The still should not be heated directly and heating should be done on a water bath. The vacuum created in the still facilitate the solvent to evaporate quickly at lower temperature which is condensed in the condenser and collected for recycling. Heating should be continued till the last traces of hexane are removed. The operation can be stopped when there is no condensation of hexane vapour in the condenser. After the complete removal of the solvent, a semi-solid wax like brown substance is left in the still which is the jasmine concrete of commerce. It is cooled and stored in glass or aluminium containers.

Crop Protection (TNAU Agritech Portal., 2014)

Insect-Pests

Bud worm (*Hendecasis duplifasciallis*): It is a green larva with a black head. It bores into closed immature flower buds and feeds on inner floral structures. Larva makes a web like pattern among adjacent buds and feed on petals. Infested flowers turn pale red in colour and fall off from the plant. Collect and destroy infested flowers along with larvae at least once in a week. Light traps are effective to attract and kill the adult moths. Spray Monocrotophos 2ml/l for the control of the pest.

Gallery worm (*Elasmopalpus jasminophagus*): It web together the terminal leaves, shoots and flower heads and feeds on them. Faecal matter can be seen attached to the silken web. Management is same as given for jasmine bud worm

Leaf webworm (*Nausinoe geometralis*): Leaves mostly in the lower bushy and shaded portions are affected. The leaves are webbed in an open and loose manner. The silk threads are seen as a cobweb on the surface of the leaves. Larvae feed on the leaves and skeletonise it. Spray dimethoate 30 EC 500 ml in 500–750 L of water/ha for the control of the pest.

Blossom midge: Spray Monocrotophos 2 ml/lit or Quinalphos 2 ml/lit to control it.

Red spider mite (*Steneotarsonemus pallidus*): Incidence of this pest is high during warm and dry weather, especially during summer. Pests feed on the under surface of the leaves which become yellow and drop off. Spray Wettable Sulphur 50 WP @ 2 g/lit or Dicofol 2.5 ml/lit to control the mite infestation.

Jasmine eriophyid mite (*Aceria jasmini*): Feeding causes felt-like hairy out growth (erineum) on the surface of the leaves, tender stem and flower buds. It makes web which appear to be a white hairy growth on the leaf surface, tender stems and flower buds. Spray wettable sulphur 50WP 5g/lit or Dicofol 18.5 EC 3ml/l.

Leaf eating caterpillar: Leaf eating caterpillar can be controlled by spraying Quinalphos 2 ml/lit.

White ants: Both nymphs and adults suck the sap from tender shoots and buds and prevent flower formation. To control, dust Lindane @ 5 g/pit to the pits before planting

Green plant hopper (*Flata ocellata*): The adult bug is green in colour with minute spot on fore wings. Both nymphs and adults feed on terminal shoots.

Jasmine leaf roller (*Glyphodes unionalis*): The green caterpillar roll the leaves and feed on them. Adult is a white moth with brown lines along the costal margin of forewings.

Jasmine bug (*Antestia cruciata*): Both nymphs and adults suck the sap from tender shoots and buds and prevent flower formation.

Diseases

Leaf Blight: It is caused by a fungi *viz., Cercospora jasminicola.* Circular to irregular reddish-brown spots (2-8 mm diameter) appear on the upper leaf surface which later become irregular covering larger leaf area. Spray of Bavistin (0.1%), Mancozeb 0.25% and Bordeaux mixture (1%) are equally effective.

Alternaria leaf spot and blight: It is caused by *Alternaria jasmini* or *A. alternata.* Dark brown spots appear on the leaves. Later the spots enlarge covering larger area causing blighting of leaves. Concentric rings can be seen the lesions. The disease also affects stem, petiole and flowers. Collect and remove fallen leaves. Spray with Copper oxychloride 0.25% or Mancozeb at 2 g/lit. from the onset of monsoon at monthly interval to control the disease occurrence.

Wilt: Wilt is caused by *Fusarium solani.* Symptoms are yellowing of lower leaves which gradually spread upwards and finally results in death of the plant. Bordeaux mixture (1%) can be drenched to soil around the plants to control the disease.

Collar rot and Root rot: It is caused by *Sclerotium rolfsii.* Plants at all stages are infected. The older leaves turn yellow first followed by younger leaves and finally death of the plant. In the root black discoloration can be seen. White strands of mycelia and mustard like sclerotia appear on the infected tissues and stem surface. Soil drenching with Copper oxychloride 0.25% is the control measure.

Leaf spot: Spraying of Mancozeb at 2 g/lit. from the onset of monsoon at monthly intervals will control the disease occurrence.

Yellowing of leaves: It is caused by 3 factors viz., iron deficiency, nematode infection, and root rot disease.

Iron deficiency: Iron deficiency can be corrected by spraying Ferrous sulphate 5 g/lit at monthly intervals until the chlorotic symptoms disappear.

Nematode: Initially test the soil for nematode infection. Apply Temik granules (10 g) near the root zone and then irrigate the field.

Root rot: Soil drenching with Copper oxychloride at 2.5 g/lit is effective to control root rot.

References

B. Hemla Naik and M. Kannan., 2012. Commercial Floriculture (FLA 202) Topic 19 – Extraction of Jasmine Concrete. e-Krishi Shiksha. http:// ecoursesonline. iasri.res.in / mod/page/view. php?id=79738

Safeena S. A, M. Thangam and N.P.Singh., 2016. Package of practices for cultivation of Jasminum sambac. Extension Folder No.77, ICAR- Central Coastal Agricultural Research Institute (Indian Council of Agricultural Research), Ela, Old Goa, Goa

Safeena S.A, M. Thangam, S. Priya Devi and N.P.Singh., 2016. A Catalogue of Jasmine Accessions in Goa, Technical Bulletin No : 58 , ICAR - Central Coastal Agricultural Research Institute (Indian Council of Agricultural Research), Old Goa, Goa

Safeena S. A, M. Thangam, S.Priya Devi AND N.P. Singh , 2017. Genetic Diversity of Jasmine and its Conservation under Coastal Humid Ecosystem of Goa. World Journal of Pharmaceutical and Life Sciences. Vol. 3, Issue 6, 116-123

Safeena S.A, Tarak Nath Saha, P. Naveen Kumar and D.V.S. Raju., 2020. Advances in cultivation of jasmine, crossandra and specialty flowers. In Training Manual (e-manual) of five days Training Programme on Advances in Commercial Floriculture jointly organized by ICAR-DFR, Pune and National Institute of Agricultural Extension Management (MANAGE) at VAMNICOM, Pune during 2-6 March 2020.

Safeena, S. A., 2018, Advances in cultivation of jasmine for loose flower production. In Training Manual of Certified Farm Advisor in Floriculture (Module-II) jointly organized by ICAR-DFR, Pune and National Institute of Agricultural Extension Management (MANAGE) during 14-28 November, 2018. Pp : 73-81

TNAU Agritech Portal., 2014. Crop Protection: Jasmine. http:// www. agritech. tnau. ac. in / crop_protection/jasmine

TNAU Agritech Portal., 2014. Export Packaging Technology for Jasmine. http:// agritech. tnau. ac. in/ export_import/jasmine/index.html

26

Marigold

B.S. Beniwal and Sonu Kumari Sheoran

Marigold (*Tagetes* spp.) is an important annual flower crop; native to Central and South America especially Mexico and belongs to family Asteraceae (Compositae). It is commonly known as "*Gainda*" in India and also called as "Friendship Flower" in United States of America. Loose flowers of marigold are in great demand throughout the year. It is commercially grown in tropical and subtropical regions of India owing to its ease in cultivation, hardy nature, wider adaptability and year round availability of single, semi-double and double flowers of vibrant colours.

Marigold is grown over an area of about 66.09 thousand hectares with production of 603.18 thousand MT loose flowers in India. Major states involved in loose flower production of marigold are Karnataka, Madhya Pradesh, Gujarat, Andhra Pradesh, West Bengal, Maharashtra, Haryana and Chhattisgarh. In India, Assam is the only state, which produces marigold for both cut as well as loose flower purpose on an area of about 0.61 thousand hectare (Anonymous, 2016a).

Marigold is a cross-pollinated flower crop and it is pollinated by insects (Kaneko *et al.*, 1990). It bears terminal and solitary inflorescence, known as head or capitulum, which possesses long and hollow peduncle with swollen apex and campanulate involucres (Setshogo, 2005). Capitulum consists of peripheral ray florets and central disc florets. Ray florets are pistillate, ligulate, sessile, epigynous and zygomorphic, whereas, disc florets are bisexual, tubular, sessile, epigynous and actinomorphic (Subrahmanyam, 1996 & Setshogo, 2005).

African marigold (*Tagetes erecta* L.) is a diploid species ($2n = 2x = 24$), while, French marigold (*Tagetes patula* L.) is a tetraploid species ($2n = 4x = 48$) with basic chromosome no. of $x = 12$ as revealed by breeding work of various researchers (Towner, 1961, Chen and Lin, 1981, Chen and Lin, 1982, Serrato-Cruz *et al.* 2000, Zhang *et al.*, 2011). *Tagetes erecta* and *Tagetes tenuifolia* are diploid species, whereas, *Tagetes patula, Tagetes minuta, Tagetes biflora* and *Tagetes remotiflora* species are tetraploid in nature (De, 2017). On the basis of floral morphology; three different types of male sterile systems i.e. apetaloid, petaloid and gynomonoecious types was observed in African marigold (Prakash *et al.*, 2016). Out of these, apetalous male sterility (femina type) is more reliable and preferred over petalous male sterility because later is more prone to break down (Goldsmith, 1968 & Janakiram, 2004). Apetaloid male sterility was controlled by single recessive gene (Prakash *et al.*, 2016). A monogenic recessive male sterile apetalous mutant was detected from *T. erecta* population by Khoshoo

(1979). Cyto-genetic male sterility is also found in marigold (De, 2017). Petaloid and gynomonoecious male sterile systems were controlled by cytoplasmic inheritance (Prakash *et al.*, 2016). In marigold, cytoplasmic genetic male sterile lines were also developed for commercial production of F_1 sterile hybrids (Pratibha *et al.*, 2000).

Importance and Uses

Marigold (*Tagetes* spp.) is known to be a versatile crop with "Bumper Harvest". It has a habit of profuse flowering, short crop duration to produce marketable flowers, wide spectrum of attractive colors, shape, size and good keeping quality. These qualities make marigold one of the most important loose flower crop grown commercially in India. Its flowers are composed of a dense arrangement of ray and disc florets that comes in off white, yellow, orange, golden yellow, primrose, mahogany, rusty red and tangerine or deep scarlet.

The entire plant of marigold is of great use to humankind in some way or the other way. Marigold flowers are extensively used for making garlands, wreath, religious offerings, flower exhibitions, insect and nematode repellants, nutrient supplement for poultry feed, extraction of essential oil and natural colour pigments for culinary, dyeing and pharmaceutical purposes. Marigold is also widely used as bed plant, pot plant and edge plant in garden for making herbaceous borders, rockeries, hanging basket and window boxes. It is also used in various dry flower arrangements.

Essential oil extracted from *Tagetes erecta* flowers has been mainly used for the compounding of high-grade perfumes. It also has anti-inflammatory, antiseptic, antispasmodic, antihaemorrhagic, astringent, diaphoretic and emmenagogue properties. This oil is very important in aromatherapy for its powerful skin healing effects and has fly repellant properties (Gupta, 2014). French marigold finds use in perfumery particularly in certain types of herbaceous fragrances like fougere and lavender. This oil is also used in aldehyde oriented tabac bases. *Tagetes minuta* oil is generally used in germicidal, insecticidal and microbicidal preparations due to the presence of toxic compound tagetone (Mohamed *et al.* 2000 and Joy *et al.* 2001). In French marigold, maximum oil content (0.800 ml/ kg) was obtained in MGH 8-2 followed by MGH 17-1 (0.717 ml/kg) extracted through hydro distillation method (Gulia *et al.*, 2017b).

Marigold is gaining industrial importance as a potential source of natural colour pigments mainly carotenoids like lutein, carotenes, mono-hydroxy pigments, di-hydroxy pigments etc. It finds applications in cosmetic, colouring and pharmaceutical industry (Gupta and Vasudeva, 2012). Carotenoids extracted from flowers are used commercially as food supplements; poultry feed additives and natural colourants in food, beverage, cosmetics and textile industries for imparting various shades of yellow, orange, and red color. These colors are safe for human consumption, unlike other artificial colors which are carcinogenic in nature. Marigold cultivars with orange flowers have higher xanthophylls as compared to cultivars having yellow flowers (Deineka *et al.*, 2007). In French marigold, maximum total carotenes (69.83 mg/100 g), xanthophylls (305.48 mg/100 g) and di-hydroxy pigments (171.55 mg/100 g) were recorded in genotype 109-1-2, while maximum mono-hydroxy pigments were reported in genotype 26-3-1 (107.82 mg/100 g) (Beniwal *et al.*, 2017a). African marigold genotype 160-8 (PN) recorded maximum mono-hydroxy pigments (121.18 mg/100g), whereas, maximum Total carotenes (77.36 mg/100g) and di-hydroxy pigments (229.18 mg/100g) were recorded in genotype 160-9-2 and maximum xanthophylls (362.03 mg/100g) were obtained in genotype 160-5 (Beniwal *et al.*, 2017b).

All carotenoid pigments like total carotene, di-hydroxy pigments and total xanthophylls were found higher at bud burst stage and their maximum content was observed on the first day of storage. Maximum carotenes were obtained in MGH 160-9-2 and mono-hydroxy pigment was present in Pusa Narangi. Chlorophyll 'a' and total chlorophyll were found higher at bud burst stage in contrary to chlorophyll 'b', which was higher at full bloom stage (Gulia *et al.*, 2017a). Marigold flower extracts have been used as feed additives to supplement the poultry feed for decades to colour egg yolks orange and broiler skins yellow (Gupta, 2014). Marigold flower is a rich source of lutein. Lutein and zeaxanthin are the predominant carotenoids of the retina and are considered to act as photoprotectants preventing retinal degeneration (Richard, 2013). Dried marigold flowers contain 0.1–0.2% dry matter of carotenoids, out of which 80% are lutein diesters (Sivel *et al.*, 2014). Orange coloured French marigold varieties contains the maximum amount of lutein (Bhattacharyya *et al.*, 2010).

Marigold also possesses nematicidal properties. It can suppress 14 genera of plant-parasitic nematodes; out of these lesion nematodes (*Pratylenchus* spp.) and root-knot nematodes (*Meloidogyne* spp.) are most affected type of nematodes (Suatmadji, 1969). *Tagetes minuta* oil concentrations (1–4 %) possess strong toxicity to eggs and juveniles of *Meloidogyne incognita* (Adekunle *et al.* 2007). These juveniles were unable to fully develop in the roots of *T. erecta* (Ploeg and Maris, 1999). Marigold roots release the chemical alpha-terthienyl, which can aid in the reduction of root-knot nematodes and other disease promoting organisms, such as fungi, bacteria, insects, and some viruses (Hethelyi *et al.*, 1986; Gommers and Bakker, 1988 and Soule 1993b).

Tagetes erecta as intercrop can efficiently reduce the levels of spiral (*Helicotylenchus multicinctus*), lance (*Hoplolaimus indicus*) and burrowing (*Radopholus similis*) nematodes in highly susceptible banana crop, whereas, Tagetes patula can be grown as an ideal cover crop for suppressing lesion nematodes (Pratylenchus penetrans and P. pratensis) in ornamentals (Wang *et al.*, 2007). Prevailing temperature can also affect nematode suppression activity of marigold. Companion planting short vine tomatoes with French marigolds throughout the growing season will slow development of white fly populations (Conboy *et al.*, 2019). A significant reduction in soil nematode population (150.25/50 ml), number of root galls (8/plant) and root nematode population (29.80/8galls) along with good vegetative growth was recorded in tomato plant with border planting of French marigold (Mali *et al.*, 2019).

The marigold flowers are widely used in folk medicine, particularly for curing inflammatory disorders (Nahak and Sahu, 2017). Healing properties of *Tagetes* species have been implemented by folk medicine for centuries (Montellano, 1986). French marigold (*Tagetes patula*) is used for preparing ethnobotanical remedies against rheumatism, stomach and intestinal problems, kidney and hepatic disorders, fever, and pneumonia (Neher, 1968, Yonzone and Yonzone, 1999).

Origin

All *Tagetes* species are endemic to the Americas and thought to be originated in Central and South America especially in Mexico (Ferraro 1955, Kaplan, 1960, Neher, 1968 & Soule, 1996). African marigold (*Tagetes erecta*), Sweet Scented marigold (*Tagetes lucida*) and Signet marigold (*Tagetes tenuifolia*) originated in Mexico, whereas, Stinking Roger marigold (*Tagetes minuta*) originated in South America (Joy *et al.* 2001). Marigold is introduced by Portuguese in India (Watts, 1893). It became popular and spread quickly because of its easy cultivation and wide adaptability to varied agro-climatic conditions. Currently marigold is well distributed and commercially grown throughout the India (Singh, 2006a).

Taxonomy

Species

Marigold belongs to Asteraceae family. The genus *Tagetes* contains about 50-56 species, out of these, African marigold (*Tagetes erecta* L.), French marigold (*Tagetes patula* L.) and Signet marigold (*Tagetes tenuifolia* Cav.) are commonly grown throughout the world due to their ornamental importance (Neher, 1966, Panero, 2007, Kaplan, 1960, Turner and Nesom, 1993, Soule, 1993a, Soule, 1996). Earlier Rydberg (1915) reported that genus *Tagetes* comprises of about 33 species of strongly scented annual and perennial herbs, later on Soule (1993a) recorded 56 species out of which, 27 species are annual and remaining 29 species are perennial in nature. African marigold (*Tagetes erecta* L., $2n = 24$) and French marigold (*Tagetes patula* L., $2n = 48$) are two important annual species, which are commercially grown in India. Other species are Signet marigold (*Tagetes tenuifolia* Cav. /*Tagetes signata* Bartl.) (Greuter, 2006+), Sweet Scented marigold (*Tagetes lucida* Cav.)[1], Lemmon's marigold (*Tagetes lemmonii* A. Gray)[1], Wild marigold or Stink weed or Muster John Henry marigold (*Tagetes minuta* L.)[1] (CABI, 2020), Licorice marigold (*Tagetes micrantha* Cav.)[1], Miniature marigold (*Tagetes minima* L.)[1] and Lesser marigold (*Tagetes filifolia Lag./ Tagetes pusilla* Kunth)[1&2]. These species have medicinal and ornamental value but not commercially grown in India.

Source:1https://plants.usda.gov/java/Classification Servlet? source = profile & symbol = TAGET & display=31& 2http://www.theplantlist.org/tpl/record/gcc-82417

African or Aztec marigold (*Tagetes erecta* L.)

It is a hardy, erect, branched and tall annual of about 90-100 cm height. It has pinnately divided leaf with lanceolate and serrated leaflets. It bears medium to large sized single or double flowers with quilled florets of vibrant colours like light yellow, golden yellow, canary yellow, bright yellow, lemon yellow, cadmium orange, deep orange, golden orange, bright orange and off white. African marigold flower oil is reddish yellow in colour, possessing characteristic marigold odour and polymerizing readily in air. This oil contains approximately 8.5% ocimene, 14.03% limonene, 21.14% linalool, 13.75% linalyl acetate and 40.38% tagetone. Leaf oil and stem oil of African marigold is greenish yellow in colour (Joy *et al.*, 2001).

French marigold (*Tagetes patula* L.)

It is a hardy, bushy and dwarf plant of about 30-45 cm height with reddish purple stem. This plant has pinnately divided dark green leaves with linear, lanceolate and serrated leaflets. It bears small to medium sized single or double flowers of yellow, orange, crimson, mahogany red, light yellow with maroon blotches, deep crimson with golden yellow margin, red, primrose and deep scarlet colour on long peduncles.

Signet marigold (*Tagetes tenuifolia* Cav.)

These marigolds produce compact plants having lacy foliage and small single flowers in cluster. They have edible flowers of yellow to orange colour. The foliage of these marigolds has a pleasant lemon fragrance. These are excellent garden plants; generally used in beds, edges and window boxes.

Wild marigold (*Tagetes minuta* L.)

It is also known as Stinking Roger or Stink weed and Muster John Henry marigold (CABI, 2020). It is an erect and woody annual with strong aroma. It has glossy green; pinnately compound leaves and lanceolate leaflets. It attains 1-2 m height with 7-15 cm long leaves with yellow to orange flowers. Strong smelling oil of *Tagetes minuta* is used as a laxative, diuretic, flavoring agent and insect repellent (Holm *et al.*, 1997). It is generally cultivated as a main crop or as an intercrop with maize, tomato and geranium for essential oil purpose (Singh and Singh, 2001). Vanphool variety of *Tagetes minuta* was found suitable for growing in North Indian plains for high yield of good quality essential oil (Kumar *et al.*, 1999).

Sweet Scented marigold (*Tagetes lucida* Cav.)

It is a perennial herb of 30-40 cm height with oblong-lanceolate and finely serrated leaves. It bears dense clusters of orange yellow flowers at terminal end. This Sweet Scented marigold is mainly suitable for border plantation in garden. *Tagetes lucida* oil is greenish yellow in colour and estragol is the major chemical constituent present in it (Joy *et al.*, 2001).

Interspecific hybrids

These are sterile, triploid hybrids resulted due to interspecific crosses between diploid and tetraploid *Tagetes* spp. They are also known mule hybrids. These plants are intermediate in vegetative and reproductive characters with medium to tall in height and double orange to red flowers.

Important varieties

African marigold

Giant Double African Orange, Giant Double African Yellow, Cracker Jack, Climax, Dubloon, Golden Age, Chrysanthemum Charm, Crown of Gold, Spun Gold, Guinea Gold, Apricot, Primrose, Sun Giant, Fiesta, Golden Yellow Glitters, Happiness, Hawaii, Honeycomb, Cerpid, Double Eagle, New Alaska, Pusa Narangi Gainda,

Pusa Basanti Gainda, Pusa Bahar, Arka Bangara, Arka Bangara-2, Arka Agni, Arka Alankara, Punjab Gainda No.1 and MDU 1.

French marigold

Red Brocade, Rusty Red, Flame, Naughty, Butter Scotch, Valencia, Marietta, Sussana, Star of India, Harmony, Pusa Arpita, Pusa Deep, Arka Honey, Arka Pari, Hisar Jaffri-2 and Hisar Beauty

Signet marigold

Golden Gem, Lulu, Pumila, Ursula, Paprika and Lemon Gem

Interspecific hybrids

Show Boat, Nugget, Red Seven Star, Red and Gold Hybrid and Show Boat

Characteristics of important varieties developed in India

ICAR- IARI, New Delhi Varieties (Sindhu, 2019)

Pusa Basanti Gainda

It is an African marigold hybrid developed by crossing of Golden Yellow variety with Sun Giant variety. It produces lemon yellow flowers of medium size after 135-145 days of sowing. This variety can be cultivated throughout India for pot cultivation and garden display purpose. It produces 20-25 t/ha fresh flowers and 75-100 kg/ha seeds.

Pusa Narangi Gainda

It is also an African marigold hybrid developed by crossing Cracker Jack with Golden Jubilee. It produces deep orange flowers with ruffled florets and flowering occurs after 125-135 days of sowing. This variety is suitable for cultivation in all over the India; but it is more popular in southern India because of its big flower size. It produces 25-30 t fresh flowers/ha and 100-125 kg seeds. Flowers are rich in carotenoids and widely used in poultry industry, food, pharmaceutical and nutraceutical industries.

Pusa Bahar

This is an open pollinated variety of African marigold with vigorous growth habit having 75-85 cm plant height. It produces compact and flattened yellow colored flowers of large size (8-9 cm). Ideal time for sowing of seeds is mid October to mid November in northern plains of India. Flowering occurs in mid January and continues up to March. Average flower yield of this variety is 50-60 flowers per plant. It is also suitable for bedding and decoration purpose in gardens.

Pusa Arpita

It is a French marigold variety with light orange flowers of medium size. It is developed by selection from heterozygous population of French marigold. Flowering starts from mid December and continues up to mid February. This variety is suitable for cultivation in northern plains of India. Its average yield is 18-20 t fresh flowers/ha.

Pusa Deep

It is an early flowering variety of French marigold with medium height (55-65 cm) and spreading habit (50-55 cm). It produces dark red and compact flowers of medium size during the month of October –November in northern plains of India. It is suitable for loose flower production during festive season with 80-90 flowers per plant and 18-20 t flowers/ha.

ICAR-IIHR, Bangalore varieties

(Source[3]: https://www.iihr.res.in/ varieties and Anonymous, 2016b)

Arka Honey

It is a photo-insensitive; dwarf variety of French marigold with spreading habit and floriferous nature having bi-coloured flowers of medium size (4.8 cm). These bi-coloured flowers have orange border of petals and their centre is dark red in colour. Early flowering initiation is observed; generally within 30-35 days after planting and it continues for 6 weeks. This variety produces flowers throughout the year under Bangalore condition. Its crop duration is of about three months and produces 5.8 t flowers/acre.

Arka Pari

It is a photo insensitive French marigold variety having orange flowers. It is also a dwarf variety with spreading canopy and good blooming habit throughout the year. Plants have medium sized flowers and complete their life cycle in about three months. Flowering starts 30 days after planting and it continues for about 9 weeks. Flower colour changes into different shades of orange depending upon light, temperature and stage of flower. Flower yield of this variety is 4.7t/acre.

Arka Bangara

It is an African marigold variety having petaloid male sterile flowers of yellow gold colour. Flowering starts in 40-45 days after planting and it continues to flower for next 65-70 days. Flowers are medium in size (5-6.5 cm). Yield potential of this variety is 18 tons/acre.

Arka Bangara-2

This variety is early flowering and mainly grown for loose flower production. It has yellow gold flowers, which are compact and large in size. It is propagated by vegetative means.

Arka Agni

It is a petaloid male sterile hybrid developed from hybridization between IIHRMGYP1 and 9-2. It has medium plant height (80-85 cm) with compact, large (7.5-8 cm) and orange flowers. Flowering starts at 40-45 days after planting and continues for 60 days. Average loose flower yield of this variety is 7-7.5 t/acre. It is multiplied by vegetative propagation.

Arka Alankara

It is also a petaloid male sterile marigold hybrid developed from hybridization between IIHRMGYP-1 and 9-2. It has medium height plant (75-80 cm) with compact, large (7- 7.5 cm) and golden yellow flowers. Flowering starts 40-45 days after planting and continues to flower for 60 days. It is a vegetatively propagated plant with high yield potential (6-6.5 t/acre).

PAU Marigold Variety (Muker *et al.*, 2016)

Punjab Gainda No. 1

It is a heat tolerant and open pollinated variety of marigold with medium height (69 cm). It bears fully double flowers of medium size with bright orange colour after 82 days of seed sowing. This variety is suitable for growing in summer season (February-March and June-July in Punjab) and produces 44.2q of flowers per acre. It possesses moderate resistance to Alternaria blight disease.

HAU varieties (Beniwal *et al.*, 2009)

Hisar Beauty: It is released from CCS HAU, Hisar in 2008-09. It is a dwarf and compact French marigold variety with 8-10 branches, its leaves are dark green with serrated margin. It produces 40-50 small to medium sized dark red flowers with yellow margin. This variety is suitable for exhibition, bedding and potting purpose.

Hisar Jaffri-2: It is also released from CCS HAU, Hisar in 2008-09. This French marigold variety is dwarf and compact with 18-22 branches. It bears dark green serrated leaf with medium sized 70-80 orange flowers per plant. This variety is suitable for exhibition and loose flower production.

TNAU variety

MDU 1: This African marigold variety is developed through selection and released in 1986. It has medium tall plants with moderate branching habit. It produces large flowers with light orange petals and 8.39 cm long stalk. An estimated yield of this variety is 41.54 t flowers/ha (Anonymous, 2013).

Climate

Marigold requires mild climate for good vegetative and reproductive growth. Profuse growth of marigold is attained between the temperature range of 18-30°C temperatures. It ceases to grow and produces very less flowers of poor quality at very high temperature. In severe winter, blackening, wilting and senescence of affected plant part occurs due to heavy frost damage. Sprouting in plants occur, if they are allowed to over winter, during spring season and produce very few flowers. Signet marigold has a stronger requirement for short photoperiod than French or African marigolds (Tsukamoto *et al*, 1968). Marigold can be grown round the year in open field except very hot and cold climatic conditions. Generally, seed sowing and transplanting of seedlings is carried out during rainy, winter and summer season.

Soil

Marigold can be successfully grown on a wide range of soils except water logged and saline conditions. Suitable soil for its cultivation is fertile, friable, sandy loam soil; rich in organic matter with good water holding, drainage and aeration capacity. African marigold flourish well in moist and fertile soil with good drainage facility, whereas, light soil is best suited for growing French marigold. Ideal soil pH should be near to neutral in reaction viz. 7.0-7.5.

Propagation

Marigold is generally propagated either by seeds or herbaceous cuttings. Marigold plants raised through seeds will look more robust and have better yield potential as compared to vegetatively propagated plants. So, commercially it is propagated by seeds. Micro propagation is rarely practiced mainly for propagation of white marigold and male sterile lines.

Propagation by Seeds

Seeds selected for nursery raising should be healthy, viable and black in colour. Freshly harvested seeds have generally 90-95 % germination rate and remain viable for about 1-2 seasons. Seeds will germinate after 5-7 days of sowing. Optimum temperature required for better germination of marigold seed is 18-30°C. 1.5 kg of marigold seed is sufficient for raising nursery of about one hectare area (Arora, 2012). Marigold seedlings can be grown in any type of container like seed box or pot on a small scale mainly for gardening purpose, but raised nursery beds should be preferred for commercial cultivation of marigold. Raised beds of 3m × 1m size with 10-15 cm height should be prepared well in advance for marigold seed sowing. Add 8-10 kg of well decomposed FYM per sq. m. and thoroughly mix it with the soil (Arora, 2012). Seeds can be broadcasted uniformly or line sowing at 8-10 cm spacing with depth of 0.5-1.0 cm should be done for better seed germination. Cover the seeds with thin layer of fine FYM or sand or leaf mould manure followed by light irrigation with fine rose can. Uniformly spread a thin layer of straw mulch over the bed after seed sowing to enhance germination process. Small amount of insecticide is applied at the border of nursery bed to avoid ant or termite attack. Nursery beds should be kept moist throughout the growing period by frequent and light watering in the cool hours of the day.

Propagation by herbaceous cuttings

It is mainly practiced to get true to type plants for their further use in various marigold breeding programmes. Some marigold varieties like Gaint African Yellow and Gaint African Orange do not set seed (Anonymous, 2017). In addition to this, several promising germplasm like male sterile lines, white marigold etc. can only be propagated through vegetative means due to lack of sufficient seed setting or seed formation or poor seed germination. Generally, 8-10 cm long herbaceous cuttings should be taken during rainy season from apical part of the plant. Nursery beds are prepared in shady areas. Plant these cuttings in pots or nursery beds filled with sand for their quick growth and easy rooting. Before planting these cuttings in the growing medium, their basal portion is treated with IBA to encourage early rooting. African marigold stem cuttings treated with 200 ppm NAA produced maximum number of roots/cutting (49.13), fresh (0.77 g) and dry weight (0.10g) of roots/cutting (Sharma, 2014). Maximum number of roots (93.44), fresh (404 mg) and dry weight of roots/cutting (34.11 mg) was recorded with 300 ppm IBA treatment and average length of root (12.23 cm) was recorded maximum with 100 ppm IBA treatment (Kaushik and Shukla, 2020), whereas, maximum root length (9.14 cm) and root spread (4.53 cm) was recorded in African marigold cuttings treated with 200 ppm IBA (Sharma, 2014). NAA treated cuttings exhibited best shoot parameters, while, IBA treated cuttings produced best root parameters (Kaushik and Shukla, 2020). Combined application of 200 ppm NAA + 200 ppm IBA on African marigold cuttings resulted in to better rooting percentage as compared to individual application of 200 ppm NAA and 200 ppm IBA (Sharma, 2014). Always keep the beds in moist condition by regular watering. Cutting starts rooting within 5-7 days of sowing and become ready for transplanting after 3-4 weeks.

Micropropagation

It is not commonly used on commercial scale for marigold propagation because of high phenol exudation from explants in initial culture establishment stage and high plant mortality during hardening stage (Kumar, 2002). White marigold possesses low seed viability and poor germination hence it is difficult to propagate through seeds. So tissue culture can be practiced for its large scale commercial propagation. Shoot tip proliferation was found suitable for its micro propagation (Misra and Datta, 1999). Some promising germplasm like male sterile lines can be maintained through vegetative means only. Round the year production and maintenance of true breeding lines can only be done through efficient tissue culture techniques (Kumar *et al.*, 2018).

Planting

Sowing time and season

Marigold can be grown throughout the year except very hot and very cold agro-climatic conditions experienced in May-June and January-February months in northern India. Generally commercial growers in India prefer to take three crops per year based on seasons i.e. rainy, winter and summer. Staggered planting at an interval of 15 days can be done from August to October or January to February for assured supply of flowers to the market during September to February or April to May.

The time of sowing and transplanting of seedlings for obtaining flowers at different seasons of year is as follows:

In plains

Season	Time of seed sowing	Time of transplanting	Peak flowering
Rainy season	May-June	June-July	September-October
Winter season	August-October	September-November	December-February
Summer season	January-February	February-March	April-May

Field preparation

Land should be free from hard and compact soil layer, clods, weeds and any other debris. Before transplanting, it should be ploughed and harrowed 2-3 times in order to make the soil friable with fine tilth.

Transplanting of seedlings

Transplanting of seedlings is done manually in the cool hours of the day (morning or evening) and established in the field easily without much mortality. Fast growing root system of marigold enable the seedlings to establish better. The one month old healthy seedlings with sturdy stalk and 3-4 true leaves are suitable for transplanting in the field. Thin and long seedlings should be avoided. Too much big seedlings are not desirable because they have lost their juvenile phase in the nursery itself. Transplanting should be done in well prepared field in the evening especially during rainy and summer seasons to avoid exposure to harsh weather, transplanting shock and to allow better establishment in cool hours of night. Soil is pressed well around root zone to avoid air pockets. Light irrigation is applied in the field immediately after transplanting for better establishment of seedlings and to reduce chances of seedling mortality. Heavy irrigation if done just after transplanting, seedlings will bend down and leaves stick with the soil which delays proper development of plant. Light irrigation before transplanting is considered beneficial in sandy soil for the better establishment of seedlings. Prepare beds of convenient size for easy handling of various intercultural practices like irrigation, weeding, pinching etc.

Spacing

Growth habit, cultivar and the soil type affect the plant density. Proper spacing between plants is required for better growth, development and flower production. Two marigold species which are commonly grown for commercial production of flowers require different spacings, for example *Tagetes erecta* requires wider spacing than *T. patula*. In general, the spacing should be 30 × 30 cm for dwarf and 40 × 40 cm for tall African marigolds. Optimum spacing for African marigold is 60 × 30 cm or 45 × 30 cm, for French marigold is 20 × 20 cm or 20 × 15 cm as recommended by DFR, ICAR at its Hyderabad centre (Anonymous, 2017). Wider spacing of 40 × 40 cm produced best results with respect to maximum flower diameter, weight of individual flower and number of flowers/plant in African marigold cv. Pusa Narangi Gainda under North Bihar agro-ecological conditions (Singh *et al.*, 2015). African marigold plants spaced at 40 x 40 cm attained maximum growth, flowering and yield in terms of plant spread (64.01 cm), number of flowers/plant (51.43), duration of flowering (53.98 days), flower diameter (6.42 cm), fresh weight of flower (9.03 g) and flower yield/ plant (460.42 g), however closest spacing of 30 x 30 cm resulted in to maximum plant height (71.76 cm) and flower yield/hectare (36.34 t) (Nain *et al.*, 2017).

Manure and fertilizers

Marigold is a hardy and vigorous crop and requires less manures and fertilizers as compared to other flower crops. The rate of nutrient utilization from soil has been reported to be 57.5 % of available N, 20.1 % P, 94.4 % K, 8.6% Ca, 23.0% Mg and 6.4% Na. Incorporate well decomposed 15-20 tonnes FYM/ hectare during the last ploughing. Apply 200 kg nitrogen, 200 kg phosphorus and 100 kg of potassium per hectare. Half dose of nitrogen, entire dose of phosphorus and potash should be applied as basal dose before transplanting. Leftover 100 kg nitrogen should be applied after one month of transplanting. A field experiment was conducted at CCS HAU, Hisar to standardize the optimum dose of nitrogen and phosphorus application in African marigold. All growth and flowering parameters were significantly influenced with every increase in nitrogen dose from 0 to 30 g/m^2 and phosphorus dose from 0 to 25 g/m^2. Combined application of 30 g N/m^2 with 25 g P/m^2 recorded maximum flower yield per plant (558.09 g) and flower yield per hectare (34.88 t) as compared to control (Nain *et al.*, 2016). French marigold cv. Hisar Beauty gave maximum flower yield when fertilized with 750 g of poultry manure per kg of sand or 750 g of FYM per kg of sand and in case of vermicompost maximum yield was found with application of 1000 g vermicompost per kg of sand (Chander *et al.*, 2015).

In marigold, higher flower yield can be obtained with application of 400 kg N, 200 kg P_2O_5 and 200 kg K_2O/ha (Joy *et al.*, 2001). African marigold cv. Pusa Narangi Gainda produced highest flower yield (15.53 t/ha) with 75:150:150 kg NPK/ha + 10t/ha FYM + *Azospirillum* + PSB @ 5 kg/ha (Kumura *et al.*, 2019). African marigold cv. Pusa Basanti Gainda plants supplied with 75% RDF + *Piriformospora indica* inoculated to seeds took minimum duration to full flowering (59.55 days) and attained maximum flower diameter (9.27 cm), fresh weight of total flowers/plant (394.47 g) and highest yield (29.06 t/ha), whereas, plants supplied with 50 % RDF + *Piriformospora indica* inoculated to seeds produced higher number of flowers per plant (53.62) (Noorjahan *et al.*, 2019).

Irrigation

Moisture stress at any stage of growth and development may hamper the productivity of marigold. Irrigation requirement of marigold varies according to variety and agro-climatic conditions. During rainy season; if rainfall is normal and well distributed, frequent irrigation is not required, but if the rain is scanty, irrigation is needed frequently. During the summer months from April to June, frequent irrigation at the interval of 4-5days is required. During winter season irrigation can be done at an interval of 8-10 days. In general, irrigate the marigold crop at weekly interval. In lighter soil more frequent irrigation is required than that in heavy soil. In sandy loam soil, weekly irrigation is necessary between September to March while during summer months between April to June irrigation at 4-5 days intervals is required.

Weeding and hoeing

Weeds are a common problem in marigold cultivation especially in rainy season. After transplanting of marigold seedlings in the field, weeds grow faster than marigold in initial stages and cover a large area in a few days. Weeding is necessary to avoid the competition with crop which causes great loss in terms of growth and productivity during all seasons. In India, weeding is done manually. During entire crop growth period, 3 - 4 manual hoeing and weeding should be done to make the soil loose and weed free.

Pinching

Pinching is done to reduce the effect of apical dominance and stimulate lateral growth of plant, which will ultimately give rise to more number of branches and hence more flower yield. It is manually practiced after 3-4 weeks of transplanting mainly in tall varieties of African marigold to encourage better plant stand with uniform lateral growth. Pinching is also beneficial in several other ways. It promotes vegetative growth and delays flowering, thus helps in growth regulation of the plant for regular supply of flowers to the market. It reduces the plant height and prevents lodging of plants during strong windy conditions in summer season.

Maximum plant spread, number of primary branches per plant, fresh and dry weight of plant was recorded in African marigold plants pinched at 2 weeks after transplanting while plant height was recorded maximum in un-pinched plants (Sheoran *et al.*, 2016). African marigold plants pinched at 4 weeks after transplanting significantly exhibited maximum growth in terms of plant spread (63.59 cm), number of flowers/plant (54.54), flower yield/plant (403.68 g) and flower yield/hectare (33.33 t) as compared to un-pinched plants (Nain *et al.*, 2017). The maximum number of flowers/plant (47.70) and flower yield/hectare (9.05 t) was recorded in African marigold plants pinched at 2 weeks after transplanting, whereas, the maximum fresh weight of flower (3.29 g) was observed in un-pinched plants of summer season (Sheoran *et al.*, 2017a). African marigold plants pinched at 2 weeks after transplanting in summer attained maximum plant spread (65.05 cm) and flower yield/plant (144.80 g), whereas, the higher plant height (84.66 cm) was observed in un-pinched plants (Sheoran *et al.*, 2017b).

Application of growth regulators

All the vegetative characters were highly influenced by higher dose of gibberellic acid in African marigold. A successive increase in GA_3 concentration from 25 ppm to 200 ppm significantly resulted in to early flowering and long duration of flowering. Number of flowers/plant (57.81), duration of flowering (87.18 days) and flower yield/ plant (639.18 g) was recorded maximum with 200 ppm GA_3 as compared to control (43.67, 80.26 days and 406.13 g) respectively (Kumar *et al.*, 2010). Application of 350 ppm dose of GA_3 produced maximum plant spread and number of primary branches/ plant, while plant height, fresh weight of plant and dry weight of plant were recorded maximum with 250 ppm GA_3 treatment as compared to control (Sheoran *et al.*, 2016). Foliar spray of GA_3 250 ppm at 4 weeks after transplanting was found beneficial in increasing number of flowers per plant (48.29) and flower yield per hectare (9.93 t) during summer season (Sheoran *et al.*, 2017a). GA_3 250 ppm dose was found best for obtaining higher plant height (88.7 cm), flower yield/plant (158.93 g), however, plant spread (65.59 cm) was observed maximum in plants sprayed with GA_3 350 ppm in summer season (Sheoran *et al.,* 2017b). Foliar spray of 350 ppm GA_3 on plants pinched at 2 weeks after transplanting resulted in maximum plant spread (68.82 cm) in summer, whereas, maximum number of primary branches/plant (16.50) was observed in winter. African marigold plants pinched at 2 weeks after transplanting attained maximum fresh weight of plant (616.67 g) and produced maximum number of flowers/plant (53.18) with 250 ppm dose of GA_3 in summer. A significant reduction in days to first flower appearance was also observed in un-pinched plants with 350 ppm dose of GA_3 in summer (Sheoran *et al.*, 2019). Ethrel showed dwarfing effect and caused late flowering of African marigold plants. A significant decrease in plant height was recorded with successive increase in ethrel concentration from 100 to 400 ppm. Maximum reduction in plant height (66.12 cm) was recorded with 400 ppm ethrel as compared to control (74.67 cm) (Kumar *et al.*, 2010). Best quality flowers exhibiting maximum shelf life of 3.67 days were produced by marigold plants with pre-harvest spray of 50 ppm BA (Benzyle Adenine) as compared to 10 ppm *Jeevamrut* and distilled water pre-harvest spray (Sharma, 2017).

Harvesting and yield

Marigold is harvested by hand plucking. Hand plucking is done because flower stalk is a hollow structure, which breaks off easily when twisted between thumb and finger. Pluck fresh and fully opened marigold flowers manually during cool hours of the day either in early morning or late evening to avoid moisture loss and maintain turgidity of flowers otherwise wilting of flowers will occur. Light irrigation should be applied before plucking of flowers so that flowers look fresh for longer period after harvest. Regular plucking of fresh flowers should be done to increase crop productivity. The best performance with respect to flower size, weight and number of flowers/plant was recorded with 1st September planting of African marigold cv. Pusa Narangi Gainda, however, the highest number of branches and plant canopy spread were attained with May and July plantings under agro-ecological conditions of North Bihar (Singh *et al.*, 2015). The yield of flower in African and French marigold varies with cultivars, cultural practices like planting, spacing, fertilization etc. Average flower yield of African marigold and French marigold varies from 11 to 18 tonnes/ha and 8 to 12 tonnes/ha, respectively.

Post harvest management

Marigold flowers are generally collected either in moist gunny bags or in bamboo baskets covered with moist cotton cloth or polythene sheet for transportation to nearby local markets during cool hours of the day preferably in early morning or late evening. Plastic crates are considered best packing materials for transport of loose marigold flowers to short distance market as later showed maximum shelf life (4 days) with minimum moisture loss percentage (6.75%) (Source[4]: https://tnau.ac.in/college-hcriperiyakulam/department-of-fp-technologies-developed/). Marigold flowers packed in cardboard boxes of size 37.5 × 16.4 × 12.5 cm and lined with polyethylene attained maximum shelf life (4.13 days) with best freshness index than flowers placed inside the cardboard boxes lined with newspaper and cellophane (Sharma, 2017).

Dry storage of cut marigold stems from harvest to shipping stage extended their vase life as compared to cut stems stored in water. African marigold stems of cv. 'Sovereign' can be stored dry for up to one week (Ahmad and Dole, 2015). Marigold cv. Pusa Narangi Gainda exhibited better shelf life (3.42-3.81 days) under cold storage (4°C) condition as compared to room temperature condition (2.40 -2.56 days) (Sharma, 2017). Fully opened flowers of marigold cv. Calcutta Gainda when packed in black polythene bags and stored in cold storage at 4°C temperature with 93% relative humidity attained greater shelf life (8 days) with better visual appearance than flowers stored at normal room conditions (3.5 days). Highest post-storage life of 2 days was also observed in black polythene packaged flowers of cold storage (Jadhav, 2018).

Cut stems of marigold cv. Double Eagle placed in either homemade or commercial floral preservatives had a longer vase life than stems placed in tap water, irrespective of duration (Ahmad and Dole, 2014). Longest vase life (15.7 days) was recorded when stems were placed continuously in vase solution containing 500 ml/litre soda (lemon or lime flavored 7 Up soda) followed by vase solution having 400 mg citric acid + 20 g sugar + 0.007 ml/litre Kathon CG (15.5 days) or pulsed for 48 hour with vase solution made of 100 mg citric acid + 20 g sugar + 200 mg/litre aluminum sulfate (15.5 days) as compared to tap water (7.9 days) (Ahmad and Dole, 2014). Flowers placed continuously for 44 hours in holding preservative solution only or along with hydrator for 4 hours had 3 days longer vase life than control flowers in deionized water (Ahmad and Dole, 2015). 0.2 mM Silver thiosulfate (STS) was found beneficial in increasing vase life of cut African marigold stems cv. Sovereign by 2.1 days as compared to deionized (DI) water (Ahmad and Dole, 2015).

All parts of marigold plant can be used for extraction of 0.02-0.08% essential oil through commercially used steam distillation method followed by absorption of distillate in benzene or petroleum ether (Joy *et al.*, 2001).

Insect-pests

Red spider mite (*Tetranychus* sp.): Attack of these mites can be easily identified by dusty appearance of plants and chances of its occurrence are more during flowering time. It can be controlled by spraying Metasystox 25E.C. or Rogor E.C. @ 1ml/liter of water (Arora, 2012).

Hairy caterpillar (*Spilosoma oblique/Diacrisia obliqua*): It is polyphagous in nature and causes drastic damage to plants by eating and damaging their foliage and flowers thus results in to significant yield loss. This caterpillar can be controlled by sprays of Nuvan 59 E.C. @ 1ml/liter of water (Arora, 2012).

Bud caterpillars (*Helicoverpa armigera* and *Phycita* sp.): These caterpillars lay their eggs on young buds. *Helicoverpa armigera* larvae feed on developing flowers, whereas larvae of *Phycita* sp. feed on heads of buds and flowers, which ultimately causes great damage to developing buds and flowers. Collect and destroy infested buds along with damaged flowers in order to reduce the spread of this bud caterpillar. (Anonymous, 2017).

Aphids (*Aphis gossypii*): Aphids mainly nymphs and adults attack on lower bases of leaves and flowers. These suck the sap from tender parts of the plant and cause their discoloration and withering. These aphids also act as vectors of some viral diseases in marigold. Spraying of Oxydemeton methyl 0.05% should be done to control these aphids (Anonymous, 2017).

Thrips: Thrips suck the sap from tender parts of plant like young leaves, buds and flowers. Distorted leaves and petals resulted due to thrips attack will ultimately turn brown. In order to control thrips attack; 2-3 sprays of Oxydemeton methyl @ 0.05% should be done at 10 days interval (Anonymous, 2017).

Diseases

Fungal diseases

Important fungal diseases of marigold are damping off (*Pythium* sp.), *Alternaria* leaf spot, *Cercospora* leaf spot (*Cercospora megalopotamica*), flower blight (*Alternaria zinniae*), wilt and stem rot (*Phytophthora cryptogea*), Collar rot (*Phytophthora* sp.; *Pythium* sp.) and *Fusarium* wilt (*Fusarium oxysporium*) (Sohi, 1983; Pawar, 1971). *Alternaria alternata, Aspergillus fumigatus* and *Curvularia lunata* were isolated and found to be pathogenic to *Tagetes erecta* and *T. patula* (Aktar and Shamsi, 2017). Leaf spot and flower blight disease caused by *Alternaria tagetica* poses a great impact on yield potential of marigold by retarding its overall growth and development. This infection leads to premature defoliation, which ultimately results in to plant death (Gurjar *et al.*, 2019).

Alternaria leaf spot and blight (*Alternaria* sp.)

Circular brownish gray spots develop on the lower leaves at the initial stage of infection. Poor vegetative growth, defoliation and reduced flower size followed by death of affected plant occurs at a later stage of infection. Alternaria leaf spot disease caused by *Alternaria tagetica* aggravated with continuous exposure to high moisture conditions for long duration (Hotchkiss and Baxter, 1983). Temperature shows remarkable effect on disease development, while no significant effect of rainfall and humidity was observed on development of this disease. The minimum disease severity was recorded in the month of January and the maximum was recorded in the month of May (Aktar and Shamsi, 2017). To control this disease 0.2% Dithane M-45 spray should be done at fortnightly intervals starting from the 1st onset of symptoms (Arora, 2012) whereas, Blitox was found efficient for controlling *Alternaria zinnia* (Mazumdar, 2000).

Septoria leaf spot (*Septoria* species)

Small, circular water-soaked spots develop on leaves, petioles, stem and calyx of infected plants at initial stage. Later on these spots became grey and their margins turn in to dark brown colour. Affected leaves become yellow and then turn brown, shriveled followed by defoliation (Shukla and Thakur, 2018). To control this disease foliar application of by 0.2% Dithane M-45 fungicide at regular intervals should be done (Aktar and Shamsi, 2015 and Bharnwal *et al.*, 2002).

Cercospora leaf spot (*Cercospora megalopotamica*)

Circular spots with ash gray centres and reddish purple to dark brown borders occur on leaves. Excessive use of fungicides lead to development of resistance of disease causing pathogens against fungicides as reported by Gangavane, (1981) and Arora *et al.* (1992). Bavistin is found to be the best fungicide for control of the Cercospora leaf spot. A bio-formulation consisting of garlic extract + cow urine + soap nut was also found effective in controlling this disease (Chandel and Kumar, 2017).

Inflorescence blight (*Alternaria zinnae*)

Large irregular blotches of light tan to dark brown colour with zonation occur on the foliage, whereas elongated lesions are formed on inflorescence (Gurjar *et al.*, 2019). Foliar spray of Dithane M-45 (0.2%) spray at regular interval can be done to control this blight disease (Aktar and Shamsi, 2015, Singh *et al.*, 2012, Bharnwal *et al.* 2002).

Wilt and stem rot (*Phytophthora cryptogea*)

It infects mainly collar portion of the plant and develops rapidly with availability of soil moisture in nursery or field conditions. This causes damping off of nursery seedlings and wilting of plants in field. French marigold and dwarf varieties of marigold are found less susceptible to the disease in comparison to African marigold (Gurjar, 2019). The disease can be controlled by treating the soil with Metalaxyl (Anonymous, 2017).

Collar and root rot (*Pellicularia filamentosa, P.rolfsii, Pythium ultimum* and *Scelrotinia slerotiarum*)

Black lesions appear on the main stem followed by rotting of root and collar region, which eventually results in to wilting of plant. In severe cases, plant death can also occur. Selection of healthy planting material and proper soil fumigation should be practiced in order to control this disease (Anonymous, 2017c). Manzate at 0.3% concentration can completely inhibit the growth of pathogen, whereas, Copper oxychloride at 0.3-0.5% concentration can control the pathogen. Use of biological antagonist like *Streptoverticillium* sp., *Trichoderma* sp. and *Gliocladium virens and* plant extracts of *Rauwalfia serpentine* can also be used to significantly control this disease (Rout and Mishra, 2008).

Damping off

It is caused by *Rhizoctoina solani* (Singh *et al.*, 2012). *Ceratobasidium* sp. as causal agent of damping off was first reported in India by Saroj *et al.* (2013). This disease more frequently occurs at the seedling stage than later stages of plant growth. Pre-emergence symptoms appear as brown necrotic spots and rings, which further extends to plumule and eventually cause pre-emergence mortality in seedlings, whereas, water soaked, brown necrotic ring appear on lower part of hypocotyls as post-emergence symptoms leading to collapse of seedlings (Singh *et al.*, 2012). Application of 2% formalin for soil sterilization prior to sowing and foliar spray of 0.2% Dithane M-45 are some effective measures to control this disease (Gurjar *et al.*, 2019). Proper drainage should be provided in the nursery bed and 3-4 years crop rotation should be followed to control the spread of this disease (Singh *et al.*, 2012).

Flower Bud Rot (*Alternaria dianthi*)

Young flower buds tend to shrivel, turn deep brown and dry up followed by rotting, while mature buds fail to open as a result of flower bud rot disease (Mondal and Chaudhari, 1976 and Arora, 2012). Blight and brown necrotic spots occur on margins and tips of older leaves, whereas, brown, scorched, necrotic discoloration of sepals, stalk and florets occurs due to incidence of this disease. Foliar spray of 0.2% Dithane M-45 effectively controls the infection of leaf and flower bud (Singh *et al.*, 2012 and Arora, 2012).

Powdery mildew (*Oidium* sp. and *Leveillula taurica*)

It is caused by *Oidium* sp. and *Leveillula taurica*, which produces tiny, whitish and superficial spots on leaves and later on whole aerial parts of plant, is covered with whitish powder having conidia and conidiophores of the pathogens (Grover, 1952 and Sreeramula, 1953). The disease can be controlled by spraying Sulfex (3g/litre of water) or 0.5% Karathane (40 E.C.) at regular intervals (Arora, 2012).

Botrytis blight or gray mold

It is caused by *Botrytis cinerea.* Gray coloured downy growth of fungus along with spores appears on infected plant parts and thus this disease is also known as gray mould. Irregularly shaped spots occur on leaves, flowers and stem. Infested flower buds fail to open and decaying of affected flowers followed by their premature fruit drop occurs. Later on, collapse of plants occurs due to rotting of their diseased stem. Remove plant debris as soon as initial disease symptoms appear and provide proper air circulation, avoid overhead watering and maintain dry foliage to control this disease. Apart from these cultural practices, use disease-resistant cultivars and apply fungicides to control the pathogen (Dhiman and Arora, 1990 and Anonymous, 2018a).

Bacterial diseases

Bacterial leaf spot of marigold (*Pseudomonas syringae* pv. *tagetis*)

This disease is seed borne and adversely affects growth as well as development of immature marigold plants. Tiny, watery and dark-green spots appear on lower leaf surface at initial stage of infection, while, diffused brownish discolouration occur on upper leaf surface after one or two days of infection (Hellmers, 1955). Occurrences of black spots on cotyledon of seedlings, necrotic leaves with chlorotic and distorted apical growth are some important symptoms of this disease. Mortality of severely infected plants occurs in some cases. To control the pathogen, healthy seeds should be used for raising marigold nursery (Gurjar, *et al.*, 2019).

Viral diseases

Cucumber mosaic virus (CMV) (Hanson *et al.*, 1951), Marigold Mosaic Virus (MMV) and Marigold Mottle Virus (Naqvi *et al.*, 1981) are important viral diseases of marigold. CMV is transmitted by seeds mainly and can be transmitted through insect vectors like *Aphis gossypii, Aphis craccivora* and *Myzus persicae.* Plants infected with viral diseases exhibit stunted growth with deformed flowers and very less yield potential. Affected leaves become yellow and show mosaic with vein chlorosis and mild curling. In severe cases, growth ceases and top necrosis occurs in plants. Another important disease of marigold is Aster yellows, which is caused by Phytoplasma. Stunted growth, week and clumpy stems, discolouration of infected leaves, deformed flowers with improper coloration are prominent symptoms of this disease (Gurjar *et al*., 2019). To control its further spread, infected plants should be completely removed and destroyed (Anonymous 2018a). Antiviral Glycoprotein isolated from *Boerhaavia diffusa* roots acts directly on viruses and effectively manages the virus (Awasthi *et al.*, 2016).

References

Adekunle, O. K., Acharya, R., Singh, B., 2007. Toxicity of pure compounds isolated from Tagetes minuta oil to Meloidogyne incognita. Australasian Plant Disease Notes, (AUS) 2:101–104.

Ahmad, I. and Dole, J.M., 2015. Optimal postharvest handling protocols for cut stems of 'Sovereign' African marigold (*Tagetes erecta*). Acta Horticulture, 1104, 303-310 https://doi.org/10.17660/ActaHortic.2015.1104.46.

Ahmad, I. and Dole, M. J., 2014. Postharvest performance of cut marigold, rose and sunflower stems as influenced by homemade and commercial floral preservatives. Turkish Journal of Agriculture and Forestry, 38: 916-925. doi:10.3906/tar-1402-31.

Aktar, M. and Shamsi, S., 2015. Blight of two species of marigold (*Tagetes*) caused by *Aspergillus fumigates* Fresenius. Bangladesh Journal of Plant Pathology, 31(1&2):1-6.

Aktar, M. and Shamsi, S., 2017. Incidence and Severity of Blight Disease of *Tagetes erecta* and *T. patula*. Bioresearch Communications, 4(1), 464-469.

Anonymous, 2013. Particulars on improved Varieties of Horticultural Crops, In: Crop Production Techniques of Horticultural Crops 2013. Horticultural College and Research Institute, TNAU, Coimbatore-641003, p.276.http://agritech.tnau.ac.in/pdf/2013/cpg_horti_2013.pdf

Anonymous, 2016a. Selected State-wise Area and Production of Marigold in India (2015-16). Ministry of Agriculture and Farmers Welfare, Govt. of India. (ON1769); websource: http://www.indiastat.com

Anonymous, 2016b. Crop Improvement. DARE/ICAR Annual Report 2015–16, p. 40.

Anonymous, 2017. Package of Practices of flower crops 2017. Floricultural Research Station, Sri Konda LaxmanTelanagna State Horticultural University, Rajendranagar, Hyderabad-30 https://dfr.icar.gov.in/Content/Pdf/aicrp%20centre%20anual%20report/HyderabadCentre.pdf

Anonymous, 2018. Managing diseases of Herbaceous ornamentals. Plant Pathology Extension Publications--Herbaceous Ornamentals http://plantpathology. ca.uky.edu/ extension/ publications#HERBACEOUSORNAMENTALS

Arora, J. S., 2012. Marigold. In: Introductory Ornamental Horticulture, 7th revised edition, 2012, Department of Floriculture and Landscaping, Punjab Agricultural University, Ludhiana (Punjab). Pub. Kalyani Publishers pp.83-90.

Arora, R. K., Kamble, S. S. and Gangawane, L. V., 1992. Resistance to metalaxyl in Phytophthora infestans in Nilgiri Hills of South India. Int. Phytophthora Newslett., U.K., 18: 8.

Awasthi, L. P., Verma, H. N. and Kluge, S., 2016. A possible mechanism of action for the inhibition of plant viruses by an antiviral glycoprotein isolated from Boerhaavia diffusa roots. Journal of Virology Antiviral Research, 5:3

Beniwal, B. S., Dahiya, B. S., Deswal, D. P. and Joon, M. S., 2009. New marigold varieties Hisar Beauty and Hisar Jaffri-2. Abst.: National Conference on floriculture for livelihood and Profitability, IARI, New Delhi. p. 38.

Beniwal, B. S., Sheoran, S., Dalal, R. P. S. and Gulia, R., 2017a. Assessment of French marigold genotypes for different biochemical constituents. Crop Research, 52(1, 2 & 3): 55-60.

Beniwal, B. S., Sheoran, S., Gulia, R. and Dalal, R. P. S., 2017b. Biochemical constituents studies in African marigold germplasm. Chemical Science Review and Letters, 6(21): 618-623.

Bharnwal, M. K., Jha, D. K. and Dubey. S. C., 2002. Evaluation of fungicides against Alternaria blight of marigold (*Tagetes* Sp.). J. Res. BirsaAgril. University, 14(1): 99-100.

Bhattacharyya, S., Datta, S., Mallick, B., Dhar, P. and Ghosh, S., 2010. Lutein content and *in vitro* antioxidant activity of different cultivars of Indian marigold flower (*Tagetes patula* L.) extracts. Journal of Agriculture and Food Chemistry, 58 : 8259-64.

CABI, 2020. Tagetes minuta In: Invasive Species Compendium. Wallingford, UK: CAB International. www.cabi.org/iscwebsource: https://www.cabi.org/isc/datasheet/52642.

Chandel, S. and Kumar, V. 2017. Evaluating fungicides and biofungicides for controlling *Cercospora* leaf spot on Marigold. International Journal of Current Microbiology and Applied Sciences. 6(5): 2072-2077.

Chander, S., Beniwal, B. S., Dalal, R. P. S. and Sheoran, S., 2015. Effect of organic manures on growth, floral characters and yield attributes of French marigold (*Tagetes patula* L.). Annals of Biology, **31**(2): 264-269.

Chen, J. F. and Lin, Y. J. 1982. Chromosome pairing in interspecific hybrids of *Tagetespatula* and *T. erecta*. Cytologia, 47: 737–742.

Chen, J. F. and Lin, Y. J., 1981. Meiotic multivalents in *Tagetes erecta* and *T. patula*. *Caryologia*, 34: 107–116.

Conboy, N.J.A., McDaniel, T., Ormerod, A., George, D., Gatehouse, A.M.R., Wharton, E., Donohoe, P., Curtis, R., Tosh, C. R., 2019. Companion planting with French marigolds protects tomato plants from glasshouse whiteflies through the emission of airborne limonene. Plos One, 14(3): e0213071. https://doi.org/10.1371/journal.pone.0213071

De, L. C., 2017. Improvement of ornamental plants -a review, *International Journal of Horticulture*, **7**(22): 180-204. doi: 10.5376/ijh.2017.07.0022.

Deineka, V.I., Sorokopudov, V.N., Deineka, L.A. and Tretyakov, M.Yu., 2007. Flowers of marigold (*Tagetes* spp) as a source of xanthophylls. Pharmaceutical Chemistry Journal, 41: 540-542.

Dhiman , J. S. and Arora J. S., 1990. Occurrence of leaf spot and flower blight of marigold *Tagetes erecta* L. in Punjab, India. Journal of Res. Punjab. Agric. Univ., 27: 231-236.

Ferraro, M., 1955. Las species Argentinas del generoTagetes. Boletín de la Sociedad Argentina de Botánica 6 (1): 30–39.

Gangawane, L.V. 1981. Fungicides resistance in crop protection. Pesticides, 15(11): 12-16.

Goldsmith, A. G., 1968. Current development in the breeding of F_1 annuals. *Hort Science*, **3**:269-271.

Gommers, F. J. and Bakker, J., 1988. Physiological diseases induced by plant responses or products. In: Diseases of nematodes. Poinar, G. O. Jr. and Jansson, H. -B. eds., Vol. I. CRC Press, Inc., Boca Raton, Florida Pp. 3-22.

Greuter, W. (2006+): Compositae (pro parte majore). – In: Greuter, W. &Raab-Straube, E. von (ed.): Compositae. Euro+MedPlantbase - the information resource for Euro-Mediterranean plant diversity. Tagetes tenuifolia Cav.. Accessed through: Euro+MedPlantBase at http://ww2.bgbm.org/euroPlusMed/PTaxonDetail.asp?UUID=8C153242-6AB6-4CC7-A9C3-796F804058A2

Grover, R. K., 1952. Folicolous fungi of Sagar. Bulletin of the Botanical Society, University of Saugar. 4: 9-13.

Gulia, R., Beniwal, B. S., Sheoran, S. and Sandooja, J. K., 2017a. Genotypic variation for important biochemical constituents in marigold. Chemical Science Review and Letters, 6(21): 544-554.

Gulia, R., Beniwal, B. S., Sheoran, S. and Sandooja, J. K., 2017b. Evaluation of marigold genotypes for growth, flowering, yield and essential oil content. Research on Crops, 18(2): 299-304. doi: 10.5958/2348-7542.2017.00051.1

Gupta, P. and Vasudeva, N., 2012. Marigold: A potential ornamental drug. HamdardMedicus, 55: 45-59.

Gupta, P., 2014. Carotenoids of therapeutic significance from marigold. Natural Products Chemistry and Research, 2: 110.

Gurjar, P., Meena, L. and Verma, A.K., 2019. Diseases of marigold (*Tagetes erecta*) and their management: A Review. International Journal of Advanced Scientific Research and Management, 4(4): 137-144.

Hanson, H. R., Weber, H. R. and Troelsen Johansan, G., 1951. Plant disease in Denmark in 1949. Annual survey of data collected by the state phytopathological service, Lyngby. T. Planteavl. 55: 1-81.

Hellmers, E., 1955. Bacterial Leaf Spot of African Marigold (*Tagetes erecta*) Caused by Pseudomonas tagetis sp. Acta Agriculturae Scandinavica. 5(1): 185–200.

Hethelyi, E., Danos, B. and Tetenyi, P., 1986. GC-MS analysis of the essential oils of four Tagetes species and the anti-microbial activity of Tagetes minuta. Flavour and Fragrance Journal, 1: 169-173.

Holm, L. G., Doll, J., Holm, E., Pancho, J.V., Herberger, J.P., 1997. World Weeds: Natural Histories and Distribution. Pub. John Wiley & Sons Inc, New York, USA.

Hotchkiss, E. S. and Baxter, L. W., 1983. Pathogenicity of Alternaria tagetica on Tagetes. Plant Disease, 67: 1288-1290.

Jadhav, P. B., 2018. Extension of the Storage-life of Marigold cv. Calcutta Gainda using Cold Room. International Journal of Current Microbiology and Applied Sciences, 7(12): 832-843. doi: https://doi.org/10.20546/ijcmas.2018.712.104

Janakiram, T., 2004. Heterosis and hybrid seed production in ornamental crops. In: Souvenir of National Symposium on Harnessing Heterosis in Crop Plants, Varanasi, pp. 143-153.

Joy, P.P., Thomas, J., Mathew S., Jose, G. and Joseph, J., 2001. Aromatic plants. In: Bose, T. K., Kabir, J., Das, P., Joy, P.P. (eds), Tropical Horticulture, vol 2. NayaProkash, Calcutta, pp 633–733.

Kaneko, J., Sugawara, N. and Kitamura, C., 1990. Relation between the number of flower visiting individuals and the number of male moths captured by sex pheromone traps in the silver Y moth (Autographa gamma L.). Ann. Report Soc. Pl. Protection North Japan, 41: 160-163.

Kaplan, L. 1960. Historical and ethnobotanical aspects of domestication in Tagetes. Economic Botany, 14 : 200-202.

Kaushik, S. and Shukla, N., 2020. Effect of IBA and NAA and their combination on the rooting of stem cuttings of African marigold (*Tagetes erecta* L.) cv. Pusa Narangi Gainda. Journal of Pharmacognosy and Phytochemistry, 9(3): 1460-1461. doi: 10.22271/phyto.2020.v9.i3x.11517.

Khoshoo, T. N., 1979. Cytogenetics in relation to plant evolution and improvement. In: Khoshoo, T. N. and Nair, P. K. (eds.). Progress in Plant Research. New Delhi, Today and Tomarrow Printers and Publishers, 2:1-74.

Kumar, A., 2002. In vitro production of male sterile lines of African marigold (*Tagetes erecta* L.) for F_1 hybrid seed production. Ph.D. thesis, P.G.School, Indian Agricultural Research Institute, New Delhi.

Kumar, K. R., Singh, K. P., Raju, D. V. S., Kumar, P., Panwar, S. and Bhatia, R., 2018. Circumventing phenolic exudation and poor survival in micropropagation of marigold. Indian Journal of Horticulture, 75(2): 273-278 doi : 10.5958/0974-0112.2018.00046.4

Kumar, R., Ram, M. and Gaur, G. S., 2010. Effect of GA_3 and ethrel on growth and flowering of African marigold cv. PusaNarangiGainda. Indian Journal of Horticulture, 67 (Special Issue): 362-366.

Kumar, S., Bansal, R. P., Bahl, J.R., Muni Ram/ Ram, M., Khanuja, S.P.S., Shasany, A.K., Darokar, M. P., Garg, S. N., Naqvi, A. A., Sharma, S., 1999. Registration of Tagete sminuta variety Vanphool for north Indian plains. Journal of Medicinal and Aromatic Plant Sciences, 21(1): 52-53 https://www.cabdirect.org/cabdirect/abstract/19991610206

Kumura, A., Pandey, G., Mishra, P. P. and Kumar, R. 2019. Effect of Integrated Nutrient Management on Growth and Flowering of African Marigold (*Tagetes erecta*) cv. Pusa Narangi Gainda. International Journal of Current Microbiology and Applied Sciences, 8(11): 1271-1278. doi: https://doi.org/10.20546/ijcmas.2019.811.150

Mali, R.S., Lavhe, N.V., Deotale, R.O. and Panchbhai, P.R., 2019. Evaluation of different marigold cultivars against root knot nematode (Meloidogyne incognita) under pot culture condition. International Journal of Current Microbiology and Applied Sciences, 8(5): 2130-2135. doi: https://doi.org/10.20546/ijcmas.2019.805.249

Mazumdar N., 2000. Epidemiological factors in relation to development of Alternaria leaf blight of marigold and fungicidal control. Plant Disease Research, 15:28-33.

Misra, P. and Datta, S. K., 1999. In vitro propagation of white marigold (*Tagetes erecta* L.) through shoot tip proliferation. Current Science,77:101-103.

Mohamed, M. A. H., Harris, P. J. C. and Handerson, J., 2000. In vitro selection and characterization of a drought tolerant clone of Tagetes minuta. Plant Science, 159, 213–222. doi: 10.1016/S0168-9452(00)00339-3.

Mondal, N. and Chaudhuri, S. 1976. Flower bud rot of marigold (Tagetes erectaL.) caused by Alternariadianthi Stevens and Hall in West Bengal. Current Science, 45(2): 75.

Montellano, B.O.de, 1986. "Aztec medicinal herbs: evaluation of therapeutic effectiveness," In Plants in Indigenous Medicine and Diet-Biobehavioral Approaches, pp. 113–127, Redgrave Publishing Company, Bedford, NY, USA.

Muker, H. S., Singh, P. and Singh, P., 2016. Proceedings of REC at SNo . 22 vide No PBG/2015/29515-615 Dt 20.12.2016, Department of Floriculture and Landscaping, Punjab Agricultural University, Ludhiana.

Nahak, G. and Sahu, R. K., 2017. Bio-controlling Effect of Leaf Extract of Tagetes patula L. (Marigold) on Growth Parameters and Diseases of Tomato. Pakistan Journal of Biological Sciences, 20: 12-19.

Nain, S., Beniwal, B. S., Dalal, R. P. S. and Sheoran, S., 2016. Effect of nitrogen and phosphorus application on growth, flowering and yield of African marigold (*Tagetes erecta* L.) under Semi-arid conditions of Haryana. Indian Journal of Ecology, 43 (Special Issue-2): 645-649.

Nain, S., Beniwal, B. S., Dalal, R. P. S. and Sheoran, S., 2017. Effect of pinching and spacing on growth, flowering and yield of African marigold (*Tagetes erecta* L.) under semi-arid conditions of Haryana. Journal of Applied and Natural Science, 9(4): 2073 -2078.

Naqvi, Q. A., Hadi, S. and Mahmood, K., 1981. Marigold mottle virus in Aligarh, India. Plant Disease, 65:271-275.

Neher, R.T., 1966. Monograph of the genus Tagetes. Ph.D. thesis, Indiana University, Bloomington, Indiana, pp 306.

Neher, R.T., 1968. The ethnobotany of Tagetes. Economic Botany, 22(4): 317–325. https://doi.org/10.1007/BF02908126

Noorjahan, K., Raja Naik, M, and Raimaiah, M., 2019. Flowering and root growth response in African marigold (Tageteserecta L.) cv. PusaBasantiGainda as influenced by Nutrients and Piriformospora indica (PGPRE). International Journal of Recent Scientific Research, 10(04B): pp.31767-31771. doi: http://dx.doi.org/10.24327/ijrsr.2019.1004.3332.

Panero, J. L., 2007. Tribe Tageteae. In: Kadereit, J.W. & Jeffrey, C. (eds.) The Families and Genera of Vascular Plants, vol. 8. Springer, Berlin, pp. 420–431.

Pawar, I. S., 1971. Addition to the deuteromycetes fungi of Kolhapur district.Journal of Shivaji University,4: 9.

Ploeg, A. T. and Maris, P. C., 1999. Effect of temperature on suppression of Meloidogyne incognita by Tagetes cultivars. Journal of Nematology, 31(4S): 709-714.

Prakash, T., Sane, A., Gadre, A. and Ghatke, M., 2016. Characterization and utilization of three distinct male sterile systems in marigold (*Tageteserecta*). Indian Journal of Agricultural Sciences 86(10): 1271-1275.

Pratibha, M., Datta, S. K. and Mishra, P., 2000. In vitro maintenance of F_1 (Tagetes) hybrid. Current Science, 78: 383-384.

Richard, L. R., 2013. Lutein, zeaxanthin and skin health. American Journal of Lifestyle Medicine, 7: 182-85.

Rout, M. K. and Mishra, B., 2008. Studies on collar rot disease of marigold and its management. Journal of Mycopathological Research, 46(2): 283-284.

Rydberg, P.A., 1915. Tagetes, North Am. Flora/Flora of North America, 34, pp. 148159.

Saroj, A., Kumar, A., Saeed, S. T., Samad, A. and Alam, M. 2013. First Report of Tagetes erecta Damping Off Caused by Ceratobasidium sp. from India. Plant Disease. 97(9):1251-1251.

Serrato-Cruz, M. A., Hernández-Rodríguez, M., Savidan, Y. and Bárcenas-Ortega, N. M., 2000. DNA content and ploidy level in Tagetes spp. using flow cytometry. Agrociencia, 34: 729-734.

Setshogo, M. P., 2005. Tagetes erecta L. In: PROTA Plant Resources of Tropical Africa / Resourcesvegetables de l'Afriquetropicale, Jansen, P. C. M. &Cardon, D. (Editors). Wageningen, Netherlands. Accessed on 10 September 2020.

Sharma, P., 2017. Studies on enhancing the shelf life of marigold cv. PusaNarangiGainda. MSc. Thesis. Department of Floriculture and Landscape Architecture, Dr. Y.S. Parmar University of Horticulture and Forestry, Nauni, Solan (H.P.) 173230.

Sharma, R., 2014 Study on the effect of auxins on rooting, growth and flowering of African marigold (*Tagetes erecta*L.) propagated through stem cuttings. M. Sc. (Hort.) thesis. I.G.K.V., Raipur.

Sheoran, S. AndBeniwal, B. S., 2017b. Influence of pinching time and gibberellic acid on growth and flower yield of African marigold (*Tagetes erecta* L.) in summer season. In: Souvenir, 7th International Science Congress (ISC-2017), Focal Theme- Widespread Research: Strengthening Nations and Spreading Happiness (8-9 December, 2017) at College of Science and Technology Rinchending, Phuentsholing, Chukkha, Bhutan. p.30.

Sheoran, S., Beniwal, B. S. and Dalal, R. P. S., 2016. Regulation of plant growth by pinching and GA_3 application in African marigold (*Tagetes erecta* L.) under different seasons. In: Abstracts (Poster Papers) of Platinum Jubilee of HSI: An International Meet 7th Horticulture Congress 2016. Doubling Farmers income through Horticulture on 15th to 18th November at New Delhi pp. 284-285.

Sheoran, S., Beniwal, B. S. and Dalal, R. P. S., 2017a. Effect of pinching and gibberellic acid on floral and yield attributes of African marigold under summer conditions. In: Abstracts of International Symposium on Horticulture: Priorties& Emerging Trends (5-8 September, 2017) at ICAR-IIHR, Banglore.p.77.

Sheoran, S.,Beniwal B.S. and Dalal, R.P. S., 2019. Seasonal regulation of vegetative and reproductive growth of African marigold by pinching and gibberellic acid. Presented and published in International Conference entitled "XIV Agricultural Science Congress: Innovations for Agricultural Transformation" from 20-23 February, 2019 at NASC Complex, Pusa, New Delhi. p.135.

Shukla, A. and Thakur, R., 2018. First report of Septoria Leaf Spot on Marigold (*Tagetes erecta* L.) from Himachal Pradesh, India. International Journal of Current Microbiology and Applied Sciences. 7(1): 1744-1748.

Sindhu, S.S., 2019. "Floricultural Research & Development of New Varieties of Flowers, ICAR-IARI, New Delhi". An International Journal of Floriculture Science & Landscaping: The Journal of the Greens and Gardens, 01(02) Pp. 29-32.

Singh, A. K., 2006a. Flower crops: Cultivation and Management. Pub. New India Publishing Agency, New Delhi, pp. 239-250.

Singh, A. K., Kumar, U. and Kumar, A., 2015. Effect of planting date and spacing on performance of marigold (*Tageteser ecta* Linn.) cv. PusaNarangi under North Bihar agro-ecological conditions. International Journal of Forestry and Crop Improvement, 6(1): 16-20. doi: 10.15740/HAS/IJFCI/6.1/16-20.

Singh, K. and Singh, K., 2001. Economics of wild marigold production and distillation in Himachal Pardesh. Indian Journal of Agricultural Economics. 56:116-128.

Singh, V., Singh, Y. and Kumar, P. 2012. Diseases of Ornamental Plants and their Management. 10.13140/RG.2.1.2250.1520. In book: Eco-friendly Innovative Approaches in Plant Disease Management. V.K. Singh, Y. Singh, A. Singh (eds.). Pub. International Book Distributors and Publisher, New Delhi pp.543-572 doi: 10.13140/RG.2.1.2250.1520.

Sivel, M., Kracmar, S., Fisera, M., Klejdus, B. and Kuban, V., 2014. Lutein content in marigold flower (*Tagetes erecta* L.) concentrates used for production of food supplements. Czech Journal of Food Science, 2014, 32(6): 521-525.

Sohi, H. S., 1983. Personal Communication on disease of marigold. I.I.H.R. Banglore.

Soule, J. A., 1993a. Systematics of Tagetes (*Asteraceae-Tageteae*). PhD. Diss. Univ. Texas at Austin.

Soule, J. A., 1996. Novel annual and perennial Tagetes. (In:) Janick J (ed). Progress in New Crops, ASHS Press, Arlington, VA. pp 546–551.

Soule, J.A., 1993b. Tagetes minuta: A potential new herb from South America. In: Janick, J. and Simon, J. E. (eds.), New Crops, Wiley, New York. pp. 649-654. https://hort.purdue.edu/newcrop/proceedings1993/V2-649.html.

Sreeramula, T., 1953. A new host Tagetes patulafor Leveillulat aurica (Lev.) Am. (Oidiopsis taurica (Lev.) Salm.). Science and culture, 18: 540-541.

Suatmadji, R. W., 1969. Studies on the effect of Tagetes species on plant parasitic nematodes. Stichting Frond Landbouw Export Bureau publicatie 47. H. Veenman Und Zonen N. V., Wageningen, Netherlands. 132p.9.

Subrahmanyam, N. S., 1996. Laboratory Manual of Plant Taxonomy, Vikas Publishing House, New Delhi p.669.

Towner, J. W., 1961. Cytogenetic studies on the origin of Tagetes patula. I. Meiosis and morphology of diploid and allotetraploidT. erecta ×T. tenuifolia. American Journal of Botany, 48: 743–751.

Tsukamoto, Y., Imanishi, H. and Yahara, H., 1968. Studies on the flowering of marigold. I. Photoperiodic response and its differences among strains. Journal of the Japanese Society for Horticultural Science, 37(3):231-239.

Turner, B. L. and Nesom, G. L., 1993. Biogeography, diversity and endangered or threatened status of Mexican Asteraceae. In: Ramamoorty, T. P., Bye, R., Lot, A., and Fa, J. (eds.). Biological diversity of Mexico: Origins and Distribution. New York, Oxford University Press, pp. 290-299.

Watt, G., 1893. A directoty of the Economic Products of India. London, 4(3): 403.

Wang, Koon-Hui, Hooks, C. R. and Ploeg, A., 2007. Protecting Crops from Nematode Pests: Using Marigold as an Alternative to Chemical Nematicides, Plant Disease, July 2007 PD-35.

Yonzone, G. S. and Yonzone, D. K. N., 1999. “Ethnobotany of Darjeeling Himalaya, India,” in ISHS Acta Horticulturae 500: II WOCMAP Congress Medicinal and Aromatic Plants, Part 1: Biological Resources, Sustainable Use, Conservation and Ethno botany, 500: 209–213.

Zhang, P., Zeng, L., Su, Y. X., Gong, X. W. and Wang, X. S., 2011. Karyotype studies on Tagetes erecta L. and Tagetes patula L. African Journal of Biotechnology,10: 16138–16144.

Web References

https://plants.usda.gov/java/Classification Servlet? source = profile & symbol = TAGET &display=31

http://www.theplantlist.org/tpl/record/gcc-82417

https://www.iihr.res.in/varieties

https://tnau.ac.in/college-hcriperiyakulam/department-of-fp-technologies-developed/

27

Lotus

P. K. Yadav and Kumari Lata

Lotus being a national flower of India, is closely associated with religion and culture. It is a perennial, aquatic crop that is grown and consumed throughout Asia. The edible parts of lotus are Leaves, flowers, seeds and rhizomes, and the flowers are also used in religious ceremonies. Lotus flowers are also, used for extraction of perfumes. It is one of the oldest plants in the world Archaeologists in China found Lotus seeds with estimated ages of 5,000 years in 1972 in Yunnan Provincelagoons.

During the pre-historic period, the South India Dravidians used lotus flowers (pundarika) as their food, much earlier than the people of the Indus Valley civilization (3000-2000 BC). However, there is no evidence of lotus being present during the Indus Valley culture of the Mohenjo Daro and Harappa period in northern India. The Rigveda reference that lotus was made by the Aryans (2000-1500 BC) wherein it is also mention that lotus can be used as food. Subsequently, lotus was mentioned in the Vedas, Upanishads, Puranas, Slokas, plays and other ancient Sanskrit literature. The lotus symbol is not only mentioned in the scriptures of Hinduism but it also occurs in the Buddhist and Jainism literature.

In the Arthasastra (350-300 BC) by Kautilya, the great diplomat during the golden age of Chandra Gupta, it is recorded that one of the varieties of sandal wood possesses the fragrance of lotus. The Mughal kings were also attracted towards the beauty of lotus. While describing Agra, the Mughal king Jehangir (1605-1627 AD) in his memories Tuzuk-i-Jehangiri, mentioned a memorable scene of a tank full of exquisite red lotus flowers, the like of which he had not seen before.

The lotus was inspired the poets, the sculptors and the artists of both ancient and contemporary periods. The lotus shapes have also been used in architecture, particularly in the decoration of ceilings of the temples of Vellore Fort and Dilwara (Mount Abu). Even during Mughal period lotus shape was represented in architecture. In Shahjahanabad city presently known as the Red Fort established by the king Shahjahan (AD 1627-58), the exclusive female apartments, the Rang Mahal, is designed as a large lotus with deftly patterned petals disburse within a square bordered frame. In the basin centre's there is a willowy stem with a silver or Indian lotus at the top from which water rushes out. The lotus was also used as a symbol of self-regeneration and rebirth.

Lotus are offered to the Gods and Goddesses in temples all over India. It is customary to offer 108 lotus blossoms for the worship of Durga on the occasion of the Durga Puja festival. Lotus stalks are used to make ornaments, like necklaces and bracelets in West Bengal.

Lotus flowers are symbolized for divineness, purity, kindness, fragrance, coolness, fertility and prosperity.

Taxonomy

The family *Nymphacaceae* has two important genera which have aquatic species with attractive flowers, e.g. water lily and lotus belonging to the genera: *Nymphaca* and *Nelumbo*, respectively. The genus *Nelumbo* has only two species, *N. nucifera* and *N. lutea*. *Nelumbo nucifera* GAERTN.; Syn. *Nelumbium speciosum* Willd., also called Sacred Lotus or Indian Lotus, originated in Asia, Persia, India (Makino, 1979).It is a potential symbol of fertility and purity and even immortality in different civilizations over the centuries. *N. lutea* is known as American lotus having yellow flowers. It is a native of America where it grows wild in the Central states of the USA, near the Great Lakes. Bythe Red Indians it was primarily being cultivated for food in the Cumberland, Tenesse and the Mississippi Rivers extended up to Connediant. All species of genus Nymphaeceae are water lilies. White Lotus lily (*Nymphaea lotus*) and Blue Lotus lily (*Nymphaea caerulea*) are native lilies represented in Egyptian art and religion. Many Eastern religions, particularly Buddhism, have adopted it as the paramount symbol of purity, paramount conscience, peace and tranquillity.

Nelumbo nucifera lacrt. syn. *Nelumbium nelumbo* Druce, is usually known as the Indian Lotus,the sacred Lotus, the East Indian Lotus,the Hindu lotus and the Chinese lotus. The lotus is native to India, Japan and China. It is generally distributed from tropical East Africa to North Eastern Australia. It is growing in lakes, ponds,rivers, and pools all over India, from the Himalayas in the north to Kanyakumari in the south.

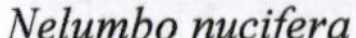
Nelumbo nucifera

Nelumbo lutea

It is large, glaucous, perennial, aquatic, herbaceous plant. Rhizomes are long about 60 to 120 cm in length, 6-9 cm in diameter. They are starchy, white to burnish-orange in colour. Cross-section shows a few large cavities adjoined by several small ones. It is fleshy, mucilaginous and slightly fibrous. This part is also edible. Leaves are large with centrally placed stalk with blue-grey colour.

Flowers are solitary borne at or above leaf level which is single. The flower is deep pink in colour, sepals are 4-5 in number, free, caducous, green, petals are obovate, obtuse, elliptic, nervulose, 8-12 cm long and 3-7 cm wide. These are multiseriate, hyogynous, anthers linear introse, numerous, yellow, 15-20 mm long, filament linear, 7-25 mm long, style short exerted, stigma terminal, carpels are many usually 10-30 in number, uniovulate, free, embedded in the large flat topped spongy receptacle thalamus resembling the roses of small watering cans, carpels maturing into nut-like achenes (seeds), nuts ellipsoidal, about 20 mm long, pericarp hard without endosperm, cotyledons thickened.

Medicinal and Uses

The lotus leaves known for their various properties like having astringent, comparable refrigerant and diuretic actions. Hence, these are helpful to cure various diseases like diarrhoea, high fever, strangury, haemorrhoids and leprosy. Seeds are helpful to stop irritation or vomiting of the digestive tract. Ripe seeds of lotus are provide a spleentonic and used for their astringent action in the seasoning of seminal emission, leucorrhoea and chronic diarrhoea. It also have a sedative action, making themuseful for heart palpitations and insomnia. The sun dried plumule or embryo of ripe lotus seeds are primarily used for insomnia, nervous disorders, high feverswith uneasiness and hypertension. The flower pods or receptacles contain proteins, carbohydrates and a little amount of Nelumbine alkaloid, used to stop bleeding andeliminate stagnated blood. Stamens are helpful to strengthen kidney function and are helpful in treatment of male sexual disorders particularly and also female leucorrhoea (Naguyen, 2001). Lotus contains large number of odour essence in leaves and flowers. In leaves, 40% of odour essence is cis-3-hexanol and in flower 75% of odour essence are hydrocarbons mainly linalool,1, 4-dimethoxybenzene-1, 8-cineoleand 4-terpineol (Omata *et al.*, 1992).

Lotus is popular for its sanctity and aesthetic value. Besides, the pods and stemsare also used as food. Its flowers are offered to gods and goddesses in temples. The lotus buds are used in flower arrangement. The flower buds are cut one or two days before opening can be transported to long distances. It is often grown. In ponds and pools for landscaping a large garden (Nguyen, 2001).

The tender leaves, petioles and flowers are eaten as vegetables. Yellowish white fibre is produced from petioles. The perfumeis also extracted from lotus flowers. Its perfume available as a blend of patchouli, benzom and styrax with phenylethyl and cinnamic alcohols in trade.

Cultivars

There are several cultivars, developed in China and Japan, which have single or double flowers of white to deep pink or red colours. There is also a cultivar having variegated leaves. The important cultivars of lotus are given here in detail.

1. **Alba Striata:** The edge of petals is striped antippedwith red colour.
2. **Alba:** This is also known as the Magnolia Lotus. The flowers are of white colour.

3. **Alba Plena:** The flowers are large in shape and white in colour. They are also double flowered.
4. **Rosea Plena:** They are double flowered. The colour of the flower is rose pink.
5. **Pekinemis Rubra:** The colour of the flower of this cultivar is rosy carmine.
6. **Shiroman:** This is a Japanese cultivar. The flower is large and double. The colour of flower is white with greenish tinge.
7. **Pygmaca Alba Plena:** This cultivar is dwarf. It has double flower with pink to red colour.
8. **Kimhiren:** The colour of the flower of this cultivator is white shaded pink.
9. **Grossberzog Ernst Ludwig:** The flower colour of this cultivar is rose purple.
10. **Pulchra:** The colour of the flower is rosy red in this case.
11. Gigantea: The cultivar is very large in size. The flower colour is rose purple.
12. **PekinensisRubra Plena:** The culuvar IS very large in size. The flower are double with rosy caramine colour.

Propagation

Sexual Propagation (by Seed)

Lotus seeds have the longest viability with age. It is estimated that up to 1,500 years old seeds were propagated and generated healthy plants and these plants were able to produce their own viable seed (Shen-Miller *et al.*, 1995). The long viability of seeds is due to the presence of extra hard, impermeable testa or seed coat. The seed propagation can be followed by the scarification of seed coat followed by incubating at 25-30°C. The seeds should submerged in water-saturated media. The choice of media depends on the application - if plants are to be transplanted then the coarse sand media is suitable for propagation. Absence of media encourages fungal growth, which borne from the inside of the seed. Plants generated by this technique are not be true-to-type because there is high degree of genetic variability in the seeds. Therefore, seed propagation should be done for selection of new varieties or identifying potential qualities may be the cross with another varieties.

Vegetative Propagation (Rhizome)

This is the most easy and effective method of lotus propagation in recent days and also produce true-to-type plants. Seed rhizomes should separate from the current crop, or a separate mother pond is constructed to supply seed rhizome material in different seasons continuously. Using this technique, a rhizome with at least two segments sealed either at the distal end by the intact node, is submerged in water-saturated media at an angle of 15° with the shoot meristem approximately 5 cm below the media surface. Meristems are left intact for further propagating material, although, if planting a crop, only one meristem is left intact. Larger rhizomes are produced more vigorous plants.

Assuming this method of plant propagation is for a crop situation or seed rhizome production, then nutrition should follow as per recommendations.

Pond Management

Design

There is a critical pond design for lotus production and, once the design constructed, it's difficult to change. That's why, it is the recommendation that civil engineers should consult these points regardingthe pond design, construction and use. Water management should be examine in the context of acquisition, storage, movement and crop pond management. A deep waterbody preferably located on higher ground, can be used for water storage. It also allows the gravity feeding of water through irrigation pipes to all ponds. Water may accumulate in each pond and be pumped back into the waterbody before being finally collected into a catchment, or the catchment pond may be second waterbody.Water may need filtering through a sedge bog and pumped back into the reservoir for example to remove nutrients. However, the amount of area under cropwould bereduced bya second reservoir. The deliberate site should have the inherent qualities likea gentle slope and rather asuitable soil type for Lotus cultivation. Attention has to be paid to the method of sealing the pond. Traditional farm dams are sealed by impervious clamping, which are unsuitable for Lotus cultivation because rhizomes can grow into them. It is difficult to harvest from hard, compacted soil and if dug, the seal of the pond may be compromised. Depending on the habit of each variety, a soft layer of soil to a depth of about 1 meter is required. Similarly, it is possible to line the pond with pond liners that are manufactured in a range of available materials. Again, if the soil on site is inappropriate, the type of soil can be selected by the grower for use. The pond liners will restrict downward growth of rhizomes and result in flat misshapen rhizomes if the soil depth is not sufficient. Consideration of the intended scale of operation will be dictated by economics; on a larger scale, the cost of construction as well as crop production and pond maintenanceis more. Further, some areas may require council or EPA approval due to significant changes to water catchment regulations on properties.

Construction

When a design has been planned and the mandatory approvals has been received and obtained, the producer should follow alike construction plan for the example given. Note that, as it only incorporates the underdeveloped elements of a specific plan, producers should always be aware to their own distinctive situation. For connecting pond system, earthmoving equipment accessibility, operators and prevailing environmental conditions at the time of construction farther consideration should given to irrigation piping. Moving soil in wet, waterlogged conditions would not be favourable situation.

- The area should be clear intended for vegetation;
- Mark the area for pond;
- Remove the topsoil and place it aside (this soilcan be used as the growing media by the growers);

- Remove the lower soil horizons to the required depth (approx. 1.2 m) and install the piping or irrigation mechanisms;
  a) Compact the subsurface soil if it is favourable to creating an impervious seal. Ensure slight slope for water flow and control; OR
  b) If this method is adopted than install the pond liner.
- Introducing water to the system of pond to test the pond's permeability and water movement (water should move frequently over the incline and not leak out - this is best done before introducing growing media);
- Drain the pond (recirculate the water to hold reservoir);
- Return thetopsoil back to the pond area (dug out) and work for necessary nutrients and soil/water ameliorants;
- Fill the pond with water to the necessary level

Cultivation

There are two common methods of planting lotus. They are by direct planting and through rhizomes which are discussed here in detail.

1. Direct Planting

Planting can be done directly into the base of the pond or the pool, particularly ina newly constructed one. Generally at the bottom of the pool there is a base of a mixture of three parts of loam and one part of well-rotten cow dung manure or farmyard manure. Coarse bone mealmay be used if cow dung manure is not available. The bottom of the pool is filled with soil and manure mixture to a depth of 15-20 centimeter. The pool should be empty at the time of planting. After planting, water just sufficient to cover the crown of the plant is allowed. After a week or ten days another 2-3 cm water is added and it is continued until the pond or pool is full which may take 6-8 weeks.

2. Through Rhizomes

Another method is to plant the rhizomes or young plants split up from the clump by division, in small baskets or pans about 35-40 cm in diameter and 20-25 cm in depth. The above mentioned compost mixture (soil and manure) is used to fill the baskets or pans. Loam, crushed charcoal and bone-meal can also be used. After planting, watering is doneand then the basket is carefully lowered and placed on bricks for about a week or ten days so that the crown remains fairly close to the surface. The plant is gradually lower when new growth appears. The planting is done in the month of March to April. Seeds are sown in small shallow pans in ordinary loamy soil. The pan is kept in a bigger container filled with water. After germination, the seedlings are transplanted into 15 cm pots. Later, they are planted in baskets, which are lowered in the pool as described earlier.

In China and Japan, lotus is grown in terraced fields as a commercial crop for edible rhizomes. However, cultivation in ponds is preferred for a better development of rhizomes that may spread up to 13m in a year. The rhizomes are harvested by digging in October. The average yield of rhizomes is about 35-40 quintals per hectare. About 10-12 kg seeds are required to produce seedlings sufficient to grow in hectare.

Flowering

Lotus flowers grow profusely in summer and rainy seasons. The flowers are produced on long stalks at or above the level of the leaves. The flower buds when picked 2 to 3 days before opening are able to withstand long distance transportation and keep well when placed in vases.

Weed Management

Sometimes aquatic weeds may grow vigorously in large ponds and completely crowd out the lotus plants. One such troublesome aquatic weed is the water-hyacinth (*Eichhornia crassipes*), which spreads rapidly through runners. Typha is another common aquatic weed. The weeds may be removed mechanically, by hand or by other means. In difficult cases, herbicides may be used like simazin (3-6 ppm) and Monuron (4-12 ppm). These herbicides when used in low concentrations do not harm fish in the pool.

Diseases and Pests

Unless we prevent the occurrence of diseases and pests we cannot get profitable yield. Common diseases and pests on lotus plant are discussed here.

Diseases

The two common diseases are leaf spots caused by *Cercospora* spp. and *Ovularia* sp. and root rot caused by *Phytophthora* spp. It is not possible to use any fungicide to control the disease because they may be harmful to fish. However, Bordeaux mixture may be sprayed on the foliage when there is no fish in the pool.

Pests

Aphids and beetles are the common insect pests. In case of infestation of aphids, spraying a strong jet of water may knock the aphids into the water where they will be eaten by fish. The infested leaves and flower stalks may be kept immersed in water for at least one or two days which will help to drown the aphids. The beetles may be either picked up by hand or dislodged by spraying water as in case of aphids.

Harvesting

Flowers: the first leaf and flower can be harvestedafter three months of planting. During summer season flower picking should be done at two days interval while in case of winter picking should be done at three days interval.The flower production is at its climax after four months of planting. The manual flower harvesting done by hand after three to four months of planting (Wikipedia).

Rhizomes for Vegetable Purpose: for cooking purpose, rhizomes mature at approximately six to nine months.The harvesting of rhizomes is done when the ponds or fields are drained and the time of early varieties varies from July to September while the late varieties harvested from October to March. It is easy to dig out the large, starch rich rhizomes from the drained soil. In small scale production, rhizomes are harvested by hand using fork like tools (Wikipedia).

Rhizomes for Transplanting: Rhizomes of lotusare harvestedat dormant stage.The growth of the plant parts above water slows down during fall stage as nutrients are directed to the underground stem. The dormant stage of plant lies between the time when the top turns yellow and wilts and the leaf bud semergence during spring. It is important to handle the rhizomes of lotus with care at the time of harvesting to prevent damage to the delicate rhizomes and growing tips.

Seed Harvesting: The seed pods are harvested after four to eight months of planting when the seeds are turning black in colour from green. The seeds are removed from seed pods and then sun drying should be done for 3-4 days after it the coat and embryo is removed.

References

https://en.wikipedia.org/wiki/Nelumbo_nucifera

Makino, Tomitaro(1979). Makino's New Illustrated Flora of Japan. The Hokuryukan Co. Ltd., Tokyo, Japan (In Japanese).

Nguyen, Q. V., 2001. Exporting lotus to Asia an agronolic and physiological study. Rural industries research and development coporation. Pp: 1-47

Omata,A., Yomogida, K., Nakamura, S., Ohta,T., Izawa,Y., Watanabe, S. (1992). The odour of Lotus (Nelumbonaceae) flower. Edited by Bhattacharyya, S.C., Sen, N. and Sethi. K.L. Proceedings of the 11th international Congress of Essential oils, fragrances and flavours. New Delhi, India, 12-16 November 1989. Vol. 4 Chemistry-analysis and structure. 1990:43-48.

Shen-Miller, J., Mudgett, M.B., Schopf, J.W., Clarke,S. and Berger, R.(1995). Exceptional seed longevity and robust growth: Ancient sacred lotus from China. American Journal of Botany 82 (11):1367-1380

www.agrifutures.com.au

28

Post-Harvest Handling of Flowers

Puja Sharma, S.R. Dhiman, Bharati Kashyap, Priyanka Sharma and Panchal Sangmesh

Floriculture is a high value industry. It is increasingly regarded as a feasible diversification option from traditional field crops due to increased per unit returns and increasing habit of 'Greeting with Flowers' during all the occasions. With increasing awareness among consumers and pressure from the artificial floral products on the fresh flower industry, and continued competition even amongst the growers, whole salers and retailers, quality is becoming increasingly important. The aspects related to visual quality of the product are the main determining indices of consumer's acceptability of flowers.

The important flower crops in the international market are rose, carnation, lisianthus, chrysanthemum, gerbera, hydrangea, gladiolus, gypsophila, bird of paradise, orchids, anthurium, tulip and lilies etc. Cut flowers have the highest economic value and are the most perishable. Their high respiration rates, rapid deterioration, and susceptibility to damage require the utmost care during postharvest handling for quality maintenance. The post-harvest losses can be both, quantitative as well as qualitative, which ultimately result in loss in terms of its monetary value.

Floriculture industry is facing main problem due to highly perishable nature of its flowers which demand for appropriate post-harvest handling techniques to maintain the quality. With changing scenario in international floriculture trade, perfection in post-harvest technology is mandatory to offer quality plant products to the consumers round the year. The post-harvest life of cut flowers and flowering pot plants can be extended relatively by a range of treatments and technologies. Quality factors that influence the post- harvest keeping quality and flower longevity are environmental factors as well as the influence of the pathogenic organisms.

Due to perishable nature of cut flowers, huge post-harvest loss is there which ranges from 30-50 per cent. Qualitative losses which include consumer acceptability of fresh produce are much more difficult to measure than quantitative losses. Quantitative losses can occur during the entire post-harvest and marketing chain due to lack of proper post harvest handling (Bhattacharjee, 2006). These losses could be on a higher side, which might be due to lack of knowledge on 'post-harvest handling' of cut flowers and proper infrastructure for post-harvest management and storage. However, all these losses can only be minimized by proper technique of harvesting, post-harvest handling, packaging, storage, marketing and value addition of flowers.

Proper post-harvest handling is an important part of modern flower production. The technology of post-harvest handling bridges the gap between the produce and the consumer, a gap often of time and distance. Post-harvest handling involves the practical application of engineering principles and knowledge of flower and flower physiology to solve this problems.

The sum total of activities carried out after harvesting of cut flowers until it reaches the end user i.e. consumer are referred to as post harvest handling. It includes various activities like optimum harvesting procedure, precooling, conditioning, pulsing, storage and transportation of cut stems etc.

Post-harvest losses of flowers can also be reduced by involving well trained people in harvesting and handling of flowers. The next most important thing is that the flowers should be harvested at optimum maturity stage and finally after harvest the cut flowers should be kept in water immediately to avoid blockage of vacuole with air. The flower then should be immediately taken to packaging house using cold chain facility. To maintain high quality of produce, storage and transportation should also be done under cold storage conditions.

Factors Affecting Post-Harvest Life

I. Pre-Harvest Factors Affecting Post-Harvest Life

Genotypic Factors: Keeping quality of flowers vary from species and cultivars, which may be due to genetic or inherent factors like differences in anatomical, physiological, physical, biochemical and genetic makeup.

Light: Light is an important factor and regulates several physiological processes like chlorophyll synthesis, phototropism, respiration and stomatal opening. Most of the commercial flowers need high intensity (3,000-8,000 f.c) and absorption of red light (625-700 nm) through a pigment phytochrome. It has been found that a decrease in available solar radiation before harvest reduces the post-harvest life of cut flowers. Low light intensity generally causes excessive elongation of flower stems and reduces stem hardening, which results in bent neck in rose and stem bending in carnation and gerbera. Similarly, excessive light intensity is also harmful and it deteriorates the flower quality. In case of lilium, the stem length reduces if grown under full light intensity.

Temperature: Higher temperature during field conditions decreases vase life and quality of cut flowers. Exposure of rose plants to high temperature reduces anthocyanin content and increases darkening of petals. A day temperature of 18-25°C has been found to be optimum for best grade carnation flowers. Increase of greenhouse temperature from 10 to 27°C for one week before harvest causes reduction of vase life of cut carnations from 4 to 7 days at ambient temperature.

Mineral nutrition: Nitrogen is an essential element for plant structure and function as it is necessary for the synthesis of amino acids, purines, pyrimidines, chlorophyll and coenzymes. The soil application of N lengthens the time taken for flower bud initiation. Field grown rose cultivars require nitrogen for better yield and longer post-harvest life in cut flowers. Similarly the availability of all other nutrients at optimum level is required for production of high quality flowers.

Irrigation: Optimum moisture level is required throughout the growing phase of the crop. Stress due to excessive or inadequate water in the growing medium reduces the longevity and quality of cut flowers. Heavy irrigation is required from the bud development stage to harvesting in all the flowers. Insufficient water during bud development leads to calyx splitting in carnation and insufficient floret opening in gladiolus.

Carbon dioxide: In general, CO_2 enrichment increases photosynthesis and decreases photo respiration. At elevated CO_2 levels, young carnation plants and mature flowers have higher dry matter content. When polyhouse is enriched with CO_2 roses with more number of petals, longer and thicker stems are produced. Chrysanthemums at a concentration of 1,000-2,000 ppm CO_2 produced longer length, more fresh weight, leaf number and increased longevity of cut flowers (Tanigawa *et al.*, 1995)

Use of agrochemicals: GA_3 is well known to promote flower formation and development in many long day and cold requiring plants (Bernier *et al.*, 1981). Application of GA_3 (50-100 ppm) improves post-harvest quality of roses by proper anthocyanin pigmentation. Pre harvest spray of BA in improves shelf life of chrysanthemum loose flowers (Sagar *et al.*, 2019).

Pests and Diseases: Export quality flowers should be completely free from pests and diseases. Hence, protection is necessary from the nursery itself so that there is no hidden infection caused by pathogens or insects which reduces the longevity as well as consumer acceptance.

II. Harvest Factors

Harvesting is the process of detaching flowers from the plant in a scientific way using sterilized equipments. For this, the maturity indices of the various cut flowers are to be known. Proper stage, method and time of harvest is of considerable importance to ensure long vase life of cut flowers. Flowers maintain a fresh appearance much longer if they are harvested at the appropriate stage of development. In general, flowers cut at more advanced phases of development have a shorter vase life than younger flowers. The optimal phase of flower development for harvest depends upon the plant species, the cultivars, the season, the distance to the market place, and the consumer preference.

Table 1: Optimum stage of harvesting of commercial flower crops.

Flower Name	Stage of Harvest
Alstroemeria (Lily of Incas)	4-5 florets open.
Anthurium andreanum	Spadix almost fully opened.
Chrysanthemum morifolium	Outer petals fully elongated.
Standard types	Open but before anthesis.
Spray types	Centre of the oldest flower fully open
Single	Open but before central disc florets begin to elongate.
Pompons and decorative	
Anemones	
Dianthus caryophyllus (Carnation)	Paint brush stage when flowers are half open
Standard cultivars	At least 2 fully open flowers
Spray cultivars	
Gerbera jamesonii (Gerbera)	Flowers open and outer two rows show shedding of pollens.
Gladiolus	1-5 buds showing colors.
Lilium spp. (Lily)	Colored buds.
Narcissus spp. (Daffodil)	Goose neck stage.
Orchids	3-5 days after opening.
Cattleyas	3-4 days after opening.
Cymbidiums	Almost fully opened flowers.
Dendrobiums	When 50-75% of the flowers open.
Oncidiums	3-4 days after opening.
Phalaenopsis	
Rosa hybrida (Rose)	First 2 petals beginning to unfold.
Red and Pink cultivars	Slightly earlier than red and pink.
Yellow cultivars	Slightly later than red and pink.
White cultivars	
Strelitzia reginae	First floret opens.
Polianthes tuberose	Buds fully developed but unopened.
Single	Maximum florets open.
Double	
Tulipa spp. (Tulip)	Half coloured buds.

Source: Nowak and Rudnicki, (1990)

Although there are well defined harvesting stages of different flowers, yet some of the flowers can be harvested at bud stage and are successfully opened by using bud opening solutions. Harvesting of flowers at bud stage is always advantageous as the bud continue to open over a long time and hence, possess long vase life. Besides, the buds are less sensitive to ethylene, easy to handle during storage and transport, economical to transport and are less prone to damage by diseases and pests. The bud harvested flowers are easier to handle and less susceptible to adverse handling conditions. Successful opening of flowers in cut stems harvested at tight bud stage has been reported by Kavita (2016) in chrysanthemum, Kuiper *et al.* (1995) in rose and Han (1992) in liatris.

Time of Harvest: Optimum time of harvest is one of the most important factor influencing flower longevity. Morning harvesting has advantages in terms of better turgidity; but at same time, flowers which may be still wet with dew are more susceptible to postharvest fungal infection. Evening harvesting has advantages in terms of higher carbohydrate concentration in flowering stems. Morning harvest is recommended for flowers which lose water quickly after picking, as in case of roses. Flowers are at full turgid in the morning, and transferring them after harvest to a floral preservative solution and then to cold storage prevents water loss and helps to preserve their quality. If, however, harvested flowers are precooled and placed directly in a floral preservative solution containing sugar, the time of harvesting is not important.

Mode of Harvest: Sharp tools should be used for removing flowers from the mother plant. The angle of the cut should be slanting, and the cut should be smooth, particularly in flowers with hard, woody stems which absorb water only through the cut surface. The harvested stems should immediately be transferred to the bucket of water to prevent air entering the cut end. The crushing of a stem at the cut should be avoided as this causes the exudation of sugar containing sap which encourages the development of micro organisms which may in turn cause stem plugging. Flowers are usually cut close to the soil to provide the longest stems possible. This is often associated with decreased water absorption by the hardened lower part of the stem, resulting in a shorter vase life. It is better to cut flowers in the less hardened portion of the stem.

Air embolism occurs when small bubbles of air are drawn into the stem at the time of harvesting. These bubbles cannot move far up to the stem, so the upward movement of water to the flower is restricted. Emboli can be removed by recutting the stems under water by ensuring that the rehydration solution is acidic (pH 3 or 4).

III. Post Harvest Factors Influencing the Quality of Cut Flowers for Export

Temperature: The high respiration rate of cut flowers, as an indicator of their rate of growth and senescence generates heat as a by-product. As in all biological systems, the respiration of cut flowers increases logarithmically with increasing temperature. The optimum temperature for storage of most of the cut flowers is 2-5 °C, just above freezing. In contrast, some tropical crops such as anthurium, bird-of paradise, some orchids and ginger show chilling injury at temperatures below 10° C.

Light: Light is essential for long distance transport or prolonged storage in alstroemeria, chrysanthemum, dahlia, gladiolus and other flowers. Similarly, high light intensity is essential for opening of tight bud cut flowers. Flowers like carnations can be stored in darkness for a longer period without affecting quality. Chrysanthemums should be illuminated during storage with a light intensity of 500-1,000 klx to protect the plants from yellowing of leaves.

Humidity: Cut flowers should be kept at 90-95 per cent relative humidity for maintaining turgidity. Flowers start showing wilting symptoms when they have lost 10-15 per cent of their fresh weight. During pre-cooling and shipment of cut flowers, the relative humidity level of the air should be maintained at 95-98 per cent. Higher air humidity reduces water loss and petal wilting.

Water Supply: Cut flowers, especially those with leafy stems, have a large surface area, so they lose water and wilt rapidly. They should be stored at relative humidity above 95 per cent, particularly during long-term storage. Water loss is dramatically reduced at low temperatures, another reason for prompt cooling. Air embolism, bacterial plugging, and poor water quality can reduce solution uptake. Water quality is very important as hard water contains minerals that make the water alkaline. Alkaline water does not move readily through stems and can substantially reduce vase life. Such water can be treated by removing the minerals using a deionizer or by making the water acid. Excessive loss of water in cut gerberas leads to bending where the flower bends from neck region and is beyond correction.

Ethylene: Some flowers especially carnations and some rose cultivars, perish rapidly if exposed to minute concentrations of ethylene. A wide range of flowers are affected by ethylene, with showing of some typical symptoms such as sleepiness of carnation and kalanchoe petals, fading and in-rolling of the corolla of ipomoea, fading and wilting of sepal tips in orchids tec.

Growth Tropism and Mechanical Damage: Negative geotropism often reduces quality in spike-flower crops like gladiolus and snapdragon because the flowers and spike bend downward when stored horizontally. These flowers should be handled upright whenever possible. While considering the aesthetic appearance of flowers, damaged flowers are not desirable.

Diseases: Flowers are susceptible to diseases, not only because their petals are fragile, but also because the sugar solution secreted by their nectarines, which is an excellent nutrient supply for even mild pathogens. The most common fungal infection during post harvest handling is gray mold (*Botrytis cinerea*). Proper greenhouse hygiene, temperature control, and minimizing condensation on flowers all reduce losses caused by gray mold.

Post-Harvest Handling Chain of Cut Flowers

Post-harvest Handling includes all the activities carried out until the flowers reach end user which is referred to as post-harvest chain. The various steps are as follows:

Conditioning of Cut Flowers: Conditioning or hardening is a process where the flowers are kept or made to stand loosely in a big container of water so the air circulates around the stem. This treatment is done to restore turgidity of cut flowers as various kinds of water stress during handling, storage and transport might have affected the turgidity of flowers. Flowers can be conditioned in the dark so that their stomata will close, reducing the amount of water loss by transportation. Conditioning is achieved by treating flowers with demineralized water supplemented with germicides and acidified with citric acid to pH 4.5-5.0 but with or without sugar.

Pre-Cooling: It is generally done immediately after harvest. Pre-cooling is done by forced air cooling or cooling to bring down the temperature to 1°C in a short period. Time between harvest and pre-cooling should be as short as possible. Forced air cooling is most effective in removing the field heat from flowers. Packed or unpacked flowers have to be pre-cooled before storage or shipment.

Impregnation: Impregnation of cut bases of flowers with high concentration of Silver nitrate ($AgNO_3$) or Nickel chloride ($NiCl_2$) or Cobalt chloride ($CoCl_2$) for 10-15 min greatly improves longevity of several flowers. Since silver travels only a short distance in the stem, bases of treated flowers should not be re-cut after treatment. Here, the Ag^{++}, or Ni^{++}, or Co^{++} ions stay at the base and protect from microbial blockages. Impregnation is better with carnation, rose, chrysanthemum, gladiolus, tuberose and orchids.

Grading: It is an important operation that facilitates sorting of flowers with similar attributes. The cut flowers are usually graded on the basis of various factors like bud diameter, stem length, degree of bud opening and aesthetic appearance etc.

In the United States, the Society of American Florists has developed recommendations concerning flower selection for various grades for some species of cut flowers only. US Grading Standards for some important cut flowers are :

A. Anthurium

Grade	Size of the Spathe (cm)
1. Extra Large	>15
2. Large	13-15
3. Medium	10-13
4. Small	8-10
5. Miniature/Tiny	<8

B. Carnation (Standards)

Grade	Flower type	Minimum flower Diameter (mm)	Minimum length overall (cm)
Blue(Fancy)	Tight	50	
	Fairy Tight	62	55
	Open	75	
Red (Standard)	Tight	44	
	Fairy Tight	56	43
	Open	69	
Green (Short)	Tight	None	
	Fairy Tight	None	30

C. Chrysanthemum

Grade	Minimum flower diameter (mm)	Minimum length overall (cm)
Blue	140	76
Red	121	76
Green	120	61

D. Gerbera

Grade No.	Stem length (cm)	Flower diameter (cm)	Flower color
1.	>60	>12	Pink
2.	50-60	11-10	Reddish orange
3.	40-50	10-9	Orange
4.	30-40	9-8	Red
5.	<30	8-7	Yellow

E. Gladiolus

Grade	Spike length (cm)	Minimum floret number
Blue (Fancy)	107 & above	16
Red (Special)	96-107	14
Green (Standard)	81-96	12
Yellow (Utility)	81 & below	10

F. Roses

Grade	Minimum Length overall (cm)	
	Hybrid Teas	Sweethearts
Blue	56	36
Red	36	25
Green	25	15

Floral Preservatives: Once cut flowers are detached from parent plant they are delinked from the supply of metabolic raw materials. Therefore, exogenous supply is essential to keep its quality and usefulness. Water is the first and foremost essential component for flowers. A source of energy deriving materials for respiration is also quite important. A clear understanding of the course of senescence is immensely helpful for establishing techniques to overcome ageing process.

Floral Preservatives are the chemical formulations used for extending the vase life of flowers. Besides increasing vase life, floral preservatives help to improve flower opening, flower size, shape and color. Many commercial formulations are now available in liquid or powder form for extending flower vase life but their constituents is not disclosed. These formulations are to be used as per the guidelines provided by the manufacturer.

Floral preservatives have two basic constituents: sugar and biocide. Sugar provides a source of additional food to the cut flower whereas the biocide checks the growth of bacteria, fungi and other microorganisms in the vase water. Besides, floral preservatives may also have other ingredients such as inorganic salts, growth regulators and anti-ethylene compounds.

Types of Floral Preservatives

Floral preservatives can be applied as (i) pulsing solutions; (ii) bud opening solutions; and (iii) holding or vase solutions.

(i) **Pulsing solutions:** Pulsing refers to short duration (16-24 h) pre-shipment or pre storage treatment. It involves loading the flowers with high concentration of carbohydrates (sucrose) to replenish the supply of food to the harvested stems. The effect of such a treatment lasts throughout the entire vase life of the flower. Pulsing is always done at room temperature.

(ii) **Bud opening solutions:** Harvesting the flowers at bud stage offers many advantages. Besides reducing crop span, it makes handling and transportation relatively very easy as compared to opened flowers. Immature buds of many flowers can be made to open in chemical solutions, referred to as 'bud opening solutions'. The components of bud opening solutions are essentially the same as those of pulsing solutions, but in case of bud opening solutions, lower concentrations (2 to 5 per cent) of sucrose are used.

(iii) **Holding or vase solutions:** The holding or vase solution can be retained with the wholesalers or retailers or the consumer to keep flowers fresh until they are sold or for the consumer to use continuously in the vase. A holding solution contains carbohydrates, germicides, growth regulators, ethylene inhibitors, mineral salts and organic acid for prolonging the life and quality of cut flowers. The level of sucrose in vase solutions is, kept very low (0.5 to 2 per cent).

Constituents of Floral Preservatives

The main constituents of floral preservatives are (i) water; (ii) sugar; and (iii) biocides. Besides, mineral nutrients, acidifying agents, anti-ethylene compounds and growth regulators are also used in floral preservatives. The brief account of these constituents is given herewith:

Water: Water content of flowers after harvest affects vase life in terms of its quantity as well as quality. Tap water can be harmful due to its high pH (alkalinity), presence of total dissolved solutes (TDS) and toxic ions. High pH promotes the microbial growth and hence, reduces vase life. Therefore, acidifying water to low pH (3.0-3.5) is advantageous because it decreases microbial growth and considerably improves water uptake by the stem. Sensitivity of different flowers to the presence of TDS in vase water varies. Vase life of rose and chrysanthemum flowers decreases at low level of TDS (200 ppm) in water whereas flowers like gladiolus show decline in vase life when TDS in water is relatively high (more than 700 ppm). The presence of sodium ions in vase water is detrimental for roses whereas fluorides are highly toxic to some flowers like gladiolus, gerbera, freesia, chrysanthemum and rose. In such cases, it is always advantageous to use distilled water.

Sugar: Sugar act as additional food source and also improves water balance of cut flowers. Sucrose is the most widely used sugar in floral preservatives. Sugar in the vase solutions act as food for the growth of detrimental fungi and bacteria, whose growth can be enhanced by organic materials from the cut stem. Substances produced by the bacteria, and the bacteria themselves can plug the water-conducting system. For this reason, flower-holding solutions should contain germicides to prevent the growth of micro-organisms. Acidic vase solutions not only improve water flow by overcoming embolism, but also inhibit bacterial growth. The optimum concentration of sugar required, however, varies from species to species.

Biocides or Germicides: Biocides are chemical compounds which are used to inhibit microbial growth in vase water as well as on the stem surface. Important biocides used for treating cut flowers are 8-hydroxyquinoline citrate, silver nitrate, aluminum sulphate, citric acid, slow release chlorine compounds and quaternary ammonium compounds.

8-hydroxyquinoline citrate (8-HQC) is very effective broad spectrum biocide. It acidifies water and also induces partial closure of stomata. This compound is used at the concentration of 200-600 ppm. Silver nitrate ($AgNO_3$) is also very effective biocide. $AgNO_3$, at the concentration of 25 ppm completely inhibits the microbial growth and also causes partial stomatal closure, thereby increasing vase life. High concentration (1000-1200 ppm) of $AgNO_3$, can also be used as short term (10 - 15 minutes) treatment $AgNO_3$, has, however, very low mobility and remains concentrated at the cut surface of the stem where it inhibits microbial growth. A lot of concern is however, being voiced in the developed countries regarding the use of silver, as being a heavy metal, it is considered to be an environmental hazard. Citric acid has been reported to improve vase life of roses, chrysanthemum, gladiolus and carnation. It acts as an acidifying agent, lowers pH of the solution and also prevents blockage of xylem vessels.

Growth Regulators: Plant growth regulators considerably affect vase life of the flowers. Growth regulators are broadly categorized in to five classes: i) auxins; (ii) gibberellins; (iii) cytokinins; (iv) inhibitors and retardants; and (v) ethylene though other compounds like salicylic acid, brassinosteroids and jasmonates are also included in this category. Single growth regulator may not play a significant role in flower senescence but interaction among different growth regulators is suggested to determine the potential life of the cut flower. However cytokinins are most frequently used in vase solutions. Cytokinins play an important role in delaying senescence of many flowers. Endogenous cytokinin content shows a pattern consistent with its presumed role in delaying senescence i.e. buds and young flowers contain high cytokinin levels, which fall as the flower ages and commences senescence (Mayak and Halevy 1970; Van Staden and Dimalla 1980). Endogenous levels of cytokinins decrease with ageing of rose flowers. Not only this, some rose cultivars with short vase life possess lower levels of cytokinins than those with long vase life. BAP in holding solution also delayed senescence of tuberose flowers. Dip treatment with BA is also reported to increase vase life of anthurium flowers, accelerates opening of buds of rose and chrysanthemum and prolongs their vase life. BA also delays the yellowing of leaves in chrysanthemum either in vase or after prolonged storage. The stems should be given pulsing treatment with BAP (100 ppm) for one hour before storage or can be held in BAP (50 ppm) solution during storage to prevent premature yellowing of leaves after storage. Thidiazuron which is having cyrtokinin like activities can effectively be used for delaying leaf yellowing in cut chrysanthemum (Kavita 2016).

Anti Ethylene Agents: The role of endogenous ethylene in causing senescence has been well documented by a range of studies reporting the dynamics of ethylene production, changes in activity of the biosynthetic enzymes (Buflcr 1984, 1986), and up-regulation of the genes encoding these enzymes (Woodson *et al.,* 1992). The key role of ethylene has been verified by studies with long-lived carnation cultivars (Wu *et al.*, 1991).

Ethylene, a hydrocarbon gas, and commonly known as a ripening hormone induces senescence in many flowers. Ethylene is present in the atmosphere in minute quantities (3-5 ppb) though its level varies with the season and the location. Internal combustion engines, ripening fruits, vegetables, pollinated flowers, and senescing leaves, diseased and damaged tissues are main sources of ethylene in the atmosphere.

The ethylene inhibitors like amino ethoxyvinyl glycine (AVG) and methoxyvinyl glycine (MVG) at concentration of 0.7 to 0.13 mM are also reported to increase vase life of carnation flowers. STS (silver thiosulphate) is a potential anti ethylene agent frequently used in vase solutins. However, with increasing awareness about reducing the use of heavy metal ions like silver a potential alternative to STS has been suggested by Mukherjee (2003) in carnation and Manikrao (2007) in alstroemeria. Pulsing of carnation in a solution containing Sucrose 10% + 8-HQC 300 ppm + BA 15ppm for 8 hrs. followed by placing the flowers in a holding solution containing Sucrose 2% + 8-HQC 150 ppm + BA 5ppm has resulted in better vase life of carnations over STS based treatments.

Flowers like Alstroemeria, Carnation, Lily, Narcissus (Daffodil) and Orchids are highly sensitive to ethylene. On the other hand flowers like Anthurium, Chrysanthemum, Gerbera, Gladiolus and tulips are relatively insensitive to ethylene.

Table 2: Symptoms of ethylene toxicity in selected cut flowers

Flowers	Toxicity symptoms
Alstroemeria	Flower malformation, petal darkening, petal abscission.
Carnation	Bud sleepiness and petal wilting.
Chrysanthemum	Slight hastening of senescence.
Gerbera	Slight hastening of senescence.
Lily	Bud blasting and petal abscission.
Narcissus (Daffodil)	Smaller flower diameters, hastening of senescence.
Orchids (Cattleya, Dendrobium, Phalaenopsis, Paphiopedilum, Vanda)	Reddish flower color, hastening of senescence.
Rose	Inhibition of bud opening, petal bluing, hastening of senescence.
Tulip	Bud sleepiness, petal bluing, hastening of senescence.

Packaging and Storage

Packaging and storage affect the postharvest life of flowers, as appropriate packaging of flowers along with pulsing may aid in keeping flowers fresh for consumer and may also extend their vase-life. Bunch of flowers previously wrapped in paper or cellophane when placed in corrugated fiber board cartons protect flowers against water loss, physical stress, and various other external conditions that may damage the flowers.

Most of the cut stems are bunched together before packaging e.g. rose, carnation, chrysanthemum etc. However some flowers require specialized packaging e.g. gerbera cut stems are individually put in polyethylene sleeves to avoid injury to the petals. Similarly the soft and shiny spathes of anthuriums are packed individually in polyethylene sleeves. The basal end of the cut stems of orchid spikes are immersed in a vial carrying food material and a biocide which enhances the vase life of cut spikes upto 15-20 days and ensures maximum opening of florets on the spike.

The flowers either bunched or individual are packaged in different materials. Some of the most common packaging materials are:

Corrugated Fibre Board (CFB) boxes: Cut flowers are generally packed in CFB boxes. Cut stems are first wrapped in materials like cellophane, polythene, newspaper etc. and finally packed in CFB boxes with holes on sides for ventilation. Corrugation can be single layered or double layered depending upon the distance of transportation. Proper cushioning material is placed inside the boxes which help to reduce shocks during handling.

Cellophane: Cellophane is generally used to wrap the cut flowers and then packed in boxes. Freshness of florets remain for longer duration due to less water loss from cellophane packed cut flowers, therefore, delaying the withering of flowers. Sisodia *et al.* (2018) observed that longevity of cut flowers of gladiolus cv. 'Punjab Dawn' was maximum when wrapped in cellophane sheets during storage. Wrapping cut flowers of roses cv. 'First Red' in polyethylene and then packing in CFB boxes improved the vase life (Bhasker and Rao, 2018).

Polyethylene: Packing of flowers is generally done in polyethylene bags of 200 gauge thickness. Shelf life and lutein content of marigold flower was recorded maximum when packed in low density polyethylene (LDPE) bags (Pal *et al.,* 2016). Shelf life of rose and Plumeria flowers was observed best in zip pouches of polythene (Kumar *et al.,* 2019). Karuppaiah *et al.* (2006) recommended packing of jasmine (*Jasminum sambac*) flowers in 200 gauge thickness polyethylene bags for extending the shelf life. Packaging of loose flowers of chrysanthemum in cardboard boxes lined with polyethylene was reported by Sagar *et al.* (2019).

Newspaper: Newspaper is generally used as cushioning material in the CFB boxes as this is cheap material.

Other Materials: Use of organic wrapping and packaging material is now recommended as these are bio-degradable. Packing of loose flowers of marigold is generally done in hessian cloth bags. Banana leaves is an alternative option for polyethylene packaging of cut flowers in India. In a study, Roychowdhury *et al*. (2011) observed that spikes of tuberose cv. 'Calcutta Double' wrapped with banana leaf exhibited maximum vase life with maximum floret opening and least amount of floret wilting. For packing of gerbera cv. 'Yanara' cut flowers, plastic sleeves with 20% ventilation and stored in CFB boxes with bedding material as tissue paper as well as newspaper can be used as best packaging materials for longer vase life (Prashanth and Chandrasekar, 2010).

Storage

Storage of flowers is also an important component in post harvest handling. It becomes utmost important in situations where there is over production and availability of flowers in the market. It ensures orderly marketing, reduced retailer's hazards resulting from unforeseen decline for demand, anticipating holidays, improved production efficiency, saving energy and making possible long term shipment. Important factors which determine the storage life of the flowers are quality of the flower to be stored, stage of harvest, temperature and relative humidity during storage, light, ethylene production and pathogens. The flowers to be stored should be of good quality. Flowers should be free from any physical injury and incidence of pests or diseases. The injured and diseased tissues are potential sources of ethylene and are also highly prone to infection with pathogens. The flowers must be cut at an optimum maturity. Some flowers like gladiolus, roses, lilies, narcissus and irises can be stored at bud stage as buds of these flowers possess the ability to open after storage.

Temperature management during storage is an important component which determines subsequent vase life of flowers. Temperature fluctuations in the cold room should be avoided. Relative humidity inside the cold room should preferably be kept high (90-95%). Low humidity may cause desiccation and browning of leaf edge with thin leaves or leaflets. Sufficient air circulation is a must to block the bad effects of high humidity. Cut flowers of rose stored at 90% relative humidity had the longest vase life, while those one kept in 60% showed the shortest longevity (Chamani and Wagstaff, 2019).

Cold rooms should be kept clean to prevent the spread of pathogens as well as ethylene accumulation by the decaying plant tissues and microbes. The storage may of flowers is done by different methods including cold storage (wet and dry cold storage), controlled atmospheric storage, modified atmospheric storage, low pressure storage, etc.

Cold Storage

Cold storage is the common method of storing cut flowers. Flowers should be harvested at the optimal stage of development, should be pre-treated with an appropriate floral preservative and pre-cooled for fast packing. Low temperature treatment during storage or shipment period reduces the entire metabolism in the tissue and slows down the respiration.

It is necessary to maintain a stable and uniform temperature during storage. In general, temperate flowers (rose, carnation and lilium etc.) are stored at 0-1°C whereas subtropical (gladiolus, jasmines, proteas and gloriosa etc) at 4-7°C and tropical flowers (anthurium, cattleya, vanda, euphorbia and bird of paradise etc.) are stored at 7-15°C. In cold storage, flowers can be stored dry or wet, depending upon the requirement.

i. Wet storage

In wet storage, flowers are stored with their base dipped in water or preservative solution for a short time. During wet storage, flowers are kept at 3-4°C temperature slightly higher than that used for dry storage. The lower-most leaves are removed from the stems in order to avoid wetting and subsequent decay. The vase water may get infected by fungi, bacteria and yeast and can cause plugging of cut stem. So, water disinfection with sodium hypochlorite should be practiced. Flowers like carnation, chrysanthemum, gladiolus and gerbera are wet stored at 4°C, roses at 2-5°C whereas anthuriums can be stored at 13°C (Singh *et al.*, 2013)

ii. Dry storage

Dry storage method is used for long term storage. In this method, fresh flowers are harvested in the morning, graded and sealed in plastic sleeves/ bags or boxes to prevent the loss of moisture. Prior to storage, flowers may be pulsed with floral preservatives containing sugar, anti-microbial and anti-ethylene compounds. Before packing in polythene sacks or boxes these should be lined with butter paper bag or news paper. Flowers should be wrapped in soft paper to absorb condensed moisture appearing on the bloom. The optimum dry storage temperature for roses is 0.5-2°C, carnation 0.5-1°C, chrysanthemum 1°C, gerbera 2°C and gladiolus 4°C (Singh *et al.,* 2013).

Table 3: Cold Storage conditions for exotic cut flowers (Lutz and Hardenburg, 1986).

Cut flowers	Temperature (°C)	Relative humidity (%)	Duration range (days)
Alstroemeria	0-4	90-95	6-10
Anthurium	12.5-15.5	90-95	3-10
Carnation	0-7	90-95	3-42
Chrysanthemum	-0.5-8	90-95	7-42
Cymbidium	-0.5-4	90-95	7-14
Gerbera	1.5-4.5	90-95	2-14
Gladiolus	0.5-10	90-95	6-8
Lilium	0-0.5	90-95	4-28
Narcissus	0-2	90-95	7-21
Rose	0-5	90-98	4-14
Tropical Orchids	15	90-95	7-28
Tulip	-0.5-2	85-95	3-42

Controlled Atmospheric Storage (CA)

The principles of controlled atmosphere storage for cut flowers are based on the close regulation of three parameters viz. temperature, oxygen and carbon dioxide. It extends storage life due to inhibited ethylene production and action, by elevated concentration. Nitrogen enrichment storages have promising result in daffodils and roses (100% N_2 and 99% N_2 + 0.5-1% O_2, respectively. The use of CA did not pick much in floriculture industry due to high cost of application and inconvenient handling of various cut flowers.

Modified Atmospheric Storage (MA)

MA storage is a less preferred storage technique in which the flowers are packed in sealed bags which leads to reduction in O_2 and increase in CO_2 levels due to respiration of the tissue. This increased level of CO_2 reduces the biosynthesis of ethylene and hence increases the flower longevity. It is a cost effective method of flower storage as no special atmospheric conditions are required in this method of storage.

Low Pressure Storage (LPS)/ Hypobaric Storage

In this storage method, flowers are stored under reduced pressure, low temperature and cooled moist air. In LPS, gaseous substances like CO_2 and ethylene produced by flowers through stomata and intercellular spaces much quicker under low pressure than at normal pressure. The desirable effect of LPS is attributed to the reduction of ethylene production at low O_2 levels and other volatile compounds. Various reports have indicated that LPS of 40-60 mm Hg pressure is helpful in prolonging vase life of cut flowers. Cut roses at 180-210 mm Hg pressure can be stored up to 3 weeks with retention of 63% of the original vase life.

Tinting of Flowers

It is a specialized technique where white flowers can be artificially altered into colours according to our wish and desire. We see blue gladiolus and tuberoses in the market, these are the tinted flowers since blue colour is not present in the natural gene pool. It is done at retailers' outlet.

Tinting is a value addition technique in flower crops like tuberose where colour pigments are absent or in light colour flowers (Sowmeya *et al* 2017). Tinting can be done with natural flowers by adding artificial colours or food colouring agents. In this the basal ends of the cut stems are dipped in the food coloring/water mixture and allowed to stand overnight. Sometimes splitting of stem lengthwise from the base is also done. The stems and leaves will also turn colors over time, but the flowers are usually the first to show the streaks, as they are in the fastest growing area.

Tinting or artificial colouring of tuberose may be a potential value addition venture (Kumari and Deb 2018). Tuberose (*Polianthes tuberosa* L.) cv. Prajwal was tinted using six different food dyes as tinting agent viz, Blue, Apple Green, Tomato red, Orange red, Lemon yellow and Rose pink at a concentration of 1%, 2% and 3% for all. The darkest shades of colours were obtained in 3% concentration while lightest shades in 1%. In crops of Tuberose (cv. Suvasini) and Gladiolus (cv.White Prosperity), 5% concentration of food dye with two hours of immersion showed a best colour shades. Tinting the spikes of tuberose cv. Suvasini at the optimum stage of harvest (1-2 basal florets open) where as in gladiolus cv.White Prosperity, harvest satge (1-2 basal florets show colour) was found to be optimum. (Kumar and Kameswari, 2014). In Gladiolus cv. 'White Prosperity' tinting was done with 5% food dyes for two hours at various stages of harvest (Kumar *et al.*, 2015). Successful tinting in gerbera, carnation and gladiolus was also reported (Sneha *et al.*, 2019).

References

Afshar M.K.,Moghadam Z.B., Taghizadeh Z., Bekhradi R., Montazeri A., Mokhtari P. 2015. Lavender fragrance essential oil and the quality of sleep in postpartum women. Iran Red Crescent Med J.17(4):e25880

Aziz, Zarith& Ahmad, Akil&Mohd-Setapar, Siti&Karakucuk, Alptug& Azim, Muhammad &Lokhat, David & Kamal, Mohammad & Ashraf, Ghulam. 2018. Essential Oils: Extraction Techniques, Pharmaceutical and Therapeutic Potential - A Review. Current Drug Metabolism. 19. 10.2174/1389200219666180723144850.

Bernier G, Klnet JM, and Sachs, RM. 1981b. The Physiology of Flowering, Vol. II (Boca Raton: CRC Press).

Bhasker V, Vijaya and Rao P. Venkata. 2018. Effect of Packaging Material on the Postharvest Life of Cut Rose cv. 'First Red'. International Journal of Current Microbiology and Applied Sciences, 7 (01): 1019-1024

Bhattacharjee, S.K. and De, C.L. 2005. Advanced Commercial Floriculture Aavishkar Publishers.vol.1

Bhattacharjee S K. 2006. Advances in Ornamental Horticulture, Vol. 6, Pointer Publishers Jaipur 302 003, (Rajasthan) India.

Bufler G. 1984. Ethylene-enhanced 1-aminocyclopropane-1-carboxylic acid synthase activity in ripening apples. Plant Physiol. 75:192–195.

Bufler G. 1986. Ethylene-promoted conversion of 1-aminocyclopropane-1-carboxylic acid to ethylene in peel of apple at various stages of fruit development. Plant Physiol. 80: 539–543.

Chadha, K. L. and Choudhury, B. 1989. Ornamental Horticulture in India. Indian Council of Agricultural Research. Pp. 191-197

Chamani, Esmaeil and Wagstaff, Carol. 2019. Effects of postharvest relative humidity and various re-cutting on vase life of cut rose flowers. International Journal of Postharvest Technology and Innovation. 6(1): 70-82

de Sousa D.P., de Almeida Soares Hocayen P, Andrade L.N., Andreatini R. A. 2015. Systematic Review of the anxiolytic-like effects of essential oils in animal models. Molecules.14:20(10):18620-60.

Göbel H, Schmidt G, Dworschak M, Stolze H, Heuss D. 1995. Essential plant oils and headache mechanisms. Phytomedicine. 2(2):93-102

Han SS.1992. Role of sucrose in bud development and vase life of Liatris spicata(L) Wild. Hort Science 27(11): 1198-1200

Karuppaiah P, Kumar S, Ramesh and Rajkumar M. 2006. Effect of different packages on the postharvest behaviour and shelf life of jasmine (*Jasminum sambac*). International Journal of Agricultural Science. 2(2): 447-449

Kavita. 2016. Studies on post-harvest handling of chrysanthemum (*Dendranthema grandiflora Tzvelve*) cultivars 'Solan Shringar' and 'Yellow Star'. M.Sc. Thesis. Department of Floriculture and Landscape Architecture, YSP University of Horticulture and Forestry, Solan. 61p.

Kuiper D, Ribot S, Van Reenen HS and Marrisen N.1995. The effect of sucrose on flower bud opening of 'Madelon' cut roses. Sci. Hortic. 60: 325-336.

Kumar BS, Kameswari PL, Pratap M and Rao PV. 2015. Optimization of stage of harvest of spikes of gladiolus cultivar 'White Prosperity' for tinting. Agriculture Science Digest- A Research Journal 2 : 89-94.

Kumar BS and Kameswari PL. 2014. Studies on standardization of dye concentration, stage of harvest for tinting and post-harvest keeping quality of tuberose (cv. *Suvasini*) and gladiolus (cv. *White prosperity*). MSc Thesis. Department of Floriculture and Landscape Architecture Dr. YSR Horticultural University, Hyderabad 185p.

Kumari S and Deb P. 2018. Effect of tinting on value addition of tuberose (*Polianthes tuberosa* L.) cv. Prajwal. International Journal of Bio-resource and Stress Management 9: 314-322.

Lutz JM, Hardenburg RE. 1986. The commercial storage of fruits, vegetables, and florist and nursery stocks. In: USDA agriculture handbook (no. 66) Washington, DC: USDA, P. 55-67.

Manikrao D R. 2007. Studies on possible alternatives of silver thiosulphate in the postharvest handling of Alstroemeria and Carnation cut flowers. Ph.D. Thesis. Department of Floriculture and Landscape Architecture, YSP University of Horticulture and Forestry, Solan. 104-105p.

Mayak S and AH Halevy. 1970. Cytokinin activity in rose petals and its relation to senescence. Plant Physiol.46:497–499.

Mukherjee P. 2003. Studies on post-harvest handling of Carnation. M.Sc. Thesis. Department of Floriculture and Landscape Architecture, YSP University of Horticulture and Forestry, Solan. 62p.

Nowak J and Rudnicki, R M. 1990. In : Post harvest handling and storage of cut flowers, Florist's Greens and Potted Plants, Portland, USA: Timber Press.

Pal Sayani, Ghosh Probir Kumar and Bhattacharjee Paramita. 2016. Effect of packaging on shelf-life and lutein content of marigold (*Tagetes erecta* L.) Flowers. Recent Patents on Biotechnology, 10 (1):103-120

Prashanth P and Chandrasekar R. 2010. Influence of pulsing and packaging materials on the postharvest quality of cut gerbera cv. 'Yanara'. Indian Journal of Agricultural Research, 44 (1) : 66 – 69

Rassem, H. H. A., Nour, A. H. and Yunus R. M. 2016. Techniques for Extraction of Essential Oils From Plants: A Review. Australian Journal of Basic and Applied Sciences.10(16): 117-127

Roychowdhury N, Chakrabarty (Das) S and Munsi P. 2011. Influence of packaging material, storage condition and storage duration on vase life of tuberose 'Calcutta Double'. Acta Horticulturae. 886: 359-364

Sagar Jagdale, Sharma Puja and Dhiman SR 2019. Studies on post harvest handling of loose flowers of chrysanthemum (*Dendranthema grandiflora Tzvelve*) cultivars 'Solan Shringar' and 'Surf'. Int. J. curr. Microbiol. App. Sci. 8(6):1472-1488

Sharma, Abhishek &Rajendran, Sasireka& Srivastava, Ankit & Sharma, Satyawati&Kundu, Bishwajit. 2016. Antifungal activities of selected essential oils against Fusariumoxysporum f. sp. lycopersici 1322, with emphasis on Syzygiumaromaticum essential oil. Journal of Bioscience and Bioengineering. 123(3)

Singh,B., Sellam P., Majumder J. and Rai P. 2014. Floral essential oils: importance and uses for mankind.HortFlora Research Spectrum,3(1): 7-13

Singh Gurdip and Maurya Sumitra. 2005. Antimicrobial, antifungal and insecticidal investigations on essential oils: An overview. CSIR. 4(3):179-192

Singh Kushal, Kumar Gunjeet, Saha TN and Kumar Ramesh. 2013. Post Harvest Technology of Cut Flowers. Pp. 26-28. Venus Printers and Publishers, New Delhi.

Sisodia Anjana, Kumawat Priyanka, PadhiMinakshi and Singh Anil K. 2018. Effect of Packaging Materials and Temperature on Post-harvest Life of Gladiouls cv. 'Punjab Dawn'. International Journal of Current Microbiology and Applied Sciences: 7(8): 889-893

Sneha M, Kukanoor L, Patil SR, Shiragur M, Naik MBN, Thippanna KS and Naik KR. 2019. Standardization of tinting techniques in gerbera, carnation and gladiolus. International Journal of Chemical Studies 7: 1153-1157.

Sowmeya S, Kumaresan S and Sanmugapriya L. 2017. Effect of Multi molours in tinting techniques in cut flowers (Rose and Carnation). Chemical Science Review and Letters 6 : 2250-2253.

Tanigawa T, Y Kobayashi, H Matsui and Y Sakai.1995: Effects of CO2 enrichment on growth and vase life of cut flowers of Dendranthema grandiflorum Ram at Kitamura. J Japanese Soc. Hart. Sci. 64(2) : 4I7 -424.

Van Staden J and GG Dimalla. 1980. Endogenous cytokinins in bougainvillea 'San Diego Red': I. Occurrence of cytokinin glucosides in the roots ap. Plant Physiol. 65:852–854.

Woodson WR, Park KY, Drory A, Larsen PB, and Wang H. 1992. Expression of ethylene biosynthetic pathway transcripts in senescing carnation flowers. Plant Physiol. 99:526–532.

Wu MJ, Van Doorn WG, and Reid MS. 1991. Variation in the senescence of carnation (*Dianthus caryophyllus* L.) cultivars. I. Comparison of flower life, respiration and ethylene biosynthesis. Scientia Horticulturae. 48:99–107.

29

Dehydration of Flowers for Value-Addition

Bharati Kashyap, Rajesh Bhalla, S. R. Dhiman, Puja Sharma and Sangeeta Kumari

Flowers and various parts of the plant have been used for decoration and to commemorate various religious, festive, matrimonial and social gathering events in day to day life. The main drawback of fresh flowers and foliage is being highly perishable and delicate which cannot retain their beauty for longer durations. Moreover fresh flowers are not available round the year which can be overcome by the dehydration and drying techniques. Since ancient times, in Egyptian pyramids, fragrant dried flowers and herb garlands were encased along with mummified bodies. Japanese people created "Oshibana", the "preserved flower art" of enduring designs to retain the exquisite beauty of the living plants centuries ago.

Later on, this craft making technique spread from Japanese to Victorian England because of their beautiful long lasting appearance. During middle ages monks dried flowers, foliage, and herbs for use in decorative motifs or for colouring their hand-printed books.

Dried flower arrangements have been popular in Europe for centuries, and as early as 1700 AD, colonial Americans used dried flowers to brighten their homes, especially during the dark winter months.

Earlier botanists practiced dried flowers in the form of herbarium for the purpose of identification of various plant species. Though drying of flower was well known even in the past but for the first time the flowers were dried commercially in Germany (Sankari and Anand, 2014).

In present scenario, dry flowers are gaining popularity as they are inexpensive, long-lasting and eco- friendly products that are available throughout the year. Dried and preserved ornamental products offer wide range of unique qualities which includes novelty, longevity, aesthetic values, flexibility and round the year availability (Joyce, 1996). Drying of plant material leads to reduced microbial activity in them so they can be stored in moisture free atmosphere and becomes free from bondage of seasons (Bhutani, 1995). Non-availability of fresh flowers and foliage round the year, while the beauty and value of dried flowers can be kept for years and they can survive cold of winter and heat of summer (Bhalla *et al.*, 2006).

Dry Flower Industry

Dry ornamentals include dried flowers, shoots, seeds, barks, fruits, nuts, cones *etc.* However in modern days any plant material with aesthetic appearance can be used for making a large number of value- added products like pot-pourries, incenses, fragrant candles, greeting cards, dry flower sticks, arrangements, bouquets, wall-pictures, floral designs, pomanders, floral wreath, paper weights, dolls, mirrors, handmade papers, hangers, soap cases and other gift items. Due to the availability of diversified agroclimatic conditions which offers wide varieties of plant materials in India, the country is a leader in the world in dry flower export. Easy and wide availability of raw materials from forests and cheap manpower, intensive craft making have greatly contributed to flourishing of the dry flower industry in India. This business can be a source of income generation to the unemployed youth, women and even the handicapped persons. Indians are famous for their creativity and craftsmanship. People can grow/ collect the material and prepare the handicraft/ floral craft in their pastime. Export of dried flowers and plants from India is more than Rs 100 crore per year. The industry exports 500 varieties of dried ornamentals to different countries. Dry flower industry is a labor-intensive and can generate ample job opportunities to thousands of unemployed men and women.

Indian Flower Industry

In India this industry is about 40 years old and was introduced by British in Calcutta for its closeness to North-East regions where plenty of exotic and diverse blooms were available in surroundings (Bhattacharjee and De, 2003).

The demand for dry flowers has increased manifold in the last decades and is increasing at an impressive rate of 8-10% annually (Singh, 2009). Indian flower industry comprises of commercial production of cut flowers, loose flowers, cut greens, seeds, bulbs and landscape plants, their marketing and most importantly production of value added products from them. Indian export of flowers is composed of 71% dry flowers exported mainly to USA, Japan, Australia, Russia and Europe (De *et al.*, 2016). USA, Singapore, UAE, Malaysia and Germany are the major exporters from India in terms of dried, dyed, bleached or otherwise prepared cut flowers and flower buds for bouquets or for decorative purposes (Anonymous, 2019).

In India, Tuticorin has three major players like Fauna International, Ramesh Flowers and W. Hogewoning while Minex in Calcutta are the largest player in dry flower production. However in global market, Botanical Resources of Australia, Sumitomo Chemical, Fauna International, KAPI, Valent, Fauna International, McLaughlin Gormley King Company (MGK), Agropharm, Beaphar UK, Novartis Animal Health, Sumitomo Chemical ,W. Neudorff *etc* are the major leading producers of value added products.

The top ten cultivated dried flower genera as "Everlasting" are *Helichrysum* sp., *Helipterum* sp., *Limonium* sp., *Nigella* sp., *Gypsophila* sp., *Delphinium* sp., *Amaranthus* sp., *Papaver* sp., *Carthamus* sp., *Rosa* sp. (Anonymous, 1989) and can be cultivated easily in India.

Apart from these, the valleys and mountains of India harbour a large number of native plants comprising of cones, ferns and grasses which are very unique and hold a great potential to be utilized for dry flower production. Above all, the agricultural waste of crops like maize, sesamum, pearl millet, sorghum, linseed, wheat etc can be utilized for making dry flower products. Although 'Dried flowers' is a lucrative business, yet lack of awareness could be another drawback in its expansion. The raw products from the wild such as cones are still being harvested through illegal means and sent to dry flower markets/ collection centres at very low rates from where the value added products are exported to bigger markets at very higher rates.

NBRI Lucknow, UHF Solan, HPKVV Palampur, ANGRAU, Hyderabad, KAU Vellanikkara are some of the institutes where work on dehydration of flowers has been taken up.

The technology for dehydration and value-addition of a large number of cultivated crops, native flora has already been standardized in the Department of Floriculture and Landscape Architecture, Dr YSP UHF, Nauni, Solan and a laboratory for demonstration of technology and the value-added products has been established which is acting as a role model for the students, distinguished guests and the farmers. Many NGO's, SHG's and farmers have shown interest in adopting the technology. A small cottage industry/ entrepreneurship can be started in the field of dried flowers with the minimum amount of rupees 10,000/ only. The dried flower products fetch good returns in the market and a single person can earn from rupees 200 to 1000/ day. Moreover, the dried flowers do not require very large area for cultivation. Farmers with limited/small landholdings can adopt this technology. The dried flowers do not have the problem of storage unlike fresh flowers which are perishable commodities. APEDA, GOI has also identified dry flowers as a potential item for export. With the sustainable utilization of all family members, farmers can easily generate 1.5 times to 5 times profit by the value addition of dry flowers and the rate of different dry flower products varies from Rs. 15 to Rs. 2000 and similarly the B:C ratio varied from 1.5 (for paper weight making) to 5.0 (for making finely designed scenery, Singh, 2018).

Techniques of Flowers Drying

In general, preferably selected species should be naturally dried along with retention of natural colour for long period of time. Deduction of moisture while retaining the unique shape, colour and texture of plant material is the goal of drying techniques. The flower dehydration technique involves reducing moisture content of flowers to a point at where microbial activity and the ageing effect come to a standstill and biochemical changes are minimized thereby cell structure, pigment level and flower shape are maintained as such. The stage of harvesting for different flower varies according to the species and the form of flower desired. The quality and appearance of dried flowers and other ornamental plant parts is greatly influenced by both the method and techniques being followed. Following table describes the optimum stage of harvesting of flowers for drying as reported by various workers (Lourdusamy *et al.*, 2002), (Jauron, 2003), (Safeena *et al.*, 2006).

Crops	Optimum harvesting stage
Helichrysum bracteatum	Harvest when outermost 1-2 layers of ray florets are perpendicular to the centre while the center is still tightly closed.
Gomphrena globosa	Both half bloom and full bloom stages are ideal.
Celosia spp	When flowers are at peak stage of bloom.
Limonium sinuatum	Harvest the flower stalk when all the cup like calyces have opened. The cream coloured petals wither and drop during the drying process, whereas bright colour calyces remain.
Gypsophila paniculata	Flowers are picked up when majority of flowers are fully opened.
Nigella spp	Harvest when the pods are firm and papery to touch.
Mollucella laevis	When the tiny flowers inside the green persistent calyx appear and becomes slightly firm.
Lagurus ovatus	When colour changes from greenish to whitish and pollens are visible
Bromus spp	When colour of spikes become green and tip of ear/inflorescence not turned prickly.
Briza minor	Colour changes from greenish to whitish
Zinnia elegans	Full bloom stage
Rosa spp	Half bloom stage

There are different methods of drying of flowers which can be broadly divided into two groups

A) Old Methods of Drying

Natural Drying

It involves the process where plant parts are allowed to dry on the plant itself naturally. In Himalayan regions, naturally dried plant parts, with beautiful fruiting shoots are *Aegle marmelos*, *Bambusa* spp, *Bauhinia retusa*, *Cassia fistula, Clematis grata, Rosa moschata, Pinus roxburghii, Picea smithiana, Sapindus mukorossii etc* while *Abrus precatorius, Aesculus indica* and *Sapium sebiferum* have beautiful seeds (Desh Raj and Gupta, 2002).

Press Drying

One of the oldest methods of drying and was first reported in 1820. In press drying the selected flowers and foliages are placed between the folds of newspaper sheets or blotting papers giving some space among flowers. These sheets are kept on one above another and corrugated boards of the same size are placed in between the folded sheets so as to allow water vapour to escape. Press drying in hot air oven at 35-39°C for 48 hours found optimum for pansy, 24 hours for drying the leaves of *Grevillea robusta Thuja occidentalis*, adiantum, nephrolepis and flowers of hibiscus, *Cassia biflora* calliandra and marigold (Kher and Bhutani, 1979) while in case of rose, carnation and helichrysum time required reported to be vary from 120, 132 and 2 hours, respectively (Gill *et al.*, 2002). Press drying in shade or oven also found suitable for maintaining two dimensional view of *Ixora* spp, *Mussaenda* spp, *Meyenia erecta*, and petals of *Bauhinia pupurea* flowers. *Viola tricolor*, *Pelargonium* spp, calendula and buttercup with soft petal stuff were amenable to press drying and highly suitable for two dimensional products like portraits, greeting cards and sceneries (Nazki *et al.*, 2012).

Vertical Hang Drying

It is the ordinary and the cheapest method where desired plant material is hung upside down with the help of rope or wire and tied to a stand. Entire process of air drying depends upon the relative humidity, air velocity, pressure, moisture in plant material and type and shape of the flowers. However the main drawback is its weather dependence and petal/foliage shrinkage which may subsequently lower down the aesthetic value of plant material.

Nigella, gypsophilla, *Achania malvaviscus*, santolina and lavender spikes can be dried by inverted vertical hanging inside room conditions away from direct sunlight (Nazki *et al.*, 2012).

Embedding Drying

It is the most efficient method of drying of plant materials in which different desiccants like silica gel, boric acid, river bed sand, saw dust, alum powder, corn granules, separately or in mixture can be used. In order to avoid shrinkage and other morphological changes which may occur due to air drying, the flowers and foliage can be embedded in drying material like; sand and silica gel (Datta, 1999). Embedded plant material can be dried in microwave oven, hot air oven, solar drier, sun drying and vacuum drying or under room conditions. For embedding drying the qualities of desiccant is very important. The desiccant should normally have the ideal size of 0.02-0.2 mm. It should be heavy to keep the plant material in its original shape, neither sticky nor costly.

Sometimes borax being hygroscopic bleaches flower petals if embedded for a longer time (Datta, 1997, Sangeeta *et al.*, 2017a). Silica gel (60-120 mesh) is the best absorbent for removing moisture from flower and foliages followed by boric acid (granules) (Desh Raj and Gupta, 2003). Flowers like rose, carnation, zinnia, chinchirenchi, gomphrena, chrysanthemum, larkspur, geranium etc. are dried well in silica gel (Smith, 1993; Chandrasekhar *et al.*, 2008; Singh *et al.*, 2004; Sharma *et al.*, 2016, Sangeeta *et al.*, 2017b).

Plant material embedded in deep containers accommodates without disturbing its shape and forms in crops like bougainvillea, candytuft, chrysanthemum, dahlia, gerbera, marigold and roses *etc.* (Bhutani, 1990 and 1995). Drying was faster with silica gel and borax in comparison to sand due to the strong hygroscopic nature of silica gel and borax which lead to rapid removal of moisture from flowers (Singh *et al.*, 2003). In case of rose, carnation, fern and silver-oak embedding in silica gel produced good results while combination of silica gel and sand combination was satisfactory (Gill *et al.*, 2002).

B) Modern methods of drying

Hot Air Oven Drying

Flowers and foliages embedded in desiccants/medium are kept in hot air oven at controlled temperature (40-50^{0}C) for appropriate time which vary from plant to plant. Standardization of drying time and temperature of some of ornamental crops has been done at NBRI, Lucknow.

The effect of embedding media, temperature and duration on drying of cut flowers of Rose *cv.*'First Red' in hot air oven revealed maximum moisture loss on embedding in silica gel and dried at 45°C for 72 hours while maximum anthocyanin content recorded when embedded in borax and dried at 35^{0}C for 36 hours (Dilta *et al.* 2014).

Colour and structure of floral parts exhibits no change, whereas total chlorophyll content significantly reduced in the dehydrated parts as compared to the control without any change in length and diameter of chrysanthemum flowers when dried in hot air oven dried at 40°C for 30-35 hours in sand as embedding media (Pandya *et al.*, 2001). *Peonia suffruticosa* and *Zinnia elegans* dried in hot air oven for 30 min. at 50°C and 40 min. at 50°C, respectively found as the suitable method of dehydration (Nazki *et al.*, 2012). The effect of dehydration treatments on overall parameters with maximum score in acroclinum flowers obtained on hot air oven embedding in silica gel as compared to others as suggested by Katoch *et al.* (2010).

Dehydration of *Rosa hybrida cvs* 'Lambada', 'Skyline', 'Ravel' and 'First Red' in hot air oven at 40°C with silica gel were more acceptable for colour, appearance and texture (Safeena *et al.*, 2006). In terms of quality parameters like colour, texture, brittleness and shape retention, the silica gel embedded flowers of *Gomphrena globosa* 'Magenta' and *Gomphrena globosa* 'White' recorded maximum visual when dried for 24 hours at 50°C in hot air oven acceptance. Longer the drying duration less is the capacity of the flowers to retain its shape (Sangeeta *et al.*, 2017a)

Microwave Oven Drying

It is quickest and efficient technique of drying of flowers. Non- metallic containers where various plant parts are embedded in different desiccants are kept in microwave oven for 2-5 minutes. This technique is based on the principle of liberation of moisture by agitating water molecules present in organic substances with the help of electronically produced heat (Bhutani, 1990).

Chrysanthemum flowers embedded in silica gel and dried in microwave oven gives the best results in terms of colour and shape retention (Bhalla *et al.*, 2006). Rose buds dried in microwave oven for 4 min. exhibited good colour and shape retention against 3 and 5 min. treatments (Dhatt *et al.*, 2007). Similarly flowers of *Calendula officinalis* cultivars ' Red Orange' and 'Yellow' embedded in silica gel and dried in microwave oven (30 sec) appeared best in terms of colour, shape and texture in both the cultivars (Biswas and Saha, 2010).

Chrysanthemum *cv.* 'Button Type Local' microwave oven dried with silica gel as medium accepted best in terms of texture and shape while for display quality flowers dried at 80% micropower level for 120 seconds (Arvinda and Jayanthi, 2004). Zinnia dried for 2 min. in microwave oven for setting time of 30 min. exhibited best results in terms of colour, shape and size (Misra *et al.*, 2009).

Nazki *et al.* (2012) evaluated the different dehydration techniques for product diversification in floral craft in some plant genera of Kashmir valley revealed microwave time of 90 sec, 240 sec, 60 sec, 30 sec, 40 sec, and 60 sec for *Rosa hybrida, Peonia suffruticosa, Dendranthemà grandiflora, Hypericum* spp, *Narcissus* spp *and Zinnia elegans*, respectively for silica gel embedded dehydration with respect to ornamental or aesthetic values.

Maximum moisture loss and size reduction observed in flowers embedded in silica gel and dried in microwave oven for 90 seconds durations while maximum carotene content in those flowers which were dried in silica gel for 30 seconds durations (Bhalla *et al.*, 2006).

In chrysanthemum (*Dendranthema grandiflora* Tzvelev.) *cv.* 'Button Type Local', microwave drying of flowers with silica gel as embedding medium recorded maximum score in terms of texture and shape while colour and overall acceptability in oven drying with white sand as embedding medium (Arvinda and Jayanthi, 2004).

Vethamoni and Nanthakumar, (2003) observed the higher percentage of moisture loss in oven drying method with a mean value of 41.4 % as compared to other methods of drying. Furthermore the percentage of moisture loss recorded greater in hydrangea flowers under oven drying methods followed by sun drying. Since the flower petals of *Hydrangea hortensis* are soft, the rate of moisture loss is greater irrespective of the situation conditions. Embedding of *Gomphrena globosa* 'Magenta' in silica gel in microwave oven for 3 min. and for 72 hours as setting duration reported best method which scored the maximum for quality parameters like colour, texture, brittleness and shape retention whereas, in case of *Gomphrena globosa* 'White' the highest score obtained in silica gel and dried for 3 min. with 24 hours setting duration (Sangeeta *et al.*, 2017b).

Maximum quality score in terms of colour, texture and brittleness to the flowers of chincherinchee embedded in silica gel and dried at 50 °C for 24 hours duration (Sharma *et al.*, 2016).

Freeze Drying

It was introduced by William Hyde Wollaston in London. Freeze drying relies on the Principle of Sublimation, whereby flowers are held under conditions of partial vacuum and low temperature will evaporate on heating without going through a liquid phase (Dubois and Joyce, 1989a). Rapid dehydration, particularly at low or ambient temperature, tends to result in better- coloured products (Chen *et al.*, 2000).

The main advantage of freeze drying is that, the products are fresh like original. During the process, freezing the flowers at (-10 °C) for at least 12 hours a vacuum pump slowly pulls the water out of the flowers as the vapour condenses as ice in another chamber. Because of this process the shape and natural colour of the flower is maintained (Murugan *et al.*, 2007).

Several cultivars of *Dianthus caryophyllus* are successfully cryo-dried and remained naturalistic in appearance after being placed in freeze drier (-20 ^{0}C) for 7 days (Bhattacharjee and De, 2003).

Rosa hybrida cvs 'Tineke', 'Golden Gate', 'Saphir', 'Roulette', and 'Rote Rose' were freeze dried for 14 days and in terms of shape and colour, the shrinkage observed in freeze drying but colour remain similar to a fresh rose (Sohn *et al.*, 2003).

In *Rosa hybrida cv*. 'First Red' maximum moisture loss and maximum sugar content noticed in flowers dried in freeze drier (Bahera, 2009).

Glycerine Drying

Being an osmotic reagent, glycerol is also used in drying process. Camelia, Ivy and Eucalyptus are good for drying with glycols (Deepthi and Santhosh, 2008). Cut stem ends are dipped or soaked in a mixture of glycerine and water where the time period of treatment varies from plant species to species. The average time taken for glycerine uptake was 2-3 weeks for foliages like magnolia and ligustrum. Genera like hedera, *Mahonia aquifolia* , magnolia, cherry laurel (*Prunus laurocerasus*) and *Sophora mollis* found suitable for glycerine drying for 2-3 weeks (Nazki *et al.,* 2012).

Glycerol and microwave preservation of annual statice maintained the flexibility, without the greasy feeling often involved in glycerol preservation (Paparozzi and McCallister, 1988).

Hiroaki *et al.* (2010) developed a protocol for the preparation of preserved flowers retaining natural color and texture of 'Moondust Velvet Blue' carnation. The three-step process consists of soaking flowers in ethyl alcohol, then soaking them in polypropylene glycol, followed by a rinse with ethyl alcohol. Some kinds of flowers processed in this manner retained their natural color and texture for at least 6 months.

Leaves preserved through glycerine are less prone to shattering and mechanical damage and are more natural in appearance. The leaves of *Peltophorum ferrugineum* and *Buxus sempervirens* treated in 40% glycerine solution by full dip method articulated minimum change in leaf area, minimum percentage of leachate, maximum score on textural quality, shape retention, brittleness and overall acceptance (Yadav *et al.*, 2018).

Dyeing

The natural colour of the plant materials may be intensified or artificial colour may be introduced by dyeing or colouring, which helps in providing the uniform colour and appearance and thus increases the saleability. Dyes are the molecules which absorb and reflect light at specific wavelengths to offer human eyes the sense of colour. Dyes can be obtained from natural substances like plants, animals or minerals or may be synthetically prepared such as Acid dyes, Basic dyes, Direct dyes, Azoic dyes, Disperse dye, Sulfur dye, Fiber reactive dye, Oxidation dyes, Mordant dyes, Developed dyes, Vat dyes, Pigment brighteners, Optical/ flouorescent brightener and solvent dyes (Kulkarni *et al.*, 2011).

In comparison to direct and acid dye colour, vat colour group of dyes scored the highest values in terms of visual aesthetic qualities and also exhibited low level of colour fading on storage (Lourdusamy *et al.,* 2002).

Dyeing of *Lagurus ovatus* where bio-colours, turmeric and lilium pollen are satisfactory, whereas henna powder, gulmehndi (wild henna), coffee, butea, punica, beet root are very poor in performance, thus cannot be recommended for dyeing Kashyap *et al.* (2016).

Fabric dyes such as Yellow fabric are suitable based on different quality parameters on dyeing of *Gomphrena globosa* L. 'White'. Indicator dyes such as Crystal Violet and Eosin Yellow gives good results with the addition of mordants such as aluminum sulphate on addition to indicator dyes (Sangeeta *et al.*, 2017c). This proved that dyeing can be improved by adding mordants to improve the contact between the dye bath solution and the plant materials as opined earlier by Joyce (1996) and Agarwal *et al.* (2007). Fastness properties viz. rubbing, washing and light fastness considerably improve involving mordants which could be attributed to affinity of mordant with dye and the plant tissue Visalakshi (2013), Samanta and Agarwal (2009).

Bleaching

In bleaching both oxidative or reductive bleaches can be used on the plant material; while former tends to break down coloured compounds, the latter tends to modify them to colourless compound (Dubois and Joyce, 1988). Oxidative (Hypochlorite, chlorite and peroxides) and Reductive bleaching chemicals (sulphite and borohydride) are used for bleaching flowers and foliages (Arulmurugan, 2007). Sodium chlorite is an excellent bleaching agent because it is relatively selective for lignin without damaging the fibre.

The L value and whiteness index of botanicals for super whiteness fixed as White Group Standard-NN-155-D as per RHS colour chart indicates maximum whiteness index in *Sesamum indicum* followed by *Luffa acutangula, Gossypium hirusutum* and *Aegle marmelos* when bleached under 30% sodium chlorite+ 5%HCl + 30% of first dip, 40% Hydrogen Peroxide Visalakshi, (2013).

Storing and Packaging

Dried products are made up of cellulose material of plant origin and are also hygroscopic and thus absorb moisture readily, thus problem of pest infestation will occur. So they should be treated with suitable biocide and packed in a water proof container (Murugan *et al*., 2007).

Dehydrated plant parts may be arranged aesthetically and covered with plastic or transparent glass to protect them from atmospheric humidity, wind and dust. (Datta, 1997; Bhattacharjee and De, 2003).

Pot-pourris are normally designed in a glass bowl or a ceramic jar or stored in colourful satin or muslin sachets, while other arrangements are designed in beautifully coloured or shaped pots (Datta, 1999).

Dried plant materials absorb atmospheric moisture and lose their shape, therefore, should be stored immediately in moisture proof containers. Different containers like glass desiccators, tin boxes, cartons, wrapped with plastic sheets or wax paper and herbarium vasculums fitted with cork sheet are used for storage of dried ornamental plant material. Small quantity of silica gel also placed at the bottom to absorb moisture.

Store separately different dried ornamental plant material and fix proper label outside with complete description of the product kept inside.

Bhalla *et al.* (2006) carried the experiment to standardize the suitable drying temperature, media, techniques and storage of flowers of chrysanthemum (*Dendranthema grandiflora* Tzvelev.) *cv.*'Nanako'. Keeping quality of dried flowers was assessed under open and covered conditions from 0 to 180 days.

Hot air oven dried flowers when embedded in silica gel and kept at a temperature of 60°C for 24 hours and microwave oven dried flowers after embedding in silica gel and kept for 90 sec exhibited maximum presentability score under covered conditions.

Maximum score (7.93) and (7.40) out of a total of 10 allotted to the two strains of *Gomphrena globosa* L. when dried in microwave oven thereafter covering with newspaper and observed after 120 days of storage duration followed by cellophane sheets covering compared to open conditions. Hence, it concluded that for aesthetic qualities and longer presentability, flowers should be dried in microwave after embedding in silica gel for 3 minutes with 72 hours setting durations (Sangeeta, 2015).

Different Value Added Products from Dried Flowers

1) **Pot pourries** Pot pourri is a sweet smelling dried floral arrangement which is a mixture of dried flowers, petals, fruit, pod, leaves, spices, seeds, root and distilled essential oil which are filled in pillows or transparent sachets. It consists of an attractive mixture of dried plant materials of different shapes, size, texture and color which makes it as a decorative item as well. The important dried ornamental plants used for making pot-pourries are roses, lilacs, lavender, pinks, hyacinths, lilies, violets, wall flowers, marigold, salvia, helichrysum, gomphrena, craspedia (bora flowers), acroclinum, rosemary and geranium etc. We can use also use fruits of koelreuteria, jacaranda, nicandra, cotton, walnut, casuarina, birch, almond, rose, acer, luffa, lotus, arjun, michelia, coix, wood roses and maize cob & citrus slices in making pot-pourries.

2) **Dry flower sticks:** Beautiful dry flower sticks can be made from dried flowers of helichrysum, statice, acroclinum, craspedia, lagurus, briza, bromus, gomphrena, fruits of tithonia, bixa, lagerstroemia, jacaranda, moti dana(*Coix lacyrma jobi*) bell cups, lotus, acacia, luffa, michelia, badam, pine cones and spongue mushroom. Elegant sticks can also be prepared by shola sheath (*Aeschynomene aspera),* maize sheath, coconut husk, cyperus, bamboo (bamboo flowers, bamboo sheath flowers, bamboo leaf triple, bamboo mat, curly coils and ting tings etc) and palms (palm spears, coco five fingers, dried palm leaves, palm lavender, palm leaf weaving, palm male chakra etc). Some of the natural materials from natures trove with long sturdy stems can directly be

used as dry flower sticks such as millet, pearl millet, statice, wild tobacco, flax, hydrangea, leek, typha, sesamum etc.

3) **Pomander:** Pomander is derived from a French word, 'Pomme d ambre' meaning an apple of amber, a ball made of perfume. In general, we make a ball from floral foam which can be covered with gomphrena, tithonia, acroclinum, helichrysum, rose hips, nicandra berries, samara of acer etc. Usually pomander is carried in a vase and may be hung down from the collar through chain or belt. Though, pomander is a decoration product but can be used to modify bad smells. We can make it perforated and add perfume sachets in it.

4) **Collages:** Collage is derived from French word, 'Coller' and literally means glue. It is an artistic floral handiwork containing pieces of coloured papers, texts, photographs, natural objects, in addition to dried flowers, leaves, roots, bark, etc. It is a typical art of novelty.

5) **Greeting cards:** Dried foliage and flowers pressed in one dimension are used to prepare greeting cards to present to friends and relatives on some social occasions. To prepare a dry flower card skeletonized leaves, pressed flowers and leaves, small thin sticks and twigs, grasses, fern and paper etc. are required.

6) **Flower arrangement:** These are used to decorate tables, shelves, corners of offices and home. To make a flower arrangement we need a vase/ bamboo basket, floral foam, dried flowers, leaves, grasses, cones, fern, twigs and pods. Helichrysum, acroclinum, gomphrena, molucella, rose, statice, chrysanthemum, achillea, ornithogalum, cocks comb, flax, hydrangea, anaphalis (buke flowers), celosia, lagerstroemia, millets, rice panicles, celastrus and abrus fruit and grasses like lagurus, briza, bromus, setaria, saccharum and leaves of cycas, peepal, magnolia,fern and oaks etc.

7) **Flower bouquets:** A flower bouquet is a collection of flowers in a creative arrangement and is normally used to decorate homes and present to greet the dignitaries. Dry flowers and grasses can be used naturally or can be dyed with various colours. Statice, hydrangea, ornithogalum, gypsophila, roses, helichrysum, anaphalis, pearl millet and grasses like lagurus, bromus, briza can be used to make a good flower bouquet.

8) **Wall pictures:** To make wall pictures we need dry flowers, fern, leaves, grass and small thin twigs. We can also frame the wall picture to maintain their beauty.

9) **Dry flower wreath** It is an assortment of flowers, foliage, twigs, grasses, fruits, cones and various other construction materials which resemble to a ring. Dry flower wreath is used to decorate doors, empty walls and for Christmas celebrations. However, fresh flower wreath is used only for paying homage to the departed soul.

10) **Dry flower ornaments:** Different ornaments like bangles, ear rings, garlands and pendants can be prepared from dry flowers. Flower bearing crops like helichrysum, gomphrena, *Abrus precatorius*, *Coix lacyrma* and rose hips can be used for this purpose.

Apart from these, products like wall hangers (gomphrena, anaphalis, millet, cones, helichrysum), dolls (maize sheath, gomphrena, ornithogalum), birds (cones), rakhi (lagurus, briza, gomphrena), mirrors (gomphrena, helichrysum), garlands (gomphrena, rose hips) and bandhanvaar (rose hips, lagurus) can also be prepared from dry flowers.

Conclusions

Dry flower industry is the major component of Indian floriculture. Owing to the availability of large variety of flora in nature which can be properly utilized from waste to useful value added products. These products are inexpensive, imperishable and eco- friendly so are gaining popularity among consumers. There are different methods of drying the flower crops. Microwave oven drying using silica gel is considered to be efficient, fastest and high reliable method for producing dried flowers. However on small scale business, air drying and press drying is also efficient method. Freeze drying is used for drying of high value crops like roses, carnation *etc.* which is helpful in preserving the natural colour. Value –addition in dried ornamentals adds in the quality and variation of products and thus increases the saleability which leads to the generation of employment to rural people particularly women.

References

Agarwal, R, Pruthi N and Singh S J S. 2007. Effect of mordant on printing with marigold flower dye. Natural Product Radiance 6(4): 306-309

Anonymous. 1989. Drying and preserving flowers and plant material for decorative use.

Anonymous. 2019. A report on dry flower industry. APEDA, New Delhi.

Arulmurugan. 2007. Bleaching of dry flowers.Website: http://www.agcom.purdue.edu/agcom/pubs.menu.html

Arvinda, K and Jayanthi R. 2004. Standardization of drying techniques of chrysanthemum (*Dendranthema grandiflora Tzvelev* cv. 'Button Type Local') flowers. Journal of Ornamental Horticulture 7(3-4): 370-375

Bahera, T B. 2009. Standardization of drying techniques in rose (*Rosa hybrida* L.). M.Sc. Dr. Y S Parmar University of Horticulture and Forestry, Nauni, Solan, Himachal Pradesh.

Bhalla, R, Moona, Dhiman S R and Thakur K S. 2006. Standardization of drying techniques of chrysanthemum (*Dendranthema grandiflora Tzevlev.*). Journal of Ornamental Horticulture 9(3): 159-163

Bhattacharjee, S K and De L C. 2003. Dried flowers and plant parts. In: Advanced Commercial floriculture. Avishkar Publishers & Distributors, Jaipur, India, pp. 162-171

Bhutani, J C 1995. Drying of flowers and floral crafts. In: Advances in Horticulture – Ornamental Horticulture. Vol. 12. Chadha K L and Bhattacharjee S K (eds). Malhotra Publishing House, New Delhi, India, pp. 1053-1058

Bhutani, J C 1990. Capturing nature, a way with flower "Everlasting". Indian Horticulture 34(4): 15-19

Biswas, S and Saha S. 2010. Standardization of drying techniques of calendula (*Calendula officinalis* L.). Journal of Ornamental Horticulture 13(4): 301-305

Chandrasekhar, R, Nirmala A, Padma M and Rajkumar M. 2008. Standardization of drying techniques of carnation (*Dianthus caryophyllus* L.). Journal of Ornamental Horticulture 11(3) :168-172

Chen W, Gast K L B and Smithey S. 2000. The effect of different freeze drying processes on the moisture content, colour and physical strength of roses and carnations. Scientia Horticulturae. 84(3/4) :321-332

Datta, S K. 1997. Dehydration of flowers and foliage and floral craft. NBRI Bulletin No. 3, EBIS, NBRI, Lucknow. 20p.

Datta, S K. 1999. Dehydrated flowers and foliage and floral crafts. In: Floriculture and Landscaping. Bose T K, Maitti R G, Dua R S and Das P. (eds). Naya Prokash, Calcutta, India, pp. 696-703

De, L.C., Rai, W., Sumanthapa, and Singh, D.R. 2016. Drying technologies of commercial flowers- an overview. Int. J. Res. Appl. Nat. Social Sci. 4(3): 111-120.

Deepthi Singh and Santhosh Kumar 2008. Dry flowers add natural splendor indoors. Floriculture Today 13:42-48

Desh Raj and Gupta P K. 2003. Standardizing dehydration technology for ornamental plant parts of shrubs from mid-hills of Himachal Pradesh. Journal of Ornamental Horticulture 6(4): 357-361

Desh, Raj and Gupta P. 2002. Naturally dried ornamentals native to outer himalayas. In: National Symposium on Indian Floriculture in the New Millenium. Bengalore, 25-27 February.

Dilta, B S, Bahera T B, Gupta Y C, Bhalla R and Sharma B P. 2014. Effect of embedding media, temperature and duration on hot air oven drying of rose (*Rosa hybrid* L.) cv. 'First Red'. Indian Journal of Applied Reseach 4(1): 67-73

Dubois, P and Joyce D C. 1988. Bleaching plant foliage. Farmnote no. 87/88. Western Australia Department of Agriculture. 4p.

Dubois, P and Joyce D C. 1989a. Drying cut flowers and foliage. Farmnote no. 10/89. Western Australia Department of Agriculture. 1p.

Gill, S S, Bakshi R and Arora S. 2002. Standardization of drying methods for certain cut flowers. Proceeding of National Symposium on Indian Floriculture in the New Millenium. Lal Bagh, Banglore. pp. 357-358

Hiroaki, I., Takahiro H, Masaki H, Katsuro M, Saori N, Youichi M and Susumu Y. 2010. A Protocol for Preparing Preserved Flowers with Natural Color and Texture. Horticulture Technology 20(2): 445-448

Jauron, R. 2003. Harvesting of everlasting flowers for drying. Iowa State University Extension and Outreach, Iowa State University of Science and Technology, USA.

Joyce, D C. 1996. Dried and preserved ornamental plant material: not new, but often overlooked and underrated. Proceeding of the Third International Symposium on New Floricultural Crops. Acta Horticulturae 454: 133-145

Kashyap B, Gupta Y C, Gupta R, Thakur P and Sharma R. 2016. Studies on dyeing and value-addition of ornamental grass Lagurus ovatus. Journal of Hill Agriculture 7(1): 28- 31.

Katoch, A, Kishtwaria J, Deshraj, Rana A and Dadwal M. 2010. Study on different methods of dehydration of acroclinum flower. Journal of Ornamental Horticulture 13(3): 223-227

Kher M A and Bhutani J C. 1979. Dehydration of flower and foliage. Extension Bulletin No. 2 NBRI, Luckhnow. 20p

Kulkarni, S.S.,Gokhale A.V.,Bodake, U.M. and Pathade, G.R. (2011): Cotton dyeing with natural dye extracted from pomegranate (*Punica granatum*) Peel. UJERT (3):135-139.

Lourdusamy, D K, Vadivel E and Manavalan R S A. 2002. Studies on critical stage of harvest of some dry flowers for dry flower production. South Indian Horticulture, 51(1-6): 241-243

Misra, P, Banerji B K and Dwivedi A K. 2009. Dehydration of zinnia (*Zinnia hybrid Benth.*) in an oven for value addition through embedding technique. Journal of Ornamental Horticulture 12(3): 222-224

Murugan, P A, Thiyagarajan G and Kumar R. 2007. Dry flower technology.

Nazki, T, Imtiyaz, Lone N H, Qadri Z A, Nelofer and Rather Z A. 2012. Evaluation of dehydration techniques for product diversification in floral craft in some genera of Kashmir valley. Journal of Ornamental horticulture 15(3-4): 259-265

Pandya, H P, Saxena O P, Nell T A and Clark D G. 2001. Preservation of chrysanthemum by drying. Acta Horticulturae 543: 367-369

Paparozzi, E T and McCallister D E. 1988. Glycerol and microwave preservation of annual statice (*Limonium sinuatum* Mill.). Scientia Horticulturae, 34(3-4): 293-299

Safeena, S A, Patil V S and Naik B H. 2006. Standardization of stage of harvest for better quality of dry flowers of rose. Journal of Ornamental horticulture, 9(3): 224-226

Samanta, A K and Agarwal P. 2009. Application of natural dyes on textiles. Indian Journal of Fibre and Textile Research, 34: 384-399

Sangeeta, K, Kashyap B, Gupta YC, Gupta R and Thakur P. 2017a. Studies on hot air oven drying of Gomphrena globosa L. International Journal of Farm Sciences, 7(1): 1-5

Sangeeta, K. 2015. Studies on drying and dyeing of Gomphrena globosa L. for value addition. M.Sc. Thesis, Dr YSP, UHF, Nauni, Solan, HP. pp.1-77

Sangeeta, Kumari, Kashyap B and Gupta YC. 2017b. Studies on microwave oven drying of Gomphrena globosa L. 'Magenta' and 'White'. Indian J. Hort., 74(1): 146-149

Sangeeta, Kumari, Kashyap B and Gupta YC. 2017c. Studies on dyeing of Gomphrena globosa L. for Value Addition. Indian Horticulture Journal; 7(1): 37-41

Sankari, A and Anand M. 2014. Process of making–waste into wealth–dry flower technology. Asian J. Hort., 9(2): 466-483.

Sharma, G, Kashyap B, Sharma BP, Gupta YC and Gupta R. 2016. Studies on hot air oven drying of chincherinchee (*Ornithogalum thyrsoides* Jacq.). Journal of Hill Agriculture 7(1): 52 – 55

Singh, A, Dhaduk B A and Shah R R. 2004. Effect of different temperature and embedding media on flower dehydration of zinnia (*Zinnia linearis* Benth.). Indian Journal of Ornamental Horticulture 61(3): 249-252

Singh, A., Dhaduk B K and Shah R R. 2003. Effect of dehydration on postharvest life and quality of zinnia flowers. Journal of Ornamental Horticulture 6(2): 141-142

Singh, D. 2018. Comparative drying method study in different flowers with respect to quality dry flower products preparation. International Journal of Agriculture Sciences 10(3): 5092-5094

Singh, H P. 2009. Floriculture Industry in India: the bright future ahead. Indian Horticulture 54(1): 3-8

Smith, R C.1993. Methods of preserving flowers. North Dakota State University Extension Service, North Dakota State University of Agriculture and Applied Science, USA.

Sohn, K H, Kwon H J and Kim E Y. 2003. Effect of drying methods on shape and colour of Rosa hybrida L. Korean Journal of Horticulture Science and Technology 3(9): 136-140

Vethamoni, I.K. and Nanthakumar, S. 2003. Standardization of technology for dry flower production at Kodaikanal. South Indian Journal of Horticulture, 50(4-6): 653-657

Visalakshi, M. 2013. Standardization of improved scientific processing techniques of dry flowers for export and dry flower market. M.Sc. Thesis, Tamil Nadu Agriculture University, Coimbatore, Tamil Nadu.

Website: http://www.techno- preneur.net/information-desk/sciencetech-magazine/2007/Dec07/Dry-flower.pdf.

Website: www.aces.edu/pubs/docs/A/ANR-1115/ANR-1115.pdf

Yadav, P, Kashyap B, Dhiman S.R., Prakash C and Gurung, A. 2018. Effect of glycerine for drying of cut foliage of Peltophorum ferrugineum. Journal of Ornamental Horticulture 21(3): 70-77

Yadav, P, Kashyap, B, Dhiman, S.R., Gupta, Y.C. and Sharma, P. 2018. Effect of glycerine for drying of cut foliage of Buxus semervirens. Journal of Ornamental Horticulture 21(1-2): 17-23

30

Importance of Cut Foliage and Fillers in Floriculture Industry

Vanlalruati, Ritu Jain, Prativa Anand, Babita Singh and Gunjeet Kumar

Floriculture has flourished with a considerable growth and a useful crop diversification option, particularly for small farmers over the past three decades. The commercial production of the cut greens has started in India in recent years and has huge demand in the market. They are suited for year round production with low investment and less care and maintenance.Cut greens or fillers are an important component of the floricultural industry and their commercial value depends on vegetative characters, such as stem length, leaf colour, density and side branching. Cut foliages (leaves and stems), which are attractive in form, colour and freshness, are lasting and in great demand. They have various other uses in making attractive fresh floral designs and floral arrangements, such as bouquets, wreaths, decoration of house interiors, etc. Evergreen plants with green, silver or variegated leaves are usually used and species with berries are now also becoming popular.Fillers are critical components in most arrangements as the flowers themselves,used to increase the volume and size of the bunch and provide colour contrast.These floral produce have various other uses in making attractive fresh floral designs and floral arrangements, such as bouquets, wreaths, decoration of house interiors, etc. Colour, freshness, stem length, freedom from pests, fragrance and vase life are important quality aspects of fillers.

In the International market, fillers accounts for about 16 percent of the total trade by volume with an annual turnover of more than Rs 800 crores including selling of ornamental leaves, filler flowers, dry and dyed flowers along with berries and capsules. On the contrary, fillers representvery less quantum with poor quality in the domestic market and are unsuitable for cut flower bouquets. Their share in the market is very low and fetches nominal rates.

Floriculture Industry World Trade and Role of Cut Greens

In India, since earliest times, use of leaves and foliage has played an important part during folk festivals, religious ceremonies and public celebrations of all kinds. In the western world and the Netherlands in particular, the floriculture industry is one of the oldest and most prosperous.The present day floral industry (with inclusion of greens) is growing fast i.e. over 10 per cent year on year. The Netherlands is trade hub for cut flowers and foliage in Europe with approximately 60% of all EU imports entering via the Netherlands. It imports greens (excluding flowers) worth around USD 250 million per year. Or in volume terms, it imports around 5 billion leaves or 15000 shipping containers of forty feet size per year. Foliage demand, however, is increasing as it gives a tremendous business opportunity since demand always exceeds supply. To meet this increasing demand, production of greens has continued to move from countries that have traditionally been consumers and growers to other relatively new producing countries such as Israel, Colombia, Ecuador, Guatemala, Costa Rica, Kenya, and Ethiopia.

Strength of Cut Foliage Industry In India

In production and export of foliage, India has certain competitive advantages in comparison to other countries:

i. Majority of cut foliages are tropical in origin. India's diverse topography with minimal temperature variations and right humidity is ideal for healthy growth of foliage plants. India can be at the forefront of cut foliage industry in the world provided farmers/growers are facilitated access to advanced technologies, innovative techniques and well developed infrastructure.

ii. The availability of wide range of inputs, coupled with low labour costs, offers India a natural advantage in growing 'greens'. It is also well suited to the economic and social conditions prevailing in the country.

iii. International logistics facilities like - availability of sea ports, availability of reefer containers, vessel frequencies, transit times, ocean freight rates to Europe base ports – are comparatively better compared with other countries.

iv. There is huge market demand for cut greens and fillers with growing urbanization, rising income and adoption of western styles by the middle class households and even small cities. Evolving domestic market will lead to increasing domestic demand.

v. India's progress in the field of Information Technology will facilitate cut foliage industry in various ways. It will make all the information instantly available on screen with a press of button and encourage exports of 'greens' to developed countries like USA, Germany, the Netherlands, France, Japan etc.

vi. Our economy is far more stable and growing much more rapidly than other countries. Self sufficiency in pre-harvest, post-harvest chemicals, irrigation systems, packing materials, etc. is the added advantage.

Uses of Fillers

Fillers are essential addition to any bouquets, vases, wreaths, flower arrangements and landscape etc.The types of fillers needed by the florists and arrangers are categorised as below:

i. **Display Flower Fillers:** These are the large flowers which are not used as main flower but used as fillers to create variation in colour e.g. Protea, Banksia, Waratah and Blandfordi.

ii. **Focal Fillers:** Some specific filler like rice flowers, wax flowers, Thryptomene serves as focal point in flower arrangements.

iii. **Small flowers:** Small flowers which increase size of the bunch, provide colour, contrast are also used as fillers. It may include seasonal/annual flowers. e.g. Achillea, *Ammi majus*, stock, delphinium, gypsophilla, statice, bells of Ireland *etc.*

iv. **Foliage Fillers:** Foliage adds contrasts and backdrop to the other flowers in bouquets and arrangements. This bulk up them with low cost materials like leaves of Asparagus, Anthurium, Cordyline, Ivy, Leather Leaf Fern, Monstera, Philodendron, Sword Leaf Fern, Tree Fern, *etc.*

v. **Tree Foliage Fillers:** Some of the trees have beautiful, exquisite foliage, which is used as fillers to provide contrasting background to the arrangements. e.g. Silver Oak, Myrtle, Eucalyptus *etc.*

vi. **Berries as Fillers:** Some of the plants (mostly shrubs) have very attractive berries which are used to add colour and contrast to the flower bouquets and arrangements. e.g Holly, Hypericum.

vii. **Dry Fillers:** Some flowers which are small in size, dried naturally and stays for a long time are used as fillers and have long lasting effect.e.g. *Ammobium sp.*, statice.

viii. **Grass Fillers:** Thin and sharp grass leaves gives better filling look when arranged with small flowers like *Briza maxima, Lagurusovatus,* Anethum and Asclepias in bouquets have long lasting effects.

Popular Flower Fillers

Filler flowers can either enhance or reduce the beauty of arrangements depending upon its usage, as they alone are not very attractive but when used with other flowers they can make enormous difference. These fillers are available in wide variety of shapes, sizes, form and colours. Adding even the simplest of filler flowers can make a bouquet look more sophisticated. Some of the popular flower fillers which can be used for flower arrangements, bouquets and decorations are mentioned below:

1. **Alstroemeria:** Also known as the Peruvian lily, originated in South Africa. The inner petals of this flower have stripes that create texture and diversity when added to bouquets. They come in a range of colours from white, yellow, pinks and purple.

2. **Astilbe:** *Astilbe chinensis*, a member of family Sexifragaceae, is native to Korea, China, and Japan and is mostly recognized by their feathery plumes of airy, dense blooms. They are available in red, white, pink, and assorted colours. It is used as flower for centrepieces, boutonnieres, and filler for wedding bouquets.

3. **Baby's Breath:** Botanically known as *Gypsophilla elegans*, is a member of family Caryophyllaceae. It is one of the most popular white filler flowers, available in single or double varieties. The flower is delicate and soft, just like a baby, and symbolizes innocence. It's also referred to as angel's breath. Tiny white flowers look finest when grouped in bunches, but it also comes in other shades like light pink. It looks best in bouquets and centrepieces, but can also be included in floral crowns. White baby's breath flowers are a familiar choice for adding mass to floral arrangements, but the flowers are easy to tint in any colour, increasing their versatility. Though traditionally used as a filler, this flower can also be used alone to create fluffy, cloud-like arrangements.

4. **Bells of Ireland:** *Molucella laevis,* commonly known as Bells of Ireland. It has a long stem lined with clusters of bell-shaped blooms which are soft green in colour. This green is perfect for adding height and texture to bouquets and centrepieces. The stems of the bells of Ireland are hollow and can be easily broken and can be ideally used for vase arrangements.
5. **Bouvardia**: It is great filler flower having clusters of star-like flowers that have a delicate scent and comes in shades of pink and white. Each stem resembles a small bouquet. It last longer in vase arrangements and makes a wonderful filler flower. Since, it is a fragrant flower, therefore, often used in wedding bouquets and corsages.
6. **Delphinium:** It is commonly known as Larkspur. It has beautiful purple and blue coloured closed flower buds which resembles to the nose of a dolphin, hence derived the name delphinium. The delphinium has multiple blooms on long stems that are similar to snapdragons. They are perfect for adding a dramatic look to any bouquet or flower arrangement.

7. **Dianthus:** Dianthus, commonly known as the carnation, is an extremely popular flower available in a plethora of vibrant colors. Mostly two species *Dianthus chinensis* and *D. barbatus* are used as fillers. The name dianthus comes from the two Greek words, "dios" and "anthos." Dios refers to the Greek god, Zeus, and anthos means flower. Thus, they are known as the "flowers of God." As regal and beautiful as they are, dianthus flowers look best as statement fillers in bouquets.

8. **Eustoma:** *Eustoma sp.*, commonly known as lisianthus is a wonderful filler flower that features anywhere between two to six budding flowers per stem. Eustoma flowers are either single-flowered or double-flowered. Both types of flowers can be found in all shades of pink, purple, white, and blue. They are usually one to three feet tall, although there are dwarf varieties that only grow up to eight inches in height.
9. **Feverfew:** Botanically it is known as *Tanacetum parthenium*. It is a medicinal plant but also act as perfect filler flower for just about any sort of flower arrangement. It has long, flimsy stems that blossom into tiny bunches flowers resembling daisies. The feverfew is a modest flower with tiny white petals that pair nicely with bright coloured flowers.
10. **Golden Rod:** Botanical name of Golden rod is *Solidago canadensis*. The variety 'Golden Baby' is an easy to grow hardy perennial that bears flat-topped clusters of golden-yellow plumes.
11. **Lily of the Valley:** *Convallaria majalis* is not a true lily. It belongs to family Asparagacreae. It has very delicate flower with tiny bell-shaped blooms and broad basal leaves. This white flower symbolizes purity, humility, happiness and sweetness. It is a perfect filler flower for delicate and elegant wedding bouquets.
12. **Medagascar Jasmine:** Madagascar Jasmine (*Stephanotis floribunda*) is a member of the family Apocynaceae. It is another plant with fragrant flowers and gorgeous foliage. It is a climber, which grows upto 10 feet height. Its flowers are also referred to as bridal veil, bridal wreath, and bridal bouquet for their popularity in wedding flower arrangements. The blooms represent marital harmony in the language of flowers, and as a filler flower, blooms contribute a significant amount of fragrance to bridal bouquets. Due to their small stems of flowers, special wiring may be required for design work.
13. **Queen Anne's lace:** *Ammi majus,* also known as Queen Anne's lace, is wonderful white bouquet filler. This elegant little flower with its lacy heads of pure white complements any colour scheme in the vase with its long sturdy stems and large full flowers. It can be used to create a perfect wedding bouquet with sweet peas, roses, berried eucalyptus. Queen Anne's lace symbolizes magic, trust, and healing.

14. **Rice Flower** – Botanical name of Rice flower is *Ozothamnus diosmifolius* previously known as *Helichrysum diosmifolium*, belongs to family Asteraceae and is native to Australia. It is also known as white dogwood, pill flower and sago bush. It has woody stem and small flowers occur in clusters of up to 100 at the ends of the branches and are usually white, but light to dark pink forms are also known. Rice flowers make a nice filler flower for garden-style designs and other arrangements.

15. **Snapdragons:** *Antirrhinum majus,* commonly known as Snapdragon is a member of family Scrophulariaceae, is possibly one of the most unique flowers. Its name represents the face of the flower that opens and closes like the mouth of a dragon. It has long, slender stalks that blossom from bottom to top in a variety of colours. They are most often used in bouquets to add height and diversity. The snapdragon is a popular filler flower, good for adding height and colour to arrangements. It is available in variety of colours such as pink, purple, yellow, orange, white, and burgundy. Snapdragons make a wonderful addition to any bouquet or centrepiece.

16. **Statice:** The statice flower, also known as the Sea Lavender. Botanically, it is known as *Limonium sinuatum* and belongs to family Plumbaginaceae. It is a beautiful flower available in white, blue, pink, purple and yellow colours, and the blooms are smaller than the stock. It is a simple yet elegant flower that symbolizes remembrance. It has long, thin stems that branch out into dense flower bunches and look best in bouquets. Statice is sometimes called the tissue flower, because of their tissue paper-like blooms. Each main stem has a spray of smaller branches with several of these tiny funnel-shaped flowers. They have a long vase life and require very little maintenance, making them popular for both fresh and dried arrangements. The meaning of statice is lasting beauty and success; however, it is also symbolic of remembrance and sympathy.

17. **Stock:** *Matthiola incana* a member of family Asteraceae is one of the most fragrant flowers. Its scent is described as both sweet and spicy. Stock flowers bloom from spring to fall and flowers range in colour from basic shades of white, pink, lavender, and rose as well as coming in deeper jewel tones of red, purple, and blue. They develop small blooms that are perfect for adding a pop of colour to any bouquet. There are dwarf varieties that only grow to be eight to twelve inches tall and larger varieties that can grow two to three feet.

18. **Sweet Pea:** *Lathyrus odoratus*, a member of family Fabaceae is a plant with very sweet scented flowers of white, red, pink, lavender and purple colour. Sweet Peas can be used in arrangements to add mass or as filler. Commonly used alone in a simple spring vase arrangement. These flowers symbolize delicate pleasures and make a great addition to any arrangement. These wonderful blossoms with their delicate petals and soft fragrance are suitable for wedding and corsage work.

19. **Thryptomene**: Most commonly two species of Thryptomene are used as a filler i.e. *Thryptomene calycina* and *T. saxicola*. It belongs to the family Myrtaceae and is native of Australia. *T. calycina* blooms in winter and the branches bear hundreds

of tiny white flowers and small spear-shaped leaves having pleasant scent. In good seasons the flowers are so numerous that they can almost completely cover the branches, whereas, *T.saxicola* bears fewer pink flowers in spring season.

20. **Wax Flower:** Geraldton wax flower, botanically known as *Chamelaucium uncinatum*, a member of family Myrtaceae is native to Australia. The tiny five-petaled blooms of wax flowers often sport a contrasting throat, packing a big style punch into a myriad of little blossoms. The blooms have good vase life and these lasts for 3 weeks in vase. Although the white, pink, or lavender flowers have little scent, the needle-like leaves have a resinous fragrance when bruised or cut. It is ideal as an accent flower, hand-tied bouquets, vase arrangements, wreaths and garlands, plus wedding designs including hair crowns, pomanders, buttonholes, etc.

21. **Winged everlasting:** *Ammobium alatus*, a member of family Asteraceae, commonly known as Winged Everlasting is native of Australia.It has small, paper-like white petals with a yellow/gold centre and stems that are silver/grey-green in colour. This simple and unique flower makes excellent filler, but can equally and effectively stand alone. Wonderful for boutonnieres, flower crowns and large centrepieces.

22. **Yarrow:** *Achillea millefolium* commonly known as Yarrow belongs to the family Asteraceae. It is perfect filler for bouquets and arrangements as the vaselife is more than 2 weeks. These native wild flowers have clustered flower heads of tiny white flowers. This flower comes in a natural bright yellow, and gorgeous painted colours, perfect to add brightness to creations.

Popular Foliage Fillers: Cut greens are generally less expensive and have longer vase life compared to the other floral components. They lend design elements and give arrangements a lush, full, balanced appearance, often add background colour, and tie together all the blooms. The quality 'greens' mainly depends on the visual appearance. The foliages of different shades, hues, gloss and texture are used in order to create mobility which is represented by leaf colour and absence of the damage.The foliage or cut greens which are used as fillers are described as follows:

1. **Anthuriums**: *Anthurium andreanum*, commonly known as Anthurium, Tailflower, Flamingo flower, and Laceleaf, belongs to family Araceae. It is of two types' flowering and foliage type, later is used as fillers in different flower arrangements.

2. **Asparagus Fern:** *Asparagus densiflorus* and *A. sprengeri*, the member of family Asparagaceae are commonly used fillers due to their very fine foliage. It is a tender evergreen perennial with bright green, ferny foliage is commonly used as an outdoor or indoor plant. Asparagus fern valued as an ornamental for its bright green, arching stems and airy foliage. Its fine foliage gives a soft or fluffy appearance and can be used to good effect for textural contrast in combination

with plants having medium or coarse-textured foliage or very large leaves. The foliage is incorporated as filler with cut flowers in arrangements.

3. **Boston Fern:** *Nephrolepis exaltata,* known as the sword fern or Boston sword fern, wild Boston fern, Boston Blue Bell Fern, tuber ladder fern, or fishbone fern, is a species of fern in the family Lomariopsidaceae native to tropical regions throughout the world. It can be used either as 'filler' in the container combination, depending on the height and form of the other plants used in the container planting. It is also used as filler in the flower arrangements to give greenery and delicate look to the arrangement.

4. **Bracelet Honey Myrtle:** *Melaleuca armillaris***,** commonly known as bracelet honey myrtle, is a plant in family, Myrtaceae, and is native to South Australia, Victoria and Tasmania in south-eastern Australia. It is a hardy, commonly grown species, often used as a fast-growing screen plant. The honey bracelet has long stems that are decorated with soft, thin leaves. Its thin stem is easily molded into any shape and can be used for just about anything. However, it is most frequently used to create flow in floral arrangements because of its long drooping stems.

5. **Cast Iron Plant***:* It is also known as Bar Room Plant, Barroom Plant, Iron Plant, Parlor Palm, Variegated Cast Iron Plant. Botanically, it is known as *Aspidistra elatior,* and is a member of family Asparageae. It is a good specimen for planting in dense shade; and slightly tolerant to salt. Prefers partial to full shade, drought tolerant, grows well on rich, well-drained soil.Its leaves are erect, long-petiolate, solitary, gigantic, ovate to lanceolate, acute, entire, glabrous with 16-20 parallel veins, dark green to reddish-purple to variegate. Leaves are used in cut flower industry as green filler in flower arrangements. Leaves should be harvested near soil level, but avoid harvesting many leaves from single plant at the same time.

6. **Chinese Fan Palm:** *Livistona chinensis* a member of family Arecaceae. This tree does best in full sun to partial shade. It prefers to grow in average to moist conditions. Chinese Fan Palm has attractive dark green foliage. The glossy lobed palmate leaves are highly ornamental and remain dark green throughout the winter. It is good filler for containers arrangement as well as flower arrangements.

7. **Cider Gum Tree:** *Eucalyptus gunnii,* a member of family Myrtaceae is a small to medium tree. The silvery-blue, rounded young leaves that give way to glaucous lance-shaped to egg-shaped adult leaves and smooth whitish-green bark. The juvenile stage heart-shaped, silvery-blue leaves are used as filler in floral bouquets and the juvenile foliage can easily be perpetuated with regular pruning. The eucalyptus "Baby Blue" is denser than its silver dollar cousin, with smaller leaves. Its color changes as it matures and morphs from a smooth

blue to a rich green. It's great for adding texture to bouquets and as an accent in floral wreaths.

8. **Dracena:** *Dracaena reflexa,* or Song of India, a member of family Asparagaceae, is an ideal house plant for cooler areas but grows well under shade when kept outside. It has a stunning variegated foliage and popularly used as fillers to sparkle and beauty in floral arrangements. 
9. **Dusty Miller:** *Jacobaea maritima,* commonly known as silver ragwort or Dusty Miller is a perennial plant species in the family Asteraceae, native to the Mediterranean region. Representing happiness and delicacy, dusty miller is a popular green filler flower. It's often used in soft, garden-like bouquets, tablescapes and wedding arrangements. It's velvety, frosted sage green leaves will be the perfect complement to petal heavy, romantic blooms like garden roses, ranunculus, and peonies. The gray and white color of the dusty millers leaves pair nicely with pastels and look great in bouquets.
10. **Good Luck Plant:** *Cordyline terminalis* is a tropical shrub belonging to the family *Asparagaceae* and commonly known as ti plant or good luck plant. The leaf colour ranging from variegated pale pink and green to purple or deep red. The shrub can reach an adult height of 10 feet. The leaves are broadly elongate, 30 inch long and 2-4 inch wide born at the top of a woody stem. In cooler climates, it grows well in container under indoor conditions.
11. **Holly:** *Ilex verticillata* commonly known as Holly or winter berry is a member of family Aquifoliaceae. It is native to eastern North America in the United States and southeast Canada. Mainly used in winters as bunches of bright red berries against fresh greenery add a fresh, natural look to wedding bouquets and arrangements. Holly traditionally symbolizes deference and domestic happiness.
12. **Hypericum:** Florist hypericum, *Hypericum androsaemum*, is prized for the beauty of its brushy habit and colorful berries. It is also called "coffee berry" in the floral industry because of the fruits' resemblance to coffee beans. Depending on cultivar, berries may ripen to green, white, yellow, peach, orange, scarlet or purplish colors, finally maturing to darker hues with some looking almost black and make wonderful accents. Hypericum berries are popular filler flowers used to complement floral arrangements, bouquets, centerpieces, and outdoor decorations at parties, weddings and other special venues. They are popular throughout the year and especially during the winter season. The stem with berries is used as filler and in boutonniere provides a pop of color against a dark background. They can easily be incorporated to most of the designs, providing texture to bouquets, centerpieces etc.
13. **Italian Ruscus:** *Ruscus hypoglossum* a member of family Asparagaceae, is native to Italy. Its small, glossy, tear shapely leaves grown in multi lateral directions make it gorgeous filler. Great focal or back drop green and adds

elegance to any arrangement. It can be used in garland, bouquets, arrangements, wreaths, boutonnieres, corsages. Besides this *Ruscus hypophyllum* (Israeli ruscus) is also used as filler as its foliage is dark green with fine texture, stiff stems, and long vase life.

14. **Ivy:** *Hedera helix*, a member of family Araliaceaeis a climber. The dark green or variegated foliage with a twisting ability due to soft stem makes it a perfect choice for filler in head crowns, wreaths, floral baskets and outdoor decorations.The leaves of ivy are unique in their triangular shape. Ivy has a longer lifespan than most types of greenery; it lasts for months, so it is an ideal choice for home decor during holiday months or celebrations.

15. **Lacy Tree Philodendron:** *Philodendron selloum* commonly known as Lacy Tree Philodendron or Tree Philodendron belongs to the family Araceae. Tree philodendron is a tropical plant with large, showy leaves. It can tolerate deep shade but prefers bright, indirect light and moist soil. It does not tolerate salt build up so flush soil regularly or used distilled water for irrigation. It can produce aerial roots that absorb water and nutrients.
16. **Leather Leaf Fern:** *Ruhmora adiantiformis,* is a member of family Dyropteridaceae. This plant grows year-round and can lasts from 7 to 21 days. The leather fern is thick and stem is strong yet flexible, extends out on the sides in a unique way. The frond branches out creates a triangular shape, can stand alone in a vase as a centerpiece or add a tropical flair to a floral arrangement. With tufted, rich green leaves that fan out triangularly, this fern provides a dramatic backdrop to the flowers in bouquet.
17. **Leucadendron Safari Sunset**: It is tall wonderful filler for accenting larger floral centerpieces and arrangements. This variety bears elongated, oval-shaped green leaves that may have some burgundy on the tips. It is highly resilient that many find easy to arrange and work with. It works well for weddings, parties, and many other events.
18. **Leyland Cedar:***Cuprocyparis leylandii* is a fast-growing, evergreen conifer with a dense, broad-columnar to narrow-pyramidal habit. This evergreen foliage is thick and lush with an attractive green hue. Dark green color foliage with subtle blue tips makes it an excellent choice for fillers. Leyland Cedar can be used in wreath or mixed in with other greens in a garland. Leyland Cedar looks especially elegant when paired with white flowers.
19. **Monstera:** *Monstera deliciosa* commonly known as Swiss Cheese Plant, Monstera, Mexican Breadfruit or Delicious Monster is a member of family Arceaee, and is wonderful decorative filler. The juvenile leaves have full leaves while the cuts begin to appear as the plant gets older, when the conditions are right. It is used alone or in combination with flowers to create unique and tropical wedding bouquets, table centerpieces or flower arrangements.

20. **Myrtle:** *Myrtus communis,* the common myrtle or true myrtle, is a species of flowering plant of family Myrtaceae. It is an evergreen shrub native to southern Europe, North Africa, western Asia, Macaronesia and the Indian Subcontinent. Myrtle is one of the most popular types of greenery because of the variety of ways that it can be used. It has long stems which are lined with glossy leaves. The thick foliage used as filler in floral centrepieces and also in wedding bouquets and home decor due to sweet fragrance of leaves.
21. **Silver Dollar Eucalyptus:** *Eucalyptus polyanthemos***is** a member of family Myrtaceae. It has one to two inch circular leaves that resemble silver dollars. It has alluring matte gray and blue foliage, making it a great pair for whites and pastels. Its thin, flexible branches similar to the ivy's make it ideal for using in wreaths and displays. The silver dollar eucalyptus is often used fresh, but can also be dried out and dyed to different colours.
22. **Silver Oak:** *Grevillea robusta,* commonly known as the Southern silky oak, Silk oak or Silky oak, Silver oak or Australian silver oak, is a flowering plant in the family Proteaceae. It has a unique type of decorative greenery with red stems that branch out into multiple green leaves. The branches vary in their length causing them to add depth to floral decor. They look best when adding accents to a bouquet or centerpiece.
23. **Sword Leaf Fern:** *Polystichum munitum* a member of family Dryopteridaceae, is a lush evergreen ground cover known for its bright green, sword-shaped fronds. The foliage is of bright green color so it really pop up in floral arrangements. It is highly suitable for line arrangements but also look great in round arrangements. It can last up to eight days with proper care. It's also available year-round.
24. **Tree Fern:** *Sphaeropteris cooperi, synonym Cyatheacooperi*, also known as lacy tree fern, scaly tree fern, or Cooper's tree fern, is a native to Australia. It is one of the easiest types of filler to use because of its extremely thin stem like ivy. The tree fern has thin, wispy branches and leaves that are often used in corsages and boutonnieres. The tree fern also works well as a solid background in a colorful bouquet.
25. **Umbrella Plant:** *Cyperus alternifolius* , a member of family Cyperaceae, is a very popular both as a house plant and pond plant. The bracts are symmetrically arranged in an umbrella formation and held atop elegant stems that sway with the breeze, giving a tropical touch to the garden. They are excellent when used in fresh or dried floral arrangements.
26. **Variegated pittosporum:** *Pittosporum tobira* "Variegata"is a member of family Pittosporaceae. It is a fast-growing shrub to small tree grown for its attractive white-edged, oval leaves. . It's beautiful variegated leaves makes it different from other greens. They are available year-round, long-lasting and can maintain its shape out of water. Variegated pittosporum is commonly paired with flowers like roses, kale, dahlias and hypericum berries. It is also available in a miniature size perfect for boutonnieres or corsages.

27. **Xanadu:** *Philodendron bipinnetifidum* xanadu belongs to the family Araceae. This plant is ideal as filler or a mass plant and can also be used to make garden borders. It is an ideal plant to grow in subtropical and tropical regions. They are very easy to grow and do not need much maintenance. These are large compact plants that offer attractive looking foliage. The leaves are deeply dissected and are lobed. These leaves are glossy in nature. Each leathery leaf will have about 15 to 20 lobes that appear distinct. Their foliage is neat and tidy and looks attractive as they mature. The best light condition for its growth is medium indirect light, but if exposed to sunlight, then its leaves will turn pale and loose its shiny green.

Grasses as Fillers: Both flowers and leaves used as fillers to add bulk to the arrangements or to the landscape. There are few ornamental grass species which are used to add beauty to the bouquets, flower arrangements and centerpieces. Few of these species are described as under:

1. **Bunny Tail:** *Lagurus ovatus* commonly known as Hare's Tail, Bunny Tail grass is a member of family Poaceae is native to the Mediterranean Basin and nearby regions. It is probably the most appealing of all the ornamental grasses. The name 'Hare's Tail' perfectly accurately describes the creamy-white flower heads, which are hare's tail-shaped, fur-like and soft to the touch. It is very good filler for dry bouquets and floral basket arrangements.

2. **Quaking Grass:** *Briza maxima*, a member of family Poaceaeais commonly known as Quaking Grass. *Briza maxima* has blue green leaves and flower heads that hang like scaly little heart shaped lockets that are tinged with pink from late spring to mid summer. It is good for dried arrangements. This plant has open, puffy, overlapping, scale-like seeds and is reddish-brown or tan, touched with green.It also pairs well with any type of berries, such as Hypericum Berries, Hanging Pepper berry and Red Ilex Berry Branches.
3. **Whorled Milkweed:** *Aesclepias verticillatta* is a member of family Aesclepiaceae. It is easy to grow in average, moist to dry, well-drained soils and flowers best in full sun but will tolerate some shade. It is resistant to deer, drought and dry soils. Seed pods are sometimes used in dry flower arrangements.

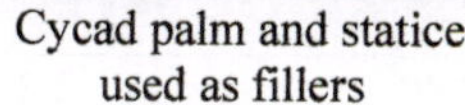

Cycad palm and statice used as fillers

Chinese fan palm and Asparagus used as fillers

Post Harvest Treatment

Some florist greens and fillers also benefit from the floral preservatives which include boxwood, leather leaf fern, ivy, *Podocarpus* etc. In addition to the floral preservatives solutions, special solutions are also used by the growers, wholesalers or retailers. Pulsing solution containing 10-20% sugars are used prior to packaging and transportation to extent the shelf-life and to aid in freshness and swimming.Floral preservatives solution that contain 1.0% sugar and biocide (200 ppm 8HQC or HQC) and acidifiers viz., 20 ppm citric acid or Al_2SO_4(10-20ppm) is recommended.

Packaging and Transportation

The bunches are tied or bundled loosely to avoid damage. They are wrapped in moisture resistant paper or placed in clean plastic sleeves. Florists greens are sometimes packed in ice, in wax-impregnated or poly ethylene film lined in the fibre board boxes. Wet newspapers or waxed paper are use to provide humidity. The society of American Florists and the Produce Marketing Associations have recommended standard fibreboard boxes to provide better staking and utilisation of pallets in transit to increase the overall efficiency in distribution and are presented below :

Industry standards packaging for fillers and florists greens

Products/sizes	Outside Dimension (2 X W X H)	
	Inches	Centimeters
Standard	41 X 5 X 7*	104 X 12.5 X 18
Florists Greens	30 X 14 X 9**	76 X 35.5 X 23
Fillers, greens	30 X 14 X 15	76 X 35.5 X 38
Gypsophylla	13 X 13 X 24	33 X 33 X 61

*Quarter Boxes

**Half boxes, centimetre dimensions are approximate

Most of the florist greens and foliages are shipped without refrigeration thereby hampering shelf-life. Use of protective packaging, precooling before loading, co-ordination at origin and destination airport and utilisation of cold storage facilities near the airport is required to minimising product loses. Refrigerated vans and containers are used for export of filler greens with adequate pre-cooling which can withstand the transit time of 2-3 weeks.

Storage

The fillers and greens if stored at higher than the recommended temperatures will lead to yellowing of leaves and bending of berries and fall-off in presence of ethylene.The recommended temperatures and approximate transit storage period for cut greens is given below:

Cut greens/ Fillers/ Foliages	Storage (^{0}C)	Temperature (^{0}F)	Approx. storage period
Adiantum	0.4	32.40	2-3 weeks
Asparagus plumosa	2.4	35.40	2-3 weeks
Asparagus sprengeri	2.4	35.40	2-3 weeks
Buxus (box wood)	2.4	35-40	-
Cedar	0	32	-
Chamaedorea	7	45	2-3 weeks
Cordyline	7-10	45-50	2-3 weeks
Croton	7-10	45-50	2-3 weeks
Dracaena	2-4	35-40	
Eucalyptus	2-4	35-40	1-3 weeks
Hedera	2-4	35-40	3-5 weeks
Leather Fern	1-4	35-40	2-4 weeks
Palm	7	45	-
Philodendron	2-4	35-40	-
Pittosporum	2-4	35-40	2-3 weeks
Podocarpus	7	45	-
Pothos	2-4	35-40	-
Salal (Lemon leaf)	0	32	2-4 weeks
Staghorn Fern	13	55	-

31

Flower Arrangement: Principles & Methods

Yeshpal Singh, Kumari Lata and P. K.Yadav

Flower arrangement is the art of using plant materials and flowers to create an eye-catching and balanced composition or display. Among all art forms, flower arrangements today take as prominent a place. The art engages ones personal involvement in welcoming guests and bringing cheer to the surroundings. It is rightly said, "Flowers are the music of the ground. From earth's lips spoken without sound."Professionally designed floral designs, arrangements or artwork incorporate the elements of floral design: line, form, space, texture, and color, and the principles of floral design: balance, proportion, rhythm, contrast, harmony, and unity.

Composition of Arrangements

There are a few compositions of flower arranging that should be considered. Once we understand these we can create so many different styles and our floral designs will have that professional look.

1. **Materials:** Would be the base for one's expression that makes a flower arrangement.
2. **Form:** This includes design, scale, balance and rhythm.
3. **Colour:** This accentuates the design and harmonizes materials with environment, helping to build the form that would be apparent in the arrangement.

1. Materials

a. **Flowers and Foliage:** These are the primary material. These come in markedly varied shapes, sizes and colour. Longer treatment before use any leaves, flowers and stems may be combined, though some need a longer treatment before use.

b. **Containers:** They must be carefully chosen. The choice of a container, which would actually compliment the arrangement and not distract from it.

c. **Pin Holders:** Help to contain the primary material in the form that is visualized by the arranger. Along with this, wire mesh could also help in holding difficult branches together.

d. **Accessories:** They accentuates the form and highlight the beauty of the flora and leaves used. E.g. Lamps, shells, stones, candles, earthenware etc.

e. **Bases:** A base can actually enhance the aesthetic appeal of an arrangement. Beautiful bases cut out of ordinary wooden planks, marbles or stone, may be carefully designed in various shapes to compliment the actual size and form of the arrangement and add to all important three-dimensional effect.

2. Forms

a. **Design:** It decides the structural pattern of the floral arrangement. It is a visualized plan about the size and shape of the floral arrangement. The design is essentially determined by the location, occasion and the materials available.

b. **Proportions:** The size of the flowers, foliage and containers should be in proportion to each other. For e.g.: a tall arrangement of long stemmed roses would be out of proportion arranged in a small vase. It would not only look top-heavy, it would be in danger of toppling over. At the opposite end of the scale short-stemmed flowers should not be placed in a large urn or they will appear lost.

c. **Balance:** This principle means that the arrangement of flowers should be symmetric like if any side of flower cluster in vase looks heavier than the other, then there is still need to do some rearrangement in cluster. We can manage this symmetry by not only adding the even things but also can use some other element of equal weight.

   Eg:- one bloom on one side and a cluster of small ones on other side or two different kinds of flowers of similar size.

   After finishing the arrangement, it should be confirm by rotating the arrangement to check from all angles.

d. **Harmony and Unity:** This principle takes care of the appropriate use of flowers for the occasion. It also takes care of the blending of colors of the flowers used in the floral arrangement and that of the container such that they contrast each other.

   It is essential that a pleasing balance and harmony be obtained between materials and containers.

e. **Rythem:** It emphasizes on directing the viewer's eye along the main line of arrangement so that the whole shape of the composition is properly followed and understood. Imperfectly balanced material will not play a rythem.

f. **Accent and Focal Point:** The central portion of the floral arrangement from where the flowers and foliage of the arrangement appear to be emerging from is known as the focal point of the arrangement. Hence, larger and dark colored flowers are usually used in the focal point of the floral arrangement. A focal point could be beautiful leaves, flowers or the accent may totally be shifted to candles, birds, statue, stones, or just beautiful twigs.

g. **Texture:** Flowers and foliage are differ in texture, so we should used different textures that go well together to give variety to our design.

3. Colour

a. **Analogous:** Schemes where two neighbouring colours of colour wheal e.g. Yellow and orange, blue, green and mauves, beige and brown are characteristically used to complement each other.

b. **Monotone:** The colour choice may even be monotone where many hues of a particular colour are utilizes in a gamut of subtle hints ranging from a rich shade to the palest of its shades.

c. **Contrasting:** The combination of primary colours or secondary colours individually on colour wheal is known as contrast colour. Contrasting colours lend brightness to drab areas. In this yellow, red and blue are contrast to each other and green, orange and purple are contrast to each other.

d. **Complimentary:** The two opposite colours on colour wheal is known as complimentary to each other. Eg: red-green, orange-blue, yellow purple etc.

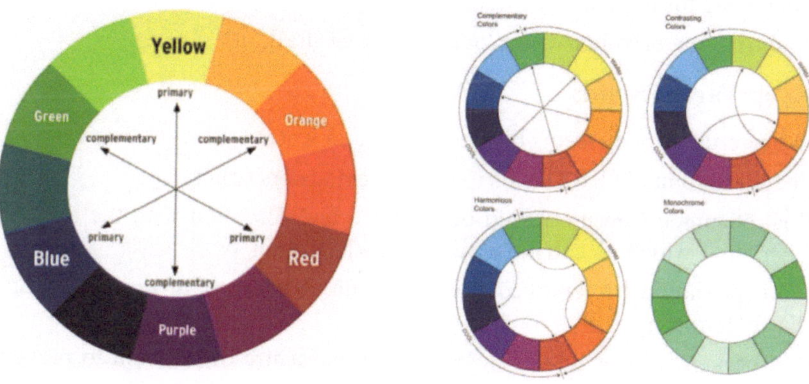

Colour Wheal

Colour combinations

General rules of flower arrangement

- Measuring of containers: the flower stem height should be one and a half to two times of the height of container.
- Centralize stems: in some case the middle flower is centralize to enhance the beauty in flower arrangement and also create balance and harmony in flower arrangement,
- Vary your stem lengths: sometimes-different length of flower stem is used in flower arrangement when done in cluster to enhance the beauty of flower arrangement with different texture of foliage.
- Avoid crossed stems: in transparent vase the crossed stem is not appears good and crossed stem also decrease the effective length of flower stalk.
- Avoid a tangled look

Cut Flower and Foliage Care

- Always pick the material late in the evening or early morning
- Carry the material in a heads down position to prevent sap from escaping and stem from drying up.
- Plunge the flowers into a bucket of cool water for 15 mins to 1 hr before use.
- A long drink of water will make the flowers/foliage last long.
- If possible, plunge material in tepid water the night before, for conditioning.
- Flowers and foliage should be separately soaked after removing 2" of the foliage from the ends as it reduces the rate of transpiration.
- One can substitute suitable foliage other than the flowers own that may wilt quicker.
- Cutting off ½" of the lowest part of the stem must be done before arranging to prevent algae formation.
- Keep the cut material in a dark place away from draft.

How To Keep Varied Stems

- Bruise or crush woody stems before plunging in water.
- Cut between joints of hard stems to make them absors water.
- Hollow stemmed flowers should be upended and force-fed with the aid of a dropper, after which its end must be sealed with cotton.
- All foliage should be removed from bleeding stems. e.g. of poinsettias and dahlias before use.
- Push a broken twig or a hard stem through the centre of a bloom to prevent drooping.

How to revive drooping flowers and prolong their lives?

- Use sprays on arrangements morning and evening to prolong freshness.
- Some delicate house plant leaves e.g. Begonias may be revitalized by laying them flat on the floor.
- Wilted flower stems may be revived by plunging them in hot water and quickly giving a subsequent cold water, drink for revival after chopping off ½" of lower stem.
- To keep water fresh copper compounds, sugar, charcoal, salt may be added to it. The best treatment however, is to change the water daily after cutting ¼" of the stem ends.

Bud Vases Arrangements

For dresser tables/ bathrooms/ writing tables/ restraunt tables etc.

1. Single Flower/Rose Bud

The thumb rule for arranging this type of arrangement:

- Leaves remain only top of the rim of the vase.
- All leaves from below the twigs that remain in water, should be removed.
- Trim off all other leaves except for one cluster behind the rose away from the vase, and one in front of it slightly covering the vase.

2. Three Flowers/Rose Buds

Carnation may be used with ferns in this picture. One twig of asparagus frames the flowers at the back to show depth, another, on the side of the flowers, and one stem in front, to complete the picture.

3. Five Flowers/Rose Buds

Notice the descending order in which the buds have been arranged. These may be from back to front or from one side to another.

The leaves, whether the flowers own, or any other, should support the flowers at the back, on one side and in front of vase.

Note: only smaller flowers are used in bud vases, annuals or perennials, from the twig of a large tree or a shrub. Branches have to be trimmed to comp lement these flowers.

Arrangements for Counters

Lobby counters/ restaurant counters/ piano top/ side board tops etc.

1. Triangular Arrangements

The tallest stem is placed in the centre, others are arranged on the sides. The last stem is lying almost horizontally pointing downwards.

The fillers below the principle "one in the centre of the two behind" e.g. Asparagus is used as the green filler and tube roses whilst glades from the effective background.

The top and lower flower are the two points of the triangle, the third point being almost rounded off.

Notice the tallest stem tilting to the left and the rest getting shorter as one descends, towards the point at the base. The last bud lies horizontally pointing sideways. The fillers are one in between two, flanked by asparagus.

2. Ovals

Oval arrangement are used for long, rectangular tables or longer counters where low arrangements may be placed in front of a picture, alongside a statue, in between lamps etc.

3. Round Bowls

A round arrangement is required to be placed in the centre of a table.

This is visible from all around the room.

Ovals

Round Bowl

Arrangements For The Banquet

1. The Fan Type Triangular Arrangements

These are one sided arrangements required to be placed against a wall, raised on the main display of chaffing dishes. Notice the two shades of peach gladiolii with their own leaves used as fillers. White and pink contrasting gladiolii are used effectively in a green vase, with green ferns forming a beautiful background. One may repeat the pink in the two front rows, or just about any two colours may be combined to make an effective display.

However, colours that pick-up the colours in carpets/walls/table frills used in the hall, look the most effective.

4. Round Arrangements

This traditional round arrangement of tube roses and roses with asparagus covering the front of a common bowl can be effectively placed in a central display.

For formal sit down meal on round individual tables; a round table arrangement in any bowl may be used. The ferns at the base of the bowl, all around will hide the bowl.

If its container, the ferns may be eliminated to show-off the bowl.

5. Eastern Style

This style commonly known as Japanese style of follower arrangement. This style of flower arrangement is not symmetric. It is asymmetric. This style impress more by the beauty of individual material not by the mass effect of flowers. In this style the materials never touch the rim of the vase. In this style the accessories like interesting brances, drift wood, pieces of bark, shells etc., are used to imitate the nature.

This is an eastern arrangement with an accent on the line. However, one must ensure that the arrangement is really not just one sided.

It has exactly the same oriental style with an equal emphasis on the shapely vase, and can be used effectively both on individual round/rectangular tables or a long table set for a larger number.

6. Hanging Arrangements

A flower basket or two hung from the ceiling in a corner; or at different heights all over the hall can be irresistibly beautiful. These provide an unusual display and give a festive look.

The number of baskets will depend on the type of function and the size of the room, and the flowers used are essentially those that can withstand a long evenins without the stems being in water.

7. Oriental Arrangements

These are frontal arrangements designed to be viewed mainly from the front, as they are always placed with their back on the walls on corner tables, buffet set ups, consoles, etc. The emphasis being on a definite line design that has been devised out of the ikebana concepts of "heaven, man and earth" fewer fillers are used in these arrangements, compared to the western style.

They are three dimensional arrangements and unlike western arrangements, they do not confirm to a rigid, conventional norm. Eastern arrangements are, therefore informal imaginative and sensitive forms of emotional expressions, and are symbolic of the varied times of the day, seasons of the year and often depict natures forms to enhance the arrangements.

8. Upright Style

A single long branch with shorter branches and flowers at the base arranged in a tall upright vase are characteristic of the nageire style.

Notice the completely different flowers, foliage and accessories in the two pictures, although they both from an upright arrangements. These are vertical forms symbolic of spring.

9. Diagonal Style

The slanting form has an essentially oriental line. Main stem is placed diagonally upward and the second stem has a downward sweep and is much shorter, placed in almost a reverse direction in a delightful bend. These symbolize a natural pouring movement of water denoting peace.

10. Crescent Shape

This shape is a semi-curving incomplete circle and can be equally effective when arranged in a tall stem vase or a low bowl.

The long stems, symbolic of a new moon in the sky.

Diagnal Style **Crescent Shape**

(*Courtesy* by http://floral.selbermachendeko.com)

11. Free Style Triangular Arrangements

This is the easiest shape to execute. The three basic lines are symbolic of "heaven, man and earth" and form the basis of most arrangements.

12. Dry Arrangements

When fresh flowers are not available, all dry materials may be used with any accessory. E.G. Bamboo, tar coal pieces and dry leaves.

Free style arrangement

Dry flower arrangement

Some other flower arrangements

References

https://www.floral-art-school.com.au
http://agritech.tnau.ac.in/horticulture/horti_Landscaping_freshflower.html
https://www.architecturaldigest.com/story/3-pro-florists-arrangements-you-can-do-at-home
http://floral.selbermachendeko.com for photo
www.cityflowers.co.in

32

Essential Oils in Ornamental Plants

Ashish S. Ghormade, Kumari Lata, Parag Sable and Sumed Thorat

All higher plants have some kinds of aroma or flavor in their various parts, which includes roots, stems, leaves, flowers and fruits. This flavor and aroma in a particular plant part due to the presence of some specific essential oils present in special glands in it (Chadha and Choudhury, 1989). Essential oils arethe concentrated liquid, which is highly concentrated and water insoluble liquid (hydrophobic) and it is also known as ethereal or volatile oils, which are fragrant oily liquid that are extracted from the various parts of the plants. Generally they are organic compounds like hydrocarbons likealcohols, aldehydes, esters, ketones and phenols. An essential oil is "essential" in case that it contains the essence of the different fragrance, and the properties of the plants from which they are derived. All though all ornamental plants have some aroma but only few of them are used commercially in perfumes, flavor and cosmetic industries. The essential oil demand by fragrance industry is 60%, flavour industry 20% and pharmaceutical industry 20%. In all the perfumes, rose oil is one of the best and costliest perfume. The first recorded reference of rose perfume is in *CharakSanghita.* The credit for discovery of rose perfume was given to Queen Noor Jehan mentioned in the book written by Md. Achem of King Jahangir's regime.

The volatile oils showed the different kinds of biological activities including the antibacterial, antioxidant, antiviral, insecticidal, etc. These oils are also used for cancer treatment, while some other has been used for the food preservations, aromatherapy, and in the perfumery industries. Thus, in the day to day life a large spectrum of usage the essential oil and aroma industry has a bright future. Yet the known applications of essential oilsare opening up ofmore areas which will benefit the industry.

Essential oilcontains on average 100 chemicals components and have myriad funcions. Some are antiseptic, antibacterial and digestive while the others are antidepressant. On the behalf of industry sources the known essential oils are more than 3,000. Some 300 out of these are usedon a regular basis for commercial purposes. The flowers commercially used for oil extraction are rose, jasmine, marigold, tuberose, plumeria, champaka, magnolia, millingtonia etc.

The essential oils production ofworld is growingannually at more than 10 per cent and at present it is estimatedat over 11 billion US dollars at about 1,10,000 tonnes valued. India is the third largest producer of essential oils in world after Brazil and USA. Among the whole jasmine oil export in the world India contributes 40% of total world export and with this share India is the second largest exporter of jasmine oil in world. The largest importer of India's floriculture produce is Europe.

History

Before knowing anything deeply regarding essential oil, we should know and understand a very good history. In the beginning, ancient Egypt was the true birthplace of essential oils, or 'aromatherapy' as they only cultivated plants for their oils and used them extensively in their religion, in cosmetics as well as for medicinal purposes. Egyptian people have been found distillation "pots" dating back about 3,500BC and most essential oils, they were produced by a type of solvent extraction method (enfleurage) using animal fat. Around the same time, China and India were exploring herbs and aromatic plants too, which would become an integral part of the Indian Ayurvedic medical system.The wisdom of the Egyptians was absorbed by the ancient Greeks; the most well-known physician of thattime, Hippocrates (c.460-377 BC) was a firm believer in treating a patient holistically and included massage in many of his therapies. Romans in turn adopted the Greek's knowledge and were great believers in hygiene to promote health –aromatic baths being especially popular. But some time went as a dark time for essential oil and after that rebirth of interest caused. A great Physician at the time, Paracelsus (1493-1541) proved his ability to his contemporaries by having great success "curing leprosy" with plant extracts. The term "aromatherapy" was originated by the French chemist and perfumer Rene Maurice Gattefosse in 1937. One incident happened with himin1910, he burnt his hand badly in his laboratory, and being the first compound available, he treated hishand with pure undiluted lavender oil and result found immediately that eased the pain as well as helped heal the hand without any sign of infection or scarring. After that, since the late 70's and early 80's a boom in the interest in Modern Aromatherapy.

Status of Floral Essential Oils in India

Damask rose is widely cultivated in Ajmer, Udaipur and in Rajasthan, Palmpur and Kullu district in Himachal Pradesh and Aligarh, Kannoj, Lucknow in Uttar Pradesh. The rose petals are also used to prepare rose water and gulkand.

India imports around 84,237 tonnes essential oils (March, 2008). India ranks 26th as Importer and 14th as Exporter of essential oil.The turnover of all perfumes in India is around Rs. 300 croreannually. Market size of Indian cosmetics is about Rs. 4500 crore (Bhattacharjee and De,2005).

Different Floral Essential Oils

Lavender oil	Lotus oil
Geranium oil	Marigold oil
Tuberose oil	Salvia oil
Rose oil	Carnation oil
Jasmine oil	Champak oil
Chrysanthemum oil	Plumeria oil

Some Popular Plant's Parts and Constitutes

Common name	Botanical name	Plant part used	Constitutes
Rose	*Rosa damascene* *R. centifolia*	Flower petals	Phenyl ethyl, citronellol
Jasmine	*Jasminum sambac* *J. grandiflorum* *J. auriculatum*	Flowers	Methyl jasmonate, cis-jasmone
Tuberose	*Polyanthus tuberosa*	Flowers	Geraniol, nerol
Geranium	*Pelargonium graveolens*	Leaves	Geraniol, citronellol
Lavender	*Lavandula officinalis*	Flowers	Linalool, Linyal acetate
Salvia	*Salvia sclarea*	Inforescene and leaves	Linalool, Linalyl acetate
Marigold	*Tagetes erecta, T. minuta*	Flowers	Tagetone, Linalyl acetate
Carnation	*Dianthus Caryophyllus*	Flowers	Eugenol, Phenylethyl alcohol
Magnolia	*Magnolia grandiflora*	Leaves	Phenols, Carbonyl compounds

These products are exported to the UK, France, Germany and Middle East countries.

Scope of Floral Essential Oils

a) **Perfume industry:** A fragrant liquid typically made from essential oils extracted from flowers and spices, used to give a pleasant smell to one's body is called as perfume. Concreteand essential oil recovery of ornamental plants:

Species	Concrete or essential oil recovery	
	%	Kg/ha of flowers
J.auriculatum	0.28-0.36	13.44-28.24
J.grandiflorum	0.25-0.32	13.85-29.42
J.sambac	0.14-0.19	1.18-15.44
Tuberose (Single)	0.08-0.11	5.5-27.5
Marigold	0.5-1.5	50-60
Rosa bourboriana	0.04	0.75-1.25
Rosa damascene	0.057-0.058	
Micheliachampaca	0.16-0.2	
Carnation	0.2-0.3	
Magnolia	0.03-0.16	

Source: Singh *et al.,* 2014

b) **Pharmaceutical industry:** There are the list of plants with their scientific name, these plant's essential oil is used to prepare different kinds of medicines as listed below.

Theraptic properties	Oils	Botanical name
Sedative	Sandal wood	*Santalum album*
	Lavender	*Lavandula officinalis*
	Chamomile	*Matricaria chamomilla*
CNS stimulant	Jasmine	*Jasminum officinale*
Adaptogen	Geranium	*Pelargonium graveolens*
Antiseptic	Geranium	*Pelargonium graveolens*
Anti-stress	Patchouli	*Pogestmone patchouli*
Carminative	Chamomile	*Matricaria chamomilla*

Source: Singh *et al.*, 2014

c) **Flavoring agent:**Many essential oils such as lemon, orange are commonly used to flavor desserts, candies and chocolate. Other, more herbal oils, such as rose and rosemary may be better suited to flavouring savoury dishes like sweets and soups. Food grade essential oils are very concentrated and are safe when used sparingly as flavouring, but should not be ingested undiluted.

d) **Confectionary:**Confectionery is a term that is generally used for those items that are rich in sugar and often referred to as a confection. Usually, confections are low in nutrient content but rich in calories. Confectionery refers to the method of creating sugar based dessert items that encompass cakes, pastries, toffees, chocolates, gum and many others. Confectioneries items are defined by the presence of sweeteners. The confectionery items are mainly divided into four main sections such as confectionery sugar, chocolate, confectionery gum and cereal bars. Most of the confectionery industries are also using artificial sweeteners as well. For the past few years, confectionary has gained huge attention amid people. Salvia is an ornamental plant, which is used as artificial sweeteners.

e) **Soaps and lotions:** Rose, jasmine, lavender essential oil etc. are the essential oils which are in demand of soaps and lotions industry. Also used as solvents in the paint and varnish industry

Essential Oils Benefits for Health

Stress and anxiety: It is estimated that approx. 43% people who have stress and anxiety use some alternative therapy to help relieve their symptoms and some essential oils can also work alongside the traditional therapy to treat the anxiety and stress.

Headaches and migraines: In the '90s, two small studiesfound that massage a peppermint oil and ethanol mixture on participants' foreheads and temples helpful to relieve headache pain (Gobel, *et al.*). Also observedthe recent studies, peppermint and lavender oil also reduces headache pain after applying it to the skin.

Sleep and insomnia: Smelling lavender oil improvesthe heart disease of patients and after childbirth also improves the sleep qualityof women.

Reducing inflammation: Essential oils may help fight inflammatory conditions.

Antibiotic and antimicrobial: The rise of antibiotic-resistant bacteria has renewed interest in the search for other compounds that can fight bacterial infections. Test-tube studies have investigated essential oils, such as peppermint and tea tree oil, extensively for their antimicrobial effects, observing some positive results.

Popular Floral Essential Oils and The Health Claims Associated with Them

Lavender: stress reliever

Rose: mood improver and anxiety reducer

Jasmine: helps in depression, childbirth, and libido

Various Method of Extraction

1. Steam Distillation

It is a type of distillation which is a separation or extraction process.This method is suitable for temperature-sensitive plant which contains natural aromatic compounds. It was a popular laboratory method for purification of organic compounds but has become outmoded by vacuity distillation. It is one of the ancient and official approved methods for isolation of essential oils from different plant materials. It is the most popular method used to extract and isolate essential oils from plant materials for use in natural products. This happens when the steam vaporizes the plant material's volatile compounds, which eventually go through a condensation and collection process.

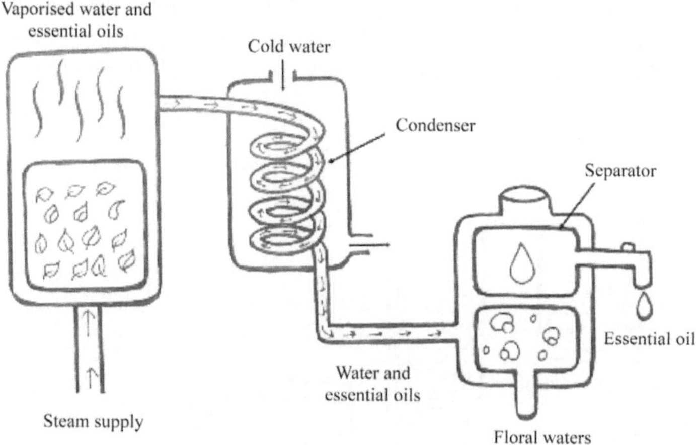

2. Water Distillation / Hydro Distillation

Delicate flowers would clump togetherwhen introduced to steam distillation process such as roses and orange blossoms, so it become more effective method to extract oil in this situation is to submerge delicate plant material in the pure boiling water. The water prevents the extracted oil from overheating. After condensation,liquids cools down and separate from each other. The remaining water which may be fragrant sometimes is referred by several names likeherbal water, essential water, hydrolate, hydrosol, floral water, or herbal distillate. The extraction principle is based on the isotropic distillation. Eg: Rose, Jasmine

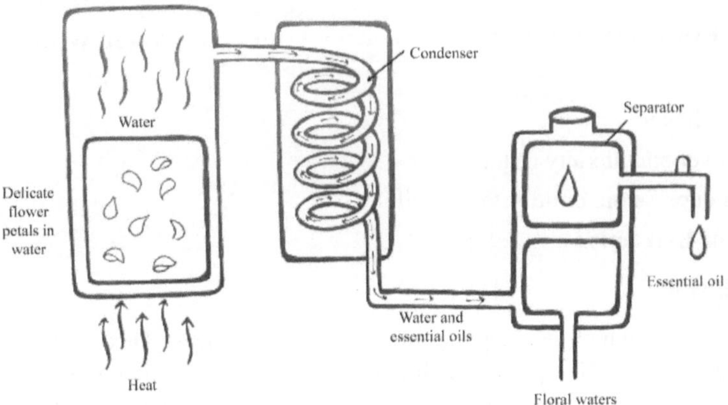

3. Solvent Extraction

The other name of this is Liquid–liquid extraction or partitioning, is a method to separate a compound based on the solubility of its parts. It assigns food grade solvents like ethanol and hexane to segregate essential oils from plant parts. It is more suitable for those plant parts which yield low amounts of essential oil, that are largely resinous, or delicate aromatics unable to withstand the pressure and distress of steam distillation. Solvent extraction method also produces a finer fragrance than any other type of distillation method. Eg: Rose, marigold, salvia, rosemary, pink lotus absolute etc.

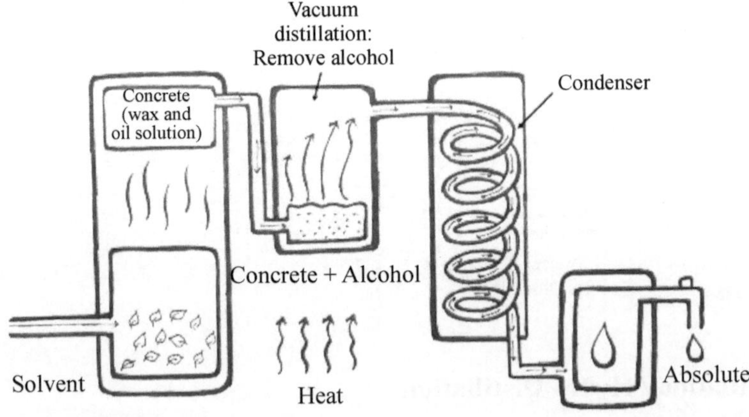

4. CO_2 Extraction

Essential oils derived from the supercritical CO_2 extraction of herbs are similar to the oils produced through distillation, which can be used in aromatherapy and natural perfumery. Eg: German chamomile flower

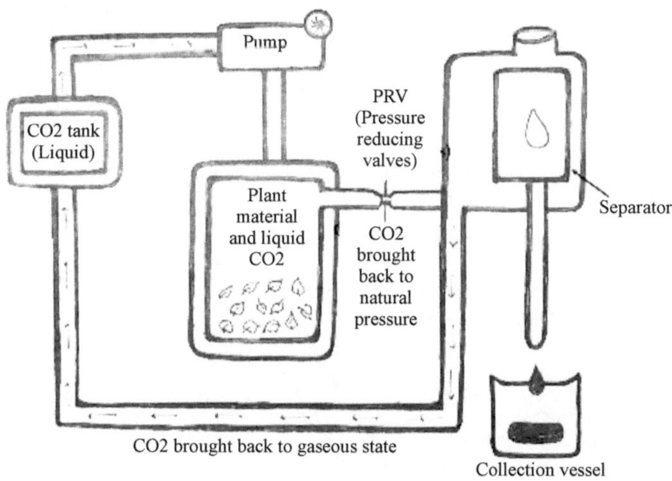

5. Maceration

These oils are also referred to as infused oils. They are extracted when carrier oils are used for solvents to extract therapeutic properties from plant material. The benefit of macerated oil as compared todistilled oil is that more plant's essence is hold in the oil, because it captures heavier, larger plant molecules than the ones captured in the distillation process. This keeps the product closer to retaining more of the plant's valuable offerings. Eg: Leaves of wintergreen

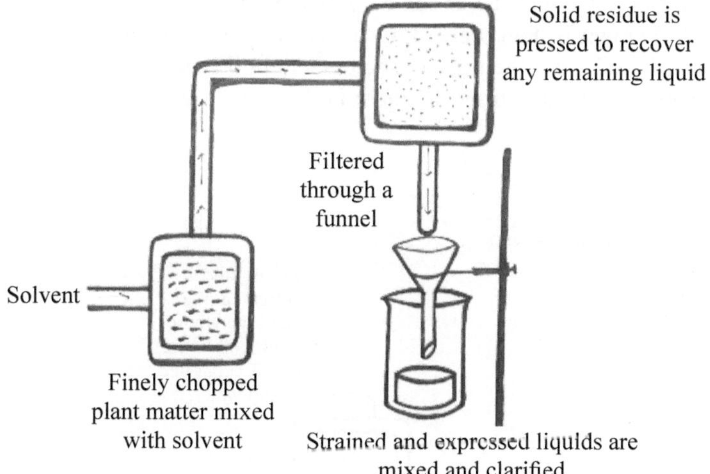

6. Enfleurage

Now a dayit is not commonly used, but it is the oldest methods of essential oil extraction which implements the use of fat. By the end of this process, either animalor vegetable fat becomes infused with the flower's fragrance compounds. The used fats in this process are solid and odorless at room temperature. The enfleurage process can be done either 'hot' or 'cold'. In both cases, the fat that is saturated with the aromais called "enfleurage pomade Eg: Tuberose

Hot Enfleurage

The fats are heated in this process and it is the only difference in this process.

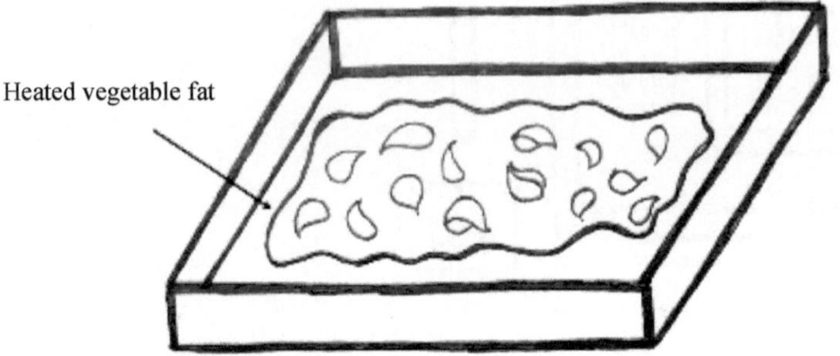

7. Cold-Press Extraction

It is also called Expression or Scarification and is suitable for citrus peels in particular. The term cold pressed theoretically means that the oil expels at low temperatures and pressure. Cold pressed method is one of the best methods to extract essential oils. It is a method of mechanical extraction where heat is reduced and minimized throughout the batching of the raw material. This method is also known as scarification method. Plant which contains oil in their outer layer can be extracted by using this technique and scrubbing. The whole plant is than pressed to squeeze the material from the pulp and to release the essential oil from the pouch. The essential oil raised to the surface of the plant material can be separated from the material by centrifugation (Rassem *et al.,* 2016).

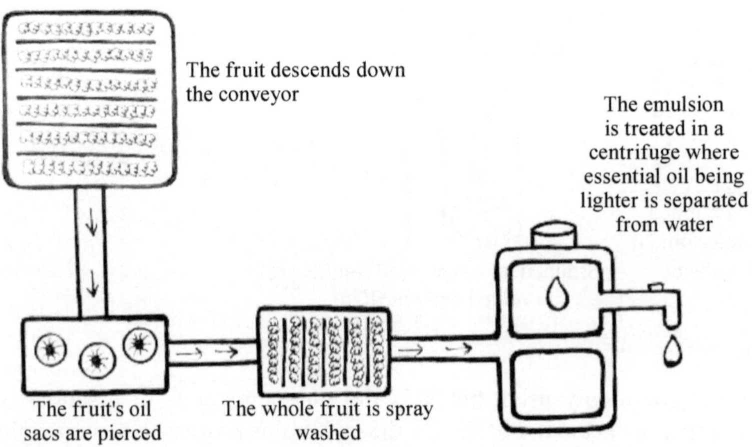

References

Afshar M.K.,Moghadam Z.B., Taghizadeh Z., Bekhradi R., Montazeri A., Mokhtari P. 2015. Lavender fragrance essential oil and the quality of sleep in postpartum women.Iran Red Crescent Med J.17(4):e25880

Aziz, Zarith & Ahmad, Akil & Mohd-Setapar, Siti & Karakucuk, Alptug & Azim, Muhammad & Lokhat, David & Kamal, Mohammad & Ashraf, Ghulam. 2018. Essential Oils: Extraction Techniques, Pharmaceutical and Therapeutic Potential - A Review. Current Drug Metabolism. 19. 10.2174/1389200219666180723144850.

Bhattacharjee,S. K. and De, C. L. 2005. Advanced Commercial Floriculture Aavishkar Publishers.vol.1

Chadha, K. L. and Choudhury, B. 1989.Ornamental Horticulture in India.Indian Council of Agricultural Research. Pp. 191-197

de Sousa D.P., de Almeida SoaresHocayen P, Andrade L.N., Andreatini R. A. 2015. Systematic Review of the anxiolytic-like effects of essential oils in animal models. Molecules.14:20(10):18620-60.

Göbel H, Schmidt G, Dworschak M, Stolze H, Heuss D. 1995. Essential plant oils and headache mechanisms.Phytomedicine.2(2):93-102

Rassem, H. H. A., Nour, A. H. and Yunus R. M. 2016.Techniques for Extraction of Essential Oils From Plants: A Review. Australian Journal of Basic and Applied Sciences.10(16): 117-127

Sharma, Abhishek&Rajendran, Sasireka & Srivastava, Ankit & Sharma, Satyawati & Kundu, Bishwajit. 2016. Antifungal activities of selected essential oils against Fusariumoxysporum f. sp. lycopersici 1322, with emphasis on Syzygiumaromaticum essential oil. Journal of Bioscience and Bioengineering. 123(3)

Singh Gurdip and MauryaSumitra. 2005. Antimicrobial, antifungal and insecticidal investigations on essential oils: An overview. CSIR. 4(3):179-192

Singh, B., Sellam P., Majumder J. and Rai P. 2014. Floral essential oils: importance and uses for mankind. HortFlora Research Spectrum,3(1): 7-13